Conference of the South African Advanced Materials Initiative (CoSAAMI-2018)

IOP Conference Series: Materials Science and Engineering
Volume 430

Vanderbijlpark, South Africa
23 - 26 October 2018

Editors:

Silethelwe Chikosha
Ronald Machaka

ISBN: 978-1-5108-7338-4
ISSN: 1757-8981

Printed from e-media with permission by:

Curran Associates, Inc.
57 Morehouse Lane
Red Hook, NY 12571

Some format issues inherent in the e-media version may also appear in this print version.

Printed by Curran Associates, Inc. (2018)

For permission requests, please contact the Institute of Physics at the address below.

Institute of Physics
Dirac House, Temple Back
Bristol BS1 6BE UK

Phone: 44 1 17 929 7481
Fax: 44 1 17 920 0979

techtracking@iop.org

Additional copies of this publication are available from:

Curran Associates, Inc.
57 Morehouse Lane
Red Hook, NY 12571 USA
Phone: 845-758-0400
Fax: 845-758-2633
Email: curran@proceedings.com
Web: www.proceedings.com

Table of contents

Volume 430

Conference of the South African Advanced Materials Initiative (CoSAAMI-2018)

23-26 October 2018, Vanderbijlpark, South Africa

Accepted papers received: 17 September 2018
Published online: 2 October 2018

Preface

Papers

Composites

Mechanical testing-Titanium

Modelling-Computational

Modelling-Smart Materials

Alloy development

Modelling-Simulation

Modelling-Numerical

Powder metallurgy

Corrosion

The 2018 Conference of the South African Advanced Materials Initiative (CoSAAMI-2018)

Abstract:

The Chairs of the Technical Committee are pleased to present the proceedings of the Conference of the South African Advanced Materials Initiative (CoSAAMI-2018). These proceedings are composed of a variety of contributions from researchers, engineering firms, consultants, nuclear industry associations, regulators and postgraduate students presented at the CoSAAMI-2018 conference held at the Riverside Sun Resort, Vanderbijlpark on the banks of the iconic Vaal River from the 23rd to 26th of October, 2018. The conference is an evolution of the Advanced Metals Initiative (AMI) series of conferences which offered a platform for researchers and students from the AMI networks, Light Metals Development Network (LMDN), Ferrous Metals Development Network (FMDN), Precious Metals Development Network (PMDN) and Nuclear Materials Development Network (NMDN) and researchers globally, to present their work on topics related to the themes of the conference. To guarantee quality, credibility and accuracy of scientific contents, all contributions submitted to CoSAAMI-2018 were peer-reviewed by a panel of capable referees in accordance with the IOP Peer Review Guidelines. As a result, out of the 93 manuscripts received, a total of 49 papers were accepted for publication. These conference proceedings and all the other scientific presentations related to the CoSAAMI-2018 must be attributed to the extraordinary efforts made by the conference's panel of peer-reviewers, the senior reviewers, the tireless conference committees, and the tremendous dedication of the CoSAAMI-2018 Conference Chair, Dr Sagren Govender. The unwavering support of the CoSAAMI-2018 sponsors is duly appreciated. Chairs of the Technical Committee with to thank firstly, Tebogo Motsai, Noluntu Muchavi, and Khodani Ramabulana who dutifully edited and cautiously formatted the proceedings manuscript by manuscript and the support of the IOP Publishing team.

The Foreword, the Table of Contents, the Steering and Organizing Committees, and the Technical are available in this PDF.

Silethelwe Chikosha, PhD
Ronald Machaka, PhD

Editors and CoSAAMI-2018 Technical Committee Chairs

ami

advanced materials initiative

FOREWORD

The Advanced Materials Initiative (AMI) was established by the three science councils working in metallurgy and materials science, viz. Mintek, the Council for Scientific and Industrial Research (CSIR) and the South African Nuclear Energy Corporation Ltd. (Necsa). It is supported by the Department of Science and Technology (DST) to facilitate research, development and innovation across the materials value chain. The goal is to target significant export income and new industries for South Africa by 2020 through the country becoming a world leader in sustainable materials production and manufacturing via technological competence and optimal, sustainable local manufacturing of value-added products while reducing environmental impact.

The key objectives of the programme are to create new industries, support existing industries, and localise technology. The AMI takes an integrated (vertical and lateral) approach across the entire value chain from resource development to materials production and the manufacture of end-products, to achieve its goal.

The AMI consists of various technology networks, each focussing on different commodities that are of strategic importance to South Africa. The AMI promotes collaborative research between the science councils, tertiary education institutions and industry, both local and internationally. Human resource development is critical for the networks to expand the country's technical capacity, develop the use of materials in new applications and more diverse industries, and develop industrial localisation.

The Conference of the South African Advanced Materials Initiative (CoSAAMI'18) is the culmination of the annual peer review mechanism of the AMI. It particularly encourages the interaction of students and developing academics with the more established academic community and industry in an informal setting to present and discuss new and current work.

The organising committee received 93 papers covering about 10 topics. Based on over 141 reviews from the technical committee and external reviewers, 50 papers appear in this proceedings. Each paper received at least 3 reviews. These proceedings will furnish scientists and engineers in the field of materials science with an excellent reference book and furthermore be an impetus to stimulate further study and research in these areas.

The success of the AMI depends completely on the effort, talent and energy of researchers in the field of materials science who have written and submitted papers on various topics, as well as the reviewers who have invested significant time in analysing and assessing multiple papers to maintain a high standard of quality for this conference. We thank all authors and participants for their contributions.

science & technology
Department: Science and Technology
REPUBLIC OF SOUTH AFRICA

CSIR
our future through science

MINTEK
Specialists in mineral and metallurgical technology

necsa
We're in your world

IOP Conf. Series: Materials Science and Engineering **430** (2018) 011003 doi:10.1088/1757-899X/430/1/011003

THE COMMITTEES OF THE 2018 CONFERENCE OF THE SOUTH AFRICAN ADVANCED MATERIALS INITIATIVE (COSAAMI-2018)

CHAIRS OF THE CONFERENCE:

Conference Chair

Dr Sagren Govender – chair@cosaami2018.co.za

Conference Co-Chair

Dr Robert Tshikhudo – cchair@cosaami2018.co.za

SECRETARY OF THE CONFERENCE:

Kelly-Anne Matthews – secretary@cosaami2018.co.za

ADVANCED MATERIALS INITIATIVE NETWORK MEMBERS:

Light Metals Development Network, CSIR
Titanium Centre of Competence, CSIR
Precious Metals Development Network, MINTEK
Ferrous Metals Development Network, MINTEK
Nuclear Materials Development Network, NECSA

AMI STEERING COMMITTEE:

Llanley Simpson	Department of Science and Technology
Wilna du Plessis	Department of Science and Technology
Robert Tshikhudo	Council for Scientific and Industrial Research
Sagren Govender	Council for Scientific and Industrial Research
Gary Partrick	Mintek
John Mone	Mintek
Jones Papo	Mintek
Keneiloe Sikhwivhilu	Mintek
Jakkie Postma	Necsa
Hertzog Bissett	Necsa

LOCAL ORGANIZING COMMITTEE

Llanley Simpson	AMI	Department of Science and Technology
Wilna du Plessis	AMI	Department of Science and Technology
Kelly-Anne Matthews	Conference Secretary	Council for Scientific and Industrial Research
Dr Sagren Govender	Chair	Council for Scientific and Industrial Research
Dr Robert Tshikhudo	Co-Chair	Council for Scientific and Industrial Research
Dr Amogelang Bolokang	Committte member	Council for Scientific and Industrial Research
Dr Annelize Botes	Committte member	Council for Scientific and Industrial Research
Dr Silethelwe Chikosha	Committee member	Council for Scientific and Industrial Research

Dr Kalenda Mutombo	Committee member	Council for Scientific and Industrial Research
Dr Ronald Machaka	Committee member	Council for Scientific and Industrial Research

TECHNICAL COMMITTEE

Editors and Chairs of the Technical Committee

Dr Silethelwe Chikosha	Council for Scientific and Industrial Research
Dr Ronald Machaka	Council for Scientific and Industrial Research

Senior Reviewers

Prof Deborah Blaine	University of Stellenbosch
Dr Annelize Botes	Council for Scientific and Industrial Research
Prof Paul Ewart	Oxford University
Dr Silethelwe Chikosha	Council for Scientific and Industrial Research
Prof Lesley Cornish	University of the Witwatersrand
Dr Ronald Machaka	Council for Scientific and Industrial Research
Dr Christopher Machio	Council for Scientific and Industrial Research
Linda Mahlatji	Council for Scientific and Industrial Research
Dr Nthabiseng Maledi	University of the Witwatersrand
Prof Reginah Maphanga	Council for Scientific and Industrial Research
Dr Walace Matizamhuka	Vaal University of Technology
Dr Ntombi Mathe	Council for Scientific and Industrial Research
Dr Kalenda Mutombo	Council for Scientific and Industrial Research
Dr Maje Phasha	Transnet Engineering (SOC)
Prof Roelf Sandenbergh	University of Pretoria
Prof Alfred Sidambe	University of Liverpool
Prof Waldo Stumpf	University of Pretoria
Prof Daniel Wamwangi	University of the Witwatersrand
Prof David Young	University of the Sunshine Coast

List of Reviewers

Nana Arthur	Council for Scientific and Industrial Research
Dr Kofi Anan	University of Pretoria
Dr Adewale Adegbenjo	University of Johannesburg
Dr Mohammed Balogun	Council for Scientific and Industrial Research
Jeff Benson	Council for Scientific and Industrial Research
Dr Amogelang Bolokang	Council for Scientific and Industrial Research
Levy Chauke	Council for Scientific and Industrial Research
Dr Gary Corderley	Council for Scientific and Industrial Research
Dr Ulyate Curle	Council for Scientific and Industrial Research
Pfarelo Daswa	Council for Scientific and Industrial Research
Dr Shahed Fazluddin	Council for Scientific and Industrial Research
Dr Romeo Fono	University of Johannesburg
Prof Kapil Gupta	University of Johannesburg
Zizo Gxowa	Council for Scientific and Industrial Research
Dr Oladeji Ige	University of Johannesburg

Marlise Jordaan	Council for Scientific and Industrial Research
Asimenye Kapito	Mintek
Desmond Klenam	Mintek
Vinod Kurup	University of Pretoria
Thabiso Langa	University of Johannesburg
Linda Lethabane	Tshwane University of Technology
Chwayita Madikizela	Council for Scientific and Industrial Research
Dr Daniel Madyira	University of Johannesburg
Dr Miyelani Maminza	Council for Scientific and Industrial Research
Velaphi Matjeke	Transnet Engineering (SOC)
Ipfi Mathoho	Council for Scientific and Industrial Research
Dr Hilary Masenda	University of the Witwatersrand
Stephen Masete	Council for Scientific and Industrial Research
Maria Mathabathe	Council for Scientific and Industrial Research
Ndumiso Mnguni	Council for Scientific and Industrial Research
Dr Rose Modiba	Council for Scientific and Industrial Research
Emma Molobi	Transnet Engineering (SOC)
Innocent M.J. Monareng	Council for Scientific and Industrial Research
Tebogo Motsai	Council for Scientific and Industrial Research
Noluntu Muchavi	Council for Scientific and Industrial Research
Tshepo Patric Nkuna	Council for Scientific and Industrial Research
Dr Babatunde Obadele	University of Johannesburg
Prof Peter Olubambi	University of Johannesburg
Dr Shaan Oosthuizen	Council for Scientific and Industrial Research
Lerato Raganya	Council for Scientific and Industrial Research
Khodani Ramabulana	Council for Scientific and Industrial Research
Munyadziwa Ramakokovhu	Tshwane University of Technology
Mandy Seerane	Council for Scientific and Industrial Research
Dr Mxolisi Shongwe	Tshwane University of Technology
Prof Charles Siyasiya	University of Pretoria
Jabu Skosana	Council for Scientific and Industrial Research
Mulalo Tambani	Council for Scientific and Industrial Research
Ghislain Tshilombo	Council for Scientific and Industrial Research
Dr Daniel Madyira	University of Johannesburg

Editing and formatting manuscript

Tebogo Motsai	University of Pretoria
Noluntu Muchavi	Council for Scientific and Industrial Research
Khodani Ramabulana	Council for Scientific and Industrial Research

The Editors and Chairs of the Technical Committee of the 2018 Conference of the South African Advanced Materials Initiative (CoSAAMI-2018) would like to express gratitude to all Committees for volunteering, support and contribution in the editing and reviewing process.

Peer review statement

All papers published in this volume of *IOP Conference Series: Materials Science and Engineering* have been peer reviewed through processes administered by the proceedings Editors. Reviews were conducted by expert referees to the professional and scientific standards expected of a proceedings journal published by IOP Publishing.

Metal matrix composite laser metal deposition for ballistic application

R Adam[1], A Botes[2] and G Corderley[2]

[1] National Laser Centre, Council for Scientific and Industrial Research, Meiring Naudé Road, Brummeria, Pretoria 0184, South Africa
[2] Light Metals, Materials Science & Manufacturing, Council for Scientific and Industrial Research, Meiring Naudé Road, Brummeria, Pretoria 0184, South Africa

E-mail: Abotes1@csir.co.za

Abstract. The armour development industry uses ceramic tiles to provide effective ballistic protection to critical areas of military vehicles. The ceramic tiles are generally attached to a ductile backing material by means of an adhesive, which often incorporates defects that are undesirable for ballistic protection. Laser cladding provides a more efficient approach for applying a ceramic material onto a ductile backing through metallurgically bonding the materials. A 3kW, IPG fiber laser system was used to manufacture test coupons to investigate the process parameters for the laser metal deposition of a tungsten carbide (chemical formula WC) metal matrix material onto an Armox 500 backing material. Results show that good fusion to the base material with limited solidification cracking can be achieved. Vickers microhardness and sound velocity results showed satisfactory results for a ballistic application.

1. Introduction

Ceramic materials offer excellent mechanical properties for ballistic applications, such as high modulus of elasticity, high hardness, high compressive strength, wear resistance and light weight [1, 2]. The primary drawback of ceramics is that they have fairly poor crack resistance, and hence they are used in combination with a ductile backing material [3]. The main purpose of the ceramic is to fragment the bullet and reduce its penetrability, whilst the purpose of the ductile backing is to absorb the kinetic energy of the bullet by plastic deformation [4, 5]. The ceramic material is conventionally bonded to the backing material by an adhesive, which may compromise the ballistic performance of the assembled component [6].

Laser metal deposition (LMD) or Laser Cladding (LC) is a process whereby material is deposited onto a substrate by essentially melting particles of the material to be deposited and the substrate with a high energy laser beam [7]. The process is a high energy, low heat input process with extremely fast cooling rates in excess of 2000° C/s. Due to the fast cooling rates, the deposited layer forms unique microstructures, e.g. extremely fine structures, which results in high hardness accompanied with high toughness [8]. The process also results in strong metallurgical bonding to the substrate, hence offering a more efficient approach that addresses the limitations of ceramic bonding by adhesive layers [7]. The main objective of this investigation is to address the limitations of bonding a ceramic material to Armox 500 by using a LMD process.

IOP Conf. Series: Materials Science and Engineering **430** (2018) 012001 doi:10.1088/1757-899X/430/1/012001

2. Experimental Procedure

A 3 kW IPG multi-mode Ytterbium fiber laser with a feeding fiber of 200 μm was used for the generation of all test coupons. A Precitec cladding head was used to give a beam spot diameter of 2 mm and 4 mm on the workpiece. WC metal matrix powder was deposited and melted onto Armox 500 substrate coupons that were milled clean and wiped with acetone. The argon carrier gas flow rate was kept constant at 2 l/min and a shielding gas flow rate of 10 l/min was also maintained.

Scanning Electron Microscopy (SEM) and Energy Dispersive Spectroscopy (EDS) evaluation of the powder used for the laser metal deposition process was performed to ensure that the powder is suitable for the process. SEM and EDS analysis was obtained using a JEOL 610 LV plus microscope.

A metallurgical evaluation was conducted by taking cross-sections through the clad layers and observing the samples using an optical microscope. A Vickers micro-hardness analysis was performed with a Zwick Roell Indentec using a 500g load and dwell time of 10 seconds, 10 indentations were performed for each sample set. Sound velocity measurements were conducted using standard non-destructive evaluation (ultrasonic testing) equipment. Initially a powder investigation was conducted to identify the powder most suitable for the laser cladding of WC with different matrix compositions onto Armox 500. The preliminary experimental parameters for the respective powders are shown in table 1.

Table 1. Experimental parameters for identifying suitable powder.

WC Powder	Spot Size (mm)	Heat Input (kJ/m)	Powder flow rate (g/min)
Powder 1 (Co, Cr Matrix)	2	180	5.3
Powder 2 (Ni Matrix)	4	180	11.4
Powder 3 (Ni, Cr Matrix)	4	180	9.7

The parameters used to generate test coupons with the identified powder are listed in table 2. The laser beam spot diameter used was 4mm.

Table 2. Experimental Parameters of identified powder.

Sample No.	Laser Power (W)	Scanning Velocity (m/s)	Line Energy (kJ/m)
C1	2000	0.010	200
C2	1500	0.010	150
C3	1200	0.010	120
C4	1800	0.017	108
C5	1500	0.017	90
C6	1200	0.017	72
C7	1800	0.010	180
C8	2200	0.010	220
C9	2500	0.010	250

3. Results and Discussion

3.1. Powder Characterization

SEM analysis of the WC powders, with varying host matrix compositions, was performed to assess the morphology and particle size of the powders. The micrographs were obtained using back-scattered electron (BSE) mode to distinguish between the WC particles and the host matrix particles. The micrographs of the respective powders are shown in figure 1. Powder 1 has uniformly distributed particles which were composed of WC, Co and Cr. For powders 2 and 3 the WC particles are dispersed within the matrix particles. The lighter particles represent the WC particles, whilst the darker particles show the Nickel based matrix particles.

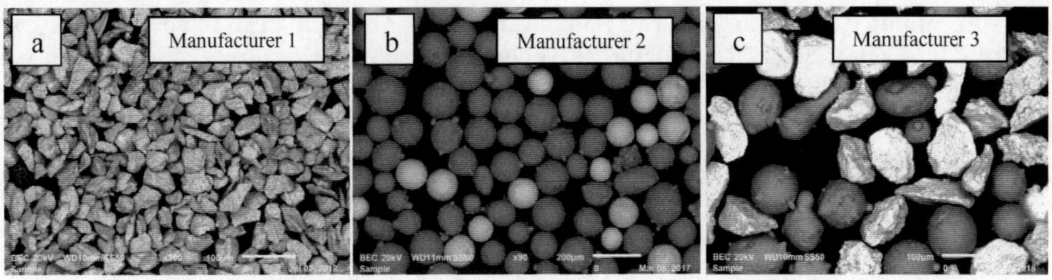

Figure 1. SEM micrographs of a) Powder 1, b) Powder 2, and c) Powder 3.

These results were confirmed by EDS analysis. From figure 1 it is evident that all the powders have different particle morphologies and different host matrices. The results of EDS analysis of the various powders is shown in table 3.

Table 3. Chemical Composition of the powders.

Powder	Main Elements
Powder 1	WC, Co, Cr
Powder 2	WC, Ni, Fe, Si
Powder 3	WC, Ni, Cr, Co, Fe

3.2. Metallurgical Evaluation

A Metallurgical evaluation was conducted by taking cross-sections through the clad layers and observing the samples with an optical microscope and SEM. Evaluation of the sample generated using a 2mm laser beam spot diameter showed that all of the WC particles were melted and that the clad layer that formed had multiple cracks and exhibited lack of fusion to the base metal. The result obtained using a 2mm beam spot diameter is undesirable for the proposed application, since the main objective was to metallurgically bond the ceramic composite to the Armox substrate while maintaining the WC particles distributed in the matrix. A 2mm beam spot diameter resulted in a high energy density which caused dissolution of the WC particles in the melt pool. The dissolution of these particles resulted in the formation of secondary phases which consequently resulted in an increase in the sensitivity to cracking [9].

Figure 2 show the micrographs of the samples generated using powders 2 and 3 respectively. The laser spot size was increased to 4mm, in order to reduce the energy density with the aim of limiting the cracking of the clad layers as well as minimizing the dissolution of the WC particles in the matrix.

In figure 2(a) the darker round spherical particles represent the suspended WC particles, similarly, the round light spherical particles shown in the cladded layer of powder 3 represents the WC particles. Evidently the distribution of WC particles suspended within the host matrix is much greater for powder 2 than for powder 3.

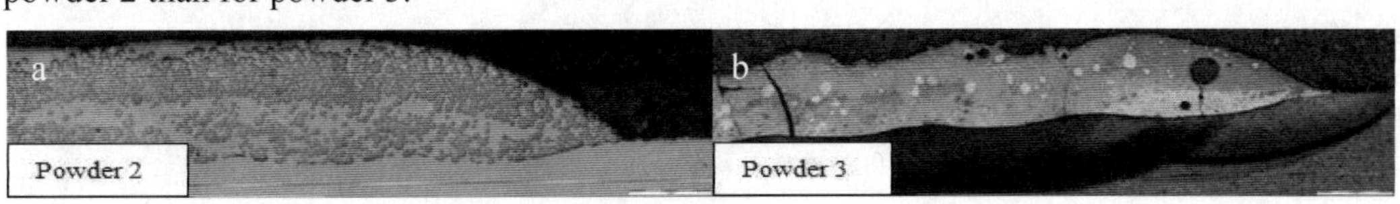

Figure 2. Optical micrograph of the cross section of the test coupon generated using (a) powder 2 and (b) powder 3.

The test coupon generated using powder 3 also shows the presence of cracking in the clad layer, which may be caused by the presence of the chromium (Cr) in the matrix. The Cr increases the precipitation of hard phases in the material, resulting in a more brittle structure, thereby increasing the material's susceptibility to cracking [9, 10]. In addition, the presence of Cr in the powder increases the WC dissolution, as observed by Liyanage et.al [11]. Another factor to consider is the formation of brittle chromium carbide precipitates within the matrix since chromium has a high affinity for carbide formation. The cracking observed is undesirable for ballistic applications since any discontinuity or free surface in the material results in the reflection of an impacting pulse as a tensile wave. If the reflected tensile wave's magnitude is larger than the tensile fracture strength of the material it will fracture [12].

Based on the results, especially the powder morphology, powder 2 appeared to be more suitable for the LMD process and the desired application. The LMD process parameters shown in table 2 were used to generate the test coupons. The resulting single clad layer heights and line energy values are provided in table 4.

Table 4. Summary of the effect of line energy (heat input) on clad layer thickness.

Sample No.	Line Energy, kJ/m	Clad layer height, µm
C1	200	388.2
C2	150	275.2
C3	120	203.5
C4	108	230.4
C5	90	197.2
C6	72	176.0
C7	180	387.8
C8	220	517.8
C9	250	593.9

The appearance of the clad layers produced is shown in the micrographs of figure 3. From figure 3 it is evident that good fusion was achieved between the clad layers and the Armox 500 substrates for the test coupons. This is a favourable outcome since it shows that the process may be used to eliminate the use of adhesives for the purpose of armour applications.

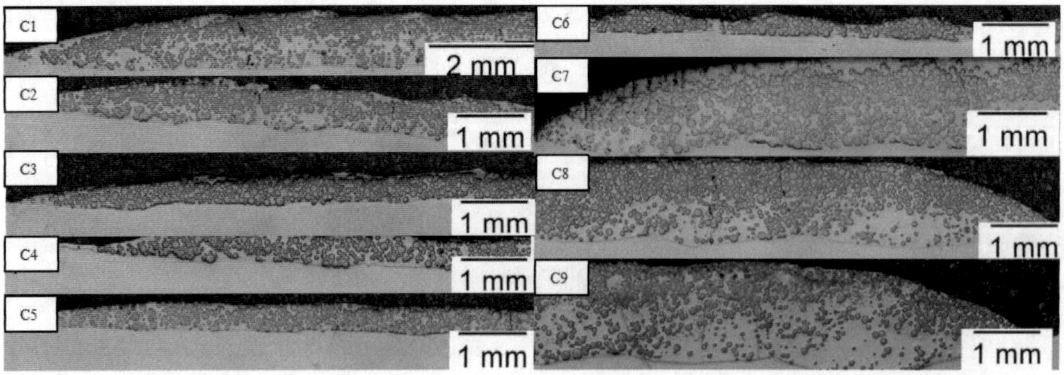

Figure 3. Optical micrographs of the cross section of the test coupons generated using powder 2.

Cracking becomes more prevalent with line energies greater than 180 kJ/which may be attributed to the increase in the WC dissolution which in turn increases the cracking sensitivity of the material [11]. Optimised parameters are those that result in good fusion, minimum WC particle dissolution, no clad layer cracking as well as minimum melt-in (dilution) into the base metal.

3.3. SEM analysis

Sample C7 was selected for SEM and EDS analysis since it produced a clad layer without cracking, together with good fusion and dispersion of WC particles in the host matrix. Figure 4 shows the SEM micrograph of the structure of the matrix and the suspended WC particles. The WC particles are suspended in a dendritic matrix. The dendritic phases are formed as a result of the rapid solidification during the laser metal deposition process. Similar results have been observed by Cheng et. al [13]. The EDS analysis confirms that the spherical particles consist of WC, whilst the dendritic matrix is composed of Ni, Si and Fe, with an increase in the concentration of Fe closer to the fusion line due to the dilution of the Armox 500 substrate.

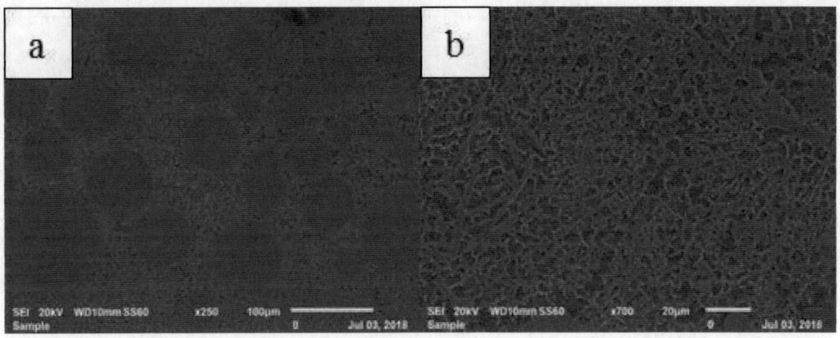

Figure 4. SEM micrograph of a) the WC particles suspended in the matrix, and b) the microstructure of the matrix.

The formation of dendritic microstructures as a result of rapid solidification has been known to increase the mechanical properties of materials [14]. The tempered martensitic microstructures have however been identified as the most suitable phases that provide the highest level of strength in armour steels. In addition, it possesses increased ductility and toughness, crucial for enhancing impact energy and strength [15].

3.4. Vickers Micro-hardness results

The Vickers micro-hardness of the particles and the metal matrix were evaluated to obtain an overview of the overall Vickers hardness profile of the clad material using the rule of mixtures (ROM), given that the composite layer consists of 60% WC particles and 40% metal matrix. A table with the average Vickers micro-hardness of the particles, metal matrix and the composite clad layer hardness is provided in table 5.

The Vickers hardness of the ceramic WC particles is significantly higher than that of the metallic matrix. This is expected since ceramics are characterized by high hardness [2]. The hardness of the matrix appears to have a linear relationship with an increase in process heat input (line energy), as shown in figure 5.

The linear relationship observed in figure 5 is attributed to the increased WC dissolution in the metal matrix with increasing heat input. The increased dissolution results in the formation of hard phases in the metal matrix, thereby increasing the overall Vickers micro-hardness of the metal matrix

[16, 17]. No relationship between the ceramic particle hardness and the heat input was observed since the particles are not changing phase during processing, i.e. melting.

Table 5. Average Vickers Micro-hardness results of the particles, metal matrix and composite clad layers based on the rule of mixtures.

Sample No.	Ave. Particle Hardness (HV)	Ave. Matrix Hardness (HV)	Composite Hardness (HV)
C1	2492 ± 375	642 ±122	1752 ± 230
C2	2646 ±313	595 ± 84	1826 ± 191
C3	2570 ±391	578 ± 56	1773 ± 236
C4	2246 ±432	548 ± 53	1567 ± 260
C5	2388 ±728	549 ± 36	1652 ± 437
C6	2579 ±176	587 ± 86	1782 ± 111
C7	2390 ±862	614 ± 58	1680 ± 518
C8	2678 ±104	642 ± 150	1864 ± 87
C9	2446 ±358	694 ± 57	1745 ± 216

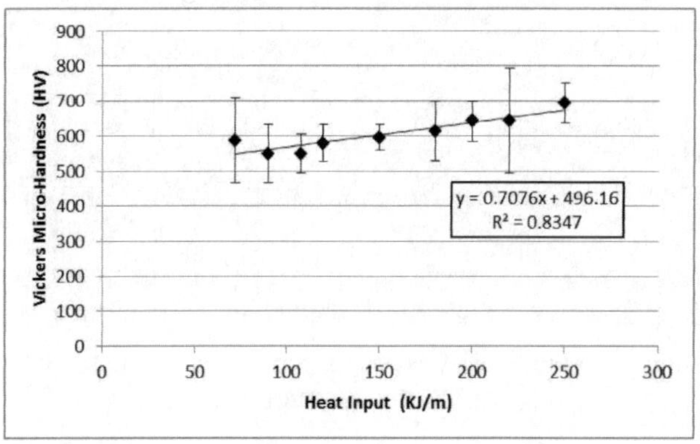

Figure 5. Graph illustrating the linear relationship between hardness of the metal matrix and heat input of the process.

The main purpose of the ceramic material in composite armour systems is to fragment the incident projectile, therefore a higher hardness can assist in increasing the abrasion of the incident projectile. Although increased hardness is useful, it should be mentioned that this may result in increased brittleness, which in turn results in an increased susceptibility to tensile cracking [18, 19]. It is envisaged that the softer metal matrix will assist in the absorption of additional kinetic energy from fragments of the projectile, whilst simultaneously providing some tensile strength to the composite material.

3.5. Sound Velocity Measurements

When a material is subjected to impact, the impact is propagated through the material as a wave pulse at a certain speed. Thus it translates to a finite time whereby the material is affected by the total applied impact. The speed, c_l of the wave is given by:

$$c_l = \left(\frac{E}{\rho}\right)^{\frac{1}{2}} \quad (1)$$

E and ρ represent the Young's Modulus and the density of the material respectively in which the wave pulse propagates [12]. The sound velocity is proportional to the energy dissipation capacity of the material, thus an increased sound velocity corresponds to increased ballistic performance [20].

The sound velocity of the clad layer was measured to be 6100 m/s. These results are comparable to the sound velocities of the ceramic materials that are currently in use by the armour industry [21].

4. Concluding remarks

The powder selection and laser cladding parameters most suitable for ballistic applications have been investigated. Results show that the powder with a Ni matrix produces clad layers with minimal cracking, even particle distribution, and good fusion to the base material. This indicates that the laser metal deposition process may be used to deposit (metallurgically bond) ceramic particles to ductile Armox backing material.

Analysis of the micrographs reveals that crack sensitivity increases as a function of heat input, and the corresponding metallic matrix forms dendritic microstructures. Vickers micro-hardness results show that the ceramic WC particles have a higher hardness than the softer metallic matrix. Additionally, sound velocity measurements show that the laser cladding process is suitable for producing materials for composite armour.

Acknowledgments

The authors would like to acknowledge Maurice Maliage (DPSS, CSIR) for his assistance with sound velocity measurements, and Dr Frikkie Mostert (DPSS, CSIR) for assistance and guidance with respect to the ballistic analysis and applications.

5. References

[1] Silva M, Stainer D, Al-Qureshi A, Montedo R and Hotza D 2014 *J. of Ceramics* **2014** 1
[2] Yadav S and Ravichandran G 2003 *Int. J Impact Eng.* **28** 557
[3] Hameed A, Appleby-Thomas G, Wood D, Hazell P and Jaansalu K 2014 *J. Phys. Conf. Ser.* **500**
[4] Walley S 2010 *Adv. Appl. Ceram.* **109** 446
[5] Kaufmann C, Cronin D, Worswick M, Pageau G and Beth A 2003, *Shock Vib.* **10**
[6] Grujicic M, Pandurangan B and d'Entremont B 2012 *Mater. Des.* **41** 380
[7] Liu, J, Yu H, Chen C, Weng F and Dai J. 2017 *Opt. Lasers Eng.* **93**, 195
[8] Rombouts M, Krieken G, Husslage W, Hofman J, and Hartgers P 2012. *Practical Guideline: Laser Cladding. Laser Applicatie Centrum: The Netherlands,* pp 1-97.
[9] Janicki D, Musztyfaga-Staszuk M 2016 *Stroj Vestn- E J. Mech. Eng.* **62** 343
[10] Amadoa J, Tobara M, Yáñeza A, Amigób V and Candelb J, 2011 Phys. Procedia **12** 338
[11] Liyanagea T, Fisherb G and Gerlicha A 2012 Wear **274– 275** 345
[12] Zukas J, Nicholas T, Swift H, Greszczuk L, and Curran D 1982 Impact Dynamics (John Wiley and Sons editors)
[13] Cheng H, Yi J, Fang Z, Dai S and Zhao X 2013 Mater. Trans. JIM **54** 50
[14] Ruan Y, Mohajerani A and Dao 2016 Sci. Rep. **6** 31684
[15] El-Bitar T, El-Meligy M, El-Shenawy M, Almosilhy A and Dawood N, 2017 JoMME **11** 214
[16] Deschuyteneer D, Petit F, Gonon M and Cambier F 2015 Surf. Coat. Technol. **283** 162
[17] Zanzarin S, Bengtsson S, and Molinari A 2015 J. Laser Appl. **27** S2
[18] LaSalvia J, Campbell J, Swab J, and McCauley JOM 2010 **62** 16
[19] Krell A and Strassburger. 2014 Mater. Sci. Eng. A **597** 422
[20] Grujicic M, Glomski P, He T, Arakere G, Bell W, and Cheeseman B, 2009 J. Mater. Eng. Perform. **18** 1169
[21] Gooch W, Burkins M and Palicka R 2000 J. Phys. IV France, Pr 9-741

Effect of milling parameters on the dispersion characteristics of multi-walled carbon nanotubes in transition metal oxides

S S Lephuthing*, A M Okoro, M Lesufi, O O Ige and P A Olubambi

Centre for Nanoengineering and Tribocorrosion, Department of Metallurgy, School of Mining, Metallurgy and Chemical Engineering, University of Johannesburg, Republic of South Africa

*senzenil@uj.ac.za

Abstract. In this research investigation on milling parameters to achieve uniform dispersion of multi-walled carbon nanotubes (MWCNTs) was conducted. High-energy ball milling (HEBM) technique was adopted to disperse 1 wt. % MWCNTs in titanium and manganese oxides with ball to powder ratio (BPR) of 10:1 at 6 and 9 hours milling times. Scanning electron microscopy (SEM), transmission electron microscopy (TEM) and X-ray diffraction (XRD) were used to characterize the as-received MWCNTs and admixed composite powders. The results indicated that HEBM was an effective route to disperse MWCNTs in transition metal oxides. In addition, the characteristics of admixed powders evidently showed that the increase of milling time effectively dispersed the MWCNTs in titanium and manganese oxide respectively, however deformation of sidewalls of MWCNTs was observed due to harsh milling conditions that resulted on non-sp^2 defects in the MWCNTs.

1. Introduction

Owing to the unique chemical, thermal conductivity, physical and mechanical properties, carbon nanotubes (CNTs) have paved way on their applications as reinforcements in functional materials such as polymers, metals and transition metal oxides (TMOs) [1–3]. Transition metal oxides have been extensively studied as promising functional materials due to their potential application on catalysis, chemical sensors, and energy conversion/storage devices. However, they have drawbacks which are low thermal conductivity, brittle nature and low fracture toughness which have motivated researchers to employ strategies in reinforcing them [3,4]. In addition, MWCNTs are the best candidate to address the inherent limitations of TMOs as they exhibit excellent mechanical properties with Young's modulus and tensile strength in the range of 0.2-0.95 TPa (5 times stronger than stainless steel) and 11-150 GPa, respectively. Additionally, they undergo plastic deformation under severe bending stress conditions without showing brittle deformation behaviour [5].

Despite these exceptional properties that MWCNTs contribute to TMOs, an effective and efficient route on synthesis of TMOs/MWCNTs composite are yet to be fully established [3]. During the synthesis of TMOs/MWCNTs composites, there are challenges associated with achieving homogenous dispersion of MWCNTs in TMOs. These challenges are due to the nanoscale dimensions and strong van der Waals forces among individual tubes, which lead to the agglomeration of MWCNTs and minimises the total surface free energy that results in poor compatibility with TMOs components [3,6–8]. Various studies on dispersion techniques of MWCNTs in metals and TMOs [9–11] have been done;

liquid phase, electrochemical deposition, thermal deposition, and High-energy ball milling (HEBM) in a bid to achieve homogeneous dispersion of MWCNTs.

High energy ball milling is a widely used technique to disperse MWCNTs, despite the effectiveness of this technique, past works have revealed that MWCNTs are subjected to harsh milling conditions that may damage the MWCNTs and eventually make them lose their unique properties [12]. When MWCNTs are damaged during a milling operation, there are tendencies of introducing vacancies and imperfections in the C–C system with broken sp2 bonds that may lead to formation of weak sp3 C–C amorphous phase which initiate sites for interfacial reactions [13]. Nevertheless, MWCNTs exhibit a significant chemical stability because of the strong graphitic basal planes, which composed of strong sp2 C–C bonds in the outer shell of the nanotubes [14]. Besides, when the outer layer become damaged unlike single walled carbon nanotubes (SWCNTs) that has single graphitic plane, the MWCNTs will still have other graphitic planes that will retain its unique properties in TMOs composites [15]. In this study, HEBM techniques were adopted with optimized milling parameters to disperse 1 wt. % MWCNTs in transition metal oxides and the dispersion characteristics of the MWCNTs were ascertain by using various advanced characterization facilities.

2. Materials and Methods

2.1. Starting materials
The starting powder materials used for this study were MWCNTs (NX7100- A -Ultra purified CNT) with average diameter of 9.5 nm and an average length of 1.5 μm, supplied by Nanocyl, Belgium. Titanium (IV) anatase (TiO_2) and manganese (IV) oxides (MnO_2) powders were 99, 6 and 99.9 % at -325 mesh particle size respectively supplied by Industrial Analytical.

2.2. High energy ball mill (HEBM)
A high-energy ball mill (Retch 400, Germany) was used to disperse 1 vol. % MWCNTs into TiO_2/ MnO_2. Dry milling was conducted by charging each TMO powder in a alumina vial (250 ml, inner diameter (ID) = 100 mm) with alumina balls of diameter 5 mm. Ball to powder ratio (BPR) was 10:1 at different time 6 and 9 hours at constant speed of 100 rpm with a relaxation time of 10 minutes, to prevent the samples from overheating. The admixed powders were collected and characterized after milling

2.3. Characterization of as received and admixed powders
Scanning electron microscopy (FESEM, Carl Zeiss Sigma) equipped with energy dispersive X-ray spectrometry (EDX) and transmission electron microscopy (JEOL-Jem 2100) were used for morphology characterisation of starting and admixed powders. X-ray diffractometer (XRD, Rigaku D/max-rB) was used for phase identification scanning using Cu-Kα (λ= 0.154 nm) radiation at a scanning rate of 1 °/min over the angular range of 20–90 °.

3. Results and Discussion

3.1. Morphology of dispersed MWCNTs in titanium oxide and manganese oxide
Figure 1 shows the SEM and TEM images of as-received MWCNTs, TiO_2/MnO_2. The MWCNTs as shown are agglomerate into clusters due to the strong Van der Waal forces between individual MWCNTs. Achieving dispersion of MWCNTs in this transition metal oxide requires suitable milling parameters to de-entangle them by reducing the effects of the strong Van der Waal forces. TiO_2 particles are spherical shaped and MnO_2 particles exhibit amorphous irregular shapes of different granules

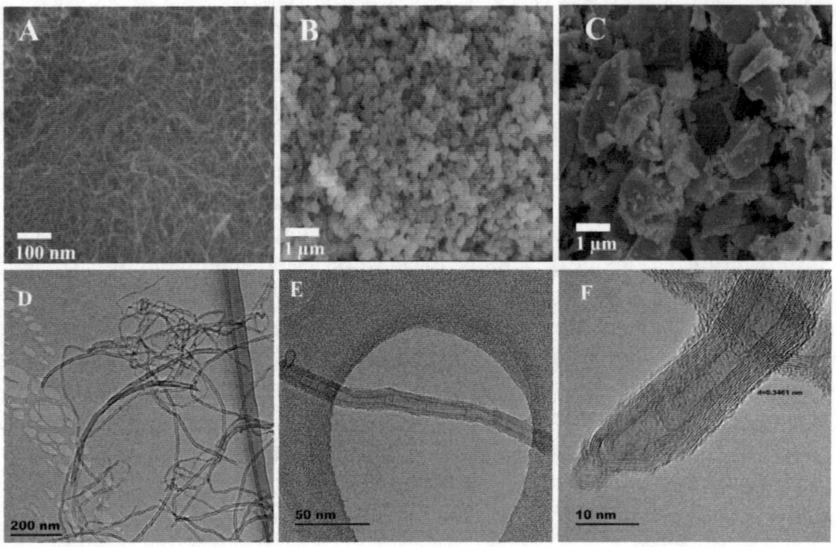

Figure 1: SEM and TEM images of the as-received powders A -MWCNTs, B -titanium oxide and C-manganese oxide, D-F indicates as-received powders MWCNTs (D) agglomerate of as received MWCNTs, (E) tubular structure of individual MWCNTs and (F) HRTEM image showing the individual walls of MWCNTs.

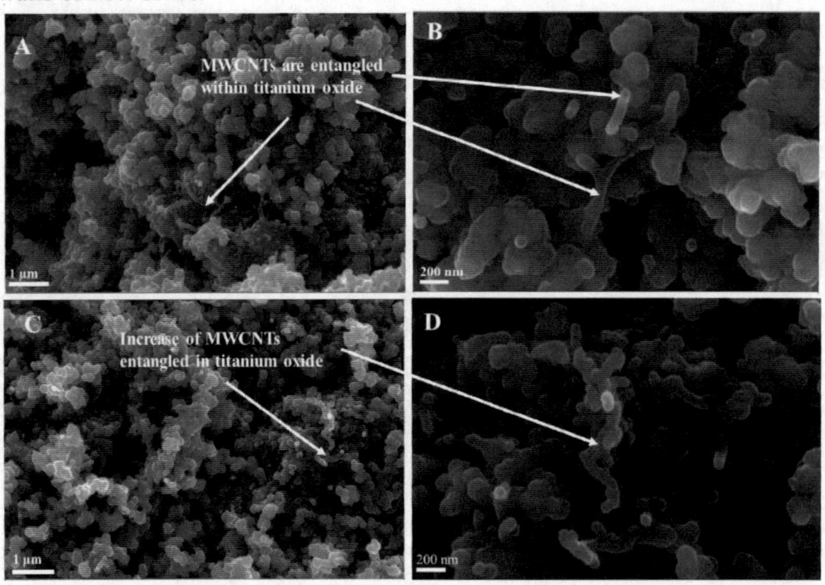

Figure 2: SEM images of admixed 1 wt. % MWCNTs in titanium oxide at lower magnification left and higher magnification right. A-B (6 hours) and C-D (9 hours).

Similarly, the SEM images of the admixed powder containing 1 wt. % MWCNTs in TiO2 after milling for 6 and 9 hours at speed of 100 rpm is shown in figure 2. The SEM images shows figure 2A&C, B&D at lower magnification and higher magnification respectively. It was observed that MWCNTs were well dispersed as the milling time increases, an interlocking behaviour between the TiO2 powder and MWCNTs was also observed and this could as a result of the bonding strength between MWCNTs and TiO2 particles. Consequently, this was further revealed at higher magnification where the MWCNTs expands in diameter in comparison to the as-received and this could be ascribed to the good bonding strength which was formed as a result of lengthy milling time.

Figure 3: SEM images of admixed 1 wt. % MWCNTs in MnO_2 at lower magnification left and higher magnification right. A and B (6 hours) and C-D (9 hours).

Likewise, similar milling parameters were used in the dispersion of 1 wt. % MWCNTs in MnO2 as shown in figure 3. From the SEM image (figure 3A), it was observed that the MWCNTs were relatively clustered on the surface of the MnO2 particles. Due to the increase in milling time at lower magnification (figure 3C), it was observed that MWCNTs detangled to an extent at the same time reducing the MnO2 particle sizes. At higher magnification (figure 3D) the images show the increase of the MWCNTs which was slightly dispersed on the surface of MnO2 oxide particles. This could be due to longer milling time where higher energy was exerted to overcome the Van Waals forces and detangle MWCNTs.

The TEM images of MWCNTs/TiO2 composite materials were shown in figure 4. In figure (4A), the TiO2 particles clumped together around MWCNTs, while in figure (4C), which shows the TEM image of the 6 hours milled MWCNTs/TiO2, the MWCNTs were embedded within the crystalline structure of the TiO2 and there no sign of wall deformation. It was also observed that an increase of milling time in figure (4G) revealed uniformly distribution of the MWCNTs, however, there was a deformation on the wall of the MWCNTs. These deformed walls can lead to the formation of non-sp2 C-C defects which are potential sites for interfacial reaction [16].

Consequently, the above TEM images in figure 5 shows the MWCNTs/ MnO_2 composite that was conducted with similar milling parameters as MWCNTs/TiO_2. It was shown that amorphous MnO_2 particles are clustered around the MWCNTs at figure (5A), however as the milling time increased in figure (5D) the dispersion of MWCNTs was observed alongside with size reduction of MnO_2 particles and this was ascribed to the lengthy milling time which was complimentary to the SEM images. Additionally, the HRTEM in figure (6C & F), shows slightly deformation of their walls, however with increase in milling time, the MWCNTs walls were highly deformed

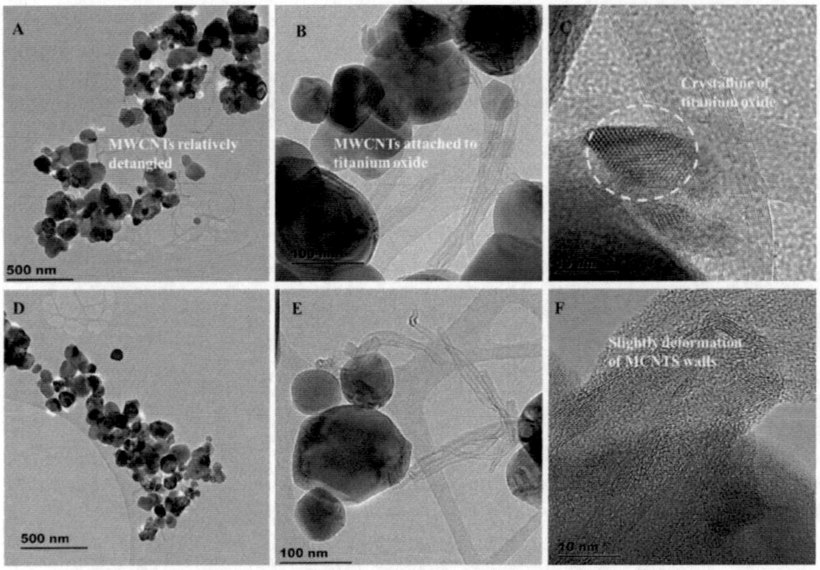

Figure 4: TEM images of admixed 1 wt. % MWCNTs in titanium oxide with varied milling time A-C (6 hours), D-F (9 hours).

Figure 5: TEM images of admixed 1 wt. % MWCNTs in manganese oxide with varied milling time A-C (6 hours), C-D (9 hours).

IOP Conf. Series: Materials Science and Engineering **430** (2018) 012002 doi:10.1088/1757-899X/430/1/012002

3.2. Phase Analysis of the CNTs in transition metal oxides

X-ray diffraction techniques was utilized to identify the various phases, structural changes and the evolution of the crystalline phases in the as-received and admixed powders. Figure 6 shows the XRD patterns of as-received and admixed powders. The peaks observed located at 25.6 and 42.4 are possibly identified to the reflection from the (002) and (100) planes of MWCNTs. Due to addition of MWCNTs to the TiO_2, the peaks observed are as follows; 25.41, 37.94, 48.2, 54.2, 55.3, 62.96, 69.02, 70.53, 75.30 and 82.55 which corresponds to (101), (004), (200), (105), (211), (204), (116), (116), (220), (215) and (303) of anatase TiO_2 powder. The MWCNTs (002) reflection overlaps the anatase TiO_2 (101) reflection in the MWCNTs/ TiO_2 composite. It was also observed that (101) peak slightly decreases with increase of milling time from which indicates that the MWCNTs are interlocked within TiO_2. There are no other phases observed which indicates that the composite materials were well crystallised [17]

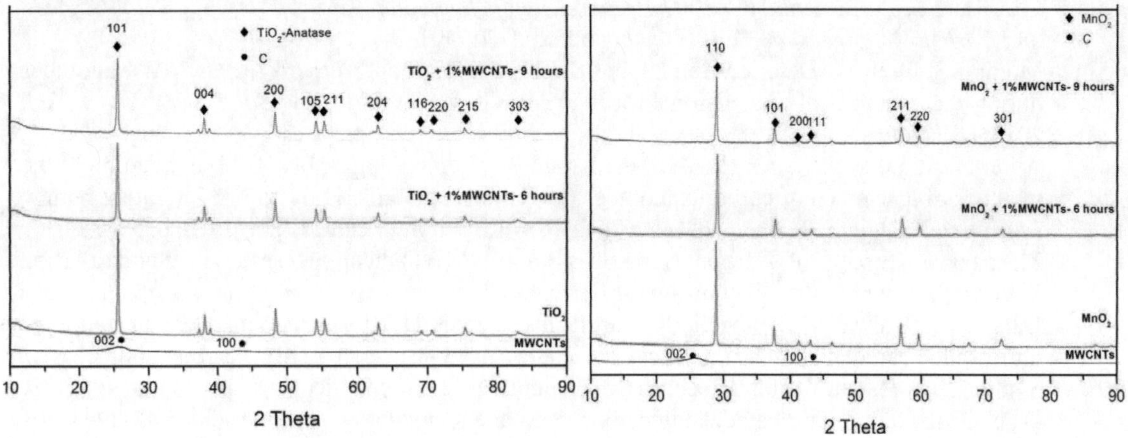

Figure 6: XRD patterns of titanium oxide and MWCNTs/ titanium oxide milled at different times (Left) and manganese oxide and MWCNTs/ manganese oxide milled at different times (Right).

Figure 6 (right) shows XRD patterns of MWCNTs and MWCNTs/MnO_2 composite material. As shown below at 25.6 and 42.4 could be the reflection from the (002) and (100) planes of MWCNTs. The major diffraction peaks which appeared as follows; 28.9, 37.6, 56.9 matched with diffraction peaks of (110), (101) and (211). The MWCNTs/MnO_2 XRD pattern showed that manganese oxide peaks were observed while the diffraction peaks from the MWCNTs were not obvious due to the uniform coating of the MWCNTs by MnO_2 layer [10]. This is complimentary with the SEM and TEM images where the increase of milling time enhanced the dispersion of MWCNTs and bonding to the MnO_2 particles surfaces.

4. Conclusion

In conclusion, the dispersion of MWCNTs in titanium and manganese oxide using HEBM technique became successful. The increase of milling time also increases the dispersion of MWCNTs in titanium and manganese oxide. However, it was observed that the irregular particles of manganese oxide powder promote the tendencies for size reduction of it particles and better dispersion of MWCNTs than titanium oxide. This is obvious on the SEM and TEM results where MWCNTs were relatively dispersed on the surfaces of the manganese oxide and with increase in milling time, the particles of manganese oxide were disintegrated. In addition, the harsh ball milling conditions of the HEBM created structural disorders such as non-sp^2 defects in the form of open edges on the sidewalls and vacancies in C–C system of MWCNTs.

Acknowledgements
I would like to thank the National Research Foundation of South Africa for their financial support.

References

[1] Gupta V, Saleh TA. Syntheses of carbon nanotube-metal oxides composites; adsorption and photo-degradation, *InTech*; 2011.
[2] Hu Y, Guo C. Carbon nanotubes and carbon nanotubes/metal oxide heterostructures: synthesis, characterization and electrochemical property, *InTech*; 2011.
[3] Zhou H, Zhang L, Zhang D, Chen S, Coxon PR, He X, et al. A universal synthetic route to carbon nanotube/transition metal oxide nano-composites for lithium ion batteries and electrochemical capacitors. *Sci Rep* 2016;**6**:1–11. doi:10.1038/srep37752.
[4] Li J, Tang S, Lu L, Zeng HC. 2007-*Preparation of Nanocomposites of Metals , Metal Oxides , and Carbon Nanotubes via Self-Assembly*.pdf 2007;**2**:9401–9.
[5] Munir KS, Wen C. Deterioration of the Strong sp2 Carbon Network in Carbon Nanotubes during the Mechanical Dispersion Processing—A Review. *Crit. Rev. Solid State Mater. Sci.* 2016;**41**:347–66.
[6] Ajiteru OA, Tshephe TS, Okoro AM, Olubambi PA, Lephuthing S, Lesufi M. Evaluation of the methods of dispersion of carbon nanotubes (CNTs) in titanium and its alloys: Literature review. *Mech. Intell. Manuf. Technol.* (ICMIMT), 2018 IEEE 9th Int. Conf., IEEE; 2018, p. 50–3.
[7] Machaka R. Effects of Carbon Nanotubes on the Mechanical Properties of Spark Plasma Sintered Titanium Matrix Composites-*A Review n.d.*
[8] Datsyuk V, Kalyva M, Papagelis K, Parthenios J, Tasis D, Siokou A, Chemical oxidation of multiwalled carbon nanotubes. *Carbon N Y* 2008;**46**:833–40. doi:10.1016/j.carbon.2008.02.012.
[9] Sun Z, Yuan H, Liu Z, Han B, Zhang X. A highly efficient chemical sensor material for H2S: α-Fe2O3 nanotubes fabricated using carbon nanotube templates. *Adv. Mater.* 2005;**17**:2993–7.
[10] Schlecht U, Balasubramanian K, Burghard M, Kern K. Electrochemically decorated carbon nanotubes for hydrogen sensing. *Appl. Surf. Sci.* 2007;**253**:8394–7.
[11] LeeáTan K. Growth of Pd, Pt, Ag and Au nanoparticles on carbon nanotubes. *J. Mater. Chem. 2001*;**11**:2378–81.
[12] Munir KS, Qian M, Li Y, Oldfield DT, Kingshott P, Zhu DM, et al. Quantitative Analyses of MWCNT-Ti Powder Mixtures using Raman Spectroscopy: The Influence of Milling Parameters on Nanostructural Evolution. *Adv. Eng. Mater.* 2015;**17**:1660–9.
[13] Ci L, Ryu Z, Jin-Phillipp NY, Rühle M. Investigation of the interfacial reaction between multi-walled carbon nanotubes and aluminum. *Acta Mater. 2006*;**54**:5367–75.
[14] Ebbesen TW, Ajayan PM. Large-scale synthesis of carbon nanotubes. *Nature* 1992;358:220.
[15] Ajayan PM, Ebbesen TW. Nanometre-size tubes of carbon. *Reports Prog. Phys.* 1997;**60**:1025.
[16] Munir KS, Zheng Y, Zhang D, Lin J, Li Y, Wen C. Microstructure and mechanical properties of carbon nanotubes reinforced titanium matrix composites fabricated via spark plasma sintering. *Mater Sci Eng A* 2017;**688**:505–23. doi:10.1016/j.msea.2017.02.019.
[17] Kiamahalleh MV, Najafpour G, Abd SS, Buniran S, Zein SHS. Multiwalled carbon nanotubes filled with transition metal oxides as supercapacitor materials. *World Appl. Sci. J.* 2010;**9**:1–7.

IOP Conf. Series: Materials Science and Engineering **430** (2018) 012003 doi:10.1088/1757-899X/430/1/012003

Evaluating extrudability of particle-reinforced 2124-Al MMCs and MMNCs within the warm working temperature range (170-280°C)

Zizo Gxowa [1, 2*], Lesley H. Chown [2, 3], Gonasagren Govender [1]

[1]Materials Science and Manufacturing, Council for Scientific and Industrial Research (CSIR), Pretoria, South Africa
[2]School of Chemical and Metallurgical Engineering, University of the Witwatersrand (Wits), Johannesburg, South Africa
[3]DST-NRF Centre of Excellence in Strong Materials, hosted by Wits, Johannesburg, South Africa

*ZGxowa@csir.co.za

Abstract. This work evaluates the warm temperature extrudability of 2124-Al alloy metal matrix composites (MMCs) and metal matrix nano composites (MMNCs) produced by powder metallurgy. Green and sintered compacts were produced by mixing and blending 2124-Al alloy powder with 5 and 10 vol. % Al_2O_3 or 10 and 15 vol. % SiC powders in a high energy ball mill, followed by ambient temperature compaction and sintering at 490°C for 1 hr. An extrusion die was designed to reduce the diameter of the compacts by 30% from 8 mm to approximately 5.6 mm, and the die set was manufactured from H13 hot work tool steel. The die and samples were heated and soaked for 20 minutes, using a high temperature furnace, before extrusion in an Instron® press. Extrusion parameters, such as temperature and speed, were informed by uniaxial compression tests performed on a Gleeble 3500® together with results from Abaqus finite element modelling. The extrusion temperature was 280°C and the speed was 7 mms^{-1}. Extrudability of the unreinforced 2124-Al was good, while the 5 vol. % Al_2O_3 reinforced MMNC and both SiC reinforced MMCs had poorer warm temperature extrudability. The composites were more difficult to extrude than the unreinforced alloy because they have higher resistance to deformation due to the hard and stiff reinforcing particles which do not deform easily. Also, this was attributed to a lack of lubrication during extrusion, which could have increased the friction and resistance to deformation, reducing material flow. Based on microstructural analysis, extrusion showed potential to improve distribution of SiC particles in 2124-Al grains. However, it seems that more deformation is required.

1. Introduction

Metal matrix composites (MMCs) are formed when an intermetallic compound, ceramic, metal or metallic alloy is embedded in a continuous metallic phase [1][2]. The metallic phase takes on the role of the ductile matrix and the phase embedded in it is the reinforcement [3]. The matrix protects the reinforcement and keeps it in an established position, while the reinforcement enhances the mechanical and physical properties of the matrix [4], such as hardness, strength and thermal stability [5].

MMCs may be continuously reinforced with fibres, or discontinuously reinforced by dispersed particles or whiskers with higher hardness in the matrix [3][6]. In MMCs, the size of the reinforcement phase is in the micron-size range, *i.e.* >100 nm [7]. When the size of the reinforcement is reduced to the nano-size range, *i.e.* 1-100 nm, the composite is referred to as a metal matrix nano composite or MMNC [7] [8].

In literature, there is a large body of work dedicated to the development of metal matrix composites (MMCs) and metal matrix nano composites (MMNCs). Achieving a uniform distribution of reinforcing particles in the matrix material continues to be the biggest challenge when fabricating discontinuously reinforced MMCs and MMNCs [9]. It has been reported that a deformation process, such as extrusion, can improve the distribution of the reinforcement in the matrix and strengthen the interfacial bonds between reinforcing particles and the matrix, ultimately enhancing mechanical properties such as hardness and strength[10].

Extrusion is a plastic deformation process where the cross-sectional area of a billet is reduced by applying a compressive force and forcing the billet to flow through a die hole with a smaller diameter [11]. The extrusion temperature is an important parameter because it influences other process variables [12]. When the extrusion temperature is increased, plastic flow in the billet is improved due to a decrease in flow stress, so the required load is lowered and deformation becomes easier [8][12]. This is the main advantage with extruding at high temperatures, referred to as hot extrusion. However, in some studies, high temperatures have been associated with defects such as surface cracks as well as poor surface finish and poor dimensional accuracy. Cold extrusion is done at ambient temperature. While the challenges experienced at higher temperatures are eliminated with cold extrusion, it has its own shortcomings [13], including high stresses due to limited plastic flow and the requirement for high capacity presses which may shorten the service life of the die set [14]. It is evident that there are shortcomings with both hot and cold extrusion, which has created an interest in intermediate warm working temperatures.

Warm working, at temperatures between 30 and 60% of the matrix material melting point, has been shown to have some advantages over both cold and hot working [15][16]. According to [17], there has been an increase in the use of warm forming technologies in the past twenty years. However, there is not much literature on the effect of warm working on deformation and properties of materials, particularly on warm temperature extrudability.

Extrudability was defined by [18] as a metal's ability to withstand deformation and indicates the ease with which the metal can be deformed. This work evaluates the warm temperature (170-280°C) extrudability of 2124-Al MMCs and MMNCs produced by powder blending in a high energy ball mill. The main focus is on the effect of warm temperature extrusion on the deformation and microstructures of these composites, as well as the distribution of the reinforcing particles in the aluminium matrix.

2. Experimental procedure

Laboratory scale extrusion experiments were done on powder metallurgically produced compacts. Green and sintered compacts were made by mixing and blending 2124-Al alloy powder with Al_2O_3 (5 and 10 vol. %) or SiC (10 and 15 vol. %) powders in a Simoloyer CM-01® 2L horizontal high energy ball mill, followed by ambient temperature compaction and sintering at 490°C for 1 hr. Uniaxial compression tests showed that a temperature of 280°C and a strain rate of 5 s^{-1} produced uniaxially compressed samples with the best integrity [19]. Results from these uniaxial compression tests were used as input data for finite element analysis (FEA) using Abaqus. According to the FEA, a strain rate of 5 s^{-1} is equivalent to an extrusion speed of 7 $mm.s^{-1}$ which was then used for the laboratory scale extrusion experiments. An extrusion die was designed to reduce the diameter of the compacts by 30% from 8 mm to approximately 5.6 mm, and the die set was manufactured from H13 hot work tool steel. The die and samples were heated to 280°C and soaked for 20 minutes, using a high temperature furnace, before extrusion in an Instron® press. The Instron® press was connected to a computer and the

Bluehill® (version 2) software was used to study and depict the materials' response during extrusion. All the samples had an initial diameter of 8mm which was inserted into the Bluehill® software prior to extrusion. The movement of the samples in the die, while being extruded as well as the applied load were monitored. The movement of the samples in the die, as a result of the applied load, gave an indication of compressive extension (mm) while the software used the applied load and change in sample diameter to calculate compressive stress. The materials' response was depicted using curves of compressive extension (mm) against compressive stress (MPa). Metallography using a Leica DM15000M® optical microscope and Jeol-6510® scanning electron microscope (SEM) showed the microstructural evolution of the unreinforced 2124-Al alloy, SiC reinforced MMCs and Al_2O_3 reinforced MMNCs during each process step.

3. Results and discussion

3.1 Microstructural evolution (blending, compaction & sintering)
Micrographs showing the microstructural evolution in the unreinforced 2124-Al and composite materials are presented in Figures 1, 2 and 3. Figure 1a shows that the particles of the 2124-Al alloy powder were spherical and, with the exception of a few agglomerated smaller particles, appear to be of similar size. The microstructure of the unreinforced 2124-Al alloy green compact was inconsistent from the surface to the core. Figures 1b and 1c show that better consolidation was achieved at the surface of the 2124-Al green compact than at the core. Figure 1c shows that there was a lot of porosity (shown by darker areas) at the core. This inconsistency in consolidation indicates that there was a density variation in the 2124-Al green compacts as a result of friction effects during compaction. Pressure gradients, caused by friction between the powder and the die walls, lead to non-uniform density distribution in green compacts [20]. The frictional force opposes the applied force and decreases the amount of pressure transmitted to the powder for consolidation [21]. The mid-height of the green compact is therefore exposed to less pressure which would explain the poorer consolidation achieved at the core when compared to the surface of the green compact.

A more uniform distribution of reinforcing particles on the aluminium alloy powder particles was achieved in the Al_2O_3 reinforced MMNCs (Figure 2a) than the SiC reinforced MMCs (Figure 3a). Figures 2b and 3b show that, in the green compacts, reinforcing particles agglomerated at 2124-Al grain boundaries which was attributed to limited plastic flow during cold compaction. So, during cold compaction, there is no mechanism for reinforcing particles at grain boundaries to migrate to the inside of the grains.

Cold compaction of 2124-Al with 10 vol. % Al_2O_3 powder was unsuccessful as the green compacts fractured into nearly horizontal slices. This fracturing was attributed to poor bonding and plastic flow due to the higher density of Al_2O_3 particles on the surface of the 2124-Al alloy powder particles. This result highlighted the need to explore other consolidation techniques, such as spark plasma sintering (SPS), for this MMNC.

It can also be observed that the white eutectic phase inside the 2124-Al alloy grains spheroidised due to incomplete eutectic dissolution (Figures 1d, 2c and 3c). The implication of this incomplete dissolution is that there would be fewer precipitates formed if the materials were subjected to an ageing heat treatment, so the hardness in the aged condition would most likely not differ much from the unaged alloy.

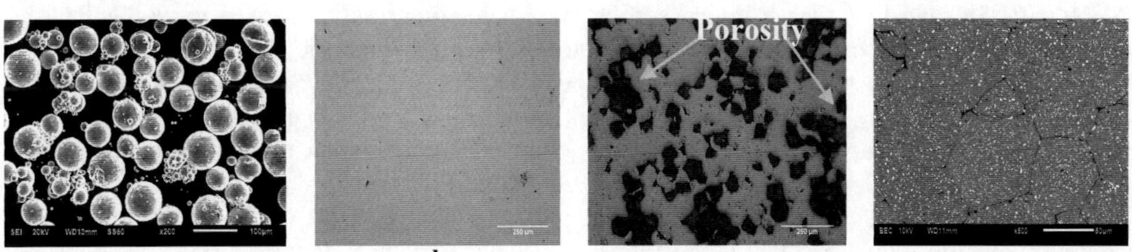

a **b** **c** **d**

IOP Conf. Series: Materials Science and Engineering **430** (2018) 012003 doi:10.1088/1757-899X/430/1/012003

Figure 1. Micrographs showing microstructural evolution in unreinforced 2124-Al from (a) as-received powder (scale bar=100μm) to (b, c) compaction (scale bar=250μm) and (c) sintering (scale bar=50μm).

a b c

Figure 2. Micrographs showing microstructural evolution in Al_2O_3 reinforced MMNCs from (a) blending (scale bar=50μm), (b) compaction (scale bar=2μm) and (c) sintering (scale bar=50μm).

a b c

Figure 3. Micrographs showing microstructural evolution in SiC reinforced MMCs from (a) blending (scale bar=50μm), (b) compaction (scale bar=50μm) and (c) sintering 9scale bar=50μm).

3.2 Extrusion

Figure 4 shows curves of compressive extension (mm) as a function of compressive stress (MPa) depicting the materials' response during extrusion. Compressive extension refers to the movement of the samples during extrusion as a result of the load being applied. The aim of the extrusion process was to apply a compressive force to the samples (*d*=8mm) and force them through a die hole of smaller diameter (*d*=5.6mm) [11]. It can be observed that the unreinforced 2124-Al alloy extruded well and plastically deformed as the dark blue curve labelled 1 in Figure 4 shows that this sample had the highest compressive extension (approximately 15mm). The sudden increase in stress for the Al alloy (the dark blue curve labelled 1 in Figure 4) shows that there was some effect of a lack of lubrication where friction between the die and sample became too high. This increased the resistance to deformation and therefore increased the compressive stress. The materials reinforced with 5% Al_2O_3, 10% SiC and 15% SiC, had very similar maximum compressive stresses at ~900 MPa. The curve shapes of the composites also show that there was limited plastic deformation before the maximum compressive stress was reached. The shape of the curves for the SiC reinforced samples in Figure 4 (curve 3: 10%SiC and curve 4: 15%SiC) are similar which suggests that the materials behaved in a similar manner during extrusion, with total compressive extensions of only 2.7 mm and 2 mm respectively; regardless of the difference in vol. % of SiC (10 vs. 15%). However, curve 2 shows that the composite reinforced with Al_2O_3 had a slightly higher compressive extension (approximately 8.3mm); implying that it had a lower resistance to deformation.

IOP Conf. Series: Materials Science and Engineering **430** (2018) 012003 doi:10.1088/1757-899X/430/1/012003

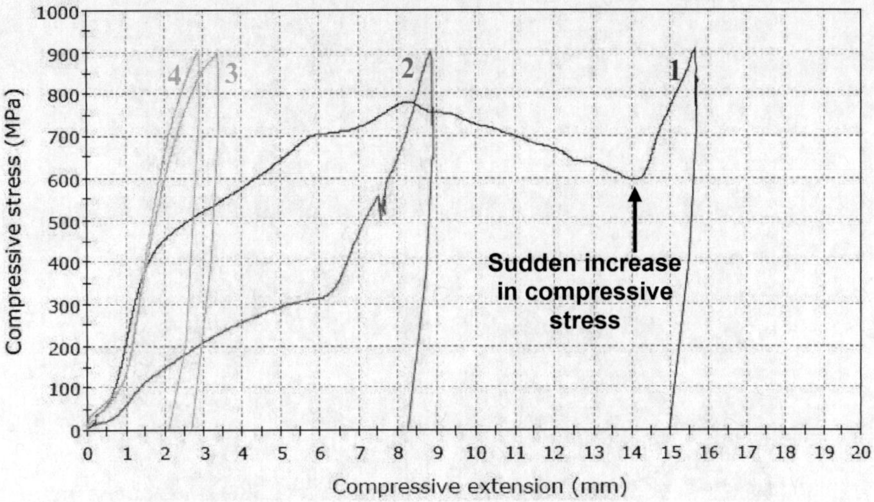

Figure 4. Compressive extension (mm) vs. compressive stress (MPa) for (1) unreinforced 2124-Al (2) 2124-Al with 5% Al_2O_3 (3) 2124-Al with 10% SiC and (4) 2124-Al with 15% SiC during extrusion.

The extruded samples were sectioned along their axes of symmetry, in the extrusion direction, and analysed. Figure 5a shows that in the unreinforced 2124-Al sample, the grains deformed and changed shape. There seems to be good bonding between the 2124-Al alloy grains but there were also many micro voids on the grain boundaries. It was inferred that these micro voids formed due to particle debonding during deformation. Some deformation was achieved in the 2124-Al with 5% Al_2O_3 MMNC since the 2124-Al grains changed shape during extrusion (Figure 5b). However, the Al_2O_3 particles concentrated on the 2124-Al alloy grain boundaries could have limited plastic flow by limiting movement of the 2124-Al grains at the grain boundaries [24].

In the 2124-Al with 10% SiC extruded sample, deformation by elongation of the matrix grains during extrusion can be seen (Figure 5c). This micrograph shows that the SiC particles lie in narrow bands in the extrusion direction. There are some areas where the matrix grains have bonded and there are a few SiC particles distributed inside the deformed grains. Figure 5c illustrates that extrusion could potentially improve matrix grain deformation and distribution of SiC particles inside the 2124-Al grains, however it seems that higher deformation (*i.e.* higher reduction in area) has to be achieved.

Figure 5d shows that the grains in the 2124-Al with 15% SiC sample deformed during extrusion. The higher vol. % of SiC led to more SiC particles concentrated on grain boundaries. While the 2124-Al grains elongated due to deformation, there were very few SiC particles dispersed in the deformed regions. It was deduced that the deformation was restricted to some extent by the high amount of SiC on grain boundaries. There was more resistance to the applied force with 15% SiC than with 10% SiC.

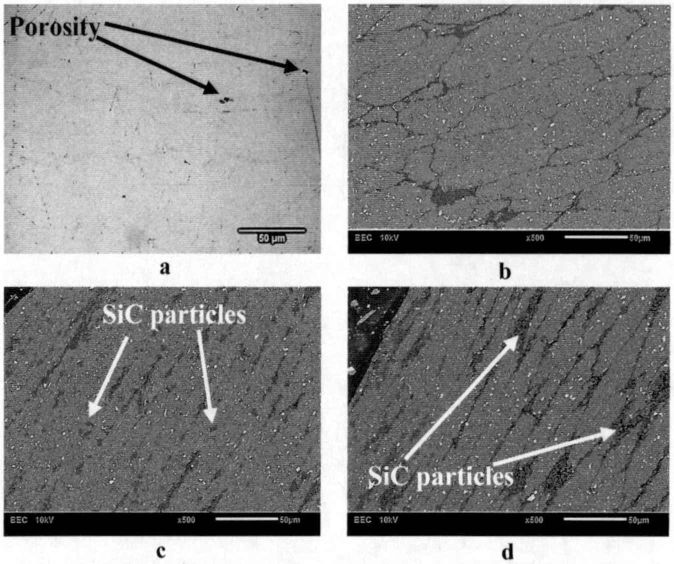

Figure 5. (a) Optical micrograph showing deformation in unreinforced 2124-Al (scale bar=50µm), (b) SEM BSE micrograph showing deformation in 2124-Al with 5% Al_2O_3 (scale bar=50µm), (c) 2124-Al with 10 % SiC and (d) 2124-Al with 15% SiC after extrusion at 280°C (scale bar=50 µm).

4. Conclusions

The extrudability of particle-reinforced 2124-Al MMCs and MMNCs within the warm working temperature range (170-280°C) was evaluated. The following conclusions were made:

1. Extrudability of unreinforced 2124-Al was good while the Al_2O_3 reinforced MMNC and SiC reinforced MMCs had poorer warm temperature extrudability due to higher resistance to deformation (owing to presence of reinforcing particles) and possibly lack of lubrication.
2. Extrusion showed potential to improve the distribution of SiC particles in the 2124-Al reinforced with 10%SiC composite; however, it seems that more deformation is required.

References

[1] Taya M and Arsenault R J 1989 Introduction *Metal Matrix Composites* (Elsevier) pp 1–8
[2] Evans A, San Marchi S and Mortensen A 2003 Metal Matrix Composites *Metal Matrix Composites in Industry* (Boston,Massachusetts: Springer,Boston,MA) pp 9–38
[3] Tumanov A . and Portnoi K . 1972 Composite materials *Met. Powder Rep.* **57** 35
[4] Agarwal B D, Broutman L J and Chandrashekhara K 2006 *Analysis and Performance of Fiber Composites*
[5] Srivatsan T S, Ibrahim I A, Mohammed F A and Lavernia E J 1991 Particulate reinforced metal matrix composites - a review *J. Mater. Sci.* **26** 1137–56
[6] Tjong S 2000 Microstructural and mechanical characteristics of in situ metal matrix composites *Mater. Sci. Eng. R Reports* **29** 49–113
[7] Giannelis E P 1992 A new strategy for synthesizing polymer-ceramic nanocomposites *J. Mater.* **44** 28–30
[8] Saravanan L and Senthilvelan T 2015 Investigations on the Hot Workability Characteristics and Deformation Mechanisms of Aluminium alloy -Al2O3 Nanocomposite *Mater. Des.* **79** 6–14
[9] Hafizpour H R, Khoeini M and Branch S 2011 Compaction of Aluminum Matrix composites reinforced with nano/micro scale SiC Particulates *J. Am. Sci.* **7** 753–9
[10] Tekmen C, Ozdemir I, Cocen U and Onel K 2003 The mechanical response of Al-Si-Mg/SiCp composite: Influence of porosity *Mater. Sci. Eng. A* **360** 365–71
[11] Samanta S K 1972 A new die profile with high process efficiency *Appl. Sci. Res.* **25** 54–64
[12] Saha P 2000 *Aluminium Extrusion Technology* (United States of America: ASM International)

[13] Palumbo G 1998 US5817193A Metal alloys having improved resistance to intergranular stress corrosion cracking 1–7

[14] Bhupatiraju M, Greczanik R and Axle A 2005 Cold Extrusion *ASM Handbook,MetalworkingBulk Form.* **14** 405–18

[15] Rao B K, Khadar M S a and Srinivasan K 1999 High temperature compression testing and determination of warm working temperature for commercial purity aluminium *Bull. Mater. Sci.* **22** 17–20

[16] Van Tyne C J 2004 Design of Forming Processes:Bulk Forming *Handbook of Metallurgical Process Design* ed L Xie, K Funatani and G E Totten (Colorado: CRC Press) pp 1–22

[17] Schubert R 2003 *Warm Forming_Principles,Applications*

[18] Kumar S 2013 Metal Extrusion *Metal Forming: Technology and Process Modelling* ed U S Dixit and G R Narayanan (New D: McGraw Hill Education (India) Private Limited)

[19] Gxowa Z, Chown L H and Govender G 2017 Warm temperature (170-280°C) uniaxial compression of SiC reinforced 2124-Al MMCs *Materials Science and Technology Conference and Exhibition 2017, MS and T 2017* vol 1

[20] Glass S J and Ewsuk K G 1995 Ceramic powder compaction *Am. Ceram. Soc. Int. Symp. Manuf. Pract. Technol.* **c**

[21] Turenne S, Godere C, Thomas Y and Mongeon P E 1999 Evaluation of friction conditions in powder compaction for admixed and die wall lubrication *Powder Metall.* **42** 263–8

[22] Turner C . and Ashby M . 1993 Experimental and Theoretical Modelling of the Isostatic Pressing of Composite Powders *Hot Isostatic Pressing* ed L Delaey and H Tas (Netherlands: Elsevier B.V.) pp 3–14

[23] Kim S H, Chung S W and Padmanaban S 2006 Investigation of lubrication effect on the backward extrusion of thin-walled rectangular aluminum case with large aspect ratio *J. Mater. Process. Technol.* **180** 185–92

[24] Casati R and Vedani M 2014 Metal Matrix Composites Reinforced by Nano-Particles—A Review *Metals (Basel).* **4** 65–83

Structural integrity and dispersion characteristics of carbon nanotubes in titanium-based alloy

A M Okoro[1,*], R Machaka[2], S S Lephuthing[1], M Awotunde[1], and P A Olubambi[1]

[1] Centre for Nanoengineering and Tribocorrosion, Department of Metallurgy, School of Mining, Metallurgy and Chemical Engineering, University of Johannesburg, Republic of South Africa.
[2] Council for Scientific and Industrial Research, Pretoria, South Africa.

*okoromo@gmail.com

Abstract. Over the years, carbon nanotubes have attracted much attention in engineering materials research due to their outstanding and superlative properties. Owing to the demands for lightweight materials with excellent mechanical and thermal properties for diverse industrial applications encouraged the incorporation of carbon nanotubes into titanium alloys. However, there are various challenges associated with the incorporation of carbon nanotubes into titanium alloys which includes; uniform dispersion and retaining the structural integrity of carbon nanotubes after dispersion. Past works have emphasized the importance of homogeneous dispersion with less structural damage to the carbon nanotubes in the various metal matrix. Therefore, this research focused on dispersion of 0.5 wt. % multiwalled carbon nanotubes in Ti6Al4V using low energy ball milling and evaluating the dispersion characteristics and structural integrity of MWCNTs in Ti6Al4V after dispersion. Various characterization techniques such as high-resolution transmission electron microscopy, scanning electron microscopy, X-Ray diffraction and Raman spectroscopy were adopted to ascertain the microstructural evolution, morphology, interfacial reaction and structural integrity of the carbon nanotubes in the Ti6Al4V powders before and after dispersion. The results indicated homogeneous dispersion of carbon nanotubes with less structural damages which are confirmed from the (ID/IG) ratio of the Raman spectra and the TEM images.

1. Introduction

The quest for materials with extraordinary properties for advanced engineering applications have unveiled the benefits of incorporating carbon nanotubes (CNTs) into metal matrix. Since the discovery of carbon nanotubes by Iijima [1] in 1991, they have accomplished unreasonable feats in materials research communities and this is due to their unique and superlative mechanical and thermal properties [2,3]. However, CNTs are recognized for their light weight (~1.7–2.0 g/cm^3), tubular morphology and ultra-high aspect ratios (~100–100,000). These have captured the interest of researchers to exploit their superlative mechanical properties (such as Young's modulus in the tera Pascal (TPa) range), strength up to 100 GPa to synthesize advanced light weight, high strength alloys and composites for aerospace and automotive industries [4]. In addition, the distinguishing properties of CNTs are generally ascribed to their seamless cylindrical morphology, quasi-one-dimensional (1D) structure, graphite-like arrangement of the carbon atoms in the shells and existence of strong sp^2 carbon–carbon (C–C) bonds in their outer shells [3,5].

Owing to the demands for engineering materials with improved properties, titanium and its alloys have attracted considerable attention for diverse applications in aerospace, petrochemical and many other industries [6]. This is due to their unique properties; relatively low densities, high-specific strength, biocompatibility, nonmagnetic properties, high melting point, corrosion resistance and adequate creep resistance at temperatures up to 400 ^{0}C [7]. However, it has been reported from past research that they possess some inferior properties such as poor hardness, stiffness when compared with steels and super nickel alloys, ease of oxidation and loss of mechanical properties at elevated temperature [7] which have encouraged the incorporation of CNTs into their structure for properties enhancements.

During the fabrication of CNTs reinforced titanium matrix composites (TMCs), homogeneous dispersion of the CNTs into the titanium-based matrix without compromising its structural integrity is a paramount challenge. Carbon nanotubes have the tendency to agglomerate into clusters within the titanium-based matrix which is due to their nanoscale dimensions and the strong Van der Waals forces among individual tubes [8]. However, it is essential to uniformly disperse CNTs in titanium-based matrix to access their unique load and heat bearing efficiency in synthesized CNT-TMCs.

In recent times, efforts have been made to homogeneously disperse carbon nanotubes in various metal matrix by adopting different dispersion techniques; solution ball milling [3], ultrasonication [9] functionalization of the CNTs [10], high energy ball milling (HEBM) [5], and gentle milling [11]. The HEBM has proved to be most efficient, effective and economical technique to uniformly disperse CNTs in the metal matrix [12]. However, other research have reported the reactive effects of harsh milling conditions of HEBM on the structural integrity of CNTs [13]. This milling technique leads to the formation of non-sp^2 phase in the C-C bonds of the CNTs resulting in the compromise of the superlative properties, weakening of the strong sp^2 bond, the formation of vacancies and open edges in the CNTs [14]. In addition, the non-sp^2 phase are potential sites for the formation of interfacial reactions at the metal-CNTs interface that promote the creation of titanium carbide phases [3]. Meanwhile, a good metal-CNTs interface is highly desirable to enhance the strengthening mechanism and properties of the resulting composites [15]. In a bid to uniformly disperse multiwalled carbon nanotubes (MWCNTs) in titanium-based matrix without compromising its structural integrity after dispersion, gentle milling approach was adopted in this work.

2. Materials and Method.

2.1 Starting materials

The starting powders utilized for this study were argon atomized, prealloyed Ti6Al4V (~25 μm particle size), supplied by TLS Technik GmbH & Co., Germany and MWCNTs supplied by Nanocycl Belgium with average diameter of 9.5 nm and length of 1.5 μm.

2.2. Powders preparation by gentle milling (GM)

A gentle mill (Retch PM 100, Germany) was utilized to disperse 0.5 wt.% MWCNTs into the Ti6Al4V matrix. This was done by charging the starting powders in a stainless-steel vial (250 ml, inner diameter (ID) = 100 mm) with stainless steel balls of two dissimilar sizes (diameter (d) = 10 and 5 mm). The various sizes of the steel balls were adopted to prevent cold welding of powder particles and enhance the collision energy introduced into the powders. The ball to powder ratio (BPR) was maintained at 5:1. In addition, the milling was carried out for 4 hours at a speed of 150 rpm with a relaxation time of 10 minutes to prevent the samples from heating up. The admixed powders were collected and characterized after milling.

2.3 Characterization of as-received and admixed powders

The morphology of the starting and admixed powders was characterized using scanning electron microscopy (Carl Zeiss Sigma FESEM) equipped with energy dispersive X-ray spectrometry (EDX). Similarly, transmission electron microscopy (JEOL-Jem 2100) was used to characterize the morphology and structural integrity of the MWCNTs before and after dispersion. The phase identification and structural changes of the powder mixtures were characterized by X-ray diffractometer (XRD, Rigaku D/max-rB) and Raman spectroscope (In-Via Raman microscope, Renishaw Plc) respectively. The XRD profiles of as-received powders and admixed powders were scanned using Cu-Kα(λ= 0.154 nm) radiation at a scanning rate of 1°/min over the angular range of 20–90°. Raman scattering was obtained in the spectral range of 200–1800 cm^{-1} on multiple powder mixtures and admixed powder at least 10 various positions with an acquisition of 50 s using a 514 nm laser (laser power = 5.63 mW). The intensity of Raman peaks and their conforming positions were acquired by deconvoluting the Raman spectra into two Gaussian peaks using WITec 2.0 software.

3. Experimentation Results and Discussion

3.1 Dispersion of MWCNTs in Ti6Al4V

The morphology of the as-received Ti6Al4V and MWCNTs powders as shown in Figure 1 (a -d). Due to the strong van der Waal forces between individual MWCNTs, they tend to agglomerate into clusters. Therefore, adequate energy is required to disperse MWCNTs into the Ti6Al4V matrix, and to de-entangle them by reducing the effects of the strong van der Waal forces.

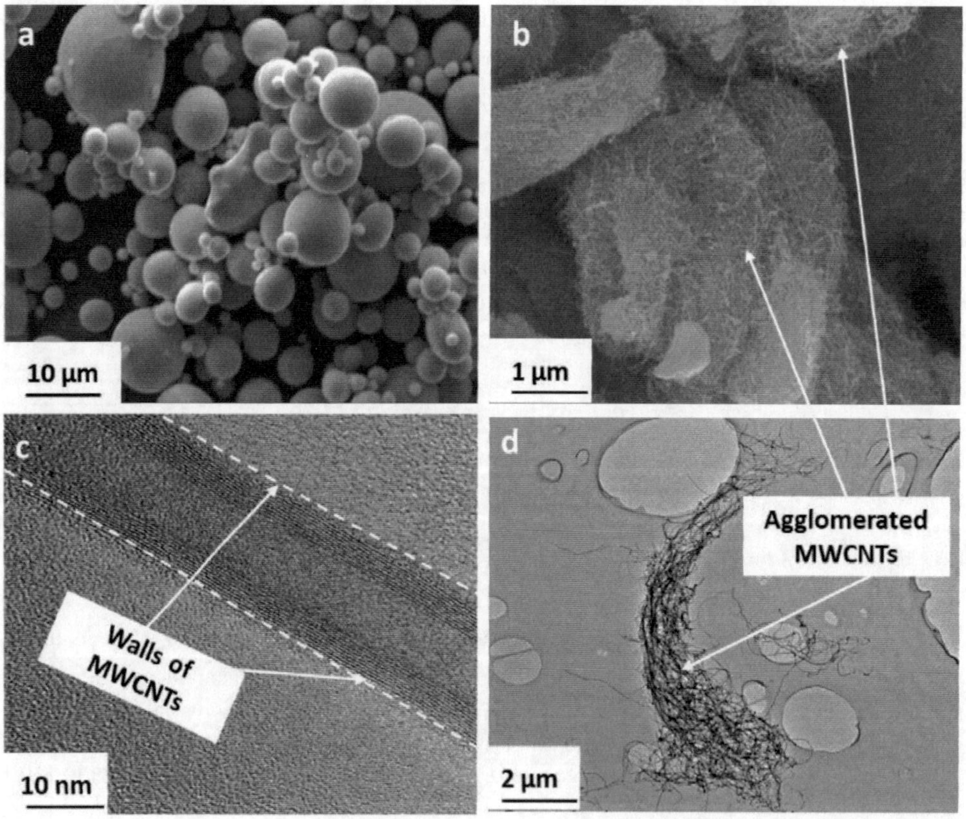

Figure 1. HRSEM (a &b) and HRTEM (c &d) images of the Powders Ti6Al4V (a) and MWCNTs (c-d).

Similarly, the SEM images of the admixed powder containing 0.5 wt. % MWCNTs in Ti6Al4V after milling for 4 h at a speed of 150 rpm is shown in Figure 1 a and b. It was observed that the MWCNTs were dispersed in the Ti6Al4V. However, some level of agglomerations was observed and this was

IOP Conf. Series: Materials Science and Engineering **430** (2018) 012004 doi:10.1088/1757-899X/430/1/012004

ascribed to the strong Van der Waal forces between the individual MWCNTs, high surface area and aspect ratios which inhibit the full homogeneous dispersion of MWCNTs in the TiAl4V [3].

Additionally, it was observed from the SEM (a & b) and TEM (c & d) images of the admixed powders in Figure 2 (a-d) that the MWCNTs were relatively dispersed and embedded in the Ti6Al4V particles, these characteristics was ascribed to the adopted milling parameters that exerted optimum energy to disperse and embed the MWCNTs in the Ti6AL4V particles. However, some level of bundling of MWCNTs were observed when the magnification of the samples was increased from 500 X in figure (2a) to 20000 X in figure (2b). Despites the clustering of the MWCNTs on some portion of the Ti6Al4V particles, the level of recorded dispersion may help to improve the load bearing capability and some other desired properties of the Ti6Al4V matrix when sintered. Moreover, slight deformation to the wall of the MWCNTs (2d) were observed and this is ascribed to the continuous shearing force exerted by the milling balls on the walls of the nanotubes during milling operation.

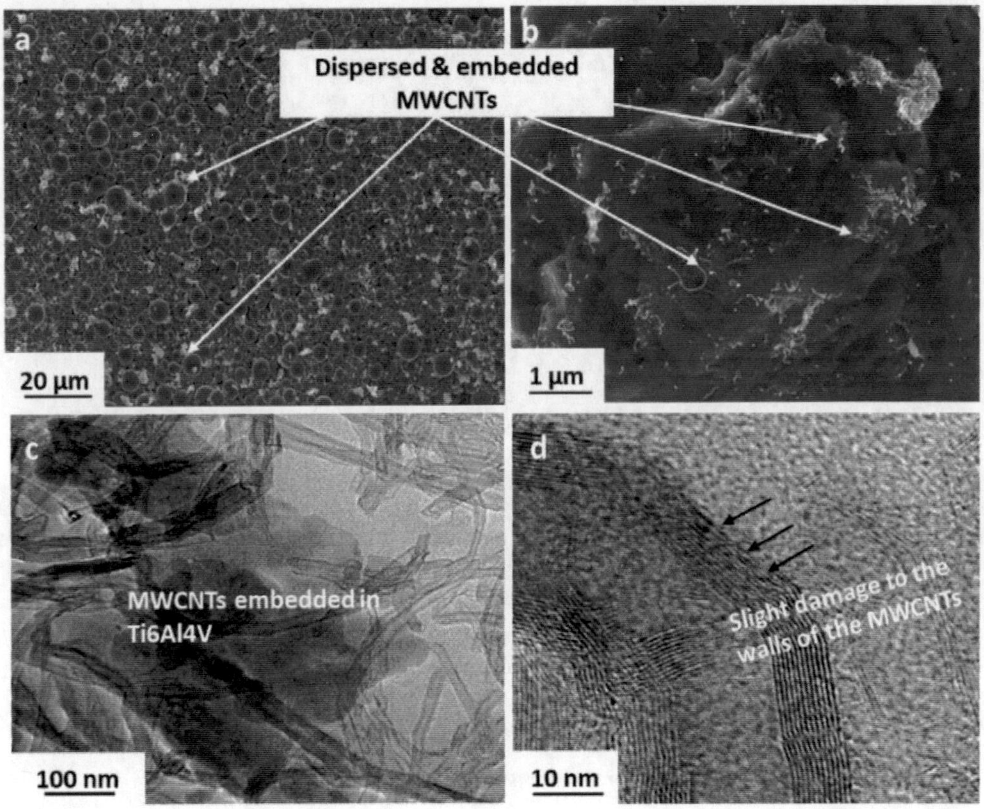

Figure 2. HRSEM (a & b) and HRTEM (c & d) images of the Admixed 0.5 wt. % MWCNTs in Ti6Al4V

3.2 Phase analysis of the MWCNTs in Ti6Al4V

X-ray diffraction (XRD) techniques was utilized to identify the various phases, structural changes and the evolution of the crystalline phases in the as-received and admixed MWCNT-Ti6Al4V. Figure 3 shows the XRD patterns of the starting powders and ball milled powder mixtures. The XRD patterns for the admixed powders depicted only slightly increased peaks of the α-Ti. Since the volume fraction of the MWCNTs in the admixed powders was significantly low (0.5 wt.%). Additionally, the XRD analysis did not identify any MWCNTs peaks in the admixed powders and this may be ascribed to the substantial difference in the mass absorption coefficient for Cu Kα radiations of Ti and carbon (C) which are 208 and 4.6 m^2/g, respectively [16]. However, XRD peaks of the admixed powder did not depict any formation of carbide phase which may likely be formed during harsh milling condition. The formation of carbide phase is usually traceable to the destruction or damage of the MWCNTs structure where vacancies are formed which act as potential sites for the formation of carbide phases (TiC).

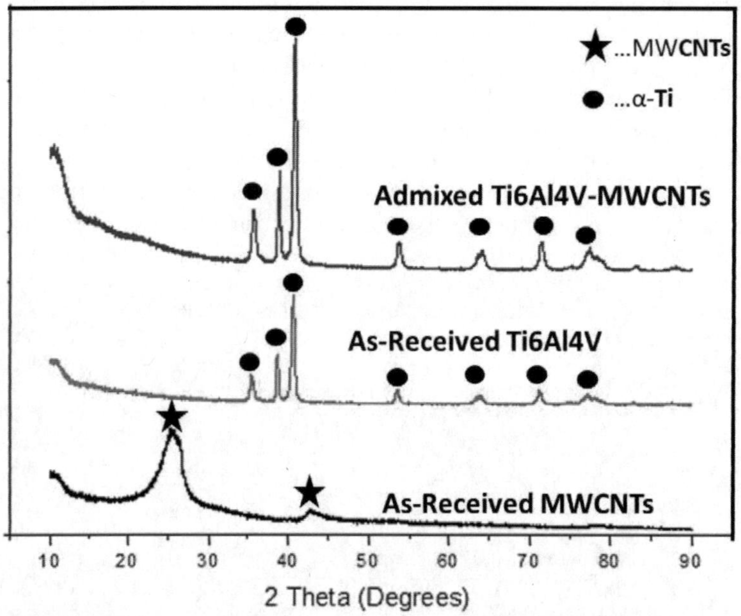

Figure 3. XRD Result of As-received MWCNTs, Ti6Al4V and admixed powders

3.3 Raman analysis of MWCNTs in Ti6Al4V

The structural integrity of the dispersed MWCNTs in Ti6Al4V was determine by Raman spectroscopic technique. This technique is used to characterize the sp^2 carbon structure of the MWCNTs in the admixed powder. The effect of processing method on the MWCNTs during dispersion is reflected on the Raman spectra.

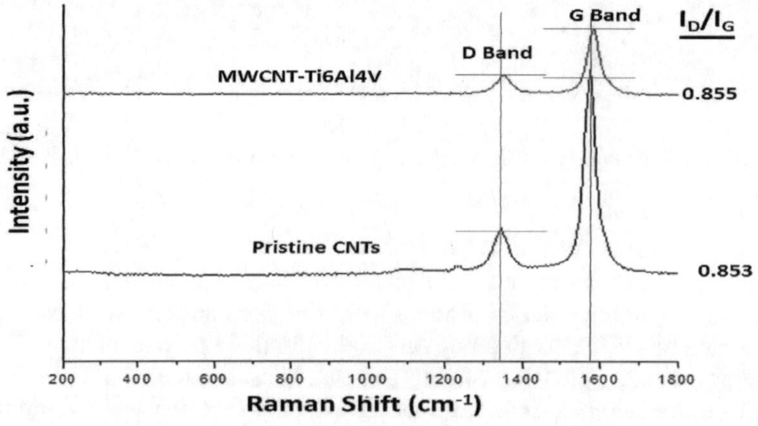

Figure 4. Raman Spectra of the Pristine MWCNTs and Admixed Powders

The Raman spectrum of the pristine MWCNTs is shown for comparison of the effects of the adopted processing parameters on the structural integrity of the MWCNTs during dispersion in Ti6Al4V. As shown in Figure 4, the two graphitic peaks of interest were observed at 1344.9 and 1575.9 cm^{-1} in the Raman spectra of the pristine MWCNTs, which represents the D band and G band, respectively. The D band depicts the defect peak and it is related to the disorder in the C-C bond which is sometimes regarded as the non-sp^2 structural defects in the MWCNTs [3]. However, G-Band represent the in-plane stretching mode of C-C bonds and it indicates the degree of crystallinity in the MWCNTs [17]. From the Raman spectra of the pristine MWCNTs and admixed powders, it was observed that the I_D/I_G ratio is 0.853 and 0.855 respectively. This is an indication that there was a slight increase in the Raman peak ratio (I_D/I_G) to 0.855 which correspond to 0.2 percent increase and a slight movement of the G band to a higher wave number of 1581.8 cm^{-1}. The increased in the Raman peak ratio is an indication of a minimal deformation in the crystalline structure of MWCNTs during dispersion which is in conformity with the TEM result where slight damaged was observed at the walls of the MWCNTs in the admixed powder. However, the level of structural damages doesn't result in the formation of titanium carbide during dispersion which would have reflected as peaks between 200 and 800 cm^{-1} on the Raman spectra of the admixed powders [3]. In addition, non-sp^2 structural defects were not formed during dispersion and this is in conformity with the XRD results where no titanium carbide phase was recorded.

4. Conclusion

The adoption of gentle milling in incorporating MWCNTs into Ti6Al4V is a viable approach for composites production since it results in relative uniform dispersion of MWCNTs with minimal structural damages to the MWCNTs. This was achieved by milling at lower energy and controlling the heating up of the samples during milling. The TEM, Raman and XRD characterization results indicated that there were less structural damages to the MWCNTs and no unwanted phase was formed during milling operation. Additionally, the SEM results revealed homogeneous dispersion with less clustering of the MWCNTs.

Acknowledgments

The authors wish to acknowledge the National Research Foundation of South Africa and the Global Excellence and Stature of the University of Johannesburg for providing financial support for this work.

Reference

[1] Iijima S 1991 Helical microtubules of graphitic carbon *Nature* 354-56
[2] Bakshi S R, Lahiri D and Agarwal A 2010 Carbon nanotube reinforced metal matrix composites - a review *Int. Mater. Rev.* 55 41–64
[3] Munir K S, Li Y, Liang D, Qian M, Xu W and Wen C 2015 Effect of dispersion method on the deterioration, interfacial interactions and re-agglomeration of carbon nanotubes in titanium metal matrix composites *Mater. Des.* 88 138–48
[4] Munir K S, Zheng Y, Zhang D, Lin J, Li Y and Wen C 2017 Improving the strengthening efficiency of carbon nanotubes in titanium metal matrix composites *Mater. Sci. Eng. A* 696 10–25
[5] Kondoh K, Threrujirapapong T, Imai H, Umeda J and Fugetsu B 2009 Characteristics of powder metallurgy pure titanium matrix composite reinforced with multi-wall carbon nanotubes *Compos. Sci. Technol.* 69 1077–81
[6] Machaka R Effects of Carbon Nanotubes on the Mechanical Properties of Spark Plasma Sintered Titanium Matrix Composites-A Review
[7] Leyens C and Peters M 2003 *Titanium and titanium alloys: fundamentals and applications* (John Wiley & Sons)
[8] Thostenson E T, Ren Z and Chou T-W 2001 Advances in the science and technology of carbon nanotubes and their composites: a review *Compos. Sci. Technol.* 61 1899–912

[9] Wang X, Lu M, Qiu L, Huang H, Li D, Wang H and Cheng Y B 2016 Graphene/titanium carbide composites prepared by sol–gel infiltration and spark plasma sintering *Ceram. Int.* 42 122–31

[10] Lacerda L, Bianco A, Prato M and Kostarelos K 2006 Carbon nanotubes as nanomedicines: from toxicology to pharmacology *Adv. Drug Deliv. Rev.* 58 1460–70

[11] Ajiteru O A, Tshephe T S, Okoro A M, Olubambi P A, Lephuthing S and Lesufi M 2018 Evaluation of the methods of dispersion of carbon nanotubes (CNTs) in titanium and its alloys: Literature review *Mechanical and Intelligent Manufacturing Technologies (ICMIMT), 2018 IEEE 9th International Conference on* (IEEE) pp 50–3

[12] Esawi A M K, Morsi K, Sayed A, Taher M and Lanka S 2010 Effect of carbon nanotube (CNT) content on the mechanical properties of CNT-reinforced aluminium composites *Compos. Sci. Technol.* 70 2237–41

[13] Kim Y A, Hayashi T, Fukai Y, Endo M, Yanagisawa T and Dresselhaus M S 2002 Effect of ball milling on morphology of cup-stacked carbon nanotubes *Chem. Phys. Lett.* 355 279–84

[14] Gill P and Munroe N 2012 Study of carbon nanotubes in Cu-Cr metal matrix composites *J. Mater. Eng. Perform.* 21 2467–71

[15] Chu K, Jia C chang, Jiang L kun and Li W sheng 2013 Improvement of interface and mechanical properties in carbon nanotube reinforced Cu-Cr matrix composites *Mater. Des.* 45 407–11

[16] Ye L L and Quan M X 1995 Synthesis of nanocrystalline TiC powders by mechanical alloying *Nanostructured Mater.* 5 25–31

[17] Delhaes P, Couzi M, Trinquecoste M, Dentzer J, Hamidou H and Vix-Guterl C 2006 A comparison between Raman spectroscopy and surface characterizations of multiwall carbon nanotubes *Carbon N. Y.* 44 3005–13

IOP Conf. Series: Materials Science and Engineering **430** (2018) 012005 doi:10.1088/1757-899X/430/1/012005

Batch adsorption study of uranium on various ion exchange resins as an alternative method to solvent extraction

J S Gama[1,*], J Barry[1], and P L Crouse[2]

[1] Applied Chemistry Department, The South African Nuclear Energy Corporation SOC Limited (Necsa), Elias Motsoaledi Street Extension, R104 Pelindaba, Pretoria 0240, South Africa.
[2] Department of Chemical Engineering, University of Pretoria, corner Lynnwood Road and Roper Street, Hatfield 0028, South Africa

*Jabulani.gama@necsa.co.za

Abstract. The project is aimed at recovering uranium from scrap material containing useful amounts of Low Enriched Uranium. The common method of recovering uranium is by using acidic or alkaline medium for dissolution, followed by solvent extraction using tributyl phosphate in kerosene. However the normal challenge with solvent extraction is the generation of enormous liquid waste which is slightly radioactive. Therefore in this project we studied the adsorption of uranium on various ion exchange resins in order to use it as an alternative method to solvent extraction. Uranium solution from the scrap materials was used, and different ion exchange resins such as Amberjet 1600 H, PPC100H, Purolite S940 and S950 chelation resin, and Purolite A500 were tested under various conditions for the adsorption of uranium. The strong acid cation resins show no selectivity between the uranium and aluminium species, both species were fully adsorbed irrespective of the pH. The strong base anion resins had good uranium adsorption attaining 100%, they also show a pH effect for the aluminium adsorption. The use of ion exchange resins, especially with anion resins showed promising results and potential for use as an alternative to solvent extraction.

1. Introduction and background

The South African Fundamental Atomic Research Installation reactor (SAFARI-1) currently uses Low Enriched Uranium (LEU) silicide fuel. Therefore during the fabrication process of fuel for nuclear research reactors, scrap material is generated. This scrap material contains un-irradiated uranium and silicon as major elements, the other elements present are traces of Mg, Fe and Mn which originate from the aluminium alloy used as cladding material. The uranium and silicon are used as starting material to produce uranium silicide (U_3Si_2), which is crushed and mixed with aluminium powder to produce a solid piece, the so-called fuel plate meat [1].

Since the introduction of LEU silicide fuel for research reactors, investigations regarding dissolution and recovery have been ongoing in order to recover the valuable uranium. For dissolution, the common methods entail acidic [2] or alkaline dissolution [3]. The common acidic routes use nitric acid, especially for the dissolution of LEU silicide fuel plates, especially those with aluminium cladding in high concentrations (3 to 6 M) [4]. However, even with higher concentrations, long dissolution times were still encountered by Touron and Cheroux (2001), whereby the cladding was removed after 3 hours using 3 to 6 M nitric acid and complete dissolution was only obtained in 6 to 8 hours [5]. However, to improve the dissolution rates, catalysts such as HF, H_2O_2, KIO [6] and Hg^{+2} have been recommended [7]. The challenge during HNO_3 dissolution is the precipitation of silicon due to its low solubility in concentrated nitric acid [3]. While in an alkaline medium, the dissolution of the uranium silicide fuel plate scrap material is normally conducted using a two-step process. The first step remove almost the entire aluminum from both the cladding and the fuel meat with either NaOH or a combination of NaOH-$NaNO_3$, followed by dissolution of the remaining U_3Si_2 particles in NaOH plus H_2O_2 at about 70 °C [8]. Alternatively, the remaining U_3Si_2 particles could be dissolved in a solution of NaClO (12% w/v) [9] or HF plus H_2O_2 [10]. Similar to the acidic medium, the comminuted particles of U_3Si_2, after the cladding has been removed, are shown to dissolve faster in the peroxide solution, according to the chemical reaction as described in equation 1. The challenge with alkaline dissolution is the decomposition of the peroxide [11].

$$U_3Si_2 + H_2O_2 + 22OH^- \rightarrow 2SiO_3^{2-} + 3UO_2 + 5.5O_2H_4^{2-} + H_2O \qquad (1)$$

Normally, after dissolution of the scrap material or fuel plate, the uranium is recovered by liquid-liquid extraction using TBP in kerosene because of its high selectivity for the uranyl nitrate species ($UO_2(NO_3)_2$) from aqueous solutions [12]. However, the challenge with solvent extraction is the generation of an enormous amount of liquid waste, which is slightly radioactive.

The aim of this work was to study uranium adsorption on various ion exchange resins, in order to use ion exchange (IE) as a possible alternative method to solvent extraction (SE). IE is regarded as an inherently cheaper method than solvent extraction [13], with the benefit of resin transformation for reuse in another cycle. IE is also used in combination with solvent extraction (Eluex process), for example it was used in high uranium concentrations of up to 900 ppm [14]. IE was used to extract uranium from leach liquor, the resin eluted with sulfuric acid then uranium was extracted using the SE process. Ion exchange resins are widely used in various areas from water softeners [15] to waste water treatment, such as the recovery of uranium from ground water [16]. There are many ion exchange resins reported in literature that are suitable for uranium adsorption from various waste streams, such as activated charcoal used to separate uranium from other alkaline metals, alkaline earth metals and lanthanide metals [17], Amberjet IRA-400 to adsorb uranium in acid leach [18], and Amberjet 1200 H (AHR) was used to study the uptake behavior of uranium from a nitric acid medium [19]. Other types of ion exchange resins used in irradiated fuel for the recovery of ^{99}Mo are TiO_2 and Al_2O_3 [20].

Ion exchange technology was previously used in earlier unpublished internal Necsa projects for the recovery of uranium from waste streams generated from purification processes via the PUREX method. Some of the resins used were Amberlite IRC-747, Diphonix, Purolite S940, and Amberjet 4400 Cl^-. These resins were used in this study as the previous studies indicated them as the best for this application.

2. Experimental

2.1. Materials and method

The fuel plate scrap material was supplied by the Nuclear Energy Corporation of South Africa (Necsa). The leaching was conducted using two separate methods. The first method used 3 M nitric acid with HF as a catalyst at 70 °C, and the second method used $NaOH\text{-}NaNO_3$ (dissolution 1) and $NaOH\text{-}H_2O_2$ (dissolution 2) at 65 to 70 °C. Screening tests of the resins were conducted using synthetic solutions of uranium, aluminium, silicon, and magnesium to test adsorption of the uranyl nitrate species $UO_2(NO_3)_2$ as well as adsorption of the other species (i.e. aluminium, silicon and magnesium). The reagents were of analytical grade supplied by LABCON South Africa. After leaching, samples were taken to measure the concentrations of the elements (U, Al, Si, Mg) using ICP-OES (Model: ARCOS FHS12) and UV/Vis spectrophotometry. The UV/Vis spectrophotometer (Biochrom Libra S60) was only used for the measurement of the uranium concentration in the range of 20 to 200 mg/L using the Na_2CO_3/H_2O_2 method. Different pH targets were attained by dilution of the pregnant leach to the desired volume and the adsorption experiments were conducted using an incubator shaker (Model: ISKO 80) at room temperature at a speed of 200 rpm. All adsorption tests were conducted using 1 g of resin with 50 mL solution for each experiment for a period of 24 h. The list of ion exchange resins used is given in Table 1.

Table 1. List of ion exchange resins used for adsorption of uranium species.

Resin name	Type	Form	Functional group	Capacity
Amberjet 1600 H	Strong acid cation	H^+	Sulfonic acid	2.4 eq/L
PPC100H	Strong acid cation	Na^+	Sulfonic acid	2.0 eq/L
Purolite S950	Chelating	Na^+	Aminophosphonic	24 g/L
Puromet MTS9500	Chelating	Na^+	Aminophosphonic	26 g/L
Purolite S940	Chelating	Na^+	Aminophosphonic	20 g/L
Puromet MTA5012	Strong base anion	Cl^-	Type I Quaternary Ammonium	1.15 eq/L
MTA8000PPSO4	Strong base anion	SO_4^{2-}	Quaternary Ammonium	1.6 eq/L
Purolite A500	Strong base anion	Cl^-	Type I Quaternary Ammonium	1.15 q/L

2.2 *Modelling of speciation*
Uranium usually exists as different species in acidic and alkaline medium, depending on various factors such as the pH. Software (Hydra/Medusa software), similar to that used by Khawassek (2018) [19] to predict uranium species under nitric acid conditions, was used to calculate the species of uranium and aluminium under HNO_3, and under NaOH plus H_2O_2, conditions. The calculated uranium species in HNO_3 corresponded to the work done by Khawassek (2018). At a pH below 4, the uranium species were found to be $UO_2NO_3^+$ and UO_2^{2+} suggesting good uranium adsorption at low pH with cation resins. However, at the same pH range of 0 to 4, the aluminium and magnesium species co-exist as cation species, namely $AlOH^{2+}$ and Mg^{2+}. The predicted uranium and aluminium anion species under NaOH-H_2O_2 conditions are $UO_2(OH)_3^-$ and $UO_2(OH)_4^{2-}$ found from pH 10 and above, while the aluminium species $Al(OH)_4^-$ appears from pH 8. This suggests a good adsorption for both uranium and aluminium using an anion resin. The complex $Na_2UO_2(O_2H)_4$, was however not predicted by the software. The adsorption percentages were calculated from the differences between the initial and final concentrations according to equation (2). Where C_0 and C_e are the initial and final concentrations of the solution.

$$\% \text{ ads} = \frac{C_0 - C_e}{C_0} \times 100 \qquad (2)$$

3. Results and discussion

3.1 *Acidic medium adsorption*
The strong acid cation (SAC) resins and chelating resins were tested for uranium adsorption from the acidic solution. The batch tests were conducted for a period of 24 hours, using 1 g sample of resin with 50 mL. The solution had concentrations of 90.65 mg/L uranium, 135.15 mg/L aluminium, 53.12 mg/L silicon, and 9.56 mg/L magnesium with a pH of 1.26. The results showed more than 90% adsorptions for both uranium and aluminium, but more aluminium was adsorbed as compared to uranium. The magnesium adsorption was always 100% for all the resins (Table 2). Silicon was eliminated since the small traces existing were not adsorbed for all the resins tested and the majority precipitated after dissolution and the precipitate was separated by centrifugation. The adsorption of both uranium and aluminium was expected from the calculations, as the $UO_2NO_3^+$ and Al^{3+} species were dominant from pH 1 to 4, however the results showed no distinct selectivity between the adsorption of the uranium and aluminium species, even though the initial aluminium concentration was higher than that of uranium. The similarity of the results, especially for the 1600 H and PPC100H resins, could be due to the similarity of the functional groups and the large capacity of 2.4 and 2.0 eq/L, respectively, while the similarity between the S950 and MTS9500 could be due to the same Na^+ form and aminophosphonic functional group.

Table 2. Adsorption percentages from acidic leach with concentrations of 18 mg/L U and 27 mg/L at pH 1.26.

Resin	U adsorption (%)	Al adsorption (%)	Mg adsorption (%)
1600 H	91.6	95.9	99.4
S940	89.8	91.3	100
MTS9500	92.5	97.4	100
PPC100H	91.1	95.7	99.7

Increasing the pH to 2 showed minor changes, with slight increases on the uranium adsorption and slight decreases on the aluminium adsorption, while magnesium adsorption remained almost similar and is not included in Tables 3. However, there was a noticeable slight increase at pH 3 with all resins reaching 100% adsorption for both uranium and aluminium species (Table 3). This indicated the effect of pH and initial concentrations for both uranium and aluminium. At low pH with higher concentration, the adsorption was about 90% and increases as the concentrations decreased when diluted to pH 2 and 3.

Table 3. Adsorption percentages from acidic leach of pH 2and 3.

Resin	pH 2 adsorption		pH 3 adsorption	
	U (%)	Al (%)	U (%)	Al (%)
1600H	96	92	100	99
S940	95	98	99	99
MTS9500	100	88	100	99
PPC100H	100	96	100	99

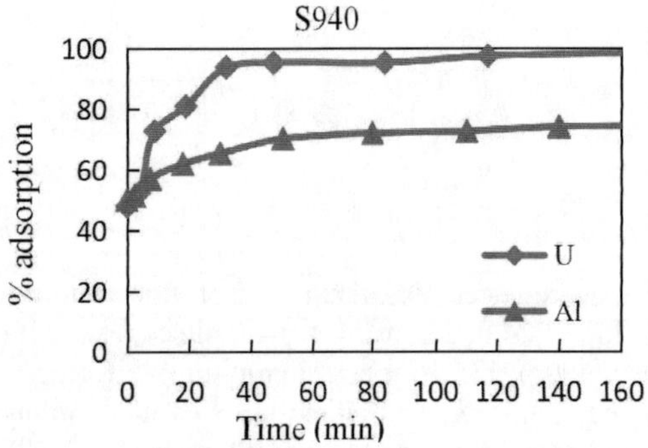

Figure 1. Effect of contact time on U and Al adsorption on S940

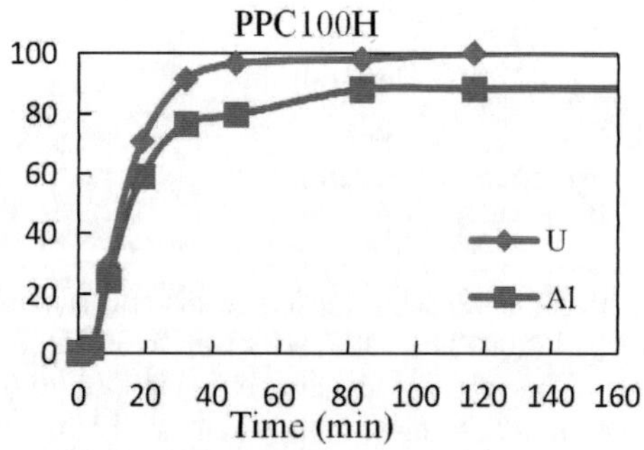

Figure 2. Effect of contact time on U and Al adsorption on PPC100H

The influence of contact time with the strong acid cation and chelating resins was determined at pH 1.26 and the study time ranged from 0 to 160 min. The solution contained uranium and aluminium concentrations of 40.46 and 59.25 mg/L, respectively. All resins showed practically similar uranium adsorption, attaining around 70% within the first 20 min. The quickest adsorption (72%) was with S940 (Figure 1), followed by PPC100H (Figure 2) and MTS9500 (Figure 3), while the 1600 H resin was the slowest, attaining 60% uranium adsorption within the same time (Figure 4). A similar uptake was also observed for aluminium adsorption, the adsorption uptake within 20 min varied from 36 to 58%. PPC1000H showed the quickest uptake (58%), followed by 1600 H (47%) and MTS9500 (41%). The S940 showed the slowest uptake for aluminium as compared to the other resins.

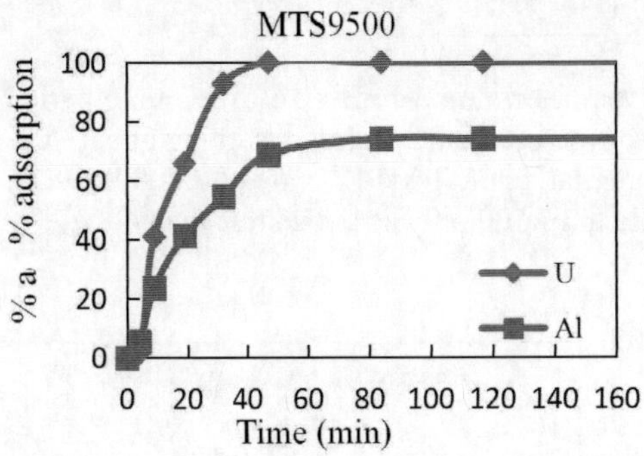

Figure 3. Effect of contact time on U and Al adsorption on MTS9500

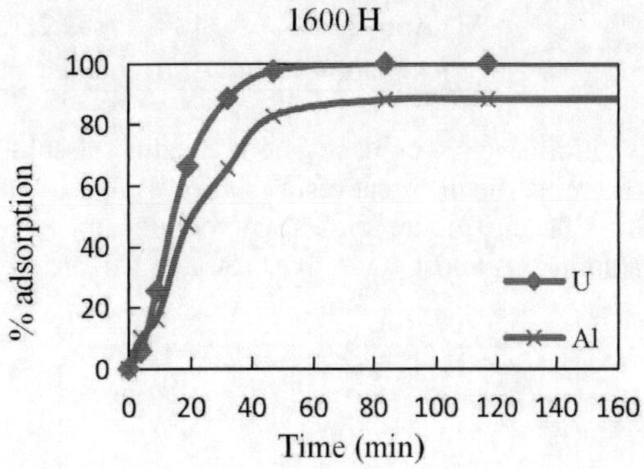

Figure 4. Effect of contact time on U and Al adsorption on 1600 H

3.2 *Alkaline medium adsorption*

The concentrations after the alkaline dissolution were 1152 mg/L U, 254 mg/L Al, 226 mg/L Si, and 10 mg/L Mg with a pH of 12.8 and about 80% of the aluminium was removed by the first dissolution. At pH 10, all resins show similar results, with 100% adsorption for the uranium species, predicted as $UO_2(OH)_3^-$ (Table 4). The aluminium adsorption was also similar with all resins attaining around 93%, and the same for silicon with all resins attaining around 85%. The two resins A500 and MTA5012 were expected to produce similar results due to the similarity in their functional groups and the Cl^- form.

Table 4. Adsorption from alkaline leach (pH 10) with 20 mg/L U, 10 mg/L Al and 30 mg/L Si.

Resin	U adsorption (%)	Al adsorption (%)	Si adsorption (%)
A500	100	93.4	85.9
MTA5012	100	93.4	85.7
MTA8000PPSO4	100	92.8	82.8

Increasing the pH to 11 and 12 had no significant change on the uranium adsorption for all three resins however; it caused a decrease in the aluminium adsorption. At pH 11 the aluminium adsorption dropped to about 66% for A500, 82% for MTA5012, and 86% for MTA8000PPSO4 (Table 5). A further decrease was observed at pH 12.6 to an average of 32% for all resins (Table 5). The low adsorption of the aluminium species was unexpected, since the Al(OH)4- species was predicted to be stable from about pH 10 to 14. The observation made was that the aluminium adsorption decreases at higher concentrations of both the uranium and aluminium. Thus, at pH 10 with low concentration of both U and Al, a complete uptake of both species was observed, the resins able to adsorb all amounts available. But at high concentrations pH 11 and 12, the aluminium adsorption dropped, suggesting that the adsorption might follow the order of adsorbing uranium species first, and only then aluminium filling the vacant sites.

Table 5. Adsorption from alkaline leach (pH 11) with 129 mg/L U, 75 mg/L Al and 58 mg/L Si.

	Adsorption at pH 11			Adsorption at pH 12		
Resin	U (%)	Al (%)	Si (%)	U (%)	Al (%)	Si (%)
A500	100	66.3	75.9	98.7	33.9	39.4

MTA5012	100	82.2	78.0	98.0	32.1	40.7
MTA8000PPSO4	100	86.2	71.3	99.4	30.6	27.1

The influence of contact time in alkaline dissolution was determined at pH 12.6 over a time period of 180 min. The different resins showed similar uptake for the uranium species, adsorbing about 90% within 30 min (Figure 5). The aluminium uptake was similar for MTA5012 and MTA8000PPSO4, both attaining almost 100% over 180 min (Figure 6), while the A500 only adsorbed less than 40%.

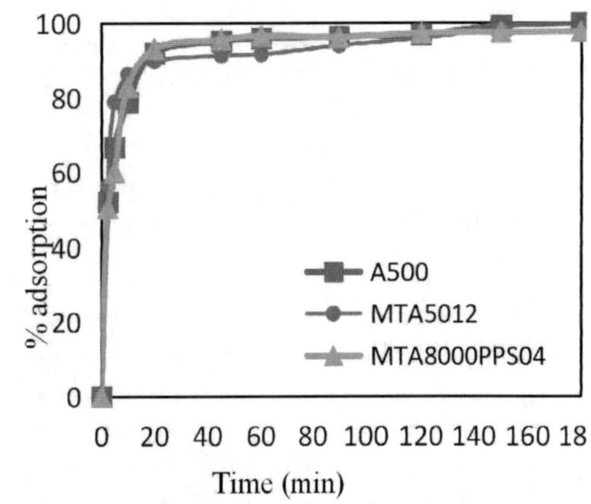

Figure 5. Effect of contact time upon U and Al adsorption with MTA8000PPSO4

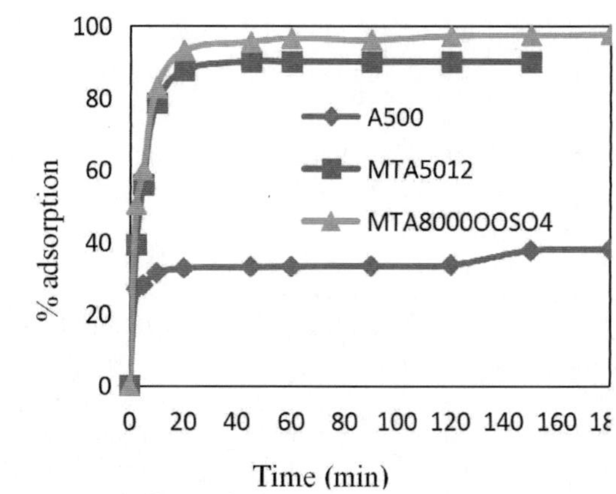

Figure 6. Effect of contact time upon U and Al adsorption with A500

4. Conclusions

The obtained results showed that the strong acid cation resins had no distinct selectivity for uranium cations, making it difficult to separate the uranium from the aluminium. The adsorption was also not sensitive to either the pH or the initial concentration of the solution. Full adsorption was obtained for both uranium and aluminium, most likely due to the high capacity of the strong acid cation. The adsorption efficiency would be determined in future work using column experiments.

However, the strong base resins, especially MTA8000PPSO4 and A500 showed good uranium adsorption with less aluminium adsorption at high pH, above 12. This indicated that it could be used with high uranium concentrations and aluminium will not be adsorbed due to a low loading capacity of about 1.6 eq/L. However, another alternative to remove all the aluminium could be further rinsing of remaining residues after the first dissolution step with the NaOH-NaNO3. The uranium after adsorption could be eluted using a solution of NaOH, NaCl or sulphuric acid, depending on the resin form. The uranium can then be recovered from the pregnant eluates by precipitation.

References

[1]. Saliba-Silva A M, Urano de Carvalho E F, Riella H G and Durazzo M 2011 Research Reactor Fuel Fabrication to Produce Radioisotopes, Brazil

[2]. Ikeda Y, Yasuike Y, Nishimura K, Hasegawa S, and Takashima Y 1995, Kinetic study on the dissolution of UO2 powders in nitric acid, J. Nuc.l Materials. 224, 266-272

[3]. Kim K, Hyun J, Lee E, Park G, Lee K, Yoo M, Song K and Moon J 2011, Recovery of uranium from (U, Gd)O2 nuclear fuel scrap using dissolution and precipitation in carbonate media, J.of Nucl. Materials. 418, 93-97.

[4]. YunFeng Z and Jing C 2008 Kinetics study on the dissolution of UO2 particles by microwave and conventional heating in 4 mol/L nitric acid Sci. China Ser B Chem 51 (7) 700-704

[5]. Touron E and Cheroux L 2001 Silicon behaviour during reprocessing of uranium silicide fuel. 5th international topic meeting on research fuel management, 9-12 March 2003, France.

[6]. Sameh A A 1995 Production of fission Mo-99 from LEU uranium silicide target materials, Symp., 18-20 May 2000, Taiwan.

[7]. Rodrigues G C and Gouge A P 1983 Reprocessing RERTR silicide fuel US Department of Energy, USA

[8]. Buchholz B A and Vandergrift G F 1995 Processing of LEU targets for 99Mo production- Dissolution of U3Si2 targets by alkaline hydrogen peroxide 18-22 September 1995 International meeting on RERTR, France.

[9]. Cristin P R, Cols H J, Bavaro R, Bronca M, Centurion R and Cestau D 2000 Production of molybdenum-99 from low enriched uranium targets, 3-8 November 2002 International RERTR meeting, Argentina.

[10]. Sameh A A 2013 Production Cycle for large scale fission Mo-99: Separation by the processing of irradiated LEU uranium silicide fuel element targets Sci and Tech. Ncl Instll Article ID 704846, 14 pages

[11]. Chen L, Dong D, Buchholz B A and Vandergrift G F 1996 Progress in alkaline peroxide dissolution of low enriched uranium metal and silicide targets, 19th International meeting on Reduced Enrichment for Resaerch and Test Reactor, 7-10 October 1996, Korea, Seoul.

[12]. Mohammed A K, Sullivan J C and Nash K L 2007 Kinetics of the complexation of UO2(NO3)2 with tributly phosphate (TBP) in xylene, solvent extraction and ion exchange, 18 5, 809-820, DOI: 10. 1080/07366290008934708

[13]. Wen Zhenqian and Niu Y 2017 Recovery of uranium from low concentrations leach liquor of acid in-situ leach IOPConf.Ser.: Mater. Sci. Eng. 191 012035

[14]. Van Tonder D and van Hege B 2007 Uranium recovery from acid leach liquors The optimisation of RIP/SX based flowsheets 2-4 September 2013 The fourth Southern African Conference on base metals, South Africa

[15]. Botez A, Dobre T, Panturu E, Filcenco-Olteanu A 2014 Uranium (VI) adsorption equilibrium on Purolite resin SGA 600U/3472 Central Europen J. Chem. 12 (7) 769-773.

[16]. Mattigod S V, Cordova E A, Golovich E C, Smith R M and Wellman D M 2010 Uranium adsorption on Ion-Exchange resins- Batch testing. US. Department of Energy, USA.

[17]. Qadeer R, Hanif J, Saleem M and Afzal M 1992 Effect of alkali metals, alkaline earth metal and lanthanides on the adsorption of uranium on activated charcoal from aqueous solutions, J. Nuc.l chemicals, 165 (4) 243-253.

[18]. Guettaf H, Becis A, Ferhat K, Hanou K, Bouchiha D, Yakoubi K and Ferrad F 2009 Concentration-purification of uranium from an acid leaching solution Phys Procedia 2 765-771.

[19]. Khawassek Y M, Masoud A M, Taha M H, and Hussein A E M 2018 Kinetics and thermodynamics of uranium ion adsorption from waste solution using amberjet 1200 H as cation exchanger, J. Rradio analytical and nuclear chemistry 315 493-502.

[20]. Stepinski D C, Krahn E O, Chung P and Vandegrift G F 2014 Design of column separation processes for recovery of molybdenum from dissolved high density LEU target (Report ANL/CSE-13/34) Argonne National Laboratory USA

IOP Conf. Series: Materials Science and Engineering **430** (2018) 012006 doi:10.1088/1757-899X/430/1/012006

Concentration of rare earth elements from monazite by selective precipitation

S M Xaba[1*], M Nete[2] and W Purcell[1,3]

[1]Department of Chemistry, University of the Free State, Bloemfontein 9300, South Africa
[2]The South African Nuclear Energy Corporation SOC Ltd. (Necsa), P.O. Box 582, Pretoria 0001, South Africa
[3]School of Chemical and Metallurgical Engineering, University of the Witwatersrand, Johannesburg, South Africa

*sbu012@gmail.com

Abstracts: Pure rare earth elements and their compounds play a unique role in a wide range of green technology and different industrial applications such as medicine, aerospace and clean energy production. The aim of this study was to concentrate and separate rare earth elements from the monazite matrix. Monazite was digested using a mixture of $Na_2HPO_4/NaH_2PO_4{\cdot}H_2O$ fluxes at 900 °C. Precipitation of rare earth elements was investigated with biphenyldicarboxylic acid and acetyl acetone chelating agents. The precipitates were re-dissolved in sulphuric acid and analysed together with the filtrate solutions using inductively coupled plasma optical emission spectrometry. Analytical results indicated selective precipitation of rare earth elements with both biphenyldicarboxylic acid and acetyl acetone. The results also revealed a negative Na^+ influence (from the flux) on the rare earth elements recovery in solution which is attributed to easily ionisable behaviour of Na which affects the flame properties of the inductively coupled plasma optical emission spectrometry. Na was removed as NaCl by addition of concentrated HCl prior rare earth elements precipitation. This step improved the rare earth elements recovery in the precipitate from 90.2(1) and 101.68(3) %, while the impurities concentrations remained below 30 %. These results indicated the successful concentration and separation of most rare earth elements from monazite by selective precipitation.

1. Introduction

The rare earth elements (REE) are defined as 15 lanthanide (lanthanum to lutetium) group elements with many similar chemical properties together with yttrium and scandium. These elements are usually divided into two groups namely the light (LREEs) and heavy (HREEs). The LREEs consist of the elements from lanthanum (La) to gadolinium (Gd) on the periodic table, while the HREE are from terbium (Tb) to lutetium (Lu) [1]. REEs and their compounds are widely used in modern technologies. For example, neodymium alloys are used in manufacturing of permanent magnets used in catalytic production of ammonia and petroleum field [2]. REEs are also in industrial applications such as in electronics, laptops, mobile phones, electrical transportation, medical, clean energy and defence [3]. Their properties improve energy efficiency and performance at higher temperatures in digital technologies [4],[5].

Monazite and bastnäsite are considered to be the most important source of LREE worldwide, while xenotime is the main source of many HREEs and yttrium [6]. Monazite is also an important source of Th and can be highly radioactive depending on the thorium (Th) quantities in the mineral. Monazite concentrate commonly contains up to 45 % REE_2O_3, 18 % P_2O_5, 8.8 % ThO_2 and 600 ppm uranium deposits [7]. South Africa is known to have reasonably large deposits of the REE with highest mineral grades which are mined at the Steenkampskraal in the Western Cape Province. The mine has a

Mineral Resource Estimate (MRE) of 605000 tons of ore with an average grade of 14.36 % containing 86900 tons Total Rare Earth Oxides (TREO) [8]. The REE minerals (mainly the cerium type monazite (Ce) - (Ce, La, Nd, Th, Y)) are mined in SA and due to their Th (and some traces of uranium (U) in significant concentration) [9] content which makes the mineral radioactive they are primarily processed near the mining site.

The radioactive elements, U and Th, can add a huge financial and operational burden in beneficiation of the REEs in monazite. The similar chemical properties between the REEs make the separation of these elements from one another highly challenging and often costly. In previous studies [10],[11],[12], REE, Th and/or U were selective precipitated from the other elements in a REE mineral samples using the different precipitating agent. Sulphuric acid (H_2SO_4) leaching or sodium hydroxide (NaOH) fusion flux methods were used to extract these elements from the mineral into an aqueous solution. From the weakly acidic solution (pH <4), REE, Th and/or U were then precipitated as hydroxides, oxalate, double sulphate, carbonate and fluoride [13].The aim of this work was to develop separation and purification methods using selective precipitation of LREE from monazite using different complexing (precipitating) agents (Figure 1). The selective precipitation of LREE, Th and/or U in the dissolved monazite sample was investigated with biphenyldicarboxylic acid and acetyl acetone as precipitating agents. Experimental factors which were investigated include the influences of acid matrix, sodium phosphate flux as well as sample:reagent mole ratios.

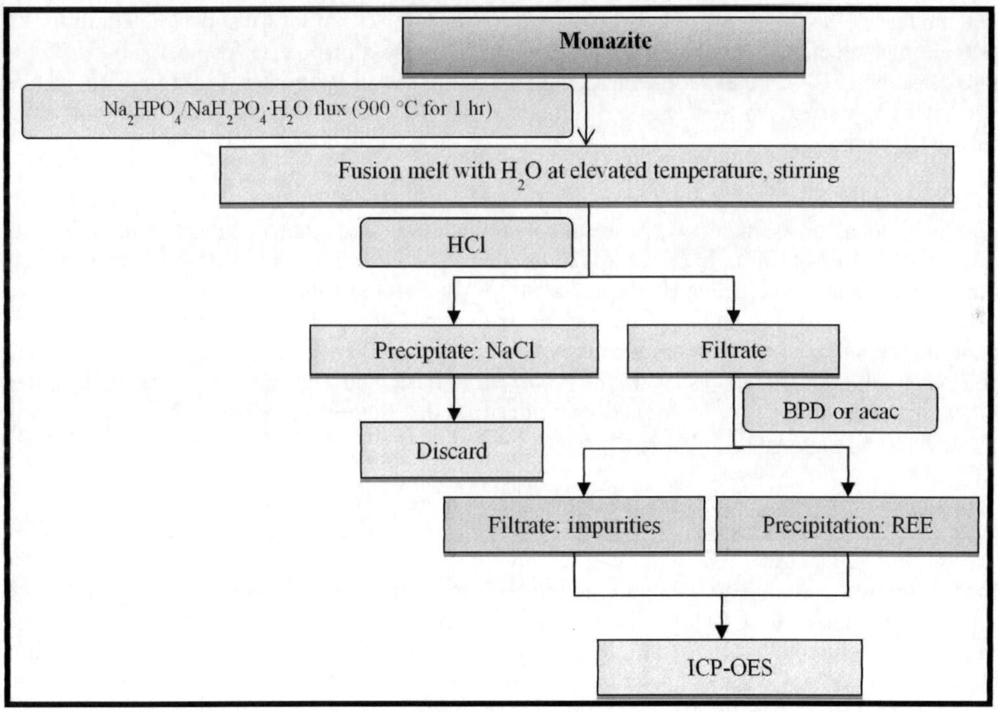

Figure 1: Scheme for the separation selective precipitation and concentration of rare earth elements from monazite.

2. Experimental

2.1 Reagents and glassware

Crushed monazite ore was supplied by the Steenkampskraal mine. Analytical grade HCl (32 %) and H_2SO_4 (98 %) and ethanol (99.5 %) were bought from Merck. Sodium phosphate dibasic (Na_2HPO_4) ≥99 % and Sodium dihydrogen phosphate monohydrate ($NaH_2PO_4·H_2O$) ≥ 98 % were purchased from Sigma-Aldrich. Acetyl acetone (acac) 99.5 % and 2, 2'-biphenyldicarboxylic acid (BPD) ≥ 98 % were bought from Riedel-de Haen AG and Merck respectively. The REE ICP multi-element standard containing 1000 mg/L (Ce, La, Pr, Y and Sm) in HNO_3 7 % v/v, was purchased from Inorganic

Ventures Inc. while a multi element standard 1000.000 mg/L (Ca, Fe and Mg) in dilute HNO_3 was purchased from Merck. Individual ICP standard solutions of 1000 mg/L Nd (in 2 % HNO_3 v/v), Si (traces in HNO_3 and HF), U (5 % HNO_3 v/v) and Th (5 % HNO_3 v/v) ICP standards were sourced from Inorganic Ventures Inc. and Ti in H_2O was purchased Merck. Grade B glass volumetric flasks and Schott Duran type beakers were bought from Merck and Sigma-Aldrich respectively. Centrifugation was performed in a rotor bought from MSE. A Shimadzu ICPS-7510 ICP-OES sequential plasma spectrometer was used for the qualitative and quantitative analysis of all elements in aqueous solutions. Ultra-pure water (conductivity, 0.00 mS/cm) was prepared in the laboratory with an ultra-reverse osmosis system bought from AJD Traders and was used for preparation of all solutions.

2.2 Sample preparations and measurements

2.2.1 Preparation of calibration curves. 2.0 g of 1:1 $Na_2HPO_4/NaH_2PO_4{\cdot}H_2O$ fluxes was accurately weighed in a platinum crucible and mixed thoroughly before been placed in a pre-heated furnace oven at 900 °C. The resultant melt was dissolved in water and quantitatively transferred to a 100.0 mL volumetric flask. Calibration solutions of 0.0, 1.0, 2.0, 3.0, 5.0, and 10.0 ppm analyte concentrations were prepared by dilution of appropriate volumes of the primary ICP-OES standard solutions. Quantitative analyses were performed at selected analytical lines of 383.826 nm for Mg, 252.852 nm for Si, 412.323 nm for La, 234 nm for Fe, 437.494 nm for Y, 410.946 nm for Nd, 318.020 nm for Th, 385.466 nm for U, 334.904 nm for Ti, 428.079 nm for Sm, 446.021 nm for Ce and 428.293 nm for Pr. The wavelength selections were based mostly on selectivity (less or no interference) and sensitivity for high detections. ICP-OES experimental conditions of 1.2 kW of RF power, 14.0 L/min coolant Ar flow, 1.2 L/min plasma Ar flow and 0.5 L/min carrier Ar gas flow were used and maintained throughout the study.

2.2.2 Dissolution of monazite using a mixture of $Na_2HPO_4/NaH_2PO_4{\cdot}H_2O$ as a flux. Approximately 0.1 g of monazite sample was accurately weighed and homogenously mixed with $Na_2HPO_4/NaH_2PO_4{\cdot}H_2O$ in a mass ratio of 1:15 sample:flux. The mixture was quantitatively transferred into a platinum crucible and heated at 900 °C in a closed oven with a periodical swirling at 15 min intervals. Visual inspection of the sample indicated a complete digestion after 60 min. The resultant melt was cooled to room temperature and dissolved in 30 mL of water. The dissolved melt was quantitatively transferred to 100.0 mL volumetric flask and filled to the mark with water. Analytical determinations were performed in triplicate to determine the elemental concentrations of the constituents in the mineral (Table 1, Expected).

2.3 Selective precipitation of REE from the solution containing the flux

0.5002 g of 2, 2'-biphenyldicarboxylic acid and 0.51 mL of acetyl acetone were separately dissolved in 100 mL in volumetric flask with ethanol and/or water, depending on which dissolves the precipitating agent completely. 10.0 mL aliquot of monazite solution (see **Section 2.2.2**) was subsequently transferred to a 100 mL beaker, in H_2SO_4 medium. 10 mL of precipitating agent was added to these mixture with a mole ratio of 1:5 sample:precipitating agent. Immediately after adding the chelating agents to a colorless solution, both reaction mixtures changed colour to yellow. The reaction mixture was stirred for 10 min and then left overnight to allow for precipitation to occur. Visual inspection of the samples on the following day indicated a white precipitate had formed in the acac solution and a yellow precipitate in BPD mixture. The solutions were separated from precipitate by centrifugation and decantation and the filtrate was quantitatively transferred into 100.0 mL volumetric flasks. The precipitates were also dissolved with H_2SO_4 and quantitatively transferred into 100.0 mL volumetric flasks. Both precipitate and filtrate solutions were filled to the mark with water. Then the solutions were analysed using ICP-OES (see Table 1).

Table 1. Metal content in the precipitation from the monazite matrix before removing NaCl.

M_xO_y(%)	Exp(s)*	Acac				BPD			
		Ppt(s)*	RSD (%)	Fil(s)*	RSD (%)	Ppt(s)*	RSD (%)	Fil(s)*	RSD (%)

MgO	0.2(3)	0.11(1)	0.83	0.25(2)	9.84	0.16(2)	7.07	0.22(1)	4.12
SiO_2	3.0(1)	1.2(1)	3.72	3.8(2)	8.28	1.3(1)	4.21	3.5(2)	8.13
CaO	8.1(1)	8.4(1)	1.84	11.1(6)	7.17	8.7(3)	3.66	11.0(6)	7.17
TiO_2	0.34(2)	0.02(8)	23.74	0.55(3)	9.56	0.03(2)	7.25	0.54(4)	11.45
Fe_2O_3	3.90(6)	2(2)	41.63	5.5(3)	6.44	0.1(3)	7.84	6.0(3)	8.96
Y_2O_3	1.59(1)	1.24(6)	3.89	0.49(6)	3.99	1.26(1)	0.34	0.44(1)	7.16
La_2O_3	8.9(1)	8.74(4)	0.50	nd	--	8.7(1)	1.15	nd	--
Ce_2O_3	17.9(1)	17.4(1)	0.60	nd	--	17.4(2)	0.85	nd	--
Pr_2O_3	0.7(1)	0.48(2)	9.17	nd	--	0.4(2)	26.92	nd	--
Nd_2O_3	6.59(3)	6.25(7)	1.00	nd	--	6.24(3)	0.51	nd	--
Sm_2O_3	0.64(7)	0.56(2)	3.43	nd	--	0.53(9)	13.82	nd	--
ThO_2	5.28(6)	4.99(2)	0.45	0.2(1)	2.57	4.96(7)	1.28	nd	--
U_3O_8	3.2(3)	2.8(3)	8.34	nd	--	2.6(3)	10.10	nd	--

M_xO_y(%) – Metal oxide percentage Exp(s) – Expected(standard deviation) Ppt(s) – Precipitate(standard deviation), Fil(s) – Filtrate(standard deviation) Tot – Total nd - not dected *Average of 3 replicates

2.4 Selective precipitation of REE from the solution without the flux.

An additional procedure was applied to reduce the salt (Na^+) concentration. 20 mL of 32 % HCl was added to the clear monazite solution (see Section 2.2.2) and then left overnight to allow for maximum precipitation of Na as NaCl. The same procedure (see the second paragraph on Section 2.3) was repeated for both precipitating agents with the 10.0 mL aliquot of monazite solution. Then precipitate and filtrate solutions were analysed using ICP-OES (see Table 2).

Table 2: Metal content in the precipitation from the monazite matrix after removing NaCl.

M_xO_y(%)	Exp(s)*	Acac				BPD			
		Ppt(s)*	RSD (%)	Fil(s)*	RSD (%)	Ppt(s)*	RSD (%)	Fil(s)*	RSD (%)
MgO	0.2(3)	0.01(1)	3.05	0.14(1)	3.57	0.01(1)	8.36	0.12(1)	6.14
SiO_2	3.0(1)	0.1(1)	4.77	2.8(2)	5.42	nd	--	0.5(4)	12.44
CaO	8.1(1)	7.2(6)	6.88	11.1(4)	4.51	10.8(5)	6.66	9.8(7)	8.20
TiO_2	0.34(2)	nd	--	0.25(2)	0.69	nd	--	0.09(1)	1.87
Fe_2O_3	3.90(6)	0.02(9)	2.18	4.25(6)	1.58	0.1(2)	4.93	4.2(2)	4.93
Y_2O_3	1.59(1)	0.72(4)	2.43	0.90(8)	4.79	0.73(3)	1.95	0.81(7)	4.26
La_2O_3	8.9(1)	8.0(2)	2.80	1.0(8)	8.49	8.64(6)	0.63	0.6(3)	2.97
Ce_2O_3	17.9(1)	17.0(1)	0.55	nd	--	17.5(2)	1.25	nd	--
Pr_2O_3	0.7(1)	0.38(5)	8.19	nd	--	0.50(2)	0.30	nd	--
Nd_2O_3	6.59(3)	6.08(1)	1.50	nd	--	6.68(5)	0.83	nd	--
Sm_2O_3	0.64(7)	0.47(2)	3.51	nd	--	0.32(3)	39.71	nd	--
ThO_2	5.28(6)	4.85(1)	0.19	0.37(6)	11.64	4.2(1)	2.82	0.3(5)	8.65

U_3O_8	3.2(3)	3.1(1)	4.53	0.11(7)	2.03	2.33(7)	2.07	0.6(1)	3.59

M_xO_y(%) – Metal oxide percentage Exp(s) – Expected(standard deviation) Ppt(s) – Precipitate(standard deviation), Fil(s) – Filtrate(standard deviation) Tot – Total nd - not dected *Average of 3 replicates

3 Results and discussion

3.1 Selective precipitation of REE from the solution with and without the flux

Two different precipitating agents were investigated for a possible selective precipitation and purification of LREE, U, Th (or removal of impurities) from monazite. Complete sample dissolution was accomplished using a flux fusion digestion with a mixture of $Na_2HPO_4/NaH_2PO_4{\cdot}H_2O$ flux. The sample/salt melt dissolves in water. Therefore, precipitation studies were investigated in water. The results indicated that both chelating agents (biphenyldicarboxylic acid and acetyl acetone) preferentially precipitated LREE, U and Th from the monazite matrix. Most of the LREEs together with U and Th were obtained in the precipitate and their recoveries (based on values in Table 2) ranged between 80.5(2) to 98.74(9) % while most of the other elements had remained in solution. However, analysis of the filtrate solution (see Figure 2) indicated higher than acceptable recoveries for the remaining elements. This observation prompted the investigation of the flux influence on the analysis of the solutions. Literature study has indicated that Na^+ ion negatively influence the flame properties of the ICP-OES and commonly enhances analyte intensities [14]. The high Na^+ concentration was reduced by addition of HCl [15] for precipitation of Na as NaCl (*cf* **Tables 1** and Tables 2). The selective precipitation was repeated on the solutions with low Na^+ content and the results are presented in Figure 4.

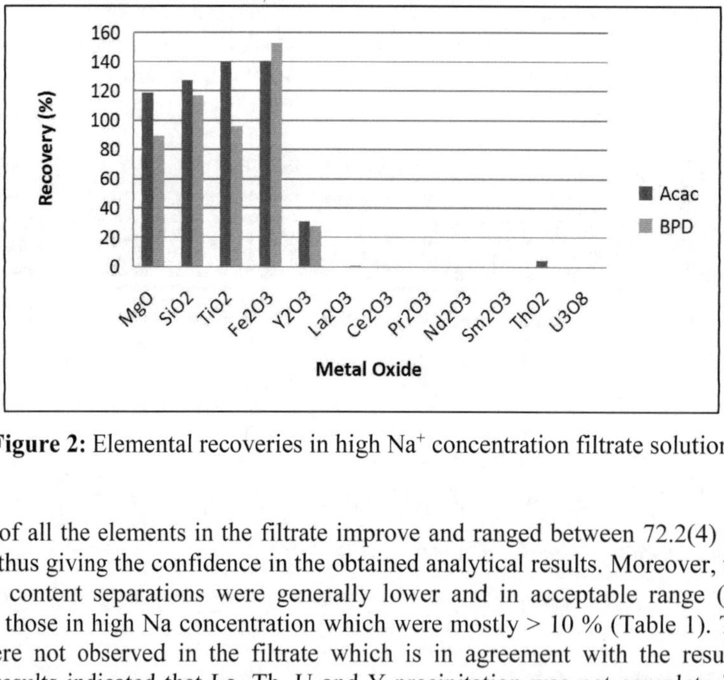

Figure 2: Elemental recoveries in high Na^+ concentration filtrate solution.

Recoveries of all the elements in the filtrate improve and ranged between 72.2(4) to 101.38(3) % (Figure 4) and thus giving the confidence in the obtained analytical results. Moreover, the RSD values in the low Na content separations were generally lower and in acceptable range (Table 2) when compared with those in high Na concentration which were mostly > 10 % (Table 1). The majority of the LREEs were not observed in the filtrate which is in agreement with the results in Table 2. However, the results indicated that La, Th, U and Y precipitation was not complete in this one step precipitation procedure. A significant concentration of Y had remained in the filtrate solution with some minor concentrations of La, Th and U.

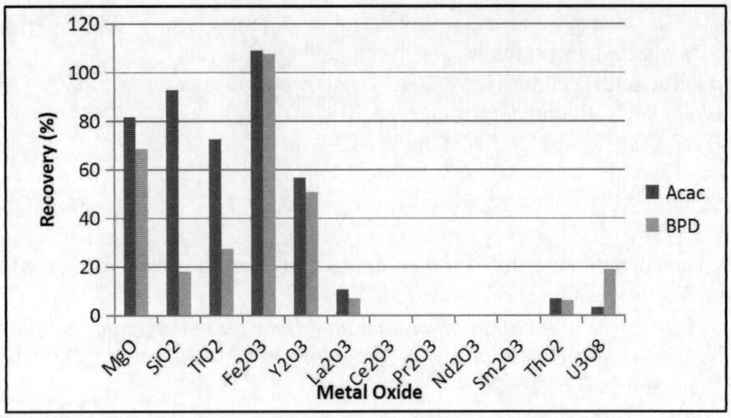

Figure 3. Elemental recoveries in low Na^+ concentration filtrate solution.

4. Conclusion

The separation of LREE from the monazite matrix using organic ligands as chelating agents was investigated in a sulphate medium. The mineral sample was digested by fusion with $Na_2HPO_4/NaH_2PO_4 \cdot H_2O$ flux and the melt was dissolved in water. Analytical results of the sodium phosphate fluxes indicated higher elemental concentrations for some elements and this led to a doubt about the accuracy of the rest of the analytes. The influence of Na on these analyses was reduced by prior precipitation of Na as NaCl using HCl and the results indicate improvement in the accuracy of LREEs, U and Th (their recoveries ranged between 72.2(4) to 101.38(3) % as well as the rest of the analytes. Results of the selective precipitation of the LREEs indicated that LREEs, U and Th were preferentially precipitated from the monazite matrix with both BPD and acac. Most of the other elements which constituted this mineral remained in the filtrate solution. This study has successfully separated the REEs from the monazite matrix. The precipitation method developed in this study can be used as a purification step for monazite beneficiation and for the concentration of low grade LREE containing minerals. The concentrated "mineral" can be used for as a starting material for the separation of REEs into their individual purer compounds.

Acknowledgements

The authors gratefully acknowledge the financial support given by the South Africa Nuclear Energy Corporation SOC Limited (Necsa) and the Nuclear Materials Development Network (NMDN) of the Advanced Metals Initiative (AMI) of the Department of Science and Technology of South Africa (DST) as well as University of the Free State for the research facility.

References

[1] Xaba S M 2015 *Method development for the quantification of selected early rare earth elements* (Bloemfontein: University of the Free State, M.Sc. thesis) p 1
[2] Jha M K, Kumari A, Panda R, Kumar J R, Yoo K and Lee J Y 2016 *Hydrometallurgy* **165** pp 2-26
[3] Joseph A 2016 *Investigating Seafloors and Oceans: From Mud Volcanoes to Giant Squid* 1st Ed (Amsterdam: Elsevier) pp 115-116
[4] Rivera V A G, Silva O B, Ledemi Y, Messaddeq Y and Marega Jr. E 2014 *Collective plasmon-modes in gain media: Quantum emitters and plasmonic nanostructures* (Cham Hiedelberg New York: Springer) p 118
[5] Bünzli J C and Pecharsky V K 2015 *Handbook on the physics and chemistry of rare earths* Vol 48 1st Ed (North Holland: Elsevier) p 288
[6] Amaral J C B S and Morais C A 2010 *Miner. Eng.* **23**(6) pp 498-503.
[7] Pike D R 1959 *The monazite deposits of the Van Rynsdorp division* (Pretoria: University of Pretoria, M.Sc. thesis) p 127

[8] Steenkampskraal rare earths, the mine [Accessed 04-05-2018]. Available from:http://www.steenkampskraalrareearths.co.za/the-mine/

[9] Sinding-Larsen R and Wellmer F W 2012 *Non-renewable resource issues: Geoscientific and societal challenges* (Netherlands: Springer) p 139

[10] Habashi F 2013 *Can. Metall. Q.* **52**(3) pp 224-233

[11] Abreu R D and Morais C A 2010 *Miner. Eng.* **23** pp 536-540

[12] Bahri C N A C Z, Al-Areqi W M, Ab-Majid A and Ruf M I F M 2016 *Mal. J. Anal. Sci.* **20**(1) pp 44-50

[13] Gupta C K, and Krishnamurthy N 2005 *Extraction Metallurgy of Rare Earths* (New York, Washington: Taylor & Francis e-Library) pp 176-177

[14] Chiweshe T T 2015 *Quantification of iridium and other platinum group metals in the presence of naturally occurring contaminants in geological ore* (Bloemfontein: University of the Free State, PhD. Dissertation) p 216-255

[15] Moila A, Chetty D and Ndlovu S 2017 *J. South. Afr. Inst. Min. and Metall* **117**(7) pp 615-621

Research and Development of Ti and Ti alloys: Past, present and future

Kalenda Mutombo

Light Metals, Materials Science & Manufacturing, Council for Scientific and Industrial Research, Meiring Naudé Road, Brummeria, Pretoria 0185, South Africa

KMutombo@csir.co.za

Abstract. This study provide an insight of established and future research and development on Titanium metal production, Titanium alloy development, Titanium powder metallurgy, and thermomechanical and manufacturing processes. In the last 60 years, research initiatives were investigating the cost effective alternative ways of producing Titanium metal from rutile, ilmenite and synthetic rutile. The very recent research focuses on the combination of pyro and hydrometallurgical process to produced synthetic rutile from ilmenite. Chloride leaching and solvent extraction will be significant strategies to reduce operating costs and protect the environment. New and developed processed for production of Titanium will be mostly focussed on direct electrowining of synthetic rutile or beach sand. Future perspectives associated with alloy development, biomedical, additive manufacturing, and Titanium powder metallurgy emerging techniques such as hydrogen sintering and phase transformation, and cold gas dynamic spraying or Titonic Kinetic Fusion are discussed.

1. Introduction

The cost of Ti and Ti alloys, product design and manufacturing process as well as emerging biomedical and additive manufacturing have driven the Titanium research and development in recent decades.

Titanium is the ninth most abundant element in the earth's crust. It is five time less than Iron and 200 times more than copper in abundance. However, its usage is 200 times less than copper and 2000 times less than Iron. This may be due to the high cost of the commercialized thermo-chemical processes such as Kroll and Hunter. Thus, the development of electrochemical and in-situ electrolysis may be promising as future production processes. However, some challenging issues such as redox-cycling, feeding, reaction kinetics, control heat balance have to be resolved before scaling-up to commercial production [1][5].

Rutile (95 % of TiO_2), and ilmenite or leucoxene (40-65 % of $FeO.TiO_2$) are the main minerals containing Titanium. Since 1981 the content of rutile in the world has decreased to less than 1 %, therefore Ilmenite is becoming increasingly important and supplying about 91 % of the world's demand for Titanium minerals. Ilmenite has extensive reserves of about 1300 million tons in beach sands and rock deposits and, therefore, ilmenite will remain the principal mineral for the production of Titanium metal in the future[1][2][4][5].

Powder metallurgy techniques which include Additive Manufacturing (AM), Metal Injection Moulding (MIM), Hot Isostatic Pressing (HIP), Cold Isostatic Pressing (CIP), Spark Plasma Sintering (SPS) or Field Assisted Sintering Technology (FAST), Gas Dynamic Cold Spray (GDCS), Direct

Powder Rolling (DPR) and Continuous Rotary Extrusion (CRE) of wire are focused on the production of near-net shapes with acceptable mechanical properties [2][3][5].

Recent research in Ti and Ti alloy has been focused on biomedical and medical applications. Superplastic Ti alloys, low cost composition Ti alloys with good ballistic and mechanical properties and near beta Ti alloys have attracted researchers in this category [5].

Thermomechanical processing and manufacturing techniques have been centred on superplastic forming sheets, forging of fuselage structures and landing gear for future aircraft and manufacturing of highly sophisticated and high performance sport cars [5].

This review therefore, aimed to identify the major drives of Research and Development (R&D) and area of growth for Ti and Ti alloys. Recent research in titanium metal production, titanium alloy development, powder metallurgy of Ti and its alloys, thermomechanical processing and manufacturing process has been reviewed.

2. Titanium metal production

Most research initiatives are currently focussing on investigating the cost effective alternative ways of producing Ti metal from ilmenite or synthetic rutile. Thermo-chemical and pyro-hydrometallurgical process routes are been researched to established the most economical process.

2.1. Thermo-chemical processes

The commercialized thermo-chemical processes such as Hunter and Kroll are very expensive and energy inefficient. Hunter Process (invented in 1910) is similar to Kroll process, however, it uses Na heated to 910 °C to react with $TiCl_4$ in two stages. The first stage reduces continuously the $TiCl_4$ to $TiCl_2$ at approximately 200 °C [1]. The second one, operating in batch, reduces $TiCl_2$ to Ti at 1000 °C.

The Kroll process (invented in 1940s) has been employed for Ti extraction since the 1950s. Its post-processing steps including multiple re-melting and product fabrications make it labour and energy intensive process. Kroll and Hunter need high grade natural rutile or upgraded synthetic rutile as feed and chlorination process to produced $TiCl_4$ [1][4].

The motivation for development of new processes for Ti metal production is mainly establishing a low-cost alternative to Kroll process. In actual research and industrial state no production process available to date has overcome both energy efficiency and scale-up challenges to replace the Kroll process.

The Armstrong® process is a continuous process for the reduction of $TiCl_4$ using Na. The process produces powder particles with a unique, low bulk density, "coral-like" morphology. Post-processing activities such as dry and wet ball milling provide a means of increasing the tap density and narrowing the particle size distribution to meet the requirements for typical powder metallurgy conversion processes. The high degree of interlocking between adjacent powder particles offers distinct advantages in roll compaction, allowing for the direct rolling of powder into titanium sheet [4][5].

2.2. Pyro-hydrometallurgical processes of Titanium

The combination of pyro and hydrometallurgical processes are recently researching ways to upgrade ilmenite to synthetic rutile. Since ilmenite is becoming increasingly important due to the depletion of rutile and, therefore, the synthetic rutile and slag are mostly used for the production of Ti metal. The development and progress on slag and ilmenite upgrading process are shown in **Table 1** [1][4].

The development of electrochemical and in-situ electrolysis have achieved some success. However, some challenging issues such as control redox reactions, feeding, reaction kinetics, control of heat balance have to be resolved before scaling-up to commercial process [1][4].

Electro-chemical processes such as direct hydrometallurgical lixiviation are advantageous in processing abundant ilmenite ores, since low energy consumption is expected and sufficiently high quality of synthetic rutile is produced. A novel sulphate process to improve leaching strategies has been developed by BHP Billiton. This process uses solvent extraction method to separate metals, reduce wastes and recycle acids. The drawback are high consumption of H_2SO_4 and production of large amount of sulphate waste and high cost for treatment. This process is very promising for commercial applications in the future[2][5][6].

Table 1. Ilmenite and slag upgrading processes.

Process	Pyro-Treatment	Leaching	Advantages	Disadvantages
Becher Sulphate	FeO oxidized to Fe_2O_3 and reduced to Fe with (C+ S) at 1200 °C	a. NH_4Cl/O_2 b. 0.5 M H_2SO_4	Diverse ilmenite type ore feed	Multi steps, high energy, emission of CO_2
Murso	Similar to Becher, but use C and H_2 and fluidized bed at 900-950 °C	20 % HCl at 108-110 °C	Efficient than Becher, easier recycling of HCl	Similar to Becher
Laporte	Oxidation in fluidise bed at 950 °C, reduction with C of Fe_2O_3 to FeO	18 % HCl	Free formation of fine TiO_2, ease leaching of FeO	Similar to Becher , but low temperature
Benelite	Reduce roasting of Fe_2O_3 to FeO using C	18-20 % HCl	One simple step	Limited type of ilmenite
Austpac	Magnetisation of ilmenite by roasting at 800-1000°C	25 % HCl	Magnetic separation, >97 % TiO_2	High acidity, magnetic Fe form
Dunn	Selective chlorination of Fe in ilmenite with Cl_2 in fluidised bed		Recycling of Cl_2 by oxidation of $FeCl_2$ to Fe_2O_3	Large Fe sulphate wastes
Kataoka	Reduction roasting of Fe_2O_3 to FeO	H_2SO_4	Low temperature and less corrosive than HCl leaching	Large Fe sulphate waste
Direct Fe leaching		HCl + Fe powder		

Direct chloride leaching has been investigated intensively, incorporating solvent-extraction to separate Ti from Fe and incidental impurities. This process is a significant strategy to reduce operating costs and protect the environment by enabling the use of low grade ilmenite and eliminate the chlorination process for production of high grade TiO_2 and/or $TiCl_4$.

Caustic leaching with high selectivity of Ti over Fe, Ti dioxide nano-technology for hydrometallurgical processing, and direct leaching of ilmenite in chloride solution and subsequent purification using solvent extraction for Ti oxide and Ti metal production are other alternative lixiviation processes that have been researched and developed nowadays [2][5][6].

Fray-Farthing-Chen (FFC) Cambridge (developed in the late 1990s) has developed a process called Metalysis®, to generate Ti powder of different morphologies and alloy chemistries. The FFC process has the capability to combine different metallurgical steps, including metal extraction, alloying, and shaping, into one step. FFC Cambridge process extracts Ti from Ti oxides at lower energy consumption via electrochemical reduction in molten salts. Its ability to produce alloys and powders, while retaining the cathode shape also promises energy and material efficient manufacturing. This process can be run with lower cost feedstocks such as Ti oxides, synthetic rutile and slags, and recycled Ti wastes. Recycling Ti scrap using FFC process will reduce the huge "buy to fly" ratio in the aerospace industry. The use of inert anodes can also improve the process sustainability by improving the current efficiency and product purity [2][5][6].

A recently developed process, Hydrogen-Assisted Magnesium Reduction (HAMR), is based on the destabilisation of the Ti–O system using hydrogen as a temporary alloying element during the magnesio-thermic reduction of TiO_2. It is a very promising process for the production of Ti metal [5].

3. Titanium alloy development

The application of Ti and Ti alloys in the field of medical and biomedical science is increasing due to its unique physical, chemical and biological properties. Ti and its alloys offer excellent biocompatibility and good osseointegration. Recent research in this area is focused on Bioactive Surface Modifications (BSM) of Ti and its alloys, in order to achieve improved biocompatibility. The current techniques that are used for BSM are intended to improve either the interaction of hard tissues with Ti and its alloys (Physio-chemistry) or the hard and soft tissues interaction with Ti and its alloy. Advanced surface modification approaches are needed to widen the possibility of applications of Ti and its alloys in medical and biomedical applications.

Low-rigidity Ti alloys containing nontoxic and allergy-free elements (Nb, Ta and Zr) are given considerable attention in biomedical applications. Low-rigidity (low Young's modulus) Ti alloys have the advantage of healing bone fractures and adapting bone properties. Ni-free superelastic and shape memory Ti alloys for biomedical applications are being actively developed. Additionally, mechanical properties such as strength, ductility, fatigue strength, fretting fatigue strength, and wear resistance have been included in the recent research [7][8][9].

New Ni-Ti shape memory alloys have been developed for intravascular devices such as stents and occlusion coil; for osteosynthesis, osteoconductivity and osseointegration. The texture and shape of the Ni-Ti device can be modified by stress and temperature variations for hard and soft tissue interactions [3],[7],[10].

Alloy development has recently been focussed on improving high temperature performance and superplasticity of α-β Ti alloy (4.5-5.5Al-3.0-5.0V-0.3-1.8Mo-0.2-1.2Fe-0.12-0.25O-0.1-0.4Si-Ti) [14]. This includes designing α-β Ti alloy with good ballistic and mechanical properties using low-cost composition, 4.2-5.4Al-2.5-3.5V-0.5-0.7Fe-0.15-0.19O-Ti [15]. The focus also includes developing a high strength, deep hardenability, and excellent ductility near-β Ti alloy for an aviation system component, including 5.3-5.7Al-4.8-5.2V-0.7-0.9Fe-4.6-5.3Mo-2.0-2.5Cr-0.12-0.16O-Ti [16].

In a joint effort, German scientists have developed a new γ-TiAl (Ti-43Al-4Nb-1Mo-0.1B at. %) which is characterized by β-solidification. This alloy exhibits a more homogenous and fine microstructure due to solidification via the bcc β-phase and can be extruded and fine grain achieved [11].

A metastable Ti-5Al-5V-5Mo-3Cr alloy has been developed recently. This alloy is expected to replace Ti6Al4V and high strength steels in the future aircraft as forged fuselage structures and forged landing gears. The new alloy offers attractive mechanical properties such as moderate strength, acceptable ductility and fracture toughness, and less residual stresses and distortion after machining, forging and heat treatment [11].

4. Powder Metallurgy of Titanium and its alloys

Powder metallurgy consists of consolidating and sintering (pressure-less and pressure-assisted) Ti and Ti alloy powders. Vacuum sintering is the mostly used process due to the minimum oxygen contamination. Unfortunately the microstructure and mechanical properties of conventionally sintered Ti powder cannot match those of wrought Ti. Thus, the conventional powder metallurgy approaches have not met the performance-to-cost ratios required to substitute wrought Ti and its alloys.

Nevertheless, the recently developed Hydrogen Sintering and Phase Transformation (HSPT) process has shown the capability of yielding microstructures that are equivalent to those of wrought Ti. This process can potentially maximise the performance-to-cost ratio by producing wrought-like Ti using press-and-sinter and conventional heat treatment processes [5].

HIP, which is the traditional TMP to close remaining porosities, will be probably replaced by a more promising low cost alternative Gaseous Isostatic Forging (GIF) process [5].

IOP Conf. Series: Materials Science and Engineering **430** (2018) 012007 doi:10.1088/1757-899X/430/1/012007

Titomic® will be commercializing the Titomic Kinetic Fusion (TKF) or Cold-Gas Dynamic Spraying (CGDS) developed by CSIRO. CGDS consists of spraying Ti or Ti alloy particles onto a scaffold to produce a load-bearing structure. The new process is able to use powders costing approximately one fifth to one tenth that of traditional AM powders, resulting in components up to 50 % cheaper, and it can produce large scale parts thirty times faster than other metal AM processes [18].

5. Manufacturing technologies

5.1. Thermomechanical processing

A method of manufacturing fine grain α/β Ti-Al-V-Mo-Fe alloy sheets that is suitable for superplastic forming has been developed. The alloy is forged and hot rolled to sheet bar, then fast-cooled from a temperature above beta transus. The sheet bar is heated between 760 °C.and 843 °C and rolled down. After reheating from about 760 °C to about 843 °C, hot rolling is performed in a direction perpendicular to the previous rolling direction to minimize anisotropy of mechanical properties. The sheets are then annealed at temperatures between 705 °C and 843 °C followed by grinding and pickling [8].

In order to increase the tensile and fatigue strength while keeping sufficient ductility, the Thermohydrogen Process (THP) has been developed. The process consists of hydrogenation of Ti below β-transus followed by air cooling, solution annealing and dehydrogenation below β-transus, solution annealing slightly above β-transus and air cooling (formation of β-phase), then ageing below β-transus followed by vacuum cooling (α precipitation hardening) [11].

The integration of design with manufacturing process and developing bioengineering have driven research in Ti and Ti alloys towards the computational materials engineering to predict fundamental phenomena focused on phase transformations, role of alloying elements and mechanical behaviour of alpha and beta phases. The comprehension through thermal modelling and compositional effects on phase transformation will be the future research focus [12].

5.2. Manufacturing processes

In manufacturing, Laser Beam Forming (LBM) technique has been established for joining of metallic components. Recent research in LBM is focused more effort on improving the laser-Ti and its alloys interaction [13].

MTU Aero Engine (Germany) has developed a linear friction welding process for the production of a new "Dual Material Ti alloy Linear Friction Welded Blisk" (DUTIFRISK). The produced linear friction welded blisks are made from a wide variety of different alloys and/or microstructures for the disk and blade [11].

A skull melting route for serial production has been developed by two companies; ThyssenKrupp Titanium GmbH®, in Germany, and Thyssen-Krupp Titanium S.p.A., in Italy. The melting process is similar to the Vacuum Arc Melting Process (VAR) [11].

Highly sophisticated, and high performance sport cars have been made by Bugatti Engineering GmbH® (Germany). The new sport car is currently the most powerful automobile with an engine performing at 1001 horsepower. The car parts are made of different Ti and Ti alloys in order to limit the weight and fulfil the specific requirement of heat shielding. The total amount of the built in parts is about 40 kg. Ti and Ti alloys used include Ti grade 1 & 2, Ti grade 1 plated with Al (leading to Ti-Al heat resistant for the exhausting system), Ti6Al4V coated with MoS (bolts suspension), Ti6Al4V bi-modal structure (Engine connecting rods), Ti6Al4V superplastically formed, β-Ti alloy (suspension springs) [11].

An erosion inhibiting coatings of thin vapour-deposited TiN/Ceramic (ERCoat) to protect compressor blades from premature loss of materials has been developed by MTU Aero Engine, in Germany. The TiN coat is applied by Physical Vapour Deposition (PVD), resulting in alternating deposition of hard and soft intermediate multi-layer structure (Pending Patents) [11].

6. Conclusion

The future of the titanium research and development remains bright, with high market demand which is driven by the growth of biomedical applications, additive and manufacturing industry. This provides the driving force for the Titanium alloy development and Titanium metal production.

References

[1] J. Gambogi, S.J. Gerdemann, Titanium Metal: Extraction to Application, Doe/Arc. (1999) 0–36. doi:10.1002/9781118788417.ch5.

[2] I. Mellor, G. Doughty, Novel and Emerging Routes for Titanium Powder Production -An Overview,*Key Eng. Mater.* **704** (2016) 271–281. doi:10.4028/www.scientific.net/KEM.704.271.

[3] Z. Abdullah, R. Razali, I. Subuki, M.A. Omar, M.H. Ismail, An Overview of Powder Metallurgy (PM) Method for Porous Nickel Titanium Shape Memory Alloy (SMA), *Adv. Mater. Res.* **1133** (2016) 269–274. doi:10.4028/www.scientific.net/AMR.1133.269.

[4] G. Crowley, How to Extract Low-Cost Titanium, *Adv. Mater. Process.* **161** (2003) 25–27.

[5] Z.Z. Fang, J.D. Paramore, P. Sun, K.S.R. Chandran, Y. Zhang, Y. Xia, F. Cao, M. Koopman, M. Free, Powder metallurgy of titanium – past, present, and future, *Int. Mater. Rev.* **0** (2017) 1–53. doi:10.1080/09506608.2017.1366003.

[6] D. Hu, A. Dolganov, M. Ma, B. Bhattacharya, M.T. Bishop, G.Z. Chen, Development of the Fray-Farthing-Chen Cambridge Process: Towards the Sustainable Production of Titanium and Its Alloys, *Jom.* **70** (2018) 129–137. doi:10.1007/s11837-017-2664-4.

[7] J. Sharan, S. V Lale, V. Koul, M. Mishra, O.P. Kharbanda, An Overview of Surface Modifications of Titanium and its Alloys for Biomedical Applications, Trends *Biomater.Artif. Organs.* **29** (2015) 176–187.

[8] M. Niinomi, Biologically and Mechanically Biocompatible Titanium Alloys, *Mater. Trans. 49* (2008) 2170–2178. doi:10.2320/matertrans.L-MRA2008828.

[9] Y. Li, C. Yang, H. Zhao, S. Qu, X. Li, Y. Li, New developments of ti-based alloys for biomedical applications, *Materials (Basel).* **7** (2014) 1709–1800. doi:10.3390/ma7031709.

[10] M. do A. Ferreira, M.A. Luersen, P.C. Borges, Nickel-titanium alloys: a systematic review, Dental Press *J. Orthod.* **17** (2012) 71–82. doi:10.1590/S2176-94512012000300016.

[11] D. Helm, O. Roder, Recent Titanium Research and Development in Germany, (n.d.). http://www.mtu.de/en/technologies/engineering_news/others/Helm_Recent_titanium_research.pdf).

[12] D. Banerjee, J.C. Williams, Perspectives on titanium science and technology, *Acta Mater.* **61** (2013) 844–879. doi:10.1016/j.actamat.2012.10.043.

[13] E.T. Akinlabi, M. Shukla, S.A. Akinlabi, Laser forming of titanium and its alloys – An Overview, *World Acad. Sci. Eng. Technol.* **71** 2012. (2012) 1522–1525.

[14] Yoji Kosaka, Patent no. 2017020607, March 2017, Titanium Metals Corporation

[15] John Fannig, Patent no. 9631261, April 2017, Titanium Metals Corporation.

[16] John Fanning, Patent no. 8454768, June 2013, Titanium Metals Corporation

[17] Yoji Kosaka, Phani Gudipati, Patent no. 8551264, October 2013, Titanium Metals Corporation)

[18] CSIRO, US patent, March 2013, Titomic Limited, Melbourne, Australia.

IOP Conf. Series: Materials Science and Engineering **430** (2018) 012008 doi:10.1088/1757-899X/430/1/012008

Development status of the CSIR-Ti Process

S J Oosthuizen* and J J Swanepoel

Council for Scientific and Industrial Research, Advanced Materials and Engineering, Meiring Naude Road, Pretoria, 0184

*SOosthuizen@csir.co.za

Abstract. South Africa is endowed with significant titanium-bearing reserves, which are concentrated to produce slag and exported as a low-value commodity, meeting about 20% of the global demand for the production of titanium dioxide pigment. Compared to the well-established titanium pigment industry, the much smaller titanium metals industry holds significant growth potential. Titanium metal is however notoriously difficult to refine from its abundant ores, and machining titanium is known to be challenging. Titanium usage is thus limited to niche applications in aerospace, defence, medical and chemical industries. Titanium powder produced at low cost promises a paradigm shift in titanium market size. Titanium powder has not been produced directly at commercial scale from titanium tetrachloride ($TiCl_4$), which is the precursor for both pigment and metal production. The Titanium Centre of Competence (TiCOC) at the Council for Scientific and Industrial Research (CSIR) has as its aim the establishment of a titanium metals industry in South Africa. To this end research and development work is being conducted towards the commercialisation of the CSIR-Ti process, a potentially cost-effective method for titanium metal production. This article discusses the history of the CSIR-Ti process for titanium metal powder production, challenges addressed, current development status and the unique products achieved.

1. Introduction

The research and development discussed herein was completed under the auspices of the Titanium Centre of Competence (TiCOC), based at the Council for Scientific and Industrial Research (CSIR) in Pretoria, South Africa. The Titanium Centre of Competence is funded by the Advanced Metals Initiative (AMI) of the Department of Science and Technology (DST). Both the AMI and TiCOC are driving the establishment of a South African titanium metals industry.

Mined titanium minerals are a mixture of rutile (TiO_2), ilmenite ($FeO.TiO_2$) and leucoxene (weathered ilmenite with a lower iron content than ilmenite). Significant processing of the minerals is required to produce the two main commodities, titanium dioxide pigment and titanium metal. Titanium is the ninth most abundant element, comprising 0.6% of the earth's crust. It is also the fourth most abundant structural material after aluminium (8.1%), iron (5.1%) and magnesium (2.1%) [1]. Recent data from the U.S. Geological Survey (2015) indicate that ilmenite accounts for about 92% of the world's consumption of titanium minerals. World resources of anatase, ilmenite and rutile total more than 2 billion tons. Titanium ore reserves are widely distributed, with the major rutile and ilmenite deposits located in China, Australia, India and South Africa. Currently, South Africa is a world-class supplier of titanium raw materials, but a minor supplier in the added-value pigment industry. South Africa is the second-largest single producer of titanium-bearing minerals in the world, but despite supplying nearly 20% of the raw materials, the country only realises about 5% of the approximately US$9 billion dollar pigment industry [2].

South Africa presently has no titanium metals manufacturing industry and limited capacity in titanium fabrication, for which all the material is imported. The market value of titanium increases dramatically as it is beneficiated. Ilmenite ore fetches about US$0.3/kg, while TiO_2 is worth some US$1/kg. Next up the scale is $TiCl_4$, which can be bought for around US$3.00/kg. Once titanium is in metal form, its price is significantly higher – titanium sponge fetches around US$12.00/kg, titanium ingots cost on average US$22.00/kg and mill product upwards of $40/kg [3]. Titanium and its alloys have been in industrial use for over 50 years. However, its applications have been restricted primarily to the aerospace industry where it is prized for its high strength and light weight, and the chemical industry, where its use is driven by its resistance to corrosion [4].

The high price of titanium metal combined with the high waste-to-final-part ratio, in aerospace termed the "buy-to-fly ratio" which is on average 8:1 by weight, constrains its use even in the aerospace sector [2,5].

Titanium's properties make it highly suitable for use in many applications, but widespread use is precluded by the relative cost when compared with competing materials [6]. Titanium is at present known to be expensive to refine, process and fabricate [1]. On the basis of its processing cost per cm^3, titanium is five times more expensive than aluminium to refine, and more than ten times as expensive as aluminium to form into ingots and to fabricate into finished products [7]. Kraft [1] indicated that this discrepancy in pricing and production volume is primarily due to the high reactivity of titanium.

Titanium has a great affinity for oxygen, nitrogen, carbon and hydrogen. The Kroll process and subsequent purification operations used for the majority of titanium production is energy, material and capital intensive. The use of titanium is furthermore hindered by long lead times of up to 12–18 months for plate, and often only aerospace companies are willing to enter into long-term offtake contracts stemming from this excessive lead time. High costs sensitivity is illustrated by Katrak, Agarwal and Loreth [8] who proposed that titanium demand should follow a '1:10:100 rule', where a cost reduction of US$1/lbs. in titanium metal sponge, followed by a reduction in mill production cost of 10% would lead to a 100% increase in the demand for titanium in non-aerospace applications. The development ideal is then the establishment of South Africa as a significant player in the higher value-addition industries of titanium metal production and manufacturing.

2. Titanium Powder

In the field of titanium, the work towards cost reduction and material savings of final products focuses on powder metallurgy, which is defined as the production, processing and consolidation of fine particles to make solid metal products. Much of the overall cost of titanium in the final application lies in machining to final configuration. Waste ratios are often very high for aerospace applications where the mill product may cost US$10/kg, but comparing material losses to the final weight of the part, the cost can be over US$500/kg [5]. According to Barnes, Peter and Blue [9] aluminium was considered to be as precious as platinum until the Bayer Process brought prices down from US$1200/kg to US$0.60/kg. Peter and Yamamoto [5] express a similar hope for titanium when they state: "The production of low cost titanium products from alternative titanium powders/sponge has the potential to enable a paradigm shift in the production and application of titanium, similar to the transformational effect of the Hall-Héroult process which enabled widespread industrial use of aluminium".

At present, powder metallurgy is based on powder from standard industrial titanium sponge. However, Sachdev, Kulkarni, Fang, Yang and Girshov [10] foresee the industrialisation of a process capable of producing titanium powder directly by a novel route that promises high-grade titanium powder at projections of US$8/kg. This cost reduction is expected to be achieved primarily through a melt-less route and smaller, low capital costs and more efficient plants. Titanium powder is prohibitively costly for most applications as it is commercially produced by crushing sponge via the extensive hydride/dehydride (HDH) process shown in Figure 1, or via melting ingots e.g. via the plasma rotating anode process.

IOP Conf. Series: Materials Science and Engineering **430** (2018) 012008 doi:10.1088/1757-899X/430/1/012008

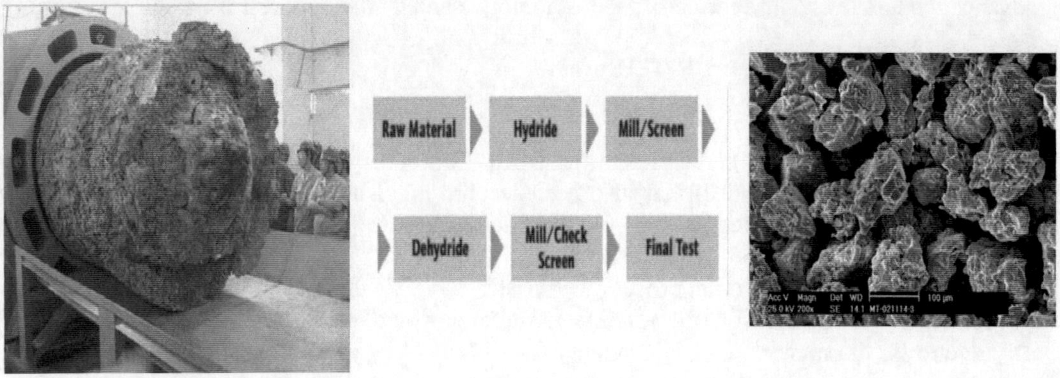

Figure 1: Steps to convert Titanium Sponge (LHS) to HDH powder.

Applications that are set to be unlocked by the availability of low-cost titanium are: automotive, heat exchangers, chemical processing, nuclear industry, condenser tubes, desalination, offshore petroleum, paper and pulp processing, additional defence applications, naval applications, liquid natural gas processing, biomedical supplies, sporting equipment and construction [5,10,11].

3. Titanium Centre of Competence (TiCOC)

A Centre of Competence (COC) is defined as a collaborative technology development consortium comprising science councils, academic institutions and industry [12]. It is guided by national priorities and relies on harnessing the resources, capabilities and synergies of all consortium partners, with clear research objectives in pursuit of products and services with commercial or public-good potential in response to market opportunity and socioeconomic challenges [12].

The development of a complete titanium value chain (production of titanium pigment, metal and downstream fabrication) has been identified as a key growth area for South Africa. This is further based on increasing world-wide demand, particularly with respect to titanium metal powder [13]. The CSIR is managing the South African Titanium Centre of Competence (TiCOC) which has as its objective the establishment of both an upstream titanium metal production industry and a downstream titanium fabrication industry [13,14]. A major goal of the TiCOC is the development and commercialisation of a novel, cost-effective and internationally competitive process for producing primary titanium metal powder in South Africa [2]. A process meeting AMI/TiCOC requirements was developed in the period 2010-12. The so-called CSIR-Ti powder process was operated initially in batch and semi-batch configuration, and a "continuous" 2kg/h pilot plant was operated in 2014-17.

4. Titanium Powder

The potential impact of a new, low-cost titanium-based industry in South Africa has been reported as follows [15]:

a) A titanium metal powder industry is estimated to have the potential to generate R3 billion to R5 billion a year, which could increase to between R10 billion and R30 billion a year once a downstream industry has been established.

b) The job creation potential of a local titanium industry includes 450 workers, engineers and technologists (on the metal production side), with about 2 000 workers, engineers and technicians in downstream component manufacturing.

c) There would also be a spillover of technologies and skills to other industries and enterprise creation opportunities in terms of inputs into the titanium value chain through the provision of goods and services. This translates into job opportunities and economic growth owing to the new businesses that will be created – all based on the beneficiation of titanium, the local production of

IOP Conf. Series: Materials Science and Engineering **430** (2018) 012008 doi:10.1088/1757-899X/430/1/012008

titanium powder and metal, and the local production of high-strength components for application in advanced industries such as the aerospace, automotive and medical industries.

5. Development status of the CSIR-Ti Process

5.1. Intellectual Property and Technology Package

The Intellectual Property (IP) pertaining to the CSIR-Ti process comprises of a substantial body of know-how and trade secrets, but IP has also been secured internationally via three patents, the initial two patents having been granted at least 18 times across various territories. These patents cover the following aspects of the proprietary process:

a) Titanium powder production process: U.S. Patent No. 8,790,441 (2011)
b) Process for the production of crystalline titanium powder: WO 2013185153 A3 (2013)
c) Electroless titanium process: UK Patent office 1716024.3 (application submitted 2017).

The CSIR-Ti process (Figure 2) utilises alkali and alkali-earth metals as reductants to produce titanium metal powder from the same commodity that is used to produce white paint pigment, titanium tetrachloride ($TiCl_4$). The by-product chloride salt is separated from the metal powder and electrolysed to recover both chlorine gas for production of $TiCl_4$, and the reductant alkali/alkali earth metal. Key aspects of process (specifically related to equipment) have been filed with the CSIR Technology Development Office as technology demonstrators. Development of technology packages relating to both 500 and 10 000 tonne per annum scale plants have been partially completed.

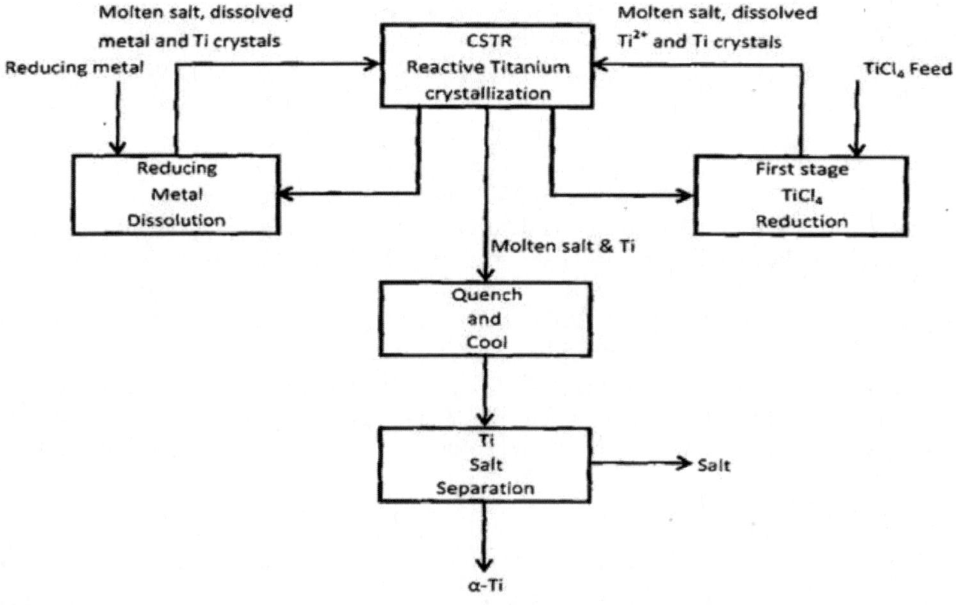

Figure 2: Block flow diagram of the CSIR-Ti process.

5.2. Process Development

The process was launched as a pilot plant (Figure 3) adopting a Continuous Stirred Tank Reactor (CSTR) configuration as per the first patents. The pilot plant was commissioned early in 2014 and operated over several campaigns until the end of 2017. The aggressive nature of the chemicals and elevated temperatures demanded by the process has provided a number of challenges [16,17]. These challenges were to a large extent overcome via sequential process developments. Numerous improvements were made to a range of systems such as electrical heating, agitation, molten salt-titanium powder slurry pumping, process control and instrumentation [17,18].

Figure 3: CSIR-Ti Pilot Plant (reduction section on right hand side).

The pilot plant met annual production targets, despite significant technical challenges which required the development of prototype equipment and novel solutions [16,17,18]. Given that several years of development effort has been expended on the process, an independent technical review of the development status was conducted in early 2018. The review panel, consisting of leading experts from industry saw merit in continuing development of the CSIR-Ti process, in both batch and continuous formats. A key outcome of the technical review was to further demonstrate the CSIR Ti-process in a scaled-up batch format as an accelerated means towards commercialization of the technology. It was also the opinion of the review panel that a batch design would attract potential investors more readily given the alignment with industry norm, given the inherent simplicity of batch processes.
Work on the continuous process is being taken forward in a new generation of reactors known as the TITAN High Flow Reactor (HFR), which targets higher production rates using a smaller reaction unit. The new reactor configuration also intends to achieve significantly better process control of parameters. Early stage powder samples have already been produced and analysed from the current TITAN V2.0 HFR.

The CSIR-Ti Pilot Plant has stimulated and intensified multi-disciplinary and inter-disciplinary research at the CSIR Pretoria campus. These disciplines include: Computer Flow Simulations and Process Modelling, Process Control and Electrical Engineering, Metallurgical Engineering and Corrosion, Mechanical and Thermodynamic Engineering and Design, Chemical and Process Engineering; and various support disciplines, such as Safety, Environmental, Risk and Project Management. The capacity developed at TiCOC can be accessed by research organisations and academic institutions by contacting the CSIR.

6. Challenges Addressed

Lower cost processes for the production of titanium metal have been a research and development goal for several decades [1,4,5,7,19], however the high temperatures and reactivity of the feed chemicals and titanium product has posed significant challenges. By developing solutions to these challenges, the CSIR has made significant progress in the demonstration of a process capable of continuously producing titanium metals powder [16,17,18]. One example is the CSIR developed molten salt slurry pump in conjunction with Lakeside Equipment, suppliers of Slurry-Busta® pumps [18]. These pumps, shown in Figure 4, were prototypes developed specifically for the high temperatures and inert gas atmosphere conditions of the CSIR-Ti process, thus requiring novel water cooled sealing systems on the rotating shafts.

The pumps operated with a single vane on the impeller as per manufacturer standard. It was found that at slurry densities approaching steady-state conditions, defined as when salt by-product and titanium powder are being produced in stoichiometric ratio, metal powder accumulation occurred behind the back plate of the impeller. Titanium powder would become compacted behind the back plate of the single vane impeller of the centrifugal pump, shown in Figure 5, which decreased performance and could lead to stoppages through seizing of the vanes. A double vane impeller was then proposed for the CSTR configuration. Nonetheless, the TITAN HFR utilises a design which may eliminate the need for vertical cantilever pumps.

Figure 4: Vertical cantilever molten salt slurry pumps (centre).

Figure 5: Single vane pump impeller with Ti metal powder visible behind the back plate.

7. Product morphology

A range of chemical and electro-chemical reactions are possible in the CSIR-Ti continuous system, and improvements to process control have allowed increased opportunity to influence product morphology. A significant breakthrough was made in the development of electrochemical sensors capable of real time monitoring of reagent concentrations and activity in the molten salt medium.

Through adjustment of reagent concentrations and process parameters, it is possible to preferentially generate certain product morphologies, ranging from amorphous to a variety of crystalline structures, or a mix of these morphologies. Typical morphologies achieved in batch and continuous operations of the CSIR-Ti process up to end 2017 are shown in Figure 6, descriptors of the microstructures were generated by CSIR staff.

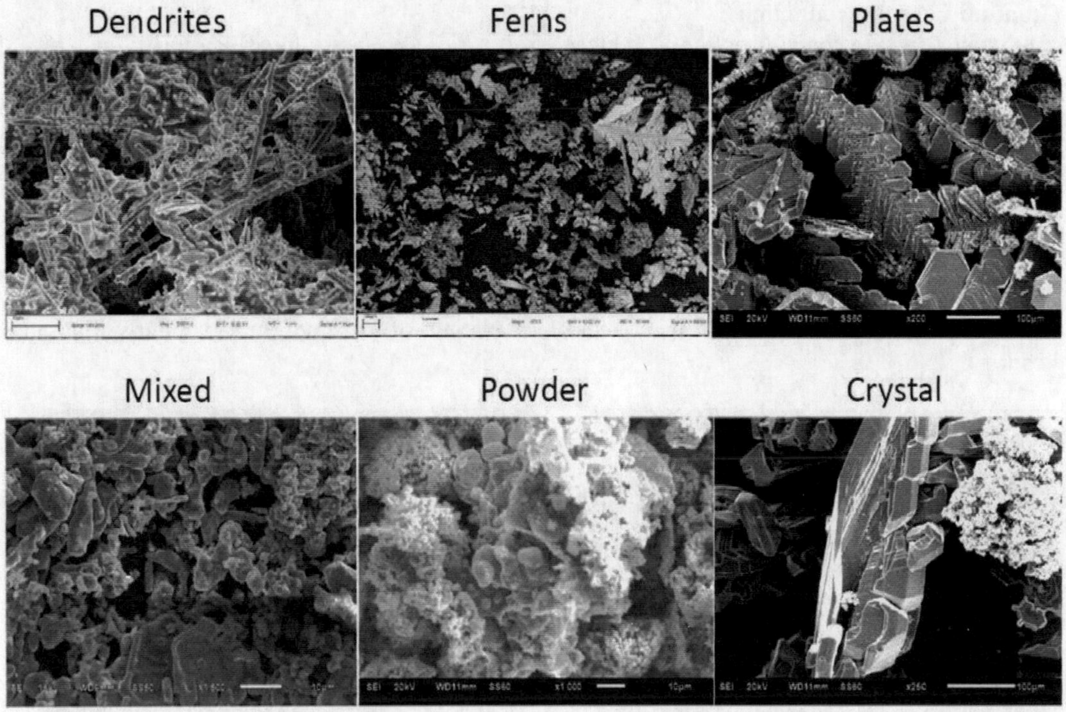

Figure 6. Microstructures of crystalline and amorphous CSIR-Ti powders produced to date.

8. Electroless Plating of Titanium

It was found that the process could be operated in a manner to grow, through contactless electrochemical plating (known in industry as electroless plating), a layer of titanium metal on particles present in the system. This was discovered when a particle of an impurity, most likely originally a particle of sand, was found to be coated with titanium metal, as shown in Figure 7a.

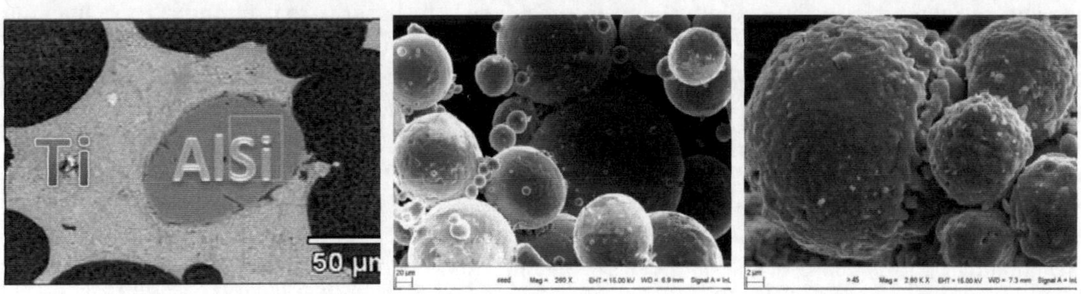

(a) (b) (c)

Figure 7: Micrographs showing electroless plating of titanium metal on particles.

The ability of the process to grow particles was further tested by feeding commercially purchased plasma spheriodised titanium powder as seed material to the process, and a scanning electron microscope (SEM) image of the seed material is shown in Figure 7b. The spherical titanium seed particles were circulated in the CSIR-Ti process, retrieved and again viewed by SEM. A new layer of titanium metal was found to have grown on the surfaces of the seed particles, shown in Figure 7c. This discovery implies the possibility of plating titanium metal on a variety of conductive and semi-conductive seed particles.

9. Titanium Crystal Production

Titanium metal sponge contains entrapped chloride salt which can cause micro and macro porosity if utilised directly, hence the industry requirement of purification through vacuum arc remelting or electron beam melting [20]. The CSIR-Ti process has shown potential for controlled growth of titanium crystals with little or no entrapped chloride. The new reactor configuration demonstrated in the TITAN HFR has delivered new product morphologies, and through altering process parameters titanium crystals were produced as fibers (Figure 8a), or as larger crystals with a hexagonal crystal structure, shown in Figure 8b.

Titanium's unique characteristics, such as the discovery of osseointegration of titanium implants which opened up a market for medical applications, have driven the creation and rapid growth of new niche markets [21]. It is the hope that electroless plating of titanium may similarly find new applications.

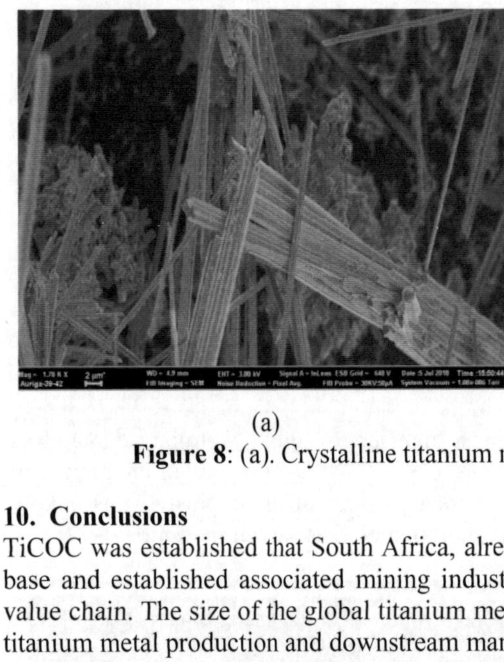

(a) (b)

Figure 8: (a). Crystalline titanium metal fibres; (b) Hexagonal titanium crystal.

10. Conclusions

TiCOC was established that South Africa, already in possession of a large titanium minerals resource base and established associated mining industry, may stand to benefit from expanding the titanium value chain. The size of the global titanium metals market is constricted by the relatively high costs of titanium metal production and downstream manufacturing. These constraints have locked titanium into a paradigm of predominantly high-end application, such as aerospace or medical components.

Research and development efforts at the CSIR aim to impact the titanium metals market by lowering metal production costs via a novel primary production process, including the utilisation of powder metallurgical processes to lower manufacturing costs of components.

The article presents non-confidential new knowledge generated in the development of the novel CSIR titanium production process, and learning gained from operation of the CSIR Ti-Powder Pilot Plant over a 3 year period (2014-2017). In this period numerous technical challenges were overcome and confidence in the process has grown significantly. The continuous CSIR-Ti process has shown the ability to produce promising new product morphologies, accelerated by the development of increasingly sensitive process control systems and a new reactor configuration. An exciting development was the demonstration of growing layers of titanium metal on seed material introduced to the CSIR-Ti continuous process, a discovery which has been added to the growing patent portfolio.

Acknowledgements

The authors would like to acknowledge the Department of Science and Technology for their continued support in the establishment of the TiCOC in South Africa. The authors also acknowledge staff at CSIR and NMISA who assisted in producing the micrographs, often at short notice, in order to help characterise products and guide the research.

References

[1] Kraft E 2004 *Summary of emerging titanium cost reduction technologies* (Vancouver, WA: EHK Technologies)

[2] Van Vuuren DS 2009 Titanium – an opportunity and challenge for South Africa. *Keynote address at the 7th Int. Heavy Min. Conf. Champagne Sports Resort, Drakensberg, 20-23 September*

[3] Roskill Information Services 2013 *Titanium Metal: Market Outlook to 2018* (London)

[4] McCracken CG, Motchenbacher C and Barbis DP 2010 Review of titanium-powder-production methods *Int. J. of Powder Met.* **46**(5) 19-26

[5] Peter WH and Yamamoto Y 2013Near net shape manufacturing of new, low cost titanium powders for industry. *U.S. Dept. of Energy Award Number: CPS Agreement 17881*

[6] Froes FH, 2012 Titanium powder metallurgy: A review - Part 1. *Adv. Mat. & Proc.* **170**(9) 16-22

[7] Seong S, Younossi O and Goldsmith BW 2009 *Titanium: industrial base, price trends, and technology initiatives* (Santa Monica, CA: Rand Corporation)

[8] Katrak FE, Agarwal JC and Loreth, MJ 1999 Cost reduction opportunities for titanium production. *Proc. of the Int. Symp. on Light Metals, held at the 38th Annual Conference of Metallurgists of CIM* 647-665

[9] Barnes JE, Peter W, and Blue CA 2009 Evaluation of low cost titanium alloy products. *Materials Science Forum* **618** 165-8

[10] Sachdev AK, Kulkarni K, Fang ZZ, Yang R and Girshov V 2012 Titanium for automotive applications: Challenges and opportunities in materials and processing *JOM* **64**(5) 553-565

[11] Wang H, Fang ZZ and Sun P 2010 A critical review of mechanical properties of powder metallurgy titanium *International J. of Powder Met.* ***46***(5) 45-57

[12] Damm O and Du Preez W 2009 Progress made by the South African Light Metals Development Network *Mat. Sci. Forum* **618** 147-154

[13] Du Preez W 2012 Forging ahead to a new South African titanium industry *Presentation at the TiCoC Planning Session, Bloemfontein, 5 September*

[14] Nel J 2012 ZrTa2011 New Metals Development Network Conference, Magaliesburg, 12-14 October 2011. *J. of the Southern Afr. Inst. of Mining and Metall.* ***112*** iv

[15] Hutton N Pandor: South Africa must develop manufacturing sector [Internet] 2012 Sep 25 Available from: http://www.mg.co.za/article/2012-09-25-pandor-south-africa-must-develop-manufacturing-sector

[16] Van Vuuren DS, Oosthuizen SJ, Heydenrych MD 2011 Titanium production via metallothermic reduction of TiCl4in molten salt: Problems and products. *J. of the Southern Afr. Inst. of Mining and Metall.* **111**(3) 141-7

[17] Van Vuuren DS, Oosthuizen SJ and Swanepoel JJ 2013 Development of a Continuous Process to Produce Ti via Metallothermic Reduction of TiCl4 in Molten Salt *Key Eng. Mat.* **551** 16-24

[18] Oosthuizen SJ, Swanepoel JJ, Van Vuuren DS 2014 Challenges experienced in scaling-up the CSIR-Ti process. *Adv. Mat. Res.* **1019** 187-194

[19] Oosthuizen SJ 2011 In search of low cost titanium: the Fray Farthing Chen (FFC) Cambridge process. *J. of the Southern Afr. Inst. of Mining and Metall..* **111**(3) 199-202

[20] Low RJ, Qian M and Schaffer GB, Chloride impurities in titanium powder metallurgy–a review *Ti 2011 – Proc. of the 12th World Conf. on Titanium*

[21] Oosthuizen SJ 2011 Titanium: the innovators' metal-Historical case studies tracing titanium process and product innovation. *J. of the Southern Afr. Inst. of Mining and Metall.* **111**(11) 781-6

IOP Conf. Series: Materials Science and Engineering **430** (2018) 012009 doi:10.1088/1757-899X/430/1/012009

Manufacturing of high added value titanium components. A South African perspective

D Dimitrov[1]*, E Uheida[1], G Oosthuizen[1], D Blaine[2], R Laubscher[3], A Sterzing[4], P Blau[4], W Gerber[5] and O F R A Damm[1]

[1]Department of Industrial Engineering, Faculty of Engineering, Stellenbosch University, South Africa.
[2]Department of Mechanical and Mechatronic Engineering, Faculty of Engineering, Stellenbosch University, South Africa.
[3]Department of Mechanical Engineering, University of Johannesburg, South Africa
[4]Fraunhofer Institute for Machine Tools and Forming Technology, Germany
[5]Aerosud Aviation, South Africa

*Dimtrov@sun.ac.za

Abstract. South Africa has significant titanium reserves and is developing strategies to not only provide this raw material on global markets, but to manufacture complex titanium alloy products and participate fully in global supply chains. Pursuing more sustainable process chains, manufacturers are constantly striving to enhance the resource efficiency of their high added value processes. Titanium alloys are used comprehensively in aerospace and biomedical applications, due to its favourable strength-to-weight ratio, corrosion resistance and biocompatibility. These properties, however, contribute to the manufacturing challenges, classifying it as a difficult-to-machine material. Various input factors affect the efficiency of a hybrid process chain. The objectives of this multidisciplinary research group were to demonstrate the reduction of material wastage by up to 50% and machining time up to 20% for selected titanium components. The results clearly demonstrate the ability to increase the technology readiness levels of the selected components and to develop competitive hybrid titanium process chains that will enable South Africa to integrate more into global manufacturing value chains.

1. Introduction

Titanium alloys have found wide application in the aerospace and biomedical industries due to their good strength-to-weight ratio and superior corrosion resistance. However, machining of complex titanium geometry (e.g. thin walled blades) is a challenging process. As titanium is difficult to shape, normal fabrication methods such as forging and machining from billets, are expensive, wasteful of raw materials, and time consuming as titanium machining is 10 times slower than aluminum. The efficiency and life service of these titanium engine blades are strongly affected by the precision of the production method [1, 2]. In the aerospace industry, buy-to-fly ratios of forged titanium parts are often only 15% to 20%, adding to the large cost for component material and machining. Blade manufacturing requires multitasking of hybrid machines such as turn-mill, which offer many technological and economic advantages.

IOP Conf. Series: Materials Science and Engineering **430** (2018) 012009 doi:10.1088/1757-899X/430/1/012009

Nowadays, the research focus shifted to enhancing the resource efficiency of the entire process chain, and is no longer driven by only high performance operations [3]. The development of near-net-shape fabrication methods is also significant in the expansion of titanium usage in aerospace applications. The developed hybrid processes chains include and are not limited to integrating of forming and additive manufacturing technologies with advanced machining. This paper provides an overview of the current-state-of-the-art manufacturing technologies for titanium components. Outcomes from a large collaborative research project on development and evaluation of different process chains for realization of prototypes of a generic blade geometry are presented and discussed.

2. Advanced manufacturing technologies for titanium components. A critical review

Press and sintering
Powder metallurgy (PM) has been earmarked as a potential growth area for titanium manufacturing due to its near net-shape capability. Metal injection molding (MIM) is widely used to manufacturing high production volumes (>10 000 per year) of small sized (<5 cm) components with complex shapes, where good dimensional tolerance is required, while additive manufacturing (AM) is used to produce individual components of complex shapes that are difficult to produce using other manufacturing processes. Figure 1 shows the different potential design for press-and-sinter processing to produce a component – a bracket. One of the distinct advantages of the press-and-sinter PM process is that it is much more cost effective than MIM or AM. A core advantage of all PM processes is the ability to produce near net-shape parts. This conducted study showed that production volumes of more than 2000 parts are required for the net-shape process to offer a more cost-efficient solution compared to traditional machining from blanks. Low cost, high quality titanium powder, along with excellent process control, are all aspects of the technology which must be achieved to attain these outcomes.

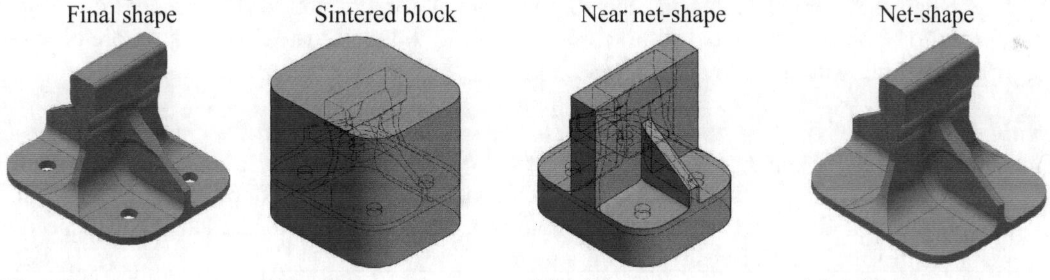

	Sintered block	Powder preforms Near net-shape	Net-shape
Material waste	89 vol%	74 vol%	1.2 vol%
Compaction requirements	simple, single level tooling	split punches, multilevel tooling	split punches, multilevel tooling requiring CNC press control

Figure 1. Comparison of various designs for press-and-sinter processing of the bracket part.

Forming
Forming applications realised from titanium and its alloys are critical structural parts, landing gears, ducts. A further important field of application for titanium alloys are the aircraft engine parts such as fan and compressor blades.

Forging. Forging can be classified into die forging and closed die forging. Advantages are low material loss in comparison with machining and the targeted change of the structure and increasing strength. A disadvantage is the lower geometric precision compared to machining processes. An example is the production of structural components where up to 95% of the material is machined. Forging is also used for manufacturing of appropriate preforms (Figure 2) for subsequent finishing

IOP Conf. Series: Materials Science and Engineering **430** (2018) 012009 doi:10.1088/1757-899X/430/1/012009

operations or finished components, respectively. It reflects the forging process as a part of an entire process route for the realization of preforms in blade manufacturing.

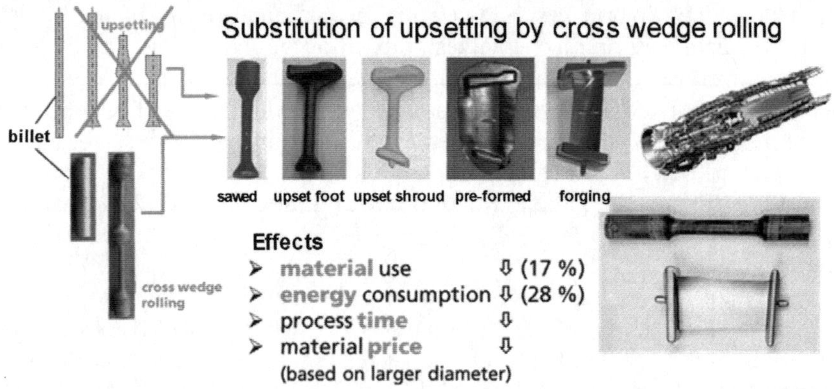

Figure 2. Process routes for preform in blade manufacturing based on forging.

Incremental sheet forming. Single point incremental forming (SPIF) application areas include rapid prototyping in aerospace and automotive industries, patient specific biomedical products, as well as components in architectural engineering. There are, however, several process issues such as poor geometric precision, high residual stresses and constrained forming limits that still need to be improved before the SPIF industrialization can take place.

A framework for understanding and analysis of the SPIF formability, forming forces and temperature limits can be found in [4]. The assembling of the three process limits enabled construction of the diagram shown in Figure 3, which can be considered as a process map for SPIF of CP grade 2 sheets. Integrating the thermal limits into the map is a novel contribution to the process knowledge; this may prove to be an expedient tool for the process analyst, to individuate design of a more efficient process when working with titanium alloy sheets [5].

Machining

The machining process is characterised by different cost factors that vary depending on the type of setup or process. The main costs, however, can be categorized into three groups, namely: machining cost, cutting tool cost and non-productive costs (Figure 4). The aim is to reduce each of the respective cost elements so that the overall costs can be minimised.

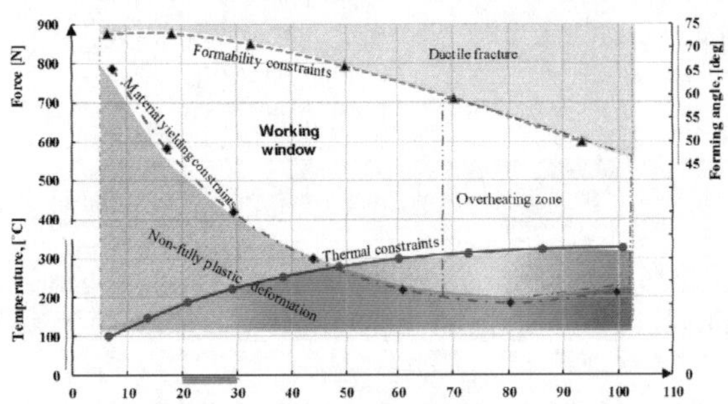

Figure 3. Schematic plot of the SPIF process map, showing the working window for the CP grade 2.

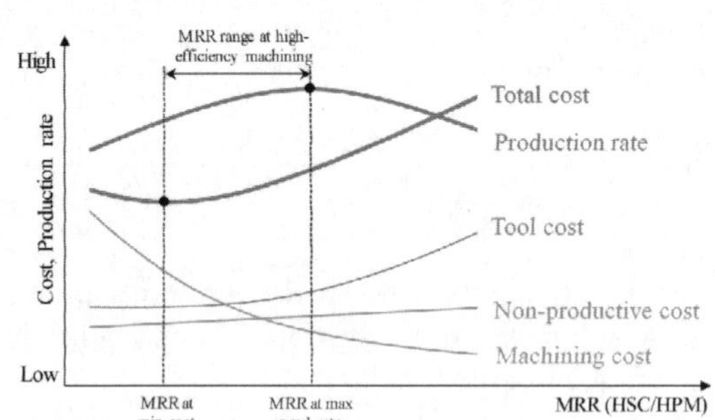

Figure 4. Interaction between main cost terms and effect of cost modelling (adapted [8]).

Research was focused on developing cost modelling approach and user interface that will allow manufacturers to improve the process production rate and cost and tool wear mapping [6]. Several demonstrators were selected from the production line of an aerospace manufacturer with the purpose of enhancing the resource efficiency of the milling process chains by utilizing the developed models. Validation has shown enhanced part accuracy, shorter machining time, and cost reductions close to 40% for producing the titanium aerospace components [7].

Additive manufacturing / SLM
Selective Laser Melting (SLM) is a powder bed additive manufacturing (AM) process whereby fine metallic powder is melted by means of a laser energy source, layer-by-layer, in order to create a 3-dimensional object. Due to the current limited capacity of metal AM machines in the country, the technology has not yet been fully adopted by industry for production purposes.

Applications in the medical industry include but are not limited to patient specific dental implants, knee and hip implants [9, 10]. Applications for the technology found in the aerospace industry are predominantly for weight reduction of low volume parts (Figure 5). For that purpose, topology optimisation can be efficiently used.

Inspection
Inspection and testing of manufactured parts typically encompass an assessment of the accuracy relative to the design and an assessment of the surface integrity. Various contact and non-contact techniques are used to assess the geometric accuracy of the part and if it is suitable for use. These must be conducted on a regular basis to maintain consistency. Surface integrity is an encompassing expression that describes the surface and near subsurface properties and state of a manufactured product. It is usually imperative that the surface integrity of high value parts be consciously engineered and quantified to ensure design performance. The two most often assessed surface integrity descriptors are topography and residual stress state.

The effect of a different cutting strategy on a typical surface integrity descriptor (residual stress) is illustrated in Figure 6 [11]. Two similar Ti6Al4V parts were manufactured at two different facilities using slightly different cutting strategies.

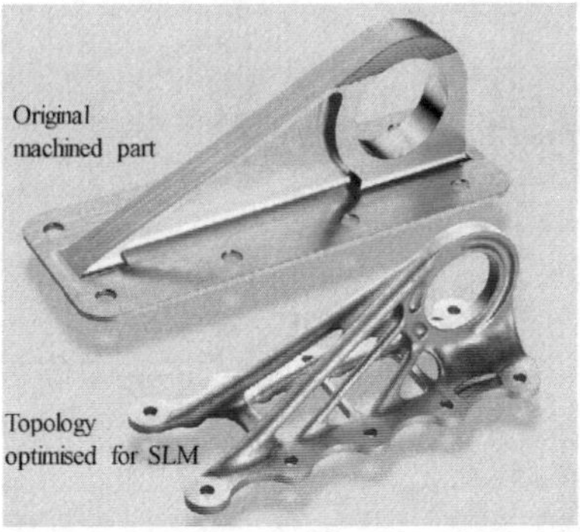

Figure 5. Cabin bracket for the Airbus A350 XWB, showing the original part and topology optimised for SLM [12].

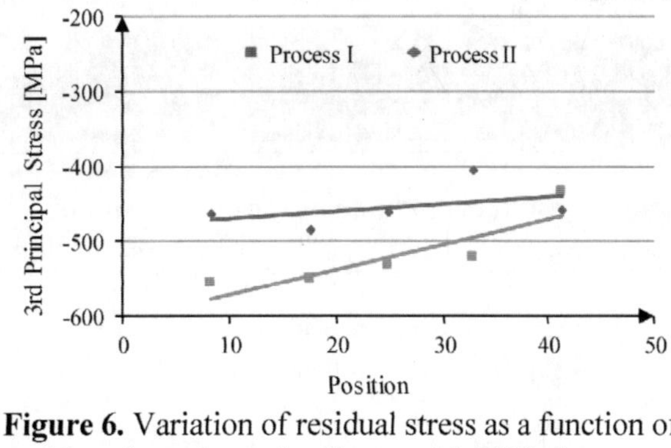

Figure 6. Variation of residual stress as a function of position along the major flange with XRD data reflecting the surface stress state [11].

Finish milling for Process II was conducted at a higher cutting speed than Process I (100 m/min compared to 50 m/min). The near surface residual stress state was measured by XRD at different positions and presented (Figure 6). This clearly shows that the process utilizing the higher finishing cutting produced a higher compressive residual stress state that may be beneficial for fatigue purposes.

3. Manufacturing of high added value components. A case study

The objective of the case study is the realization of prototypes of a generic blade geometry based on the selected manufacturing routes:

- Machining from a whole, i.e. billet

- Forming (for preform realization) + machining
- Additive manufacturing (for preform realization) + machining

The reference basis was the conventional machining process. Within this project, the complete process routes were developed in detail, including single process steps. Another aspect which played an important role was the achievement of a high Technology Readiness Level (TRL) for the manufacturing route.

Process route "Machining"

Machining of high-added value titanium blades is a challenging process. The main reason for this is not only the exotic material, but largely the complex geometry and tight tolerances required. The blades are usually very thin and high in length, making the clamping and machining especially difficult. Introducing of multitasking machining that comprises several machining operations in a single platform, such as turn-mill, can respond to the complexity and energy saving growing demands [13].

The turn-mill technology was used to produce rotor blades from Ti6Al4V rod-type workpiece. The blade design obtained from a reputable manufacturer of power generation was used as a capability demonstrator. The blade thin and varying thicknesses over its specified height add to the machining and clamping complexity. Figure 7 depicts the turbine blade on the turn-mill machine, showing the workpiece, the clamping system and the volume ratio of material used. Three fan blades were machined of Ti6Al4V pellets ($\emptyset 110, 350\ \mathrm{mm}$). Figure 7(c) highlights the large waste of material by volume ratio. This also can be improved by using more efficient start-up workpieces geometry as well as better clamping strategy.

Figure 7. Machining of turbine fan blade out of mill product showing the clamping on both main and sub-spindle: (a) the starting billet; (b) the machined blade before the offcut; (c) material usage.

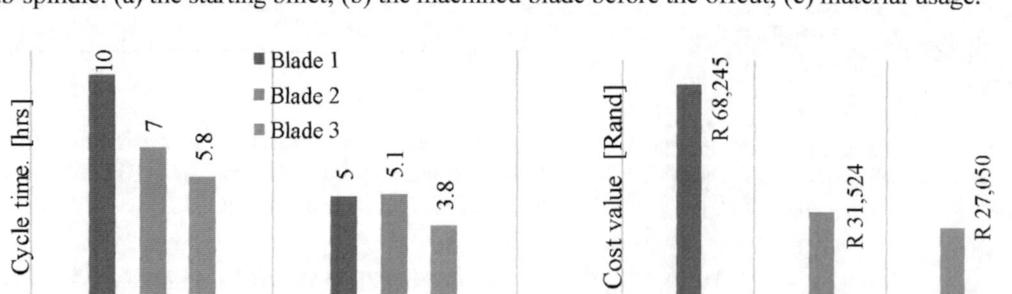

Figure 8. Optimisation of the machining operation: (a) reduced cycle time; (b) reduced costs.

The iterative-like approach followed, that is stimulus, or adapting of the process over the course of machining enabled learning and optimise the process over the three samples. Reduction of machining time and cost was obtained using various milling strategies, and different tool types. Figure 8(a) illustrates a comparison of the actual machining time between the three iterations (blades) from both the roughing and the finishing operations. The assembled costs of the machining hours, tools and material are shown in Figure 8(b). Comparatively, significant reductions of the total costs have been realised between the first and last samples.

Process route "Forming + machining"
Based on an analysis of the generic blade geometry, the rough process route that was identified as a promising variant for the realization of the preform comprises cross wedging, upsetting and forging as shown in Figure 9(a).

Concerning the subsequent finishing process, the blade geometry was slightly changed to optimize the clamping strategy. Based on successfully performed feasibility studies using FE simulations, required tools, jigs, etc. were realized and the prototyping process was started. Finally, it was possible to realize preforms based on the selected forming route. Figure 9(b) illustrates the final forging step for the realization of the preforms.

The next process step was the machining of the blade root which is the basis for workpiece clamping and referencing. The machining of air foil geometry was realized using a 5-axis turn-mill-centre. The root part was mounted on the main spindle. Therefore, a special clamping adapter was developed for adapting the prismatic root on the three-jaw chuck. With the opposite spindle the counter holder was gripped. Chucking from both sides gave high stiffness to the part clamping and provided the ability to machine the entire blade surfaces in a single clamping setup. Finally, the air foil area was finished based on two step strategy consisting of pre-finish milling and finish milling.

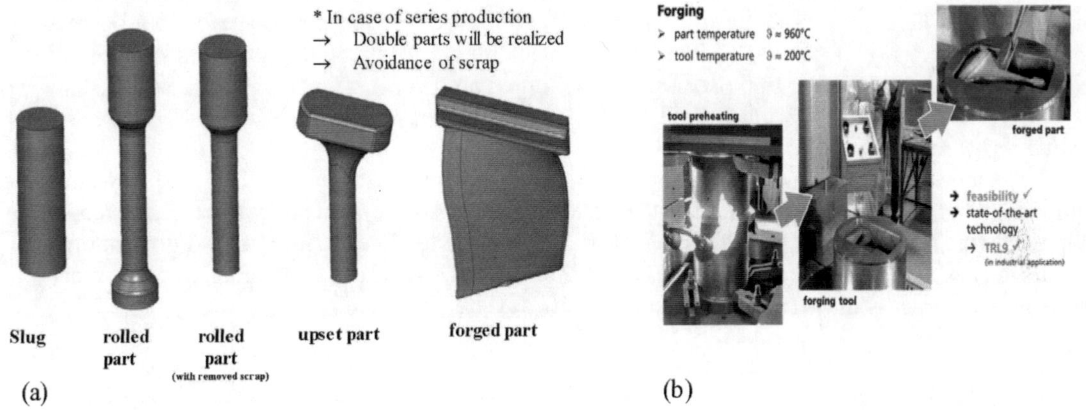

Figure 9. Process route for preform realization: (a) based on intermediate part geometries; (b) proof of feasibility for modified preform geometry.

Process route "Additive manufacturing + Machining"
Additive Manufacturing is another promising process route for realization of blade preforms. An advantage is not only the realization of (near) net-shape contours. It is also possible to realize internal structures, i.e. cooling channels can be integrated. To realize an appropriate preform for subsequent machining processes, the working steps performed included: development of adequate pre-part, development of additive manufacturing strategy, development of finish machining strategy and prototyping.

Similar to the process chain described above, the preform geometry was also modified in this case in order to ensure an optimal clamping condition in the machining process. Only the air foil area was realized by additive manufacturing to increase the efficiency. The blade root was realized by machining from a solid titanium bar. This approach enabled significant reduction of part volume to be built and a precision root as reference base for subsequent machining. Selective Laser Melting was used as a suitable process for the realization of the air foil area. The parts are generated directly from

IOP Conf. Series: Materials Science and Engineering **430** (2018) 012009 doi:10.1088/1757-899X/430/1/012009

titanium powder. The structure is built up in layers. For an increased efficiency three blades were built up simultaneously in one additive job (Figure 10).

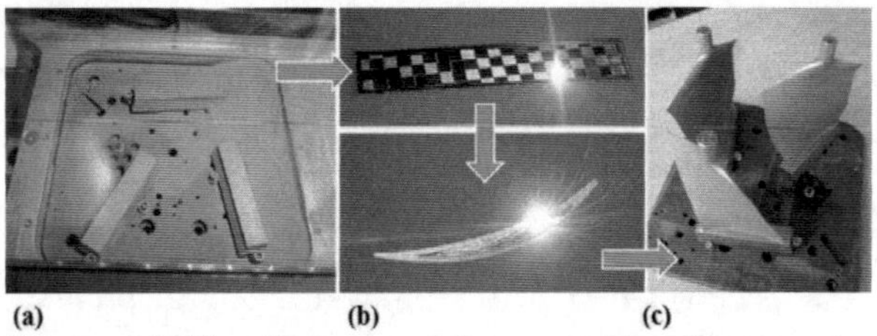

Figure 10. The AM route: (a) pre-machined root parts mounted on SLM-table; (b) middle- additive building process; (c) SLM-table with AM pre-parts.

Table 1. Optimizing AM process time with batch size.

Part number per job	3	6	10	20
Light exposure per part, hr	7,9			
Light exposure total, hr	23,8	47,6	79,3	158,7
Layer lamination, hr	16,5			
Time per part hr	**13,4**	**10,7**	**9,6**	**8,8**

Measurements of the first preforms showed a torsional distortion of the air foil geometry. This geometric error can be mainly attributed to residual stresses because of the thin walled geometry and the local heating during the additive manufacturing process. A beneficial tool to solve this problem is a new simulation code, which is still under development. The process can be simulated regarding the thermal and residual stresses. Consequently, these stresses can be minimized and the SLM process optimized.

The minimization of the process time is another challenge. In comparison to the laser time which increases linear with the number of parts, the time for powder layer lamination is not dependent on the part numbers. Therefore, an increasing number of parts to be realized per job leads to a decrease of process time per part in general (Table 1). The machining and clamping strategies of the preform realized by SLM were similar to those used for the forged preform.

Evaluation of process routes

For the planned benchmark of the different routes, issues such as material use, lead time, tooling equipment, and finally manufacturing costs were considered. A first interesting aspect is the amount of titanium which is needed for the realization of the parts (see Figure 11). Furthermore, the manufacturing costs were determined. Beside of the material, criteria such as required tools, machine-hour rates and process times (including auxiliary times) were considered.

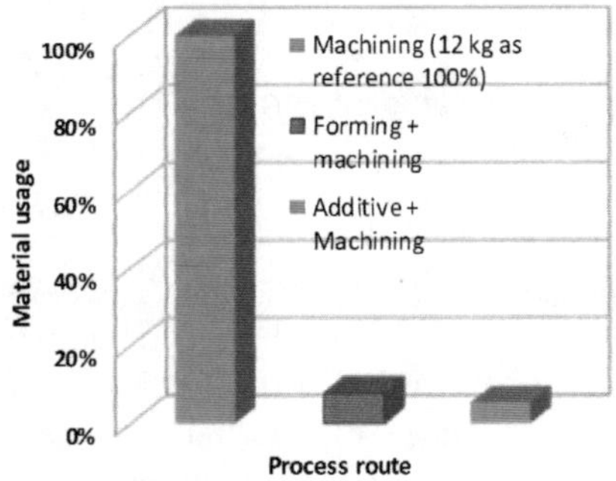

Figure 11. Comparison of material usage for realization of blade geometry between the process routes.

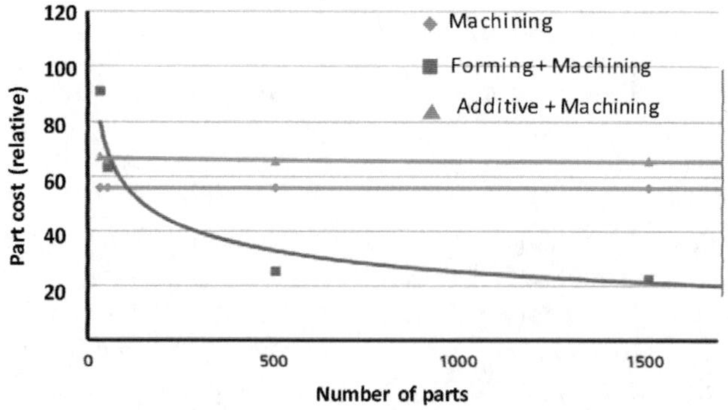

Figure 12. Relative cost for a titanium turbine blade demonstrator part for different manufacturing process chains and part volumes.

The diagram in Figure 12 shows the development of part costs dependent on the part quantity. Based on these calculations, it is possible to see that the complete machining of a blade is suitable for small part numbers. Above a part number of about 65, the alternative process route "forming + machining" is gaining importance. Due to the cost of titanium powder, the parts realized using the process route "additive manufacturing + machining" are characterized by the highest costs.

Therefore, in summarizing the advantages and disadvantages of the process routes considered, machining provides high flexibility, low material utilization ratio, and is therefore suitable for low demand. In the process route "forming + machining", part costs are considerably reduced with increasing production volume. However, it has low flexibility with an expected tool life of 1500 (tool reworking expected) parts which renders this process route suitable for medium to high demand. For "additive manufacturing + machining" the high flexibility and material utilization with additional function integration in tooling is advantageous, but the feasibility of this route is challenged by high residual stresses in as-built parts and elevated powder costs.

Regarding the achieved results, the used processes, the used machine equipment and the state-of-the-art in blade manufacturing it can be summarized that TRL 9 was achieved/proven for the process route "forming + machining". In the case of the route "additive manufacturing + machining" further optimization loops regarding the guarantee of dimensional and geometrical accuracy are required. However, the principal feasibility was proven and TRL7 was achieved.

4. Resource efficient production of high-added value components. A South African perspective

The research outlined in this paper forms an integral part of the broader DST R&D led industrialization strategy, which aims at the growth of local industrial capacity and activity, as well as value addition to South African minerals. The work to date has been successful in developing core competencies in downstream manufacturing technology building blocks, such as high efficiency machining of titanium (and aluminium), additive manufacturing, powder metallurgy, forging, and materials/component characterization. More importantly, these building blocks have been combined in different manufacturing process chains to investigate the optimum configurations for particular component types. This is a differentiating competence that can provide a competitive advantage to local manufacturers of high value titanium products for the fast-growing aerospace as well as the more established biomedical markets in particular.

The importance of this ability to design and implement different process chains is illustrated in Figure 12 above. This graph is significant in that it is based on validated data that was derived from manufacturing processes that are within the locally developed competence base.

Evidently, different process chains and production volumes yield very different cost curves and part costs. If local manufacturers were not able to have access to such data and competence, then clearly they would not be able to compete in the global marketplace; and local value addition to titanium minerals would be limited to raw materials such as powder, or semi-finished products like plate and sheet.

In this context it is important to note that for the current powder bed additive manufacturing technologies, the titanium powder accounts for less than 10% of the actual AM component cost. Hence, the cost of the titanium powder is not a significant cost driver in the hybrid additive manufacturing chain. (Powder cost nevertheless becomes more significant for AM processes with much higher material deposition rates. Such processes/machines are currently under development, e.g. the local Aeroswift programme). Moreover, there is a global oversupply of titanium metal sponge, which has seen sponge prices fall to historical lows [14].Titanium and titanium powder are thus commodities, albeit with a relatively high value, and the local manufacture of high value parts and products must form a key part in a sustainable titanium industrialization process.

Competencies and technologies are of little value, however, unless they can be transferred and implemented in industry. The work described in this paper has been successfully demonstrated at a TRL level of 5 – product related, which is still some way away from commercial implementation.

For any new technology or process, the key requirement for successful industrialization is a motivated user (MU), who is able to develop the technology to TRL level 9 in collaboration with the technology partner. Ideally, the MU should be involved in the technology development process to ensure a seamless progression. Once the technology has been successfully demonstrated and adopted by the MU, it can then be industrialized further by the MU and/or diffused into broader industry. From an industrialization perspective, it is thus important to progress the hybrid process chain technologies described in this work from TRL 5 to TRL 9, in collaboration with a MU.

5. Conclusions

The most important outcomes and conclusions from the presented research can be summarised as follows:

1. The initial target to demonstrate reduction of material wastage by up to 50% through implementation of preforms and near-net shape manufacturing technologies (thereby reducing the required amount of machining to achieve the final component geometries) has been clearly proven, in fact substantially exceeded.
2. The second target—reduction in machining time up to 20% has been clearly proven too. This becomes possible through implementing of more efficient cutting and clamping strategies. The using of preforms brings additional advantages in this regard.
3. The TRL challenge:
   - The lack of an end user partner forced the project to differentiate between component readiness levels and process readiness levels.
   - Component readiness levels cannot exceed TRL 5 in the absence of a "real" component.
   - Process readiness levels can be proven all the way to TRL 9.
4. The importance of the results represented by the graph (Figure 12) is far reaching, in the sense that, it is validated data which was derived from processes within our reach. The data contained in the source document shows that the material price of the additive powder is not a cost driver in the hybrid additive chain. (This means further that the availability of SA powder is not a risk to this method). Moreover, this data can be used to make design and manufacturing decisions in an industrial environment.

Finally, this project clearly illustrated the value of a multidisciplinary research group. The ability to develop and demonstrate in parallel stimulates and enables cross cutting problem solving. Of a particular importance hereby is also the fact that industry have direct access to R&D institutions.

Acknowledgement

The presented results reflect to a large extent work carried out within the project "Resource Efficient Project Chains for Titanium Components in Aerospace, Automotive and Medical Applications", which forms part of the research platform of the TiCoC. The authors express their sincere gratitude to the DST for funding and supporting the research over several years.

References

[1] Tung C and Tso P 2011 *Int. J. Ind. Manuf. Eng.* **1** 172–7
[2] Lu D, Liu J, Zhao W, Lu B, Wu D, Song D, Xue F and Cheng B 2017 *J. Manuf. Sci. Eng.* **139** 111015
[3] Mose C and Weinert N 2015 *Robot. Comput. Integr. Manuf.* **34** 44–51
[4] Uheida E H 2016 *Development and optimisation of incremental sheet forming of titanium grade 2—Process mapping* (Dissertation, Stellenbosch University)
[5] Grün P A, Uheida E H, Lachmann L, Dimitrov D and Oosthuizen G A 2018 *Int. J. Adv. Manuf. Technol.* 5–7
[6] Oosthuizen G A 2010 *Wear characterisation in milling of Ti6Al4V – A wear map approach* (Dissertation, Stellenbosch University)
[7] Conradie P, Dimitrov D and Oosthuizen G 2016 *Procedia CIRP* **46** 412–5
[8] Neugebauer R, Drossel W, Wertheim R, Hochmuth C and Dix M 2012 *Procedia CIRP* **1** 3–16
[9] Santos E C, Shiomi M, Osakada K and Laoui T 2006 *Int. J. Mach. Tools Manuf.* **46** 1459–68

[10] Murr L E, Quinones S A, Gaytan S M, Lopez M I, Rodela A, Martinez E Y, Hernandez D H, Martinez E, Medina F and Wicker R B 2009 *J. Mech. Behav. Biomed. Mater.* **2** 20–32
[11] Dimitrov D M, Laubscher R F, Sterzing A, Conradie P J T, Oosthuizen G A, Blau P, Schmidt G, Hochmuth C, Styger G and Zachäus R 2016 *Procedia CIRP* **45** 155–8
[12] GmbH C L 2017 Voraus ! Topologisch optimierte Bauteile in der Luftfahrt - Ahead ! Topologically optimised components in aviation
[13] Calleja A, Fernández A, Rodríguez A, De Lacalle L N L and Lamikiz A 2015 *Proc. Inst. Mech. Eng. Part B J. Eng. Manuf.* **229** 1324–36
[14] TBGA The Barnes Group Advisors, Personal communication

IOP Conf. Series: Materials Science and Engineering **430** (2018) 012010 doi:10.1088/1757-899X/430/1/012010

Development and Process Verification of a Linear Friction Welding Platform for Small Axis-Symmetrical Ti6Al4V Components

N T A Mohlala[1,*], D G Hattingh[1,*] and W Rall[1]

[1]Nelson Mandela University, Department of Mechanical Engineering
*narishe.mohlala@mandela.ac.za
*danie.hattingh@mandela.ac.za

Abstract. This paper reports on the developments of a study done to establish the feasibility of implementing Linear Friction Welding (LFW) as a joining technique for small axis-symmetrical Ti6Al4V samples. This work will attempt to facilitate the manufacturing of high-integrity small near-net-shape components for aerospace, automotive and medical applications. LFW is a solid-state welding technique that utilises frictional heat generated by the rubbing of surfaces under an axially applied load, thereby forming a weld at temperatures below the beta transus point. The technique is advantageous as it has the potential to reduce defects normally associated with conventional welding of this material. This paper will describe the development of an experimental platform, which will facilitate the evaluation of the influence of selected process parameters and their influence on joint integrity. Process parameters that will form the basis of this study include axial load (weld interface force), consumed length (axial shortening), oscillating amplitude and frequency. Welds made on this platform will allow for the characterisation of the microstructural change across the weld zone or regions exhibiting grain refinement, linking this to the mechanical properties of the joint region. The study will create an understanding of how process parameters can be manipulated to achieve optimum joint properties or assist in eliminating weld defects. Knowledge generated will form the basis for developing a bench-top LFW research platform, with a reliable closed-loop response system that will aid in studying the effects of welding parameters on joint integrity of small titanium components.

1. Introduction

Joining and welding technologies remain fundamental to manufacturing and engineering, consequently the ability to form durable and strong connections between materials enables the fabrication of items that we rely on daily. Friction Welding(FW) is a solid state welding technology were welding takes place below the melting temperature of the material, this makes it a more attractive and practical joining technique for alloys that present difficulties to be joined by fusion welding [1]. Titanium alloys are alloys of choice for aerospace applications, cryogenics, automotive, bio-medical, petroleum and power generation; this due to their high specific strength, good fatigue, creep, fracture toughness properties and excellent strength to weight ratio [2]. Numerous FW techniques exist and include; rotary friction welding (RFW), friction stir welding (FSW), friction hydro pillar process (FHPP) and linear friction welding (LFW) with the last being of interest for welding symmetric and non-symmetric components. RFW has been a prevalent research field as opposed to LFW but it is limited to axi-symmetrical (rotational) welding. LFW is carried out by a relative reciprocating motion, in one plane of two components rubbing against each other under a certain pressure (10 - 200 MPa) at a high linear oscillating motion (200 - 500 mm/s) [3]. LFW is predominantly used in blisk fabrication, therefore significant research has been done on developing the application for this technique, especially were welding takes place below the beta transus temperature (~ 980^{0}C) of titanium [4]. Welding below the beta transus assist in alleviating welding defects which include; high thermal strain cycles, solidification segregation (macro-segregation) and micro-segregation),solidification cracking, contamination cracking, hydrogen embrittlement and sub-solidus (ductility dip) cracking porosity, which are inherent during fusion welding as a result of complex alloying and heat treatments [2] experienced during the manufacturing process of Ti6Al4V.

2. Literature Review

2.1. Background on LFW

The concept of LFW can be traced back to 1929, but proved doubtful because of difficulties in generating linear reciprocating motion at the time [5]. In the 1960's, patents were presented on the linear reciprocating motion and not the friction welding process [6]. A purpose built electro-mechanical actuated LFW machine was commissioned in 1990 by The Weld Institute (TWI Ltd), which facilitated initial research and development aero-engine manufactures. This technology was initially initiated by the need to repair severely damaged blades and developments in near-net shape blisks manufacturing. Companies like Pratt & Whitney, MTU aero engines and Rolls Royce were assisted by these developments to commercialize production processes through LFW [1]. To date, Boeing Co. and United Technologies Corporation hold patents that protect certain applications and welding methods or tooling concepts of the process [8]. From literature, it appears that the fundamental principles of LFW are not patented.

The physics of the LFW process relies on heat generated by friction at the contact surfaces until a deformation-temperature is achieved which results in the formation of metallic bond[1].The process has several phases as shown in Figure 1(a).

- Phase 1: Dry contact (no motion or oscillation).
- Phase 2: Dry friction with solid-to-solid contact, oscillation starts and axial load is applied. Initial flash forms but no coalescence between the components exists.
- Phase 3: A plasticized layer forms between the two components, while the oscillation frequency and axial load remain constant. Increased flash formation is observed accompanied by axial shortening (consumed length).Resulting in a formation of a fused flash indicating coalescence between the components.
- Phase 4: Rapid deceleration, oscillatory frequency reduced to zero and axial load increased consolidate weld (forging action).

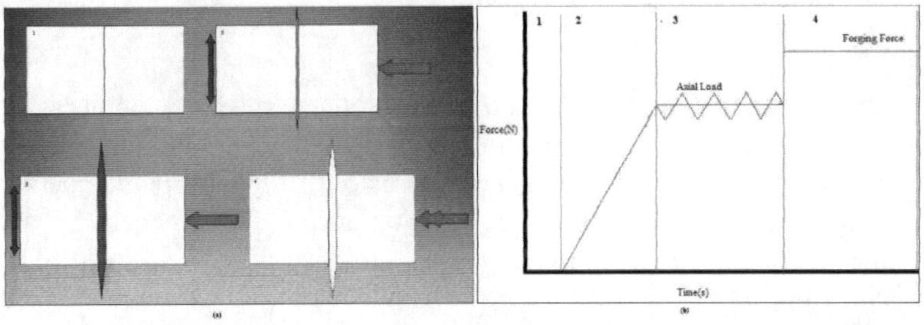

Figure 1. Schematic of the LFW process sequence and proposed process "Force Foot" print of LFW. The schematic illustration to the right of Figure 1 gives a proposed "Force Time" relationship for linear friction welding. The experimental work proposed for this platform includes the characterisation of a LFW "Force Foot Print" as was proposed by Hattingh et.al for friction stir welding in 2000 [15].

The joining by friction are divided in to four main stages, as original identified by Vairis and Frost [1].

- Stage 1: Heat generation by columbic friction between the rubbing surfaces. Asperity contact exists between the two surfaces, and as heat is generated, the asperities soften and deform, increasing the true area of contact between the two parts. (Columbic refers to friction generated by relative motion between bodies rubbing against each other)
- Stage 2: Transition stage, during which the true contact surface becomes 100% of the sample contact surface. This transition is accompanied by an increase in the force required to oscillate the parts. True area is the area revealed as the asperities are removed due to the oscillatory motion.
- Stage 3: Equilibrium, the shear force reaches a steady state value and the initial lengths of the joint components shortens through the generation of ash. The plastic zone gets progressively larger during this stage.
- Stage 4: Deceleration stage, the two specimen seize motion relative to each other and they are aligned. Depending on the application, additional forging pressure may be required.

2.2. Power input and modeling of LFW

The maximum rubbing velocity is determined by assuming sinusoidal motion of the oscillating work-piece, so the displacement, x can be approximated by (1) [8].

$$x = sin\,(\omega t) \tag{1}$$

Where (α), is amplitude of oscillation and (ω), is angular velocity which equates to $2\pi f$, (f) is the frequency of oscillation. Differentiating (1) with respect to time approximates the rubbing velocity of the oscillating specimen:

$$\dot{x} = \alpha\omega\, cos\,(\omega t) \tag{2}$$

Therefore, the maximum rubbing velocity:

$$\upsilon = \alpha 2\pi f \tag{3}$$

In-plane force:

$$F_p = \mu F_n \tag{4}$$

Where (μ) is coefficient of friction and F_n is the axial load. The heat generated at the interface per cycle is estimated byQ_o:

$$Q_o = F_p\upsilon \tag{5}$$

The process parameters to be used in this study will be derived using a trial and error approach using parameters from industrial scale welds using guidelines from work done by McAndrew et al [6].

2.3. Aptness of Titanium to LFW

Materials that exhibit conducive high temperature properties especially compressive yield, shear strength and low thermal conductivity are well suited to LFW [1]. The low thermal conductivity allows for good retention of frictional heat generated as a result of high temperature mechanical properties where the heat generated is confined at the interface [1]. Heat generation at the interface is influenced by; frequency, amplitude and frictional pressure, therefore the synergy must be carefully controlled to achieve a uniform weld and a good welded joint [1]. Consequently making titanium alloys with their good high temperature mechanical properties and low thermal conductivity ideally suitable for this type of joining process. Mill annealed Ti6Al4V grade 5 samples will be utilized in this study.

LFW can be used for a wide variety of materials, including certain aluminides and metal composites with low temperature mechanical properties and high thermal conductivity. However, high input processes forces are generally required to achieve sufficient heat to produce sound welds [13].

Alpha alloys exhibit creep resistance superior to β alloys subsequently they are preferred for high temperature applications [12, 14]. The absence of a ductile-to-brittle transitions, a feature of β alloys, makes α alloys suitable for cryogenic applications. Beta alloys have excellent harden-ability, and respond readily to heat treatment. Solution treatment is the common thermal treatment followed by aging. Titanium alloys exhibit a high chemical affinity for oxygen, which increases with an increase in temperature, coupled to a growth in surface oxide layer [3]. Temperatures exceeding 500^0C, result in reduced oxidation resistance of titanium at a greater rate and the alloy is highly vulnerable to embrittlement by O, N and H as they dissolve interstitially in titanium [3]. LFW of titanium should be done in total vacuum or inert environments. For the purposes of this study argon (Ar) will be the inert gas of choice. Good shielding improves the weld integrity of the joint, 200 ppm is the desired amount of O_2 remaining in your medium to achieve optimal welded joints [9].

3. Linear Friction Welding Platform Development

3.1. Platform Design Methodology and Process Variables

The design of this Linear Friction Welding (LFW) platform required the identification of key process variables (constraints) fundamental to LFW, these include; generating oscillatory motion, ramp up time and deceleration. Hence, an evaluation was done to determine the feasibility of using existing equipment to satisfy the above mentioned process parameters, where the platform will be attached or adapted during operation. The Zwick and Instron fatigue platforms were the two cyclic platforms at the Department of Mechanical Engineering considered as base equipment for integration of the LFW research platform. The type of dynamic motion of interest, in this case, is controlled cyclic loading from a frequency and displacement point of view. With further investigations into available platforms, it was realized that the Instron platform was the best suited from a process control point a view as it inherently provided precise control during the ramp-up and deceleration cycles. Consequently, the developed platform was designed with a view of attaching it to the Intron platform.

The main features of the Instron that dictated the adaption of the welding platform was the top and bottom jaw configuration, with the later been attached to the actuator that generates oscillating motion during fatigue tests, dictating how the platform is going to drive the oscillatory component. With this in mind the top jaws was envisaged to support the axial load applied on the non-oscillatory component as discussed as in section 2.1. The maximum travelling distance between the two jaws was of particular importance as it determined the maximum height of the platform, while facilitating the adaptation of the platform while maintaining functionality.

3.2. Process Parameters and Control

The welding platform has to meet the process parameters of LFW which includes the following; axial load, consumed length (axial shortening), oscillating frequency, oscillating amplitude and forging force. LFW is a self-regulating process from an interface temperature point of view, therefore a precise control and monitoring of the process during welding is fundamental to ensure that repeatable and defect free weld can be produced.

3.3. Axial Load and Forging Force

A constant axial load is required on the non-oscillatory component before deceleration (stage 4), consequently a load cell is required to continuously measure the applied axial load throughout the process. The load cell will translate the applied axial load via a control loop to relay the measured signal to the servomotor in an effort to control the applied load. Load control is essential during welding as the material at the interface undergoes a repeated heating, plasticise and collapse cycle, resulting in the components inability to support the initially applied axial load continuously. Hence, the term self-regulatory, as this process continues until the applied load stabilizes or exhibits minute fluctuations. Which will be a clear indication of steady state and coalescence between the components is achieved.

Due to the complicated design of the non-oscillatory chuck, an off-the-shelf load cell was not available to fit into this chuck. Therefore, a load cell was designed to meet the requirements of this stage and overcome platforms space constrains by using strain gauges. Altair *Hypermesh* was used to ensure that the deformation is linear and increase the sensitivity of the load cell as depicted in Figure 2(a).

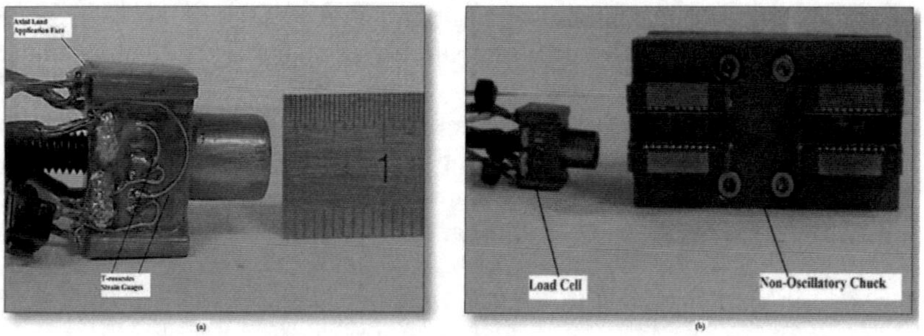

Figure 2. Load cell and non-oscillatory chuck

Figure 2 (b) shows the load cell and non-oscillatory chuck prior to and after assembly, a T-rossetes with three measuring grids as depicted in Figure 1. This design was in particularly guided by space restriction on the load cells face and non-oscillatory chuck prior to assembly.

3.4. Consumed length (Axial Shortening)

The consumed length is of particular interest because it can be related to the observed flash between the components, which aids with expelling surface contamination during plasticisation (stage 2) of the welding process, therefore ensuring a clean weld fusion zone (oxide free).

4. Discussions

Figure 3 depicts the LFW research platform for small sample developed after considering the above mentioned process variables and parameters. The top and bottom clamps will be mounted in the jaws of the Instron from where the cyclic motion will be introduced via the Instron actuator. The servo-motor will be used to drive the non-oscillatory chuck. The servo-motor will apply the required torque to maintain a constant axial load during the welding process. Trapezoidal nut and lead screw were used to connect the servo-motor to the non-oscillatory chuck with the aid of the coupling. The trapezoidal nut is mounted on the non-oscillatory chuck housing to ensure that the servomotor can slide forward and backwards, ensuring that the above mentioned fundamentals of LFW are satisfied.

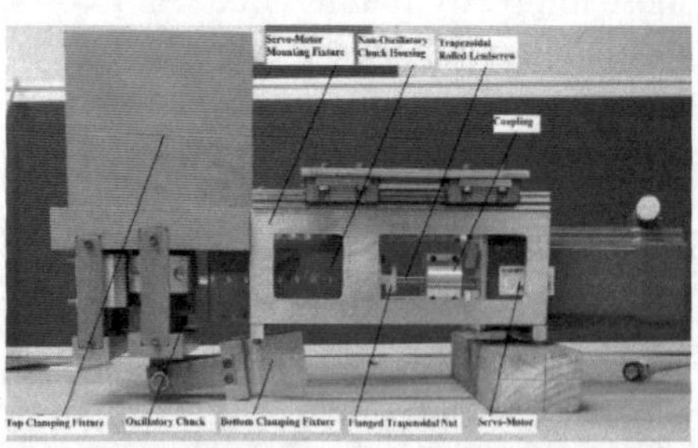

Figure 3. Linear Friction Welding Platform

References

[1] Bhamji I, Preuss M, Threadgill P, and Addison A, Solid state joining of metals by linear friction welding: a literature review. Materials Science and Technology, 27(1):2-12, 2011.

[2] Cengel Y. A and Ghajar A. Heat and mass transfer (a practical approach, SI version). McGraw-

Hill Education, 2011.
[3] Donachie M. J,Titanium: a technical guide. ASM international, 2000.
[4] Javadi Y, Akhlaghi M, and Najafabadi M A. Using nite element and ultrasonic method to evaluate welding longitudinal residual stress through the thickness in austenitic stainless steel plates. Materials & Design, 45:628-642, 2013.
[5] Maurya R and Kauzlarich J. Bonding apparatus {friction welding by reciprocal motion. Patent nos. US3420428-A, DE1552871-A, CA844858-A, 1969.
[6] Midling O and Grong. A process model for friction welding of al mg si alloys and al sic metal matrix composites. haz temperature and strain rate distribution. Acta Metallurgical et Materialia, 42(5):1595-1609, 1994.
[7] Nunn M. Aero engine improvements through linear friction welding. In 1st International Conference on Innovation and Integration in Aerospace Sciences, pages 4-5. CEIA, Queen's University Belfast Northern Ireland, UK, 2005.
[8] Shtrikman M. Linear friction welding. Welding international, 24(7):563-569, 2010.
[9] Tsikayi D. S. Friction Hydro Pillar Riveting Process of TI-6AI-4V Titanium Sheet. PhD thesis, Nelson Mandela Metropolitan University, 2015.
[10] Vairis A and Frost M. Modelling the linear friction welding of titanium blocks. Materials Science and Engineering: A, 292(1):8-17, 2000.
[11] Vill V. I. Friction welding of metals, volume 1. American Welding Society; trade distributor: Reinhold Pub. Co., 1962.
[12] A. H. Volume. 2: Properties and selection: Nonferrous alloys and special-purpose materials. ASM international, pages 1770-1886, 1990.
[13] Wanjara P and Jahazi M. Linear friction welding of ti-6al-4v: processing, microstructure, and mechanical-property inter-relationships. Metallurgical and Materials Transactions A, 36(8):2149-2164, 2005.
[14] Welsch G, Boyer R, and Collings E. Materials properties handbook: titanium alloys. ASM international, 1993.
[15] Hattingh D, Blignault C, Van Niekerk T, and James M. Characterization of the influences of fsw tool geometry on welding forces and weld tensile strength using an instrumented tool. Journal of Materials Processing Technology, 203(1-3):46-57, 2008.

Development of a welding platform and tool for the study of weld and process parameters, during continuous friction stir welding of AA6082-T6 sheets

P P Chikamhi*, D G Hattingh and B Dreyer

Department of Mechanical Engineering, Nelson Mandela University, Port Elizabeth 6001, South Africa

*s211174920@mandela.ac.za

Abstract. Bobbin tool friction stir welding is a relatively novel technology whose application, despite its benefits, is still limited due to unfamiliarity presented by less published literature. Advantages associated with the bobbin-tool technique lie imbedded in the resultant double-sided processed zone, of somewhat rectangular cross section, along the joint line. Currently, the joint integrity benefits are overshadowed by high setup costs associated with slightly more complex tool and platform designs of bobbin-tool friction stir welding. To largely exploit and optimise the technique, there is need for an in-depth understanding of interaction between welding parameters and response variables. This comprehension could assist with process optimisation, reproducibility, automation and possibly process economic feasibility. This paper proposes design considerations with regards to development of a continuous solid-state welding platform and tool, for instrumentation of process output variables. As an instance, upon tool and platform development, calibration and verification, data acquisition of weld forces developed during bobbin-tool friction stir welding, as a function of process time, can then be implemented to enable analysis. Feasibility of the proposed methodology is then left for evaluation in future work. Thus, analysis of weld forces can be facilitated by the design and development of an instrumented fixture and a bobbin friction stir welding tool, in joining AA6082-T6 aluminium plates.

1. Introduction

Friction Stir Welding (FSW), is a welding technique both invented and patented by The Weld Institute (TWI) in Britain, in the 1990s that involves the traversing of a rotating and wear-resistant weld tool along the weld interface of two metals [1, 2]. Dynamic contact friction between the tool and workpiece material and the resulting material plastic deformation, generate heat required to soften, stir and join the material, in solid state via a weld nugget. Material stirring, facilitated by the tool's material mixing features, is done under the effect of a weld-consolidating forging force. Forging force continuously applied to the material by the welding machine spindle through the welding tool during welding, is counter-acted by the support given by a backing anvil or plate. A typical FSW tool consists of a pin, or probe, and either one shoulder in the case of conventional FSW (CFSW) or two shoulders in the case of self-reacting tool FSW (SRFSW). The latter FSW variant, SRFSW, is alternatively known as bobbin tool FSW (BFSW). Figure 1 illustrates these two approaches, along with the accompanying tool configurations.

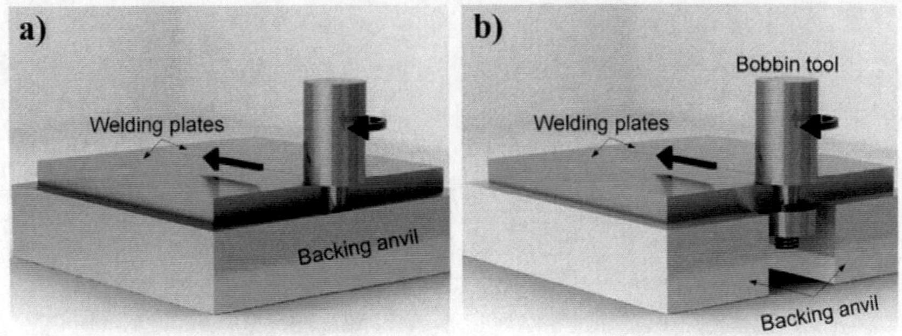

Figure 1. Illustration of two FSW variants; (a) CFSW (b) BFSW. [3]

Owing to its simple tool design, CFSW remains dominant in both published literature and industrial applications. Sued MK and Pons DJ [4] report that CFSW is best suited to welding thicker materials and is associated with the potential of root flaws due to incomplete tool pin penetration. Unlike the conventional tool, the bobbin tool design is characteristically more complex, owing to the increased number of variables affecting weld quality, especially for thin materials [4]. Presently, less is understood about the complex nature of these relationships existing among the variables and resulting weld quality. Furthermore, a study carried by Sued MK et al. reveals that tool features and tool dimension heuristics are not directly transferrable from CFSW to BFSW, not without compromising process variables and tool part functionality [5]. Therefore, literature is yet to develop a model linking weld variables like weld speed and tool geometry to response variables like weld forces and the resulting weld quality. These limiting factors combined with additional set-up costs, commensurate with complex platform and tool design and development, consequently delay and discourage ready adoption of BFSW.

CFSW is generally associated with the potential of root flaws formation, unbalanced heat input and high process forces (vertical), requiring rigid workpiece support and clamping. Conversely, BFSW eliminates the net vertical forces along with the need of a backing plate at the weld interface. Process forces on fixtures and the FSW machine itself are reduced. Moreover, risks of root flaws and unbalanced weld distortions are eliminated. Balanced heat input profiles, higher peak temperatures and cooling rates are attainable, at a low heat input and an improved weld processing speed [6]. Finally, finer microstructure formed in the stir zone of BFSW samples, result in higher hardness values and slightly higher tensile strength. FSW in the past decades, has been successfully implemented in automotive, aerospace and shipbuilding industries. FSW has been attaining increasing application where non-ferrous metals have replaced steel as structural material, [1, 7]. This adoption of FSW as a preferred alternative with non-ferrous metals can be explained by the fact that, other welding techniques like MIG, TIG and spot welding, are either difficult or impossible to implement, for such metals. Despite the seemingly successful and widespread implementation of FSW, there remains a space for improvement, to further optimise process efficiency and fully exploit its benefits, using BFSW techniques. Focus of this paper forms a subsection of an ongoing project focused on automatic control of the continuous BFSW of long aluminium alloy sheets, as used in the marine industry. This is to be accomplished by making use of a short bed bolt-on extrusion feeder providing workpiece feed to a rotating spindle, using the TWI's floating-bobbin tool concept and an FSW machine, as illustrated in Figure 5. The paper seeks to address the implementation of a system that enables analysis of variation of weld variables, especially separation forces, with process parameters; tool geometry, tool rotational and welding speed. Design considerations accompanying platform and tool development, for resistance to and instrumentation of weld forces, are examined.

2. Experimental Setup

2.1. *Workpiece Material*

Whilst almost any aluminium alloy can be used as substrate, AA6082-T6 plates 3mm thick were considered as base metal (BM), with workpiece dimensions of surface area of 110 mm x 400 mm each, in accordance with the ASTM B 557M-02a standard tensile testing sample sizes. Table 1 and Table 2 show respectively, the chemical composition and mechanical properties of the plates to be welded. AA6082-T6, being a high strength and light weight 6xxx series alloy, is often used for structural purposes. Its typical applications include highly stressed applications in trusses, bridges and transport systems. As such, high specific strength coupled by corrosion resistance [7], make AA6082-T6 preferable for marine applications.

Table 1. Chemical composition of AA6082-T6 (wt. %).

Elements	Si	Fe	Cu	Mn	Mg	Cr	Zn	Ti	Al
Base Metal	0.92	0.445	0.05	0.61	0.75	0.05	0.142	0.065	Bal

Table 2. Mechanical properties of AA6082-T6.

Material	Yield Strength (MPa)	UTS (MPa)	Elongation (%)	HV (0.5 kg)
Base Metal	255	318	9	95

2.2. *BFSW tool development*

2.2.1. Tool features. On the developed platform, to manufacture butt welds, bobbin tools were designed and developed. Tool configurations shown in Figure 2 and Figure 3 as Tool 1 and Tool 2, whose features and dimensions are displayed in Table 3 were adopted. Both tools feature scrolled bottom and upper shoulders and a 2.9 mm cylindrical, threaded and tri-flat probe. In consultation with existing literature [2, 5, 6, 8-12], effort was made to formulate an optimal bobbin tool design, to enhance the overall process efficiency and resulting butt weld quality. As an illustration, Casalino et al. [13] report that there is need for shoulder diameter optimisation, especially with regards to heat transfer and material flow. This is proposed due to the significant influence certain geometry features have on thermal cycles (peak temperature) and the required process power and torque. Thus, special tool features selected were recommended for better material mixing, essential for defect tolerant welds. Shoulder scrolls minimize flash formation and eliminate the tool tilt angle [12] whilst flats and threads promote material flow and circulation. Owing to tool (shoulders and probe) size limitations, Sued MK et al. assert that these special tool features, are difficult to fabricate and are rarely used in welding of thin materials [4]. In practice, when incorporated into design, additional costs are incurred. A case in point is the cost of fabrication, which rapidly escalates with the reduction of tool size and the complexity of features. K110 tool steel was selected as the bobbin tool material because of its good compressive strength, dimensional stability in heat treatment, adhesive and abrasive wear resistance, all necessary for high integrity welds.

Table 3. BFSW Tools 1 and 2 geometrical parameters.

Tool	Shoulder diameter, D (mm)	Probe diameter, d (mm)	Probe thread pitch, p (mm)
1 (D_1, d_1, p_1)	14	6	0.8
2 (D_2, d_2, p_2)	17	8	0.8

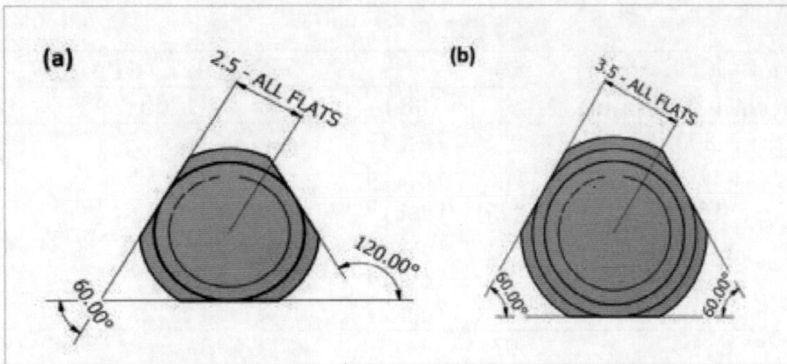

Figure 2. BFSW tool probes' profiles: (a) Tool 1 (b) Tool 2.

2.2.2. Floating mechanism. The main components of the bobbin tool are as shown in Figure 3, a tool holder (sleeve) and the welding tool. The welding tool is attached to the tool holder, via a slot, a slot key and a floating mechanism. The floating mechanism comprises of two (compression and tension) mechanical springs, of identical compliance. The floating capability of the bobbin tool enables mechanical auto-alignment of tool with workpiece during welding, eliminating the need for accurate setup procedures and sophisticated force control systems [2]. Essentially, the floating feature ensures continuous contact between tool shoulders and the plate surfaces during welding. Misalignment could possibly cause the tearing away of the thin plate weld material or the excessive formation of flash.

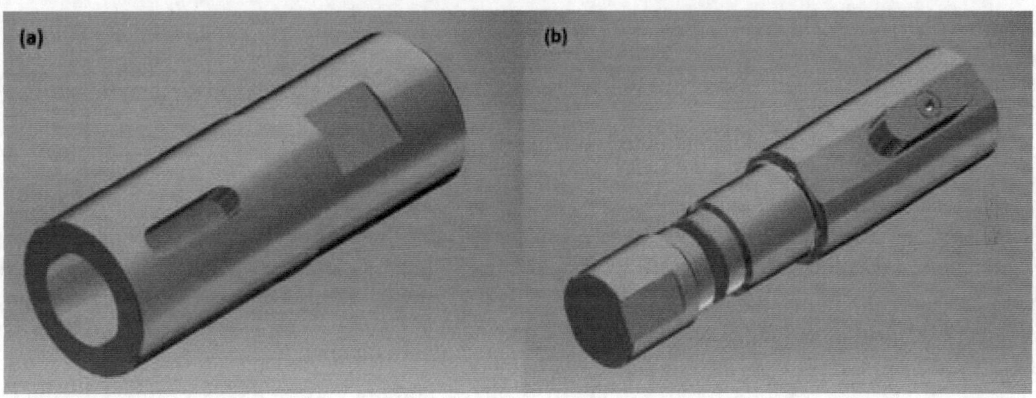

Figure 3. BFSW tool components: (a) Tool holder (b) Tool.

2.3. *Design of Experiments.*

To facilitate data acquisition and analysis of process variables, in future work, welds will be done on a specialised PDS Rotary friction welding machine, in position-control mode, according to the test matrix shown in Table 4. Preliminary tests to establish the weldability window, with respect to weld quality and weld variables will be conducted first, to confirm the test matrix selected. Table 5 summarises control variables and corresponding magnitudes where applicable, to be used in these experiments. Table 3, Table 4 and Table 5, show that the main process variables considered in the tests would be (1) tool geometry, (2) weld speed and (3) tool rotary speed. It can also be noted from Table 4 and Table 5 that focus of the study would be limited to fast welds in a bid for process optimisation for industrial applications. Additionally, literature associates faster travel speeds with high integrity and reduced weld forces. Fast welds also naturally result in cold welding conditions and short processing times.

IOP Conf. Series: Materials Science and Engineering **430** (2018) 012011 doi:10.1088/1757-899X/430/1/012011

Table 4. Test matrix.

Tool	Speed, v (mm/min)	Speed, n (RPM)		
		800	**1200**	**1600**
1	800	Test 11	Test 12	Test 13
1	1200	Test 14	Test 15	Test 16
1	1600	Test 17	Test 18	Test 19
2	800	Test 21	Test 22	Test 23
2	1200	Test 24	Test 25	Test 26
2	1600	Test 27	Test 28	Test 29

Table 5. Control variables.

Variable	Value
Workpiece material	AA6082-T6
Workpiece thickness	3.0 mm
Workpiece curvature	0.0 deg
Weld length	400 mm
Weld condition	Cold, Fast
Dwell time	5.0 s
Start-up parameters: Feed, Speed	1.75 mm/min, 1400 rpm
Tool tilt angle	0.0 deg
Tool pin flats, shoulder scrolls	3, 3
Pin thread depth	1.0 mm
Pin thread pitch	0.8 mm
Pin flat depth	0.5 mm
Shoulder gap/pin length (@ 3.33% interference)	2.9 mm

2.4. BFSW Platform development

2.4.1. Design and Instrumentation. Due to the tool-workpiece interaction, transverse, longitudinal and axial mechanical forces are generated. In BFSW, the resultant axial force experienced by the spindle is approximately zero since equal and opposite axial forces contained between the two shoulders cancel out. Therefore, only longitudinal and transverse forces are expected to show a marked variation with process variables. A mild steel welding and Load-cell platform shown in Figure 4 was designed and developed to meet the various force measurement and clamping requirements during BFSW. One backing plate was mounted on linear guides, as the other was constrained, to effect a sliding contact between one backing plate and the common base plate. Done in a way that supports only infinitesimal movement whilst rigidly clamping the workpieces in place, the micro-strains generated during welding would be detected by three force transducers. Each of the three transducers were installed against the sliding backing plate, along the weld direction. Rigid clamping resists plate separation forces during welding, whose effects compromise weld quality and whose magnitudes can be measured and ascertained, under different welding conditions. Force transducers therefore, measure the transverse forces experienced by the clamping system on the platform, during welding. Transducers consist of full-bridge configuration gage installations of four active gages, two in compression and two in tension, during force detection. This configuration is more immune to apparent strain arising from temperature variations during welding. Each gage is connected as a resistor in the Wheatstone bridge circuit, registering any strain variation as voltage variation. The strain variation can then be mapped to the corresponding force variation, from calibration readings. Other accompanying weld forces, namely longitudinal and axial, can be measured by the PDS machine's internal sensors.

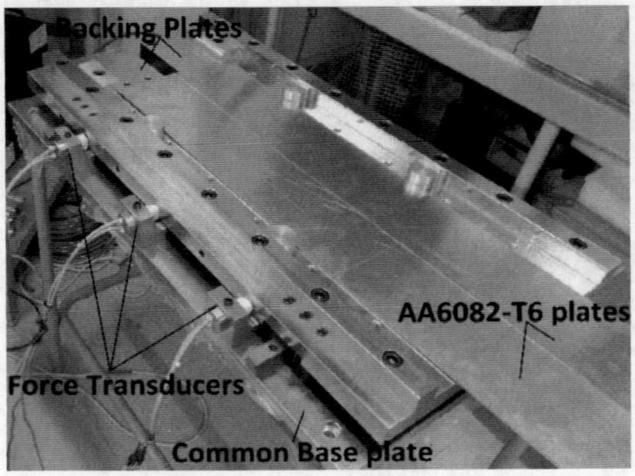

Figure 4. BFSW platform

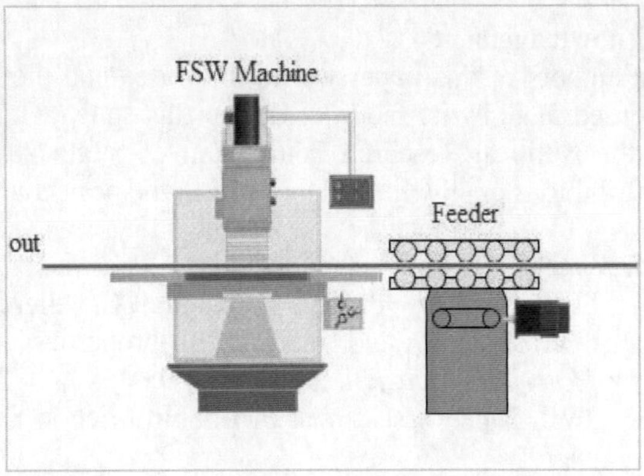

Figure 5. Continuous BFSW of aluminium sheets, using a bolt-on Feeder.

2.4.2. Data Acquisition. Weld forces data acquisition can be provided for by the SoMat eDAQ field computer and the PDS machine's data logging system. Calibration of the platform is achieved by an application of dead weight loads to the Figure 4 platform and noting corresponding gage strain readings.

3. Discussions and conclusions

Weld forces encountered in practice during FSW, are enormous enough to warrant concern. As a result, clamping forces to counteract them and arrest the undesirable effects of their development are necessary for every friction stir welding operation. For instance, separation forces must be resisted accordingly, to avoid compromising the weld quality through enlargement of the weld gap during welding. This often limits the implementation of continuous FSW, hence drastically reducing the efficiency and feasibility of the overall process. Sued MK et al. postulates that process settings such as clamps, support arrangements and shoulder gap create compression, vibrations and heat distributions, hence influence weld quality. For that reason, implementation of continuous FSW requires an address and investigation of weld forces and the complex relationships that exist among them, weld variables and weld quality, to enable efficient clamping and fixturing during welding. Such invaluable information would assist with process optimisation, automation and adaptive control. This observation is also concurred by Forcellese et al. [14]. An important consequence of these gains would be improved productivity through reduced processing and setup time. Other possible benefits would be improved weld quality and better process reproducibility. Additional benefits, particular to the methodology pursued and illustrated in Figure 5, may include easy handling of material before and after welding whereby rolls can be used for wrapping and storage of material. The developed platform and bobbin tool facilitate investigation into these issues.

Although less is known concerning the relationships governing variation of the weld forces and the weld quality with weld variables, artificial intelligence can be used to aid analysis and model such, making predictions possible. As data of this variation becomes available, implementation of continuous BFSW may be facilitated by use of derived control algorithms to control actuators providing workpiece feed and clamping. There yet remains many possibilities of process implementation and employment, to fully reap benefits of the FSW technology. It is the opinion of the authors that continuous bobbin friction stir welding presents a new methodology for achieving higher production rates for solid state butt welds and as such platform and tool development facilitating its investigation is necessary. BFSW process heuristics can also then be formulated, adding to the existing limited body of knowledge.

Acknowledgements
The authors of this paper would like to extend their appreciation to eNtsa, for the research capacity afforded them by its facilities during this study. This work is based on the research supported in part by the National Research Foundation of South Africa (Grant Numbers: 113407). Also, the expert contributions of all who collaborated in the project are acknowledged.

References
[1] Scialpi A, De Filippis LAC and Cavaliere P. 2007. Influence of shoulder geometry on microstructure and mechanical properties of friction stir welded 6082 aluminium alloy. *Materials & Design*; 28(4):1124-9.
[2] TWI. 2008. Assessment of Bobbin Friction Stir Welding for the Joining of Aluminium Alloys. *TWI*.
[3] Esmaily M, Mortazavi N, Osikowicz W, Hindsefelt H, Svensson JE, Halvarsson M and et al. 2016. Bobbin and conventional friction stir welding of thick extruded AA6005-T6 profiles. *Materials and Design*; 108(2016):114-25.
[4] Sued MK and Pons DJ. 2016. Dynamic Interaction between Machine, Tool, and Substrate in Bobbin Friction Stir Welding. *International Journal of Manufacturing Engineering*; 2016:14.
[5] Sued MK, Pons D, Lavroff J and Wong EH. 2014. Design features for bobbin friction stir welding tools: Development of a conceptual model linking the underlying physics to the production process. *Materials & Design* (1980-2015); 54:632-43.
[6] TWI. 2012. Further Development of the FSW Floating-Bobbin Technique. UK: TWI.
[7] Krasnowski K, Hamilton C and Dymek S. 2015. Influence of the tool shape and weld configuration on microstructure and mechanical properties of the Al 6082 alloy FSW joints. *Archives of Civil and Mechanical Engineering*; 15(1):133-41.
[8] Colligan KJ, O'Donnell AK, Shevock JW, Smitherman MT and Hostetter GJ. Friction Stir Welding of Thin Aluminium Using Fixed Gap Bobbin Tools. 9th *International FSW Symposium*; 15-17 May 2012; Huntsville, AL2012.
[9] Threadgill L.P, Ahmed M, Martin J, G. Perrett J and Wynne B. 2010. The Use of Bobbin Tools for Friction Stir Welding of Aluminium Alloys. 1179-84 p.
[10] Hattingh DG, Blignault C, van Niekerk TI and James MN. 2007. Characterization of the influences of FSW tool geometry on welding forces and weld tensile strength using an instrumented tool. *Materials Processing Technology*; 203(2008):46-57.
[11] Khan NZ, Khan ZA and Siddiquee AN. 2015. Effect of Shoulder Diameter to Pin Diameter (D/d) Ratio on Tensile Strength of Friction Stir Welded 6063 Aluminium Alloy. *Materials Today*: Proceedings; 2(4):1450-7.
[12] Zhang YN, Cao X, Larose S and Wanjara P. 2012. Review of tools for friction stir welding and processing. *Canadian Metallurgical Quarterly*; 51(3):250-61.
[13] Casalino G, Campanelli S and Mortello M. 2014. Influence of Shoulder Geometry and Coating of the Tool on the Friction Stir Welding of Aluminium Alloy Plates. *Procedia Engineering*; 69:1541-8.
[14] Forcellese A, Simoncini M and Casalino G. 2017. Influence of Process Parameters on the Vertical Forces Generated during Friction Stir Welding of AA6082-T6 and on the Mechanical Properties of the Joints. *Metals*; 7(9):350.

Process optimization of rotary friction welding of Ti-6Al-4V alloy rods

MC Zulu[1]* and PM Mashinini[1]

[1]University of Johannesburg, Department of Mechanical and Industrial Engineering Technology, Doornfontein Campus, Johannesburg, South Africa

*mthobic91@gmail.com and *mmashinini@uj.ac.za

Abstract. This paper presents an investigation on the performance of rotary friction welded 25.4 mm diameter Ti-6Al-4V alloy rods. The process parameters used are friction and forging pressure, rotational speed, forging time and upset distance. The heating time was determined by the amount of friction pressure and rotational speed utilised. Visual examination of preliminary weld joints was done which revealed lack of bonding, weld oxidation and discolouration. The final weld joints were examined for mechanical properties. The results revealed that friction welding process parameters have significant influence on the weld joint properties and weld joint integrity. The hardness and ultimate tensile properties of the weld joints varied with the variation in rotational speed and axial pressure. The tensile properties of the weld joints were higher than the parent material at lower rotational speed or higher axial pressure. The weld width was proportional to relative motion and inversely proportional to axial pressure.

1. Introduction

Joining of similar and dissimilar materials has always been a challenge in conventional welding processes. This challenge instigates because of deterioration in mechanical properties and change in the microstructure of the material at the weld joint. As a result, the weld joint becomes a point of failure during the transverse tensile test and the component loses trustworthy for the critical applications. This challenge has always been a nightmare for fabricators even though many investigations and developments have been done to improve the efficiency of conventional welding processes, such as arc welding [1]. In many cases, deterioration of mechanical properties is a consequence of melting that happens at the weld interface as well as low temperature segregation caused by filler material and its poor selection during welding process [2-3].

Rotary friction welding (RFW), a solid-state welding process, is one of friction welding processes that maintains or have improved properties at the weld interface without the use of filler material. This is because this typical welding process occurs at a temperature below melting point of parent material being joined therefore no melting occurs. On the other hand, since there is no filler material being utilized, the process eliminates defects associated with poor selection of filler material and undesirable effects of low temperature segregation [2-4]. RFW can be used to join many similar and dissimilar materials without any difficulties, even materials that are considered not weldable by the use conventional methods or techniques.

RFW process consists of two joining workpieces, one placed at a stationary fixture and the other at a rotating fixture. The process is performed by rotating one part relative to the other along their common axis. This is done simultaneously with the application of axial pressure to enhance rubbing as a result of an increased in friction coefficient. The relative motion together with axial pressure increases interface temperature. This temperature is proportional to the axial pressure and inversely proportional to rotational speed [5-7]. When the interface temperature is high enough to induce plastic state, plasticized material at the interface is displaced out through the edges. This is a very important phase as all impurities are ejected out, leaving clean material from both workpieces. The final increased axial pressure is applied to permit consolidation after stopping relative motion. In this stage too, plasticized material is pushed out in a form of a flash. The weld interface is allowed to cool while consolidating. Finally, the axial pressure is quickly removed and the welding is complete. The weld joint formed consists of distinct weld zones as a result of microstructure variation [8-9].

Although this process produces a high-quality type of weld joint, there are major difficulties associated with poor selection of process parameters. Firstly, metal surfaces of workpieces are rough at atomic scale. Therefore, when low axial pressure is applied only fewer sections of the interface get into contact. Secondly, high axial pressure increases the number of contact surfaces, but oxides layer and occluded gas present in all metals under normal atmospheric conditions are encountered. On the other hand, higher relative motion or lower axial pressure increases the volume of material affected by heat, while lower relative motion or higher-pressure results in a visible physical deformation [1, 11-12]. However, these difficulties can be avoided if much attention is paid on the selection of process parameters based on the geometry of the workpieces and the properties of the material to be welded.

2. Welding Ti-6Al-4V under different conditions

Titanium grade 5 (Ti-6Al-4V) is one of material that is regarded difficult to weld using conventional welding processes. This is because this metal is highly reactive to the environment (air) at elevated temperatures. Therefore, much attention has to be paid on shielding to prevent defects associated with atmospheric gases. Rotary friction welding of Ti-6Al-4V has been performed widely under different process conditions or parameters. It has been reported that the variation in process parameters has a significant influence on the properties of friction welded Ti-6Al-4V. Poor selection of process parameters results in a weld joint with poor mechanical properties. Dalgaard [7] reported that an increase in relative motion increases heat density, which in turn decreases friction coefficient resulting in a low heating rate. The research also reported that the heat density is inversely proportional to axial friction pressure. Dinaharan et al [3] reported that at higher rotational speed the coefficient of friction is lower resulting in longer heating time, high heat conductivity rate and later amount of material pushed out in a form of flash. Tolvanen [13] mentioned that cooling rate is the very important factor that determines the integrity and properties of component at the interface. The researcher also revealed that the cooling rate is proportional to pressure and inversely proportional to rotational speed.

Although other researchers believed that variation in process parameters have significant influence on the weld joint properties, Vill [14] believed that rotational speed is the least sensible process parameter and axial pressure does not have significant effect if it is not varied over a wide range. This was also supported by Yate [15] on the inertia friction welding of Ti-6Al-4V. However, Da Silver [1] and North [16] opposed this revealing that low relative motion reduce weld width therefore increases the tensile properties of weld joint. Munchen [17] reported that an increase in axial pressure reduces heat input, which results in higher cooling rate and hardness. Although researchers have different opinions regarding the effect of process parameters on the weld joint quality, it is important to understand the role of each process parameter in the production of weld joint.

The aim of this research study is to optimize the rotary friction welding process of Ti-6Al-4V under various conditions and to evaluate the effect of variation in process parameters. This will be achieved by keeping certain process parameters constant while varying others.

3. Experimental procedure

Welding platform

The platform utilized for the welding of Ti-6Al-4V rods of 25.4 mm diameter was Process Development System (PDS) for friction welding process, located at Nelson Mandela University in Port Elizabeth, South Africa. The welding platform is fully automated with a computer interface that allows the switch between different process parameters and data collection.

Process parameters and Mechanical Testing
Rotational speeds of 1600, 1900, 2300 and 2700 rpm, and heating pressures of 25, 40, 60 and 80 MPa were utilized during rotary friction welding of Ti-6Al-4V rods, which was conducted on the laboratory temperature under argon gas shielding. Forging time, forging pressure and upset distance were kept constant at 25 s, 95 MPa and 2 mm respectively. The heating and braking times were determined using process cycles recorded for each welding process. The range of heating pressures and rotational were selected based on the analyses of preliminary welds.

Tensile strength and micro-hardness were the mechanical properties that were tested on the welded specimens. The tests were performed on the welded and parent material samples. The tensile test samples were prepared using ASTM E8/E8M-13a standard while hardness test samples were prepared using ASTM E92 standard. The tensile test speed of 2 mm/min and a pre-load of 2 MPa were used. The tensile testing was carried out at room temperature using Zwick/Roell Z250 tensile test machine, with the load capacity of 250 kN. Hardness testing was determined on the samples using Timc Vickers micro-hardness tester. The hardness profile was performed at different locations of weld interface. The indentations were made over a distance of 10mm with a spacing of 0,5mm. The load of 300gf for a dwell time of 10 seconds was used.

4. Results and discussion

Process cycle
Figure 1 illustrates the rotary friction welding process cycle of Ti-6Al-4V that was recorded at a rotational speed of 1600 rpm and 40 MPa heating pressure. The different process parameters used as well as the responses are graphically represented against the welding time. The process is made out of three stages (as clearly marked with the letters in Figure 1) and these stages are heating stage - A, braking stage - B and forging stage - C. heating stage is divided into initial phase and constant heating phase.

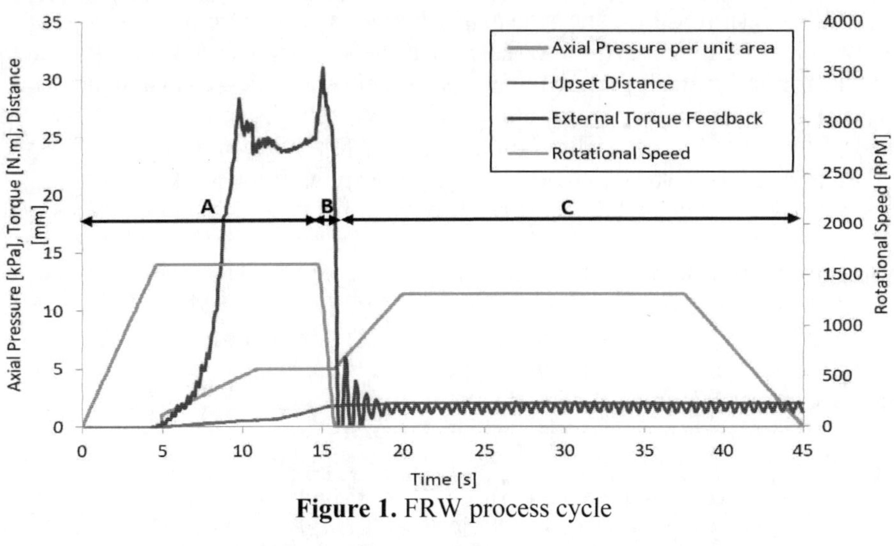

Figure 1. FRW process cycle

The pre-determined value of rotational speed (1600 rpm) was reached before the initial contact of the workpieces. This was done to reduce the friction coefficient at the initial contact. The predetermined value of heating pressure (40 MPa) was reached during the initial phase of the heating stage. This value was reached through a constant increment known as the ramp rate. The initial contact of workpieces causes a torque to increase and reach its first peak value. This was due to a higher friction coefficient as a result of low temperature on the joining surfaces. As the heating stage progressed, the interface temperature increased resulting in a decrease in friction coefficient, which in turn caused a sudden decrease in torque to some extent, as shown in Figure 1. The interface temperature was inversely proportional to the coefficient of friction and rotational speed. This was backed up by Dalgaard [7] on the Linear Friction Welding (LFW) of Titanium alloys. The torque decreased further to some extent that was considered as an equilibrium torque as the heating stage progressed further. This may be correlated to a high temperature being reached at the interface causing a further decrease in the wear rate and friction coefficient. This may also be linked to the interface material getting softer and pushed out because of heat and pressure. The torque oscillated along its equilibrium value for a short period and gradually increased at a minimum gradient through the remainder of the constant heating stage. This was due to the variation of interface temperature as a result of self-cooling process of Ti-6Al-4V. In addition, an increase axial shortening was observed throughout the constant heating phase. This was linked to interface material being soft and pushed out as a result of relative motion and applied pressure. The braking stage was reached at the end of heating stage. At the braking stage, the drive motor was dis-engaged and brakes were applied. This was done to eliminate the relative motion before consolidation. The torque value increased yet again and reached its second peak value that was the maximum torque of the welding cycle. This was caused by an increase in rotation resistance. Additionally, an increase in axial pressure to a pre-determined forging pressure resulted in a rapid increase in axial shortening. This was achieved by pushing more soft material out in the form of a flash. A forging pressure of 95 MPa was applied at the end of braking stage to the stationary workpieces. This resulted in a further increase of axial shortening to a pre-set value of 2 mm. Forging pressure was maintained at a pre-determined value for a period of 20 seconds to allow consolidation and self-cooling of material affected by the heat. At the end of forging stage, pressure was quickly removed and Ti-6Al-4V weld joint was formed. Similar process cycle was obtained by Palanival et al [3] during friction welding of Titanium tubes.

Effect of process parameters onTi-6Al-4V weld joint
Rotational speed is one of the sensitive process parameters in rotary friction welding of Titanium alloys. It has an extensive influence on the weld joint quality if it is varied over a wide range. The efficiency of RFW is improved by using a relatively low rotational speed, as shown in Table 1. The observation on friction welded Ti-6Al-4V revealed that the shearing effect at the weld interface was replaced by polishing action at a higher rotational speed. This was attributed to a minimum coefficient of friction at a higher rotational speed resulting in a low heating rate that allows for higher heat propagation rate along workpieces axially. Lower heating rate prolonged heating time, which in turn caused heat loss resulting in a greater amount of energy being used for welding process. In addition, the weld joint produced at higher rotational speed had wider width as a result of heat dissipation. The tensile and hardness tests revealed that the weld joints of higher rotational speed were weaker while those of low rotational speed were improved. This was concluded after having failure points within the weld zone for high speed (as shown in Figure 2A) and outside of the weld joint for lower rotational speed (as shown in Figure 2B). This was attributed to the minimum temperature gradient as a result of higher heat dissipation rate. Higher rotational speed demotes the cooling rate, widen weld width and hence reduces mechanical properties of the weld joint.

Table 1. Tensile and hardness test results at different speeds and pressure

	Rotational Speed (rpm)				**Axial Pressure (MPa)**				**Parent material**
	1600	1900	2300	2700	25	40	60	80	
Tensile Strength (MPa)	1035	1033	1030	1024	1022	1022	1031	1034	1030
Micro Hardness (HV)	375.9	372.3	370.8	365.0	346.5	351.6	359.2	360.2	336

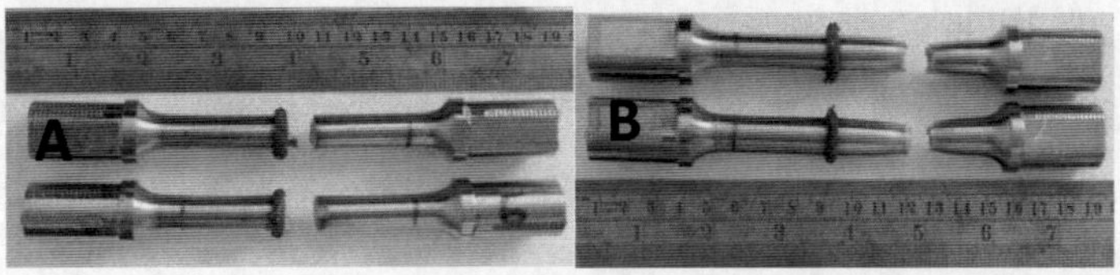

Figure 2. Tensile test specimens of higher (A) and low (B) rotational speeds.

Another sensitive process parameter is the axial pressure. Its regulates the weld joint temperature gradient, friction power required and upsetting. Axial pressure can be varied over a wide range depending on the materials properties as well as the geometry of workpieces to be welded. The main function of axial pressure is to maintain intimate contact of the surfaces, preventing oxidation and enhancing consolidation. When low axial pressure is applied fewer sections of the interface get in contact. This prolongs welding time, increases weld defect possibilities and promotes heat propagation along the workpieces consequently, larger volume of material gets heated and a large weld joint is produced. The mechanical properties of weld joints of light axial pressure exhibit poor mechanical properties, as shown in Table 1. This is caused by lack of bonding and weld defects present within the weld joint. Therefore, the weld joint becomes a point of failure during tensile test of welded specimens, as shown in Figure 2A. Low axial pressure reduces shortening rate, increases heat propagation rate, increases weld defects and increases weld width.

Results and characterisation of weld joint.

The analyses of welded Ti-6Al-4V samples revealed a change in microstructure across the weld joint. The weld nugget (WN), thermos-mechanically affected zone (TMAZ) and heat affected zone (HAZ) where three distinct weld zones that were obtained during microstructure characterization. The same results were reported by Dalgaard [7]. A basket-weave microstructure with complete recrystallization was obtained at WN, as shown in Figure 3A. The refined alpha grains were well distributed within transformed beta grains. This was caused by recrystallization occurred as a result of temperature elevating beyond beta-transus temperature. Cooling from the temperature above beta transus resulted in very fine acicular grains within the WN. In addition, very fine martensitic alpha grains were more visible at the weld joints produced at lower rotational speed and higher axial pressure. This was due to the faster heating and cooling rate as a result of high temperature gradient between weld joint and cold base material. The microstructure of TMAZ is almost similar to that of WN. This is because this region is exposed to extreme temperature for a short period of time, therefore allowing partial dynamic recovery, although it is not physically active on the welding process. This region contains very fine grains with the presence of acicular alpha grains, as a result of extreme temperature exposure, which may possible be resulted in recrystallization, as shown in Figure 3B. Furthermore, the microstructure of the HAZ contains visible alpha grains within the fully transformed beta phase. The grains of this region were elongated and arranged in a radial flow pattern, as shown in Figure 3C. This may be attributed to the nature of RFW and heat propagation, which may have caused a loss of torsional stiffness at this location.

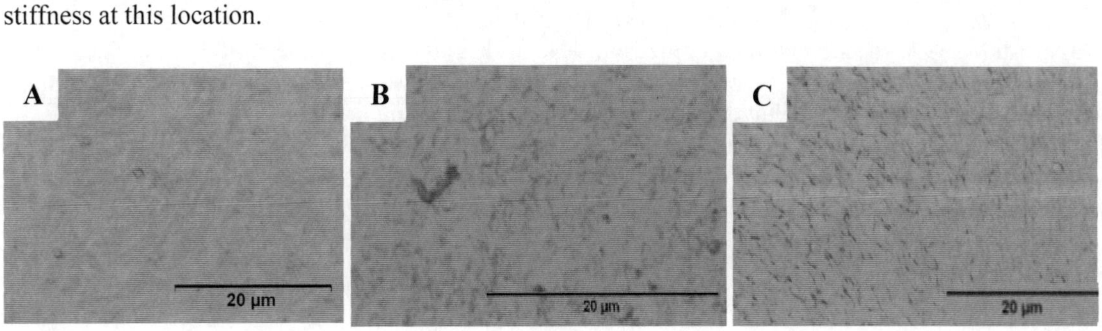

Figure 3. Microstructure of WN (A), TAMZ (B) and HAZ (C)

Conclusion

Titanium rods of 25.4 mm diameter were successfully welded at different rotational speeds and axial pressures. The results revealed that axial pressure and rotational speed had significant influence on the quality of the weld joint and weld integrity. The increase in rotational speed reduced the mechanical properties of the weld joint as a result of reduced friction coefficient. Rotational speed was proportional to weld width and inversely proportional to welding time. The weld joint obtained had weld nugget, thermo-mechanically affected zone and heat affected zone. The weld nugget has basket-weave microstructure with very fine acicular grains.

Acknowledgements

The authors would like to extend their gratitude to the staff members of eNtsa-innovation through technology at Nelson Mandela University for the great contribution toward the success of this research study. Authors also like express their profound gratitude to Prof. Hattingh, Mr Brown and the staff of Mechanical and Industrial Technology at University of Johannesburg for their kindly assistance and knowledge in friction welding. Moreover, a great thanks to Mr Mukhawana for his assistance to conduct analyses. Lastly, we like to extend our sincere gratitude to National Research Foundation (NRF) for funding this research study.

References

[1] Da Silva A A 2006 An *Investigation on the Structure/Property Relationships of Solid State Welding Processes in a Titanium Matrix Composite Alloy (Ti-6Al-4V)* (Duisburg-Essen: University of Duisburg)
[2] Tech S 2001 *Friction welding process* http://www.nctfrictionwelding.com/process
[3] Palanival R, Laubscher R F, Dinaharan I and Hattingh D G 2017 *Int. J. of pressure vessels and piping* **154** 17-28
[4] Bohme D, Appel L and Cramer H 2009 *Joint of Titanium/Titanium/Steel by Friction Welding with continuous drive* (Munchen: Germany)
[5] Moarrefzadeh A 2012 J. of Mech Eng. **1** Issue 1
[6] Sahoo R and Samantaray P 2007 *Study of friction welding* (NIT: Rourkela)
[7] Dalgaard E C 2011 *Evolution of microstructure, micro-texture and mechanical properties in linear friction welded titanium alloys* (Montreal, Canada: McGrill University)
[8] American Welding Sociecty 1989 *Recommended practices for Friction welding* (New York: American National Standards Institute (ANSI))
[9] Amecrican Welding Society 1989 *Friction welding process* (Miami: Florida). Vol 2. pp. 740-63.
[10] Mech4study 2017 *Friction welding : Principle, work, types, application, advantages and disadvantages* http://www.mech4study.com/2017/04/friction-welding-principle-working-types-application-advantages-and-disadvantages (Accessed 07 June 2017)
[11] Dippenaar R J, Reid M H and Dehghan-Manashadi A 2005 Effect of mictrostucture morphology on the mechanical propeties of titanium alloys *J. of Physics* **1**1-4
[12] Linnert G E 1994 *Welding Metallurgy Carbon and Alloy Steels* **1**1-940
[13] Tolvanen S 2016 *Microstructure and mechanical properties of Ti6Al4V welds produced with different processes* (Sweden, Gothenburg: Chalmers University of Technology)
[14] Vill V I 1972 *Friction welding of Metals* (New York, New Jersey: AmericanWelding Society)
[15] Yates A 2015 *The effect of microstructure on mechanical properties in inertia welding titanium 6-4* (Birmingham: University of Birmingham research archive)
[16] Li Z, Maldonando C, North T H and Alsthuller B 1997 *Mechanical and Metallurgical properties of MMC friction welds* (Welding journal) **76** 9 367-373
[17] Munchen V S 2000 *Joint of Titanium/Titanium/Steel by Friction Welding with Continuous Drive* (Innovative technologies for joining advanced materials – tima09) **1** 1

IOP Conf. Series: Materials Science and Engineering **430** (2018) 012013 doi:10.1088/1757-899X/430/1/012013

On Machinability of Titanium Grade 4 under Minimum Quantity Lubrication Assisted High Speed Machining

T N Mathonsi[1], R F Laubscher[1], K Gupta[2]

[1]Department of Mechanical Engineering Science, University of Johannesburg, Auckland Park-2006, Johannesburg, Republic of South Africa
[2]Department of Mechanical and Industrial Engineering Technology, University of Johannesburg, Doornfontein-2028, Johannesburg, Republic of South Africa

kgupta@uj.ac.za

Abstract The wonder metal Titanium and its alloys are prime candidate for various automotive, biomedical and aerospace applications due to their good strength-to-weight ratio, bio-compatibility and corrosion resistance. Titanium and its alloys are known as difficult-to-machine materials i.e. their machining is challenging. The experimental work reported in the present paper attempts to enhance the machinability of Titanium Grade 4 under the influence of minimum quantity lubrication at high speed conditions. In this work a total of twenty seven experiments has been conducted based on full factorial design of experiment technique. Cutting speed, feed rate, and depth of cut are varied at three levels each and the values of important MQL parameters are fixed. The effects of machining parameters on surface roughness are discussed. Machining at optimum combination of parameters resulted in precision finish with maximum roughness value 2.16 μm and maximum tool flank wear value 0.201 mm. The research results reveal the superiority of MQL over conventional wet cooling to successfully machine Titanium Grade 4 at high speed conditions with sustainability.

1. Introduction

Titanium is known as wonder metal and used in various biomedical, aerospace and industrial applications in commercially pure and alloyed forms [1-3]. The special characteristics that make titanium superior to other materials are biocompatibility, high strength to weight ratio, and corrosion resistance etc. But, it is also known as difficult-to-machine (DTM) material and its machining is challenging. It possesses poor machinability that results in excessive tool wear, high consumption of energy and resources, deteriorated work surface quality, and high machining cost [3]. Titanium Grade 4 is a reliable material for various industrial and bio-medical applications such as heat exchangers, vessels, condenser tubing, dental implants, surgical instrument etc. It undergoes extensive machining operations (i.e. turning, milling, drilling, and grinding etc.) while manufacturing products for the aforementioned applications. The high speed conditions (important to achieve high productivity) make the machining extremely difficult. Conventional wet cooling is not effective at high speed machining conditions especially for titanium and its alloys, because wet cooling cannot manage the extensive heat generated at high speed conditions and fails to do effective lubrication [3-6]. It results in extensive tool wear, deterioration in work surface integrity, and high machining cost with environmental footprints. There are some ways such as using sustainable cooling and lubrication; optimum machining parameters; appropriate tool materials, geometries and coatings; and employing hybrid techniques etc. to make the machining of difficult-to-machine materials such as titanium and its alloys easy [3-6].

Sustainable cooling and lubrication techniques are very helpful to reduce the consumption of energy and cutting fluids as well as friction and heat generation in the tool-chip interface and thereby tool wear, and to maintain the work surface quality [5-7]. One of the important sustainable lubrication techniques is minimum quantity lubrication where tiny droplets of green lubricants mainly synthetic esters, vegetable oil, fatty acids and alcohols are sprinkled over the tool-chip interface in the machining zone [7]. A detailed literature review conducted by the authors reveals that there are many articles where MQL assisted machining of various materials have been reported [8-12], but there is a lack of research conducted on the MQL assisted high speed machining of titanium Grade 4.

The research work reported in this paper fills the gap and discusses the experimentation details, effect of machining parameters on an important machinability indicator such as surface roughness, and optimization to secure the best set of machining parameters for enhanced machinability of Grade 4 titanium.

2. Experimental details

Titanium Grade 4 at high speed conditions has been machined on CNC turning centre using rhomboid shaped carbide tool inserts. Producut make MQL device has been used to supply green lubricant which is a blend of natural, synthetic and sulphurized esters. The experimental setup used in the present work is shown in Figure 1.

Full factorial design of experiment methodology has been used to design the experiments at various combinations of machining parameters. The full factorial design includes all possible combinations of input parameters and hence the most appropriate experimental results [13]. Three important machining parameters namely cutting speed, feed rate and depth of cut at three levels each have been selected based on some preliminary experiments, machine settings and literature review. A total of twenty seven experiments have been conducted to analyze the effect of the aforementioned three machining parameters on machinability indicators i.e. maximum surface roughness and tool flank wear. The details of variable and fixed input parameters are given in Table 1.

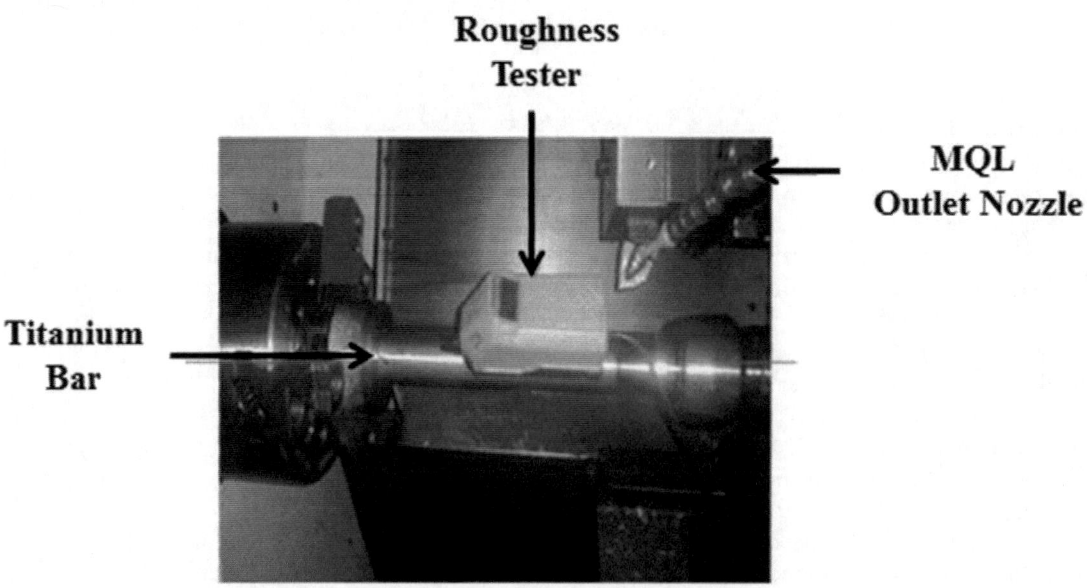

Figure 1. Experimental setup

Table 1. Details of input process parameters

Machining Parameters	Unit	Level		
		-1	0	1
Cutting Speed 'V_c'	m/min	200	250	300
Feed Rate 'f'	mm/rev	0.1	0.15	0.2
Depth of Cut 'd_c'	mm	1	1.5	2
MQL parameters	Flow rate 70 ml/hr	Air pressure 4 bar	Nozzle distance 30 mm	
CP Ti Grade 4: Carbon- 0.012% + Iron-0.206% + Hydrogen- 0.0028% + Nitrogen- 0.0028% + Oxygen- 0.3 % + Ti- balance Yield strength- 480 MPa				

3. Results and discussion

Table 2 presents the values of average surface roughness and tool flank wear corresponding to all twenty seven set of experiments. Figures 2-4 depicts the effect of machining parameters mainly cutting speed, feed rate and depth of cut on maximum surface roughness. The statistical analysis of variance (ANOVA) study found feed rate and cutting speed as the most significant parameters affecting surface roughness and tool wear respectively.

Table 2. Experimental results corresponding to twenty seven combinations of machining parameters

Expt. No	V_c (m/min)	f (mm/rev)	d_c (mm)	Maximum Surface Roughness R_{max} (μm)	Tool Flank Wear T_w (mm)
1	300	0.1	1	2,25	0,55
2	250	0.3	1	7,08	0,11
3	300	0.2	1	4,83	0,17
4	200	0.1	2	3,06	0,14
5	250	0.3	2	8,31	0,32
6	250	0.3	1.5	7,83	0,17
7	300	0.1	2	2,67	0,19
8	300	0.3	1	6,48	0,14
9	250	0.2	1.5	5,34	0,15
10	200	0.3	1.5	7,89	0,16
11	300	0.3	2	8,04	0,35
12	200	0.3	1	7,32	0,09
13	200	0.1	1	2,79	0,28
14	300	0.3	1.5	7,11	0,18
15	300	0.1	1.5	2,46	0,32
16	250	0.2	1	5,16	0,14
17	300	0.2	2	5,37	0,27
18	250	0.1	1.5	2,82	0,30
19	200	0.2	1	5,58	0,08
20	300	0.2	1.5	5,13	0,14
21	250	0.1	1	2,52	0,24
22	200	0.2	1.5	5,70	0,09
23	250	0.1	2	2,94	0,17
24	250	0.2	2	5,55	0,18
25	200	0.1	1.5	2,97	0,10
26	200	0.2	2	5,79	0,09
27	200	0.3	2	8,46	0,47

As shown in Figure 2, increase in cutting speed leads to improve surface finish. High surface roughness at low speed is due to the formation of built-up edge on tool tip which deteriorates the work surface quality including surface finish [14-15]. At high feed rates, large amount of chips flow on the cutting edge of the tool which results in high friction that causes high roughness. Therefore, increase in feed rate increases surface roughness (See Fig. 3). As depicted in Figure 4, there is a straight line relationship between depth of cut and surface roughness i.e. increasing depth of cut leads to increase surface roughness. It is due to the fact that material removal is increased at high depth of cut which contributes to increase the cutting force, and high cutting forces cause fracture of tool particles that form voids on the surface being machined and hence increase the surface roughness [16].

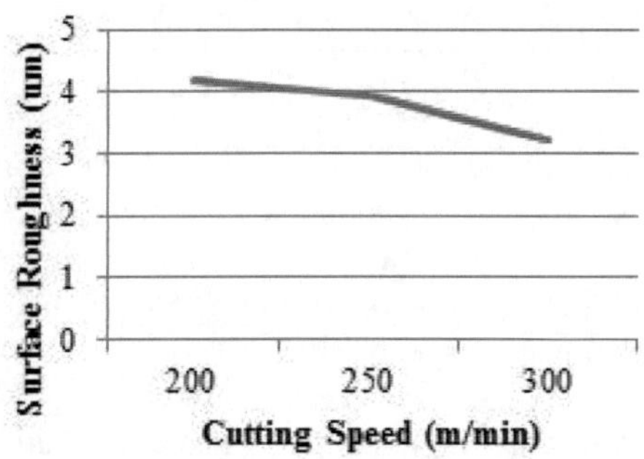

Figure 2. Effect of cutting speed on maximum roughness.

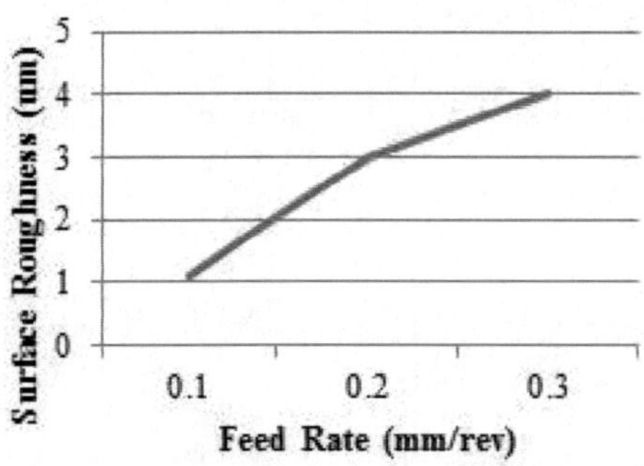

Figure 3. Effect of feed rate on maximum roughness.

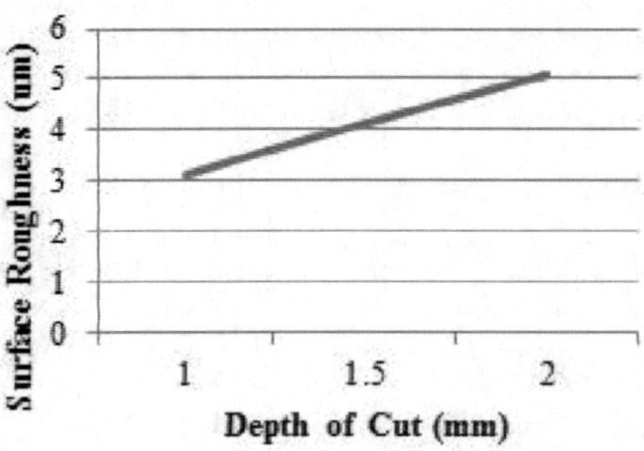

Figure 4. Effect of depth of cut on maximum roughness.

Similarly, the effects of machining parameters on other machinability indictors such as tool wear, energy consumption, and cutting forces have been investigated and reported elsewhere. After a detailed analysis of the results, it has been found that the variation trends of all machinability indicators are different and a trade-off exists that requires multi-performance optimization to simultaneously secure the best values of all machinability indicators at single set of machining parameters.

A statistical optimization technique Desirability analysis [17] with 'smaller the better' type of desirability function has been used for multi-performance optimization. Table 3 presents the optimum values of input machining parameters as well as responses predicted by desirability and the actual values obtained by confirmation experiment conducted at the optimal parameters. The results of the confirmation experiments validate the predictions and ensure significant improvement in surface finish and reduction in tool flank wear at a single set of machining parameters. Figure 4 depicts the SEM image of flank wear of the tool used to machine titanium grade 4 at optimum parameters.

Table 3. Results of confirmation experiments for optimization

Optimal parameters	Predicted optimal responses		Optimal responses at confirmation expt	
Machining condition	R_{max} (µm)	T_w (mm)	R_{max} (µm)	T_w (mm)
Cutting speed-Feed-Depth of cut 250-0.1-1	2.546	0.307	2.158	0.201

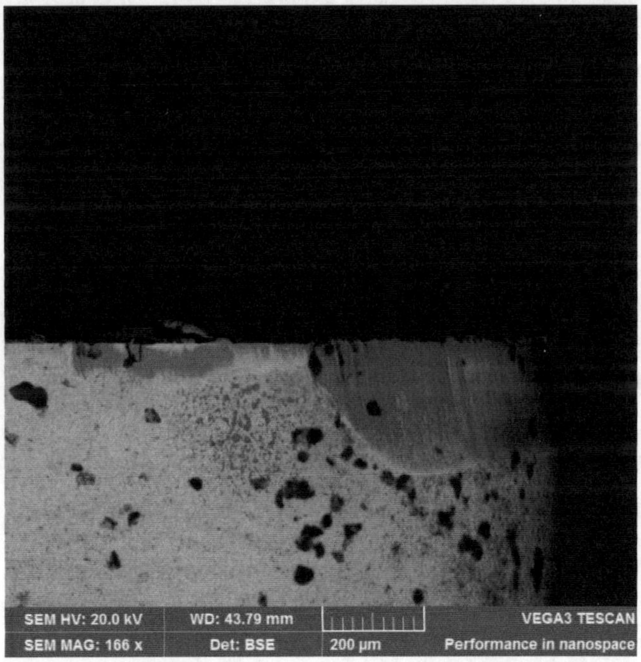

Figure 5. SEM image of tool flank wear used to machine Ti Grade 4 at optimum parameters

4. Conclusion

This paper reports various important aspects of a detailed investigation conducted on the MQL assisted high speed turning of titanium grade 4. The following conclusions can be drawn from this investigation:

- Increasing feed rate and depth of cut increase the surface roughness, whereas increasing cutting speed decreases the roughness.

- Machining at optimum parameters i.e. cutting speed- 250 m/min, feed rate-0.1 mm/rev, and depth of cut- 1 mm resulted in optimized values of maximum surface roughness- 2.16 μm and tool flank wear- 0.201 mm.
- Tool flank wear value 0.2 mm is under the recommended ISO limit (0.6 mm).
- Cutting speed and feed rate were found to be the most significant parameters.

A comparative study of titanium grade 4 machining in different cooling and lubrication conditions, MQL assisted machining of other titanium grades, and life cycle analysis of the titanium machining process etc. are the possible scope for future research in this area.

References

[1] Pramanik A 2014 *Int J Adv Manuf Technl* 70 919–928
[2] Donachie MJ 2000 *Titanium: a technical guide* (ASM International: Ohio)
[3] Davim JP 2014 *Machining of titanium alloys* (Springer: Heidelberg)
[4] Chauhan SR, Dass K 2012 Mat and Manuf Proc 25 (5) 531-537
[5] Gupta K, Laubscher RF 2017 *Proc. IMechE, Part B: J Eng Manuf* 231-14, 2543-2560
[6] Gupta K, Laubscher RF 2016 *In Proc. of International Conference on Competitive Manufacturing*, Stellenbosch (South Africa) pp 211-217
[7] Gupta K, Laubscher RF, Davim JP, Jain NK 2016 *J Cln Prod* 112 (4), 3320-3330 (DOI: 10.1016/j.jclepro.2015.09.133)
[8] Sun J, Wong YS, Rahman M et al. 2007 Mach Sci Technol 10, 355–370
[9] Cai XJ, Liu ZQ, Chen M et al. 2012 *Proc IMechE, Part B: J Eng Manuf* 226-11, 1784–1792
[10] Gao K, Qi L, Yu D et al. 2014 In Proc of the first symposium on aviation maintenance and management, Xian, China Springer
[11] Nath C, Kapoor SG, DeVora RE et al. 2012 *J Manuf Proc* 14, 452–459
[12] Okada M, Hosokawa A, Asakawa N et al. 2014 Int J. Advan Manuf. Technol. 74, 1255–1266
[13] Montgomery DG 2009 *Design and analysis of experiments* (John Willey & Sons)
[14] Grzesik W 2017 *Advanced Machining Processes of Metallic Materials* (Elsevier)
[15] Rao DC, Rao DDN, Srihari P 2013 Procedia Eng 64, 1405-1415
[16] Astakhov V 2007 Int J Advan Manuf Technl 7-8, 631-640
[17] Derringer G, Suich R 1980 J. Quality Technl 12, 214–219

IOP Conf. Series: Materials Science and Engineering **430** (2018) 012014 doi:10.1088/1757-899X/430/1/012014

A study of the machinability of Ti-6Al-4V compacted powders

M T Tambani[1, 2,*], C Machio[1], D M Madyira[2], E H Uheida[3] and G A Oosthuizen[3]

[1]Light Metals, Materials Science & Manufacturing, Council for Scientific and Industrial Research, Meiring Naudé Road, Brummeria, Pretoria 0185, South Africa. [2] Department of Mechanical Engineering Science, University of Johannesburg, South Africa. [3]Stellenbosch University, South Africa

* mtambani@csir.co.za

Abstract. Machining is an integral shaping process for most titanium alloys based components. However, the ease of machining (machinability) of Titanium and its alloys is generally poor because the alloys have low thermal conductivity, high strength at high temperatures, which both combine to speed up the tool deformation and wear. The poor machinability of the titanium alloys contributes to their high cost, which prevents their widespread industrial application. In this study, this challenge of poor machinability is investigated for the most common titanium alloy, Ti-6Al-4V. The envisaged approach was to machine and characterise the machinability of the alloy in the green state. Green compacts of blended powders of Ti-6Al-4V were pressed at room temperature. To enable machining, a minimum green strength to handle clamping and withstand mechanical forces during machining was required. The green strength of this study was varied by changing the compaction load and the binder during pressing. The compacts were subjected to a drilling operation where the tool type, feed rate and the cutting speed were varied. The machinability was evaluated by measuring average edge breakout and the drilling forces. Results indicate that high strength steel (HSS) twist drills, low feed rate and high speeds generally reduce machinability of the Ti-6Al-4V compacted powders. The drilling forces were independent of the green strength of the compacts.

1. Introduction

Ti-6Al-4V is a lightweight metal alloy with a high specific strength and high corrosion resistance. Despite the alloy superior properties, it has not found widespread use due to the high inherent cost of producing components[1]. The cost is from extraction as well as manufacturing, which almost always includes machining steps. Ti-based alloys are known to have low machinability due to the inherent properties of low thermal conductivity and high chemical reactivity. During machining of Titanium and its alloys, temperature at the tool/chip interface rapidly increases and the cutting tool quality deteriorates quickly [2].

Powder metallurgy (PM) has the potential to reduce the cost of Ti-based alloys because near net shape components that require minimal machining are possible. PM also, offers better material utilization because of the reduced machining. The buy to fly ratio of powder metallurgy parts is way less than the traditional method of producing parts of wrought and casting. The simplest PM technique, which also offers the most advantages, is press and sinter (P&S). In P&S, features that are perpendicular to the pressing directions such as holes, threads and undercuts cannot be accommodated during compaction; they have to be introduced on the part after sintering by machining [3]. Such features are prime candidates for green machining.

Green machining refers to the production of features in compacted powder (also called green bodies). This work evaluates the machinability of Ti-6Al-4V alloy green bodies. It is hoped that machining the green bodies/compacts can improve machinability because of their low green strength. This has the potential to increase productivity and prolong tool life. Green machining requires compacted powders with adequate green strength to permit the green compacts clamping and also withstand the machining forces [4]. The green strength also affects the quality of the drilled holes: low green strength leads to poor surface finish and an allowance of average breakout that occur by pulling of powder particles [5]. The average breakout is one the distinct criteria that is used to characterize the machinability of the green compacts. In this study, the machining operation used was drilling as it is one of the most widely used manufacturing operations. Another criterion to access the machinability is the measuring of the cutting forces that are generated during the metal removal process. The resultant cutting force is generally resolved into three components, namely feed force (Fx), thrust force (Fy) and cutting force (Fz). The cutting forces acting on the tool are an important characteristic of machining; they can be used to indicate tool and workpiece deflections [6]. The strength of the material strongly influences the cutting forces. Other factors that affect cutting forces are cutting speed, feed rate, depth of cut, cutting fluids and tool geometry [7], [8].

2. Materials and Methods

The material used was blended elemental Ti-6Al-4V made by blending hydrogenated, dehydrogenated titanium (Ti HDH) with particle sizes of less than 45 μm (325 mesh) with a 60Al40V master alloy. A one weight percent of polymeric binder was used as an admixed lubricant. The base powder, master alloy and the polymeric binder were mixed dry by blending in a tubular mixer for an hour to make a homogeneous mixture. Cold compaction of the blended powder was done using an Enerpac 100T laboratory hydraulic press. Lubrication was applied on the walls of the die to minimize friction between the powder particles and the die walls. A cylindrical die of 43 mm diameter was used to press 10 mm thick samples. The compaction pressure used for this study was 430 MPa and 600 MPa. Some samples were cured in air at 200 °C for 1hr. The green strength of all the samples was determined using Brazilian disc test according to ASTM D3967 standard. Table 1 shows the green density and green strength of the uncured and cured Ti-6Al-4V samples.

Table 1. Properties of green compacts used.

Compaction pressured	Cured/ Uncured	Density (%)	Green strength (MPa)
430 MPa	Uncured	72 ± 1	12 ± 2
430 MPa	Cured	73 ± 1	15 ± 2
600 MPa	Uncured	76 ± 1	20 ± 3
600 MPa	Cured	78 ± 1	27 ± 1

Drilling experiments were performed on a CNC Leaderway V650 milling machining. In these experiments, through holes of 3mm diameters were drilled using two types of drills: High Strength Steel (HSS-Co) tool and a carbide tool. The characteristics of the drill tools are provided in Table 2. The drilling speeds used were 2000 rpm, 4000 rpm and 6000 rpm, and the feed rates were 0.02 mm/rev and 0.06 mm/rev. For each parameter, 3 holes were made. Air was used as the cooling medium for the tools during drilling. Tool wear was regarded as insignificant.

Table 2. Characteristics of drilling tools.

Material	Coating	Helix angle (◦)	Point geometry	Point angle (◦)
High strength steel-Co	TiAlN	35	Split	120
Solid carbide	Coated	30	Split	118

The surface integrity was analyzed by measuring the average width of breakout near the outlet edge. The outlet edge is the edge that is last drilled by the cutting tool, whereas the inlet edge is the edge where drilling starts. Micrographs were captured using a stereo microscope and analyzed using Image pro software. The thrust force and torque during drilling were determined only for the HSS drill bit using a speed of 6000 rpm and 0.06 mm/rev. The force was measured using a Kistler piezoelectric dynamometer.

3. Results and discussion

Figure 1 shows the variation of the size of breakout zones versus drilling speed, feed rate, compaction pressure and drilling tool. For both the uncured and cured specimens, the size of breakout zones generally increased with cutting/drilling speed, indicating that as the cutting speed increased, the mechanical bonds of the green compacts increasingly failed to hold the particles together. Even though intuitively, productivity can be increased by using higher speed, the results here show that very high speeds can lead to high scrap rates due to the size of breakouts. Furthermore, regardless of the type of sample, breakout zones were generally larger for samples drilled with the HSS drill bits. There is a possibility that the HSS tool reacted with the samples forming bonds that acted over a larger area.

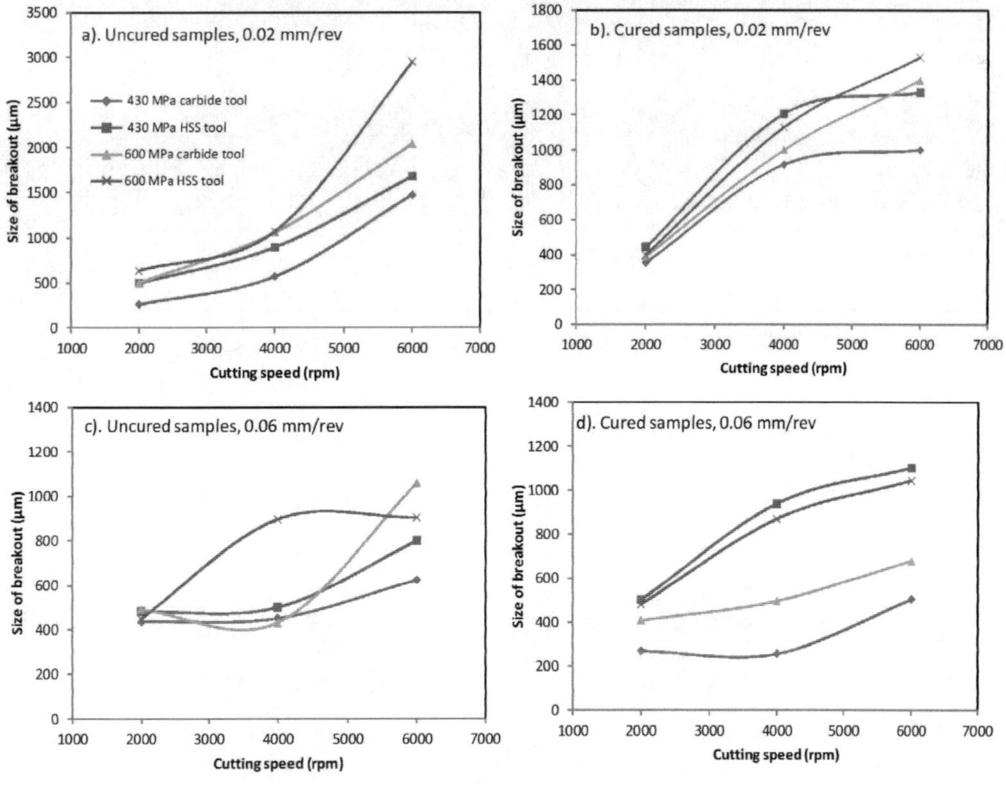

Figure 1. Average breakout of Ti-6Al-4V samples: a) Uncured, 0.02 mm/rev, b) Cured, 0.02 mm/rev, c) Uncured, 0.06 mm/rev, d) Cured, 0.06 mm/rev. Legend in Figure 1a) applies to all graphs.

Concerning the compaction load, it appeared that the samples compacted at 430 MPa had smaller break-out zones than the samples pressed at 600 MPa. This, according to most literature, was a counter-intuitive observation given the green strength was lower for the former samples pressed at 430 MPa, and a higher green strength is associated with smaller breakouts [7–9]. However, the observation tallies with others that argue that a high green strength is not beneficial because the stronger mechanical bonding between powder particles leads to bigger breakouts [5]. Characteristic break-out zones are shown in Figure 2, for uncured and cured samples drilled using HSS and similar parameters.

The effect of feed rate on the size of breakouts was very pronounced at high speeds, where increasing the feed rate from 0.02 mm/rev to 0.06 mm/rev reduced the size of the breakouts (Figure 1a) and c) (for uncured) and Figure 1b) and d) (for the cured) samples).

This observation was in line with literature [2], where a low feed rate is associated with an increase in temperature of samples, leading to loss of green strength and accelerated particle removal. In the current study, it is possible that the low feed rate led to an increase in temperature and caused the polymeric binder to over-cure, leading to a brittle network.

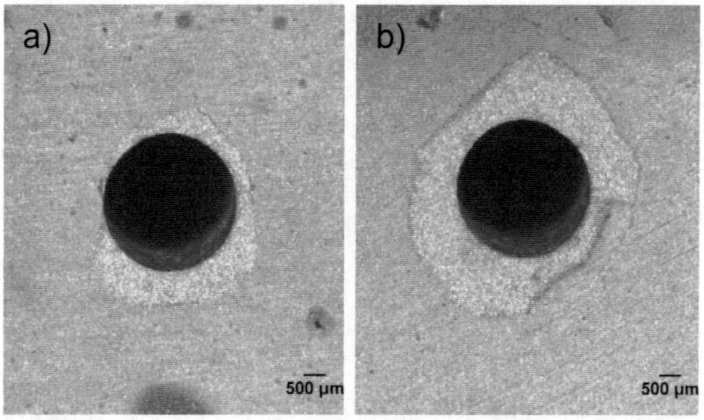

Figure 2. Micrographs of Ti-6Al-4V showing the average breakout on samples drilled using a HSS tool at 0.06 mm/rev and cutting speed of 2000 rpm: a) cured sample b) uncured sample.

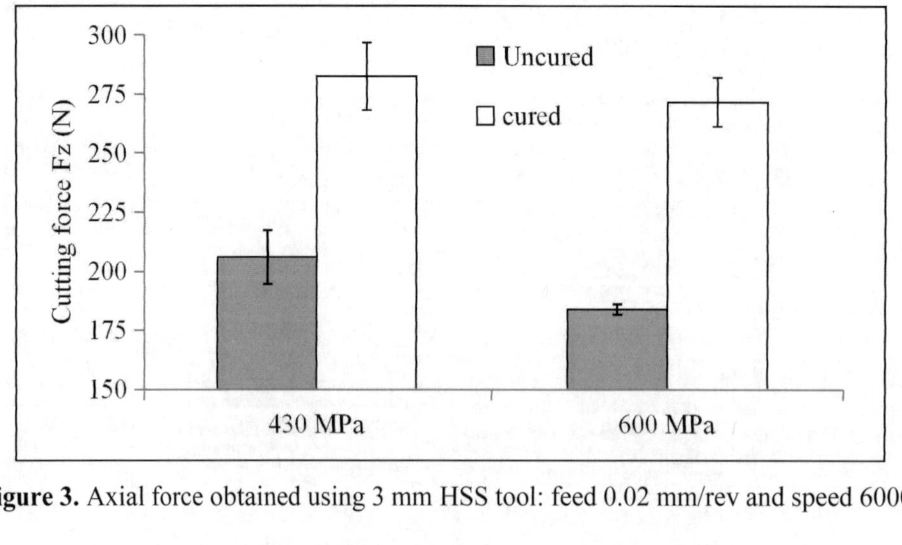

Figure 3. Axial force obtained using 3 mm HSS tool: feed 0.02 mm/rev and speed 6000 rpm.

Figure 3 presents the cutting forces measured when drilling with a HSS tool at a speed of 6000 rpm and feed rate of 0.06 mm/rev. The cutting forces for the uncured samples were lower than for the cured samples. It is highly possible that the uncured polymeric binder partly acted like a machining lubricant, hence lowered the cutting forces. The least cutting force was obtained on uncured samples pressed at 600 MPa, while the highest force was obtained for the cured samples, regardless of the compaction load. The similarity of cutting forces for the cured samples, which had different green strengths, indicates that the green strength had only a marginal role in determining the cutting forces. It is possible that the high cutting forces obtained for the cured samples were caused by clogging of twist drill by the machined powder, as has been observed in other alloy systems [10].

4. Conclusion

It may be concluded from the present work that the interaction between the strength and the machinability of Ti-6Al-4V green compacts is not straightforward, but dependent on the drilling parameters. The cutting speed displayed significant impact on machinability where high cutting speeds reduced machinability. The effect of feed rate was muted at the lowest cutting speed, but increasing feed rate at high speeds reduced size of breakouts and improved machinability. HSS twist drills marginally reduced machinability compared to carbide twist drills. The cutting forces were highest for the cured samples, regardless of the compaction load, indicating that it was independent of the strength of the compacts.

Acknowledgements

The authors would like to give gratitude to CSIR for funding the project, Mr M. Grober and Mr S. Masete for their technical expertise.

References

[1] Wang H, Zak Fang Z and Sun P 2010 a Critical Review of Mechanical Properties of Powder *Int. J. Powder Metall.* **46** 45–57

[2] Robert-Perron E, Blais C, Pelletier S and Thomas Y 2007 Machinability of green powder metallurgy components: Part I. characterization of the influence of tool wear *Metall. Mater. Trans. A Phys. Metall. Mater. Sci.* **38** 1330–6

[3] Robert-Perron E, Blais C, Pelletier S and Thomas Y 2007 Drilling of high quality features in green powder metallurgy components *Mater. Sci. Eng. A* **458** 195–201

[4] Liu T, J.Danaher and R.Guo 2003 Green machining for conventional P/M process *Japan Soc. powder Metall.* **50** 832–6

[5] Degnan C C, Kennedy A R and Shipway P H 2007 Relationship between physical structure and machinability of green compacts *Powder Metall.* **50** 14–20

[6] Igarashi N, Sonoda Y, Take R and Terai H High-Quality High-Productivity Manufacturing of Variable Valve Timing Parts by Green Machining 48–53

[7] Robert-Perron E, Blais C, Thomas Y, Pelletier S and Dionne M 2005 An integrated approach to the characterization of powder metallurgy components performance during green machining *Mater. Sci. Eng. A* **402** 325–34

[8] Cimino T M and Luk S H 1995 Machinability evaluation of selected high green strength PM material *J. Chem. Inf. Model.* **53** 1689–99

[9] Chagnon F, Tremblay L, St-laurent S and Gagné M 1999 Improving Green Strength to Enable Green Machining *Soc. Automot. Eng.*

[10] Vitiello a. and Prisco U 2009 Evaluation of drilling parameters effects on machinability of PM materials using ANOVA *Powder Metall.* **52** 164–71

IOP Conf. Series: Materials Science and Engineering **430** (2018) 012015 doi:10.1088/1757-899X/430/1/012015

The effect of specimen geometry on tensile properties of titanium alloy metal sheet.

M S Masete[1,*], N S Muchavi[1,2] and S Chikosha[1]

[1] Council for Scientific and Industrial Research (CSIR), Pretoria, South Africa
[2] Department of Engineering Metallurgy, University of Johannesburg, South Africa

*SMasete@csir.co.za

Abstract. Determination of tensile properties of materials is required to reliably predict performance of engineering materials in service and for design purposes. Tensile properties are also used as a measure of product quality. Various testing standards recommend specific specimen sizes for reporting of tensile properties. This recommendation is normally a challenge in materials development where the required specimen size is not achievable due to component size limitation. In addition, the materials developer still wants to compare their tensile results with the reference materials produced from standard sizes or produced from different geometries. It is well documented that tensile properties are affected by the specimen geometry (gauge length and cross sectional area). It is therefore the purpose of this study to investigate the effect of specimen geometry on tensile properties. In addition, this study seeks to determine suitable specimen geometry required to yield comparable tensile properties obtained from other specimen geometries. Various dog bone specimen geometries were produced from a titanium alloy metal sheet and tensile properties determined. Yield strength, Ultimate tensile strength and, % elongation were compared. Yield strength and ultimate tensile strength were not affected by different specimen geometries. However, % elongation differed for different geometries.

1. Introduction

It is common practice during development of engineering materials to determine tensile properties. These properties are required to qualify materials, predict performance and sometimes improve on properties of currently available materials.

Even though determination of tensile properties is well known, certain considerations need to be adhered to for the results to be acceptable. Various testing codes prescribes geometries and test conditions in order for the results to be valid and comparable. During material development, one normally seeks to compares the new results with the existing ones to determine if improvement has been achieved or not. Various testing standards and authorities [1][2] allow such comparison only if all the conditions are similar i.e. same testing conditions and same sized specimens. The ISO 6892-1 standard [1] prescribes a gauge length of either 50 mm or 80 mm while ASTM E8 standard [2] prescribes a gauge length of 25 mm, 50 mm or 200 mm for tensile properties evaluation. In addition both standards [1][2] require that the gauge length to width ratio (L/W) be either 4 or 5. The specimen geometry is reported to have negligible effect on the Yield Strength (YS), Ultimate Tensile Strength (UTS) and Young's modulus, while ductility is greatly affected [1][2]. The ASTM E8 standard [2] suggests keeping the gauge length over square root of area ($L/\sqrt{A}$), the same to reduce the effect of specimen geometry on ductility while ISO 6892-1 [1] suggest the gauge length over width (L/W) to be constant. Depending on material availability, prescribed specimen sizes are often not achievable. This would then imply that the material developer cannot confidently compare the new tensile results with the known one.

Various studies have been carried out to try and address this difficulty [3–6]. Han et al [3] compared tensile properties of either steel or aluminium alloys for plate type specimens having different gauge length and width. The results showed that YS, UTS and Young's modulus were not affected by the change in specimen geometry. However, % elongation (% EL) differed by around 10% for different gauge lengths. Even though Han et al [3] showed that % EL was greatly affected by specimen geometry, there was no suggestion on how this problem could be overcome if one wants to compare % EL of differently sized specimens. Strnadel and Brumek [6] kept the $L/\sqrt{A}$ the same and equal to about 5.56 to remain compliant with the standards. The results showed that ductility (% EL) decrease with increasing cross sectional area. However, the authors [6] did not show whether keeping the $L/\sqrt{A}$ equal to 5.56 reduced the effect of specimen geometry on % EL. Yuan et al [5] used a specimen with a gauge length of 25 mm and width of 6 mm as specified by the ASTM E8 standard [2] to investigate the effect of specimen thickness on tensile properties. As Yuan et al [5] increased the specimen thickness, the $L/\sqrt{A}$ reduced leading to an increase in % EL. The YS and UTS were not affected by the change in thickness. The ASTM E8 standard [2] gives no exact requirement for thickness of plate specimens hence a study by Yuan et al [5]. Al-bakri et al [4] also found that for specimens with similar gauge length and width, as thickness increases, ductility increases.

Literature agree that specimen geometry affects tensile properties especially % EL. Various considerations such as keeping the L/W equal to 5, keeping the $L/\sqrt{A}$ similar, and having a gauge length of at least 25 mm, were put forward in ensuring that tensile properties are in compliant with the standards. However, none of the studies clearly showed how one can design a sub-size specimen to ensure similarity of tensile properties especially % EL. The question that is not answered is whether the L/W ratio of about 4 to 5 is a must for compliancy. In addition, can one get similar tensile properties from specimens with gauge length much smaller than 25 mm? The purpose of this study was then to investigate the effect of specimen geometry (gauge length, specimen width, specimen thickness) on tensile properties of plate type specimen. In addition, the aim of the study was to propose suitable geometries required to compare with tensile properties obtained from specimens of other sized plate specimens. The results of this study would be beneficial to material developers that use non-standard and sub-size specimens to confidently compare their tensile results to those of other researchers using completely different geometries.

2. Experimental procedure

2.1. Materials

The Ti-6Al-4V alloy metal sheet 1.1 mm thick was used for this study. Electric discharge machining was used to cut tensile specimens of various geometries indicated in Table 1. Geometry A and B were cut to have different gauge length over area ratio ($L/\sqrt{A}$), while geometry C and D were cut directly according to ASTM E8 standard [2]. It should be noted that ASTM E8 standard [2] kept only the L/W ratio the same for geometry C and D. All specimens were cut parallel to the rolling direction as shown in Figure 1.

Table 1. Dimension of specimens tested.

	Gauge length	Width	$L/\sqrt{A}$	L/W	Total length
Geometry A	18	2.5	10.9	7.2	40
Geometry B	27	2.5	16.3	10	50

Geometry C	32	6.25	12.2	5.12	62
Geometry D	60	12.5	16.2	4	200

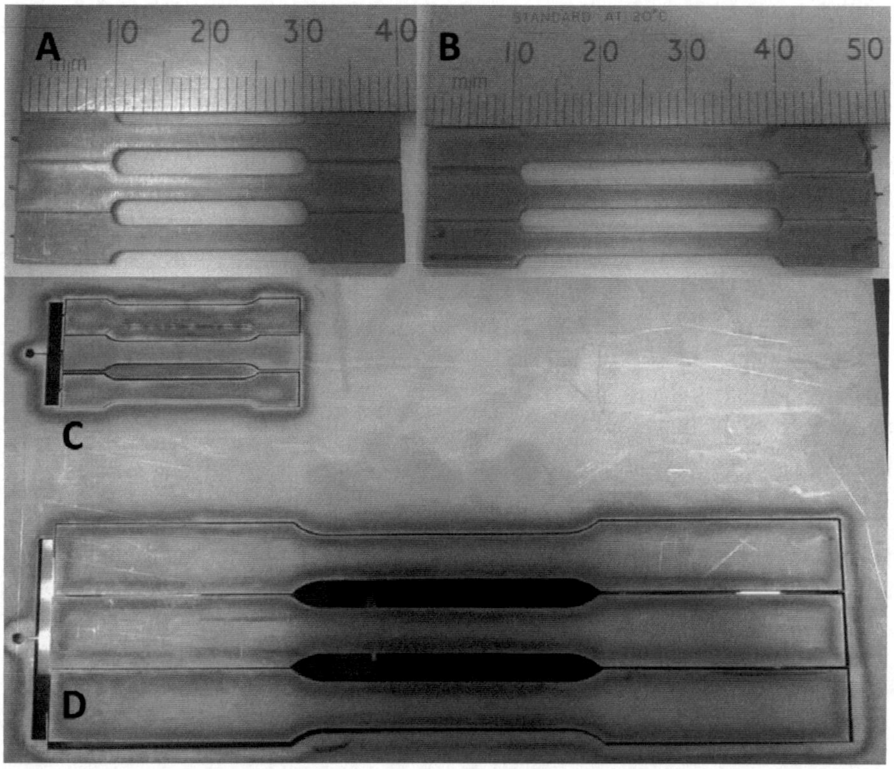

Figure 1. Specimens (Geometries A to D) used for tensile testing.

2.2. Tensile testing

An Instron^{TM} 1342 tensile tester fitted with a 50 kN load cell was used to perform tensile tests at room temperature and at a constant crosshead speed of 0.5 mm/min. The tensile strain was measured using an extensometer attached to the gauge section of the test specimen. The extensometer was then removed at a strain of 1.5% to prevent damage during specimen fracture. Testing continued until the specimen fractured. Three specimens were tested for each geometry type. The 0.2 % YS and UTS were automatically determined by the Bluehill 2 software after testing. The gauge length of the specimen was measured before and after testing to calculate the % EL.

3. Results and discussion

Table 2 shows tensile properties of the geometries tested.

The change in YS from the lowest value (geometry D) to the highest value (geometry A or B) is 2 %. This change show that there were no significant differences between the YS of all the geometries tested. Therefore, the change in gauge length or gauge area did not influence YS. Similarly, UTS increased by 3 % from geometry D to geometry A. Therefore, UTS was not significantly influenced by the change in specimen geometry. These results are consistent with those reported from literature that specimen geometry has a negligible effect of YS and UTS [3–6].

There was a 20 % increase in the value of % EL from the lowest value (geometry D) to the highest value (geometry A). Therefore, specimen geometry had a significant influence on % EL. However, specimens having similar L/√*A* ratio (geometry B and D) had % EL values differing by 6 % which is lower than the 10 % normally accepted as a significant change for destructive tests. In addition, Figure 2 shows that geometry B and D had overlapping error bars indicating that the % EL values were not significantly different.

Table 2. Tensile properties.

	(L/√*A*)	YS (MPa)	UTS (MPa)	% EL
Geometry A	10.9	969±2	1040±20	14.1±0.7
Geometry B	16.3	969±0	1039±3	12.2±1.0
Geometry C	12.2	964±6	1024±6	13.1±0.8
Geometry D	16.2	953±3	1012±9	11.4±0.2

Geometry C was a direct sub-size of geometry D as prescribed by ASTM E8 standard [2] but the % EL varied by 14 %. The error bars of geometry C and D were not overlapping indicating significantly different % EL values. This indicates that similar L/√*A* ratio leads to similar % EL values than similar L/W ratio. Therefore the L/W ratio used by ASTM E8 standard [2] to decide on sub-size specimen geometry D relative to geometry C was not appropriate if one wants to compare % EL of differently sized specimens. However, ASTM E8 standard [2] still warns about the need for the L/√*A* to be kept similar. Therefore, it has been established in this article that one cannot always keep both the L/W ratio and L/√*A* ratio the same for the plate type specimens. This consideration becomes even more challenging when the resultant specimen thickness cannot be varied at will.

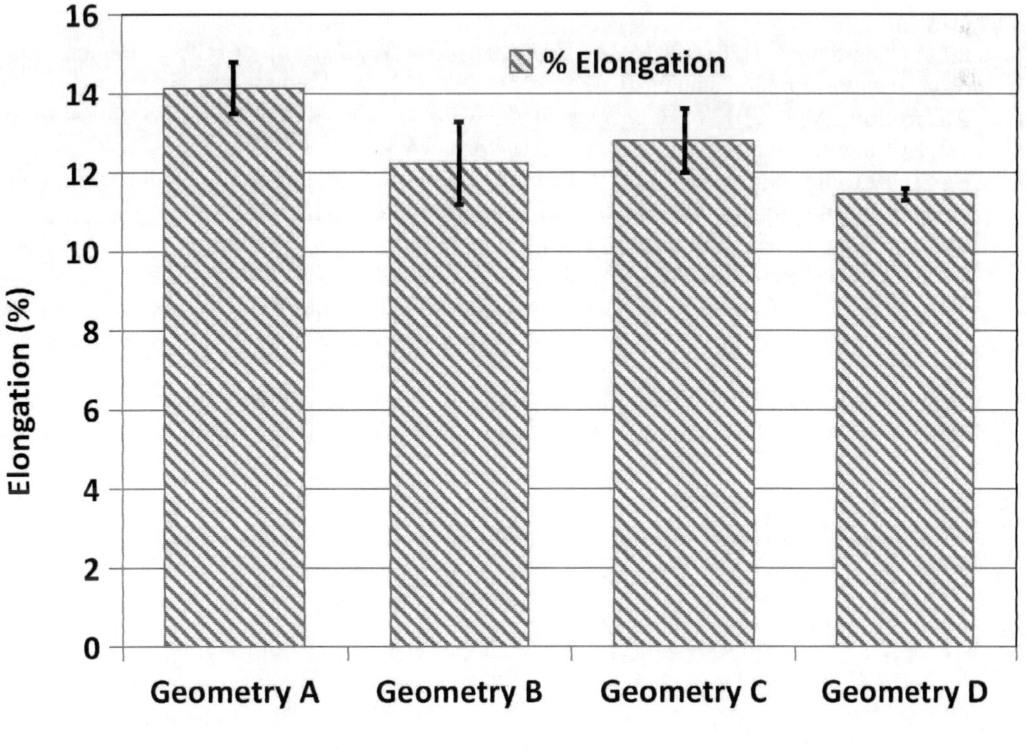

Figure 2. % Elongation for various specimen geometries.

Consistent with the literature [2][4][5], the lower the value of the L/$\sqrt{A}$ ratio, the greater the % EL as shown in Figure 2.

The results of the current study showed that the L/$\sqrt{A}$ has the most influence on determining similarity of tensile properties than the L/W ratio. This implies that one can use sub-size specimens and still get tensile results comparable to larger conventional standard specimens. Therefore, one can design specimens with comparable tensile properties as long as the dimensions of the reference specimens are known and the L/$\sqrt{A}$ is kept constant.

4. Conclusions

The following conclusions were drawn from tensile testing of plate type specimens with different geometries:

- Yield strength and ultimate tensile strength were not significantly affected.
- Keeping the L/W the same was not adequate to yield similar percentage elongation values.
- Similar L/$\sqrt{A}$ produced similar percentage elongation values.
- Tensile properties of specimens with different geometries can be compared by keeping their L/$\sqrt{A}$ the same.

Acknowledgements

The authors wish to acknowledge the South African Department of Science and Technology (DST) and the Council for Scientific and Industrial Research (CSIR) for sponsoring this study. The assistance of Chris McDuling during tensile testing is greatly appreciated and Martin Williams for his advice in specimen design.

References

[1] ISO Standard 6892-1:2016 2016 *Metallic materials - Tensile testing - Part 1: Method of test at room temperature* (BSI standard)

[2] ASTM E8/E8M-11 2013 *Standard test methods for tension testing of metallic materials* (ASTM International, West Conshohocken, PA, USA.)

[3] Han L, Reynolds N, Dargue I and Williams G 2009 The Effect of specimen dimensions on obtained tensile properties of sheet metals *Key Eng. Mater.* **410–411** 481–91

[4] Al-bakri A A, Sajuri Z and Kamal A 2016 Tensile and fracture behaviour of very thin 304 stainless steel sheet *J. Teknol.* **9** 45–50

[5] Yuan W J, Zhang Z L, Su Y J, Qiao L J and Chu W Y 2012 Influence of specimen thickness with rectangular cross-section on the tensile properties of structural steels *Mater. Sci. Eng. A* **532** 601–5

[6] Strnadel B and Brumek J 2013 Effect of tensile test specimen size on ductility of R7T steel. *22nd International Conference on Metallurgy and Materials* (Brno, Czech Republic) pp 2–7

IOP Conf. Series: Materials Science and Engineering **430** (2018) 012016 doi:10.1088/1757-899X/430/1/012016

Interatomic potential parameters for Li-Cl-Ti interaction

A F Mazibuko[1,*], H R Chauke[1], P E Ngoepe[1] and C R A Catlow[2]

[1]Materials Modelling Centre, School of Physical and Mineral Sciences, University of Limpopo, Private bag X1106, Sovenga 0727, South Africa
[2]Kathleen Lonsdale Materials Chemistry, Department of Chemistry, University College London, 20 Gordon Street, London WC1H 0AJ, UK

[*]andile.mazibuko@ul.ac.za

Abstract. Alkali metals and alkali earth metals can be used as reducing agents of titanium halide in titanium production. Despite South Africa's position as being the major raw titanium material producer, titanium production is low and expensive as a direct consequence of the outmoded technology that is used in its extraction from raw materials such as the Kroll process. In this study, computational modelling techniques were employed to simulate the conditions for LiCl that will be suitable for generating a large quantity of metallic titanium in pure and powder form. We used a combination of density functional theory and molecular dynamics, employing FHI-aims, DL_POLY and GULP to characterize LiCl in a solid and molten form. The derived potentials reproduced the LiCl structure to within 1% in agreement with experimental data. More importantly, the melting temperature was deduced from the diffusion coefficient as 800 K which is closer to the experimental melting point of 878 K. Furthermore, the interaction of Ti-Li, Ti-Cl and Li-Cl-Ti were tested and gave reasonable results to set an environment for titanium clusters. The new pair potentials were deduced as Ti-Cl: D_e=0.400 a_0= 1.279 r_0=2.680 and Ti-Li: D_e =0.730 a_0=1.717 r_0=2.000. The findings of this work will contribute towards the development of alternative ways of titanium production in a continuous and less expensive processes.

1. Introduction

Titanium has many applications in various industries such as in catalysis, medical and aerospace, where the fabrication of components with excellent corrosion and high temperature performance are significant [1] . Titanium clusters have shown great potential since they exhibit different properties compared to that of bulk structure. This has prompted a great interest in this field and thus extensive studies on Ti clusters are being performed [2, 3]. It has been reported that studies of titanium clusters are important to understand the evolution and growth regarding geometry and electronic properties [3]. In previous work, investigations by Zhao et al. [4] , showed that the Ti clusters follow a pentagonal growth behavior that favors close-packed structures. Sun et al. [5] reported on the tetrahedral-, octahedral-, and icosahedral sequences being some of the size-dependent growth patterns. These structures or geometries can be obtained by either adding or removing atoms from certain basic motif or positions [5] . This is the underlying fundamental studies that may inform the process of titanium production.

Titanium powder can be produced commercially through thermochemical processes by reducing a precursor with a reductant which is either sodium (Na) or magnesium (Mg) [1] . However, the

IOP Conf. Series: Materials Science and Engineering **430** (2018) 012016 doi:10.1088/1757-899X/430/1/012016

reduction reaction occurs extremely fast and produces crystalline particles that tend to aggregate [6]. In other work, Seon and Nataf formed titanium powder by lithiothermic reduction of titanium tetrachloride ($TiCl_4$) in a molten salt reactor using a high-energy mixer that served to disperse undissolved lithium and gaseous $TiCl_4$ into the molten salt. This procedure may also break up any lumps of titanium that might form [7].

In this paper, we explore alternative means to gain understanding and knowledge on the titanium formation processes. We report the fundamental investigation using an alternative lithium chloride (LiCl) salt which has a similar structure as TiCl. We have performed molecular dynamics calculations on the LiCl structure using a combination of interatomic potentials from the CATLOW-library of potentials [8] . We also discuss the derivation of potentials for Ti in the LiCl system. A good reproducibility of Li-Cl, Ti-Cl, Cl-Cl interactions was achieved. These potentials are transferable to TiCl since all interactions have been considered.

2. Computational methodology

The initial configuration of LiCl was created from GULP [9] using a combination of potentials from the CATLOW-library that were firstly developed and validated. Molecular dynamics simulations were performed using DL_POLY code. A radial cut-off of 6.0 was used in a simulation of 216 atoms of LiCl system. The timestep was set at 0.001ps and the thermal data computed every 120 000 steps. The NVT ensemble using the Nosé-Hoover thermostat and barostat was run for 50000 steps to bring the system to equilibrium. Furthermore, pair potentials were deduced using the density functional theory FHI-aims [10, 11] code for the Ti-Cl and Ti-Li interactions. This was achieved by simulating a lithium ion, a chlorine ion and then a Li-Cl cube between two stable and symmetrical titanium clusters (Ti_7). The distance between the clusters and the Li^+/Cl^- was varied and the binding energies were calculated using the equation:

$$E_b = \frac{E_{FHI} - 2E_{Ti} - E_{alk}}{2},$$

where E_b is the binding energy, E_{FHI} is the energy from FHI-aims, E_{Ti} and E_{alk} are respectively, the energies of Ti and Li/Cl. The Morse-potential was used to investigate the interactions i.e. fit potentials.

3. Results and discussion

3.1. LiCl structure

The configured structure was firstly validated by comparing the lattice parameters of the LiCl structure with the experimental data. We observed that the interatomic Buckingham potentials used reproduced the structure to within 1% in agreement with experimental data. A lattice parameter of a= 5.049 Å was obtained and it compared well with the experimental value as 5.070 Å [8].

Elastic constants

The large dispersion about the behavior of C_{11} and C_{12} with respect to experiment, is mainly related to the correlation between the two quantities, both computationally and experimentally [8]. For C_{44}, the situation is different in that the dependence of C_{44} on the lattice parameter is very small, much smaller than for C_{11} and C_{12} [8]. In Table 1, the value of C_{12}=C_{44} suggests that the Cauchy relation holds and that the material can be well described by pair-potentials. The calculated and experimental results agree as shown in Table 1.

Table 1. Calculated (Calc.) and experimental (Exp.) LiCl elastic constants C_{11}, C_{12}, C_{44} in GPa. Experimental data is from Ref. [12, 13].

	C_{11}		C_{12}		C_{44}	
System	Calc.	Exp.	Calc.	Exp.	Calc.	Exp.
LiCl	69.140	63.200	26.152	23.800	26.152	27.6

3.2. Thermodynamic properties

The mean square displacement and the diffusion coefficient were determined to understand the motion of Li^+ and Cl^- ions in the simulated system and the trajectories of these ions also traced.

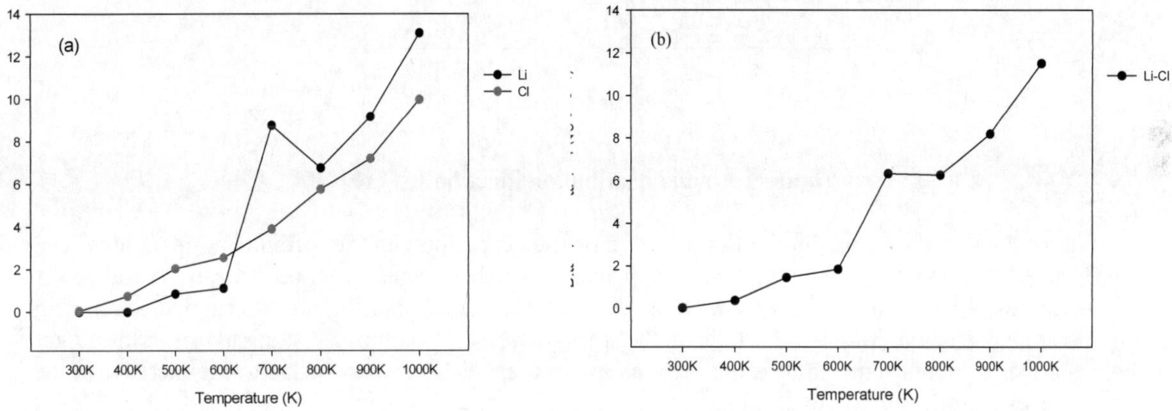

Figure 1. Diffusion coefficient ions (a) comparison between Li^+ and Cl^-, and (b) average for Li-Cl ions.

From the Figure 1(a-b), the diffusion coefficient depends exponentially on the temperature. The melting point was found to be 800 K which is closer to the experimental value of 878 K (605 °C -614 °C) [13]. This difference might be due to a slight mismatch in the potential parameters which have an important effect on the melting point, in this case an insignificant difference in the structural parameters was observed.

Radial distribution function.

In anion-cation radial distribution functions, the maxima and minima occur where there are corresponding minima and maxima hence the observation of a well-defined main peak at ~3.13 Å for all the graphs depicted [13].

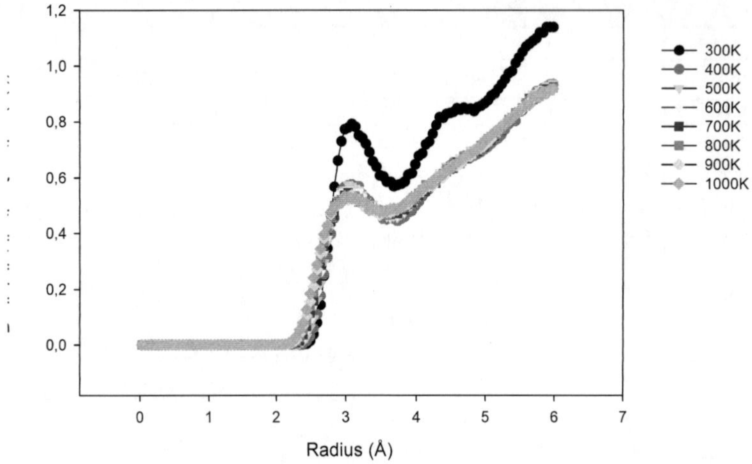

Figure 2. Radial distribution function for LiCl.

Figure 2 shows the radial distribution function of LiCl corresponding to different temperatures. The peak at 300 K shows that the structure is still intact, which is what we expect to observe at lower temperatures. Also, the height of the first peak of the radial distribution curve decreases with temperature increment while the width of the peak increases to almost a straight line. This might indicate that at higher temperatures the interaction between Li-Cl is weaker due to the increase in the kinetic energy of the atoms suggesting melting. Furthermore, the broader second peak at 300 K shows that there are still observable interactions of the atoms in the system, but this peak starts disappearing around 800 K, implying that the structure is beginning to lose its morphology as atoms begin to be randomly distributed.

3.3. Fitting pair potentials for the interaction of Ti in LiCl systems

The comparison between the DFT results generated using FHI-aims and Morse-potential results.

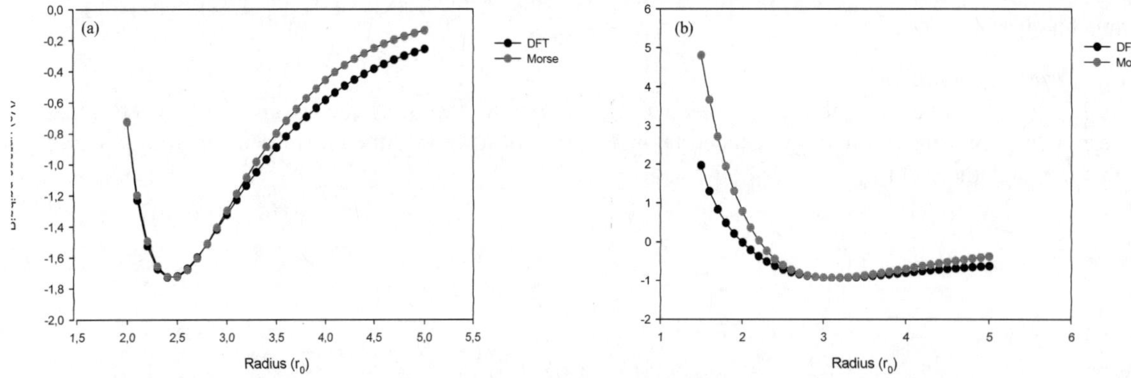

Figure 3. The interaction between (a) Ti-Cl and (b) Ti-Li for the DFT and Morse potential.

IOP Conf. Series: Materials Science and Engineering **430** (2018) 012016 doi:10.1088/1757-899X/430/1/012016

Figure 3(a-b), shows the interaction of the Ti-Li and Ti-Cl deduced from binding energies calculated from FHI-aims energies. The DFT results show similar trends with the Morse-potential curves and this observation provides guidance on which potential parameters will be best suited to describe the interactions of the system.

Table 2. Morse pair potentials.

	Ti-Cl		Ti-Li	
	Initial	Final	Initial	Final
D_e	0.400	0.400	0.033	0.730
a_0	1.279	1.279	0.857	1.717
r_0	2.681	2.681	3.875	2.000

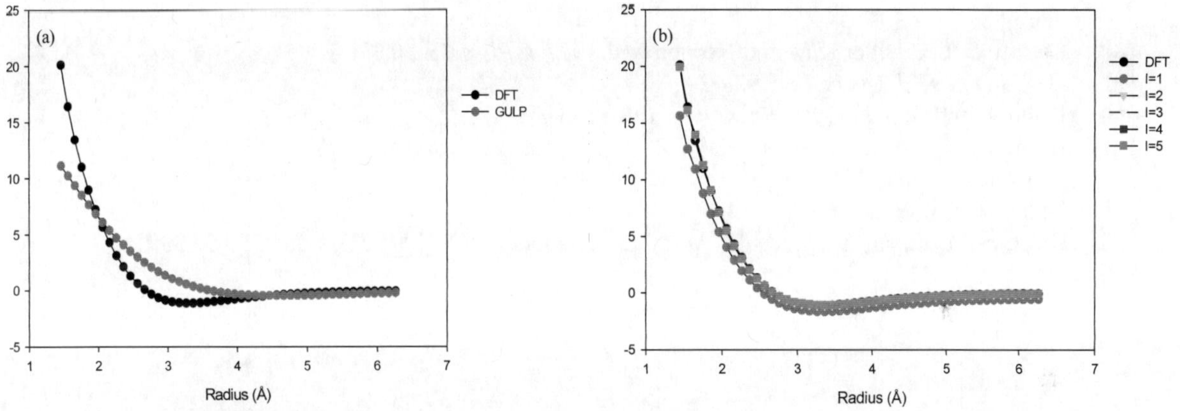

Figure 4. Energy Vs Radius curve for the Li-Cl-Ti system. (a) Comparison of DFT and GULP fitting, (b) fitting of Morse potentials from adjusting the D_e and r_o parameters and comparing with DFT.

In Figure 4(a), the fitted Ti-Cl and Ti-Li parameters were noted when the final sum of squares was close to zero. However, it was observed that the GULP curve shows that the parameters used for fitting are not the best ones since there is no alignment between the DFT and GULP curves from Figure 4(a). The Ti-Cl parameters were unchanged, while those of Ti-Li were changed since these were far from the final sum of squares of close to zero. Thus, an adjustment in the D_e and r_o parameters to better align the graphs was done. Hence, Figure 4(b) shows the fitting of Morse pair potentials and the parameters are given in Table 2.

4. Conclusion

The derived potentials reproduced the LiCl structure to within 1% in agreement with the experimental data. The elastic constants are also in agreement with experimental observation whilst still obeying the Cauchy relation. It was predicted that the melting temperature was 800 K which is closer to 877 K [13] . More importantly, from Figure 4 we observed the DFT and Morse curve alignment. This suggests that the derived potentials could be used to set an ideal environment for exploring the growth and evolution of the titanium clusters in a LiCl solution.

Acknowledgements
The research is part of the collaboration between the groups of H.R.C. and P.E.N. University of Limpopo (UL) and C.R.A.C and S.M.W. University College London (UCL). This work made use of the facilities of UCL Faraday and UCL Research computing platforms services. The authors wish to recognize the support of the South African Research Chair initiative of the Department of Science and Technology and the National Research Foundation

References
[1] Fang Z Z, Paramore J D, Sun P, Chandran K S R, Zhang Y, Xia Y, Cao F, Koopman M and Free M 2017 *Int. Mater. Rev.* 1743
[2] Alonso J A 2000 *Chem. Rev.* **100** 637
[3] Fernando A, Weerawardene K L D M, Karimova N V and Aikens C M 2015 *Chem. Rev.* **115** 6112
[4] Zhao J, Quin Q, Wang B, Wang J and Wang G 2001 *Solid State Commun.* **118** 157
[5] Sun H, Ren H, Wu Z and Xu N 2015 *Comput. Theor. Chem.* **1062** 74
[6] Noda T 1965 *J.M.* **17** 25
[7] Seon F and Nataf P "Production of metals by metallorthermia" US Patent 4725312 16 February 1988
[8] Catlow C R A, Diller K M and Norgett M J 1977 *J. Phys.* C. **10** 1395
[9] Gale J D 1997 *J. Chem. Soc.* **93** 629
[10] Kohn W and Sham L 1965 *Phys. Rev.* **140** 1133
[11] Blum V, Gehrke R, Hanke F, Havu P, Ren X, Reuter K and Scheffler M 2009 *Comput. Phys. Commun.* **180** 2175
[12] Haussiihl S 1960 *Z. Phy.* **159** 223
[13] Prencipe M, Zupan A, Dovesi R, Apra E and Saunders V R 1995 *Phys. Rev.* B. **51** 3391

IOP Conf. Series: Materials Science and Engineering **430** (2018) 012017 doi:10.1088/1757-899X/430/1/012017

Computational chemistry study of zirconium monomers in low acid concentration aqueous solutions

D B Jansen van Vuuren[1], H M Krieg[1], C G C E van Sittert[1,*]

[1] Research focus area: Chemical Resources Beneficiation, North-West University, Potchefstroom, Republic of South Africa

*cornie.vansittert@nwu.ac.za

Abstract. Zirconium and hafnium have to be separated prior to their use in nuclear energy reactors. Therefore, a lot of research has been done on the solvent extraction of zirconium. However, most of this research involved screening studies that focused on finding suitable solvent extraction processes, rather than attempting to gain insights into the extraction mechanisms. An important part of understanding extraction mechanisms is knowing which zirconium species are present in aqueous solutions since these are the species with which extractants react. In this study, we used density functional theory based molecular dynamics calculations to model zirconium monomers in low acid concentration solutions. Specifically, we modelled aqua and hydroxo zirconium complexes using an explicit solvent approach. By calculating radial distribution functions for each of these zirconium complexes, using either 10, 20, 30, 40 or 50 explicit solvent water molecules, we were able to determine that at least about 30 solvent water molecules are needed to model the first two solvation spheres of a zirconium ion during an explicit solvent approach. We anticipate that this work will be useful for further explicit solvent based computational chemistry studies on the solvent extraction mechanisms of zirconium and hafnium.

1. Introduction

During the production of nuclear energy, uranium undergoes nuclear fission and in the process produces thermal neutrons [1]. These liberated thermal neutrons radiate between uranium atoms, which are held in so-called fuel rods inside a nuclear reactor, causing further fission of other uranium and sustaining the nuclear reaction [1]. The heat given off by this nuclear reaction causes water, which is also present in the nuclear reactor, to rise as steam. This rising steam causes turbines to rotate and produce electrical energy.

Zirconium and hafnium have important applications in nuclear energy reactors. Since zirconium is transparent to thermal neutrons, it is used to make the casing of the uranium fuel rods [2]. Conversely, hafnium absorbs thermal neutrons and is therefore sometimes used to make control rods which are used to moderate nuclear reactions [2]. Since zirconium and hafnium have opposing thermal neutron capturing properties, these two metals have to be separated prior to their use in nuclear energy reactors. A promising candidate technology for the separation of these two metals is solvent extraction [3].

A lot of research has been published on the solvent extraction of zirconium and hafnium [3]. When poring over this published research it becomes apparent that most of it was done by adopting an experimental screening approach. This approach involved choosing an extractant on the basis of novelty. Then doing a sequence of solvent extraction experiments in which the extractant's concentration was varied in a stepwise manner. Although this approach can lead to the discovery of suitable extractants for the solvent extraction of zirconium, it seldom yields insights into the mechanisms that underpin how these extractants extract zirconium. An important part of understanding these mechanisms is knowing exactly which zirconium species are present in an aqueous solvent since these are the species with which extractants react. Besides for understanding the extraction mechanisms, knowing exactly which zirconium species are present in an aqueous solution could also be applied to the selection, or design, of extractants for zirconium.

Although the aqueous speciation of zirconium is not fully understood, it has been experimentally established that in low acid concentration aqueous solutions zirconium can be present as either a monomer, dimer or tetramer [4]. In this study, we used computational chemistry to model the zirconium monomer in such aqueous solutions. In principle, it is possible that monomeric zirconium exists as an aqua or hydroxo complex in such aqueous solutions, and therefore we focused on both cases. Specifically, we modelled aqueous zirconium complexes surrounded by water molecules; an approach is known as explicit solvent modelling [5]. Given these systems, we did density functional theory based molecular dynamics (MD) calculations in order to calculate so-called radial distribution functions (RDF). In simple terms, these RDF indicate how oxygen and hydrogen atoms are spatially arranged around a zirconium ion.

We calculated the RDF for systems that contained added either 10, 20, 30, 40 or 50 explicit water solvent molecules. Using these RDF we determined how many water molecules are necessary to model the solvation sphere around a zirconium complex when using an explicit solvent approach. Following this, we made predictions regarding the structures of aqua and hydroxo zirconium complexes. These predictions include coordination numbers of the zirconium complexes and the extent to which the aqua and hydroxo ligands undergo protonation or deprotonation. We anticipate that this work will serve as a foothold for future computational chemistry studies on the solvent extraction of zirconium; especially studies that also use an explicit solvent approach.

2. Method

All the modelling presented in this paper was done within the paradigm of density functional theory, using the DMol [6] code [6,7] as implemented in Biovia's Materials Studio 2018 computational chemistry software package. Specifically, we used the PBE functional [8]. As for the basis set, the $DMol^3$ code implements numerical basis sets. In this study, we used the DNP basis set, which is an unrestricted basis set in which each core orbital is represented by a single primitive, while each valence orbital is represented by two primatives plus a third primitive. This third primitive is a p function for hydrogen, and a d function for all other elements. Zirconium's core electrons were represented by a relativistic core potential.

Since the systems that we modelled consisted of multiple molecules, interatomic dispersion interactions had to be accounted for. We did this using the Grimme DFT-D scheme [9]. In addition to modelling the solvent environment around the zirconium complexes explicitly, we used an implicit solvent model to model the solvent environment around the entire solute-explicit solvent cluster. For this purpose we used the COSMO implicit solvent model [10], with a dielectric constant of 78.54, to represent a water solvent environment. The reason why we added this implicit solvent model was to stabilize the cluster's charge.

Regarding the MD calculations, we calculated NVT trajectories using a time step of 1.0 fs. The temperature was controlled at 298 K using the Nose-Hoover thermostat [11,12]. We equilibrated each system by doing 1000 step geometry optimization calculations, followed by 2 ps MD calculations. After this equilibration procedure, we did 3 ps MD calculations and used the resulting trajectories to create RDF.

We created the RDF by writing a Python script. Given a trajectory, this script samples the coordinates from every configuration in the trajectory. These sampled coordinates are then used to create a histogram of either Zr-O or Zr-H distances, using a bin width of 0.01 Å. The histogram is normalized and used as the RDF for the system whose trajectory was subjected to the script.

3. Results and discussion

We began by doing geometry optimization calculations on $Zr(H_2O)_8^{4+}$ and $Zr(OH)_8^{4-}$. For these calculations, we used an implicit solvent model without adding explicit solvent molecules. Both structures converged to eight coordinate complexes. Thus, when using a purely implicit solvent model hydroxo zirconate was predicted to exist as an eight coordinate structure. However, as will be shown later, when explicit solvent molecules were present, hydroxo zirconate was predicted to be either a six or seven coordinate structure, depending on how many explicit solvent molecules were present. This observation highlights the fact that different approaches to modelling the solvent environment can result in markedly different predictions.

Using the geometrically optimized $Zr(H_2O)_8^{4+}$ and $Zr(OH)_8^{4-}$ complexes, we created new structures by placing either 10, 20, 30, 40 or 50 additional water molecules around each of these complexes. This resulted in ten new systems which contained varying amounts of explicit solvent molecules; five of these structures had a charge of 4+, while the other five had a charge of 4-. Since the 4+ systems each had 8 H_2O ligands to begin with, these systems ended up with either 18, 28, 38, 48 or 58 total water molecules after adding explicit water molecules. Similarly, the 4- systems ended up with a total of either 18, 28, 38, 48 or 58 oxygen atoms. These ten systems were equilibrated, before doing further MD calculations to create Zr-O and Zr-H RDF for each systems. Figure 1 shows a representative example of what these structures looked like after equilibration.

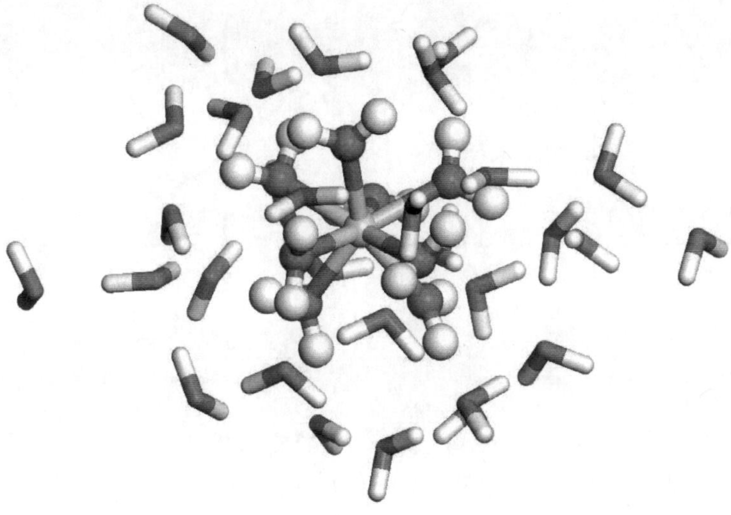

Figure 1. The equilibrated structure of $Zr(H_2O)_8^{4+}$ in the presence of 30 explicit solvent water molecules. Zirconium, oxygen and hydrogen atoms are shown in teal, red and white.

The results for the Zr-O RDF are shown in figure 2. For each of the ten systems, we integrated the first two Zr-O RDF peaks and multiplied the integral by the total number of oxygen atoms in the system. This gave us the average number of oxygen atoms that corresponds to each peak. For instance, the first peaks for the 4+ systems, which are all centred at 2.3 Å, each corresponds to an average of 8 oxygen atoms. After considering the Zr-H RDF, which is discussed later, we interpreted these oxygen atoms as belonging to aqua ligands that are coordinated to the zirconium centres. The results of these kinds of integrations of the Zr-O RDF are shown in figure 3.

When considering figures 2 and 3, it seems that in addition to any aqua or hydroxo ligands coordinated to zirconium, at least about 30 water molecules were necessary to model the first and second solvation spheres of a zirconium ion. This is clearly apparent when considering the left side of Figure 3. As the number of water molecules in the 4+ systems increased, so too did the number of oxygen atoms corresponding to the second Zr-O RDF peak. However, the number of oxygen atoms corresponding to this second peak seems to plateau for systems that contained 38, 48 or 58 water molecules. This suggests that for the structures that contained 18 or 28 water molecules, the second solvation sphere had not been "filled up".

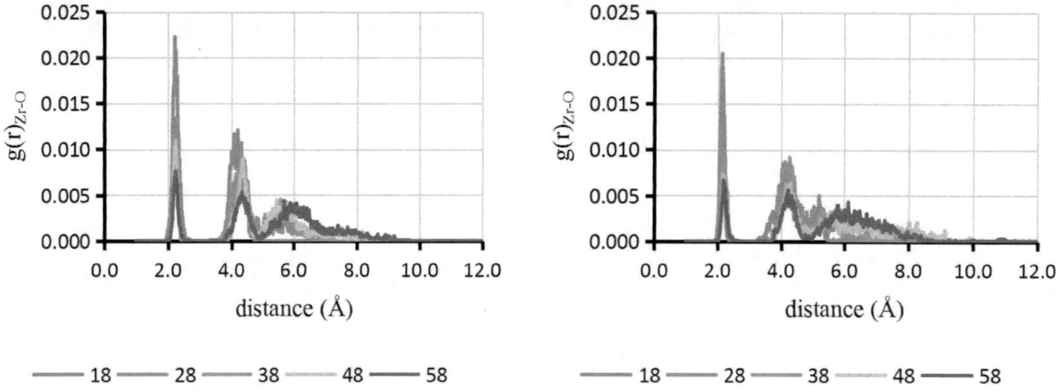

Figure 2. Zr–O RDF for each of the 4+ (left) and 4- (right) systems.

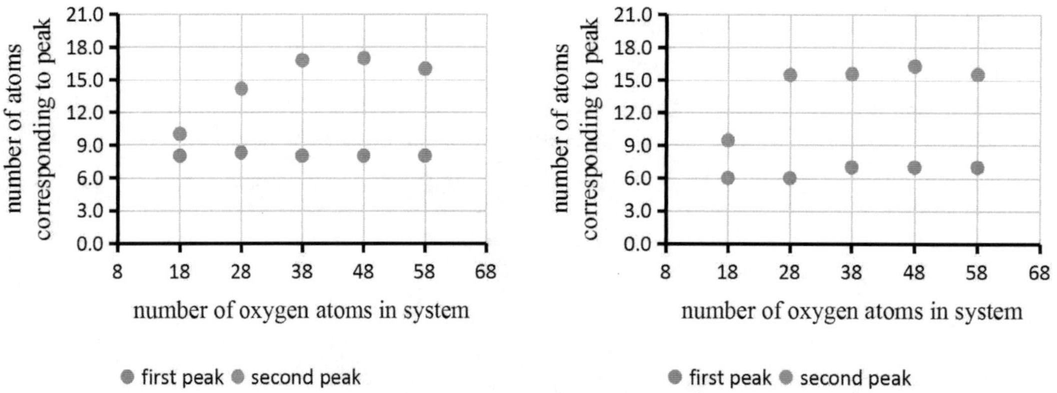

Figure 3. The number of oxygen atoms corresponding to the first and second Zr-O RDF peaks of the 4+ (left) and 4- (right) systems.

A similar argument can be made for the right side of figure 3 when considering the number of oxygen atoms corresponding to the first Zr-O RDF peak. For systems that contained 10 or 20 additional water molecules, the first peaks corresponded to an average of 6.0 oxygen atoms each. Conversely, for systems that contained more explicit solvent water molecules, the first peaks corresponded to an average of 7.0 oxygen atoms each. Therefore, it seems that the number of explicit solvent molecules influenced how the first and second solvation spheres of the zirconium ion were predicted to be.

Since its clear that 30 explicit solvent molecules were necessary to model a zirconium ion's first two solvation spheres, the rest of the discussion will focus only on the two systems that contained 30 explicit solvent molecules. However, the arguments that follow can also be made for the systems that contained 40 or 50 explicit solvent molecules.

We now discuss the Zr-H RDF, which is shown in figure 4, and the number of hydrogen atoms corresponding to the first two Zr-H RDF peaks, shown in figure 5. For the 4+ system, the first Zr-H RDF peak corresponds to 16.0 hydrogen atoms. Given that this system's first Zr-O RDF corresponds to 8.0 oxygen atoms, it seems that for the 4+ system zirconium's first solvation sphere consisted of eight water molecules. Furthermore, when considering the relative positions of these Zr-O (2.3 Å) and Zr-H (3.0 Å) RDF peaks, it becomes clear that these eight water molecules were aligned in such a way that their oxygen ends pointed towards the zirconium centre. This interpretation is consistent with the notion that these water molecules were coordinated to the zirconium centre.

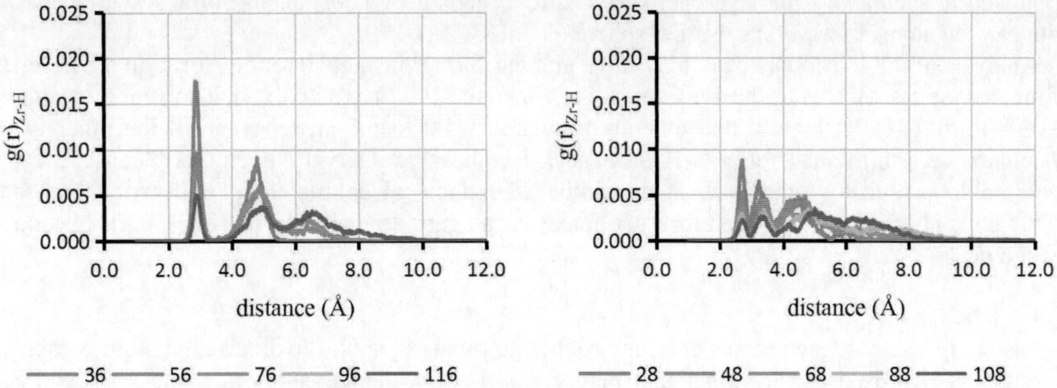

Figure 4. Zr–H RDF for each of the 4+ (left) and 4- (right) systems.

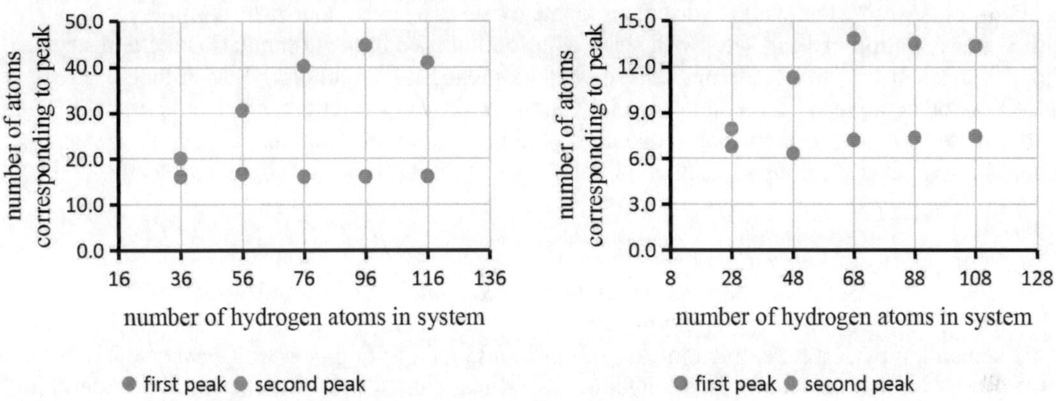

Figure 5. The number of hydrogen atoms corresponding to the first and second Zr-H RDF peaks of the 4+ (left) and 4- (right) systems.

When considering the second RDF peaks of the 4+ system, we see that the Zr-O RDF peak (4.5 Å) is closer to the zirconium centre than the Zr-H RDF peak (5.0 Å). This suggests that the water molecules in the second solvation sphere also had their oxygen ends pointed towards the zirconium centre. An explanation for this orientation is that hydrogen bonds were present between the hydrogen atoms in the first solvation sphere and oxygen atoms in the second solvation sphere.

Lastly, we discuss the right sides of figures 4 and 5. The first Zr-H RDF peak of the 4- system correspond to 7.19 hydrogen atoms, while the first Zr-O RDF peak corresponds to 7.0 oxygen atoms. Therefore, it seems that the first solvation sphere of the 4- system consisted of approximately 7 hydroxy groups. Similar to the 4+ system, the Zr-O RDF peak (2.2 Å) is closer to the zirconium centre than the Zr-H RDF peak (2.9 Å). This suggests that these hydroxo groups were coordinated to the zirconium centre. Observing that the average number of hydrogen atoms corresponding to the first Zr-H RDF is greater than 7.0, it seems that some of these hydroxo ligands were protonated for a portion of the 3 ps MD trajectory.

The second Zr-H RDF peak of the 4- system corresponds to 13.8 hydrogen atoms. To be clear, the 4- system's second Zr-H RDF peak is not analogous to the 4+ system's second Zr-H RDF peak. The 4+ system's second Zr-H RDF peak (5.0 Å) is centred further away from the zirconium centre than its first (2.3 Å) and second (4.5 Å) Zr-O RDF peaks. Conversely, the 4- system's second Zr-H RDF peak

(3.5 Å) is centred between its first (2.2 Å) and second (4.4 Å) Zr-O RDF peaks. A possible explanation for this may be that the water molecules in the system's second solvation sphere was orientated in such a way that hydrogen atoms in the second solvation sphere formed hydrogen bonds with oxygen atoms in the first solvation sphere.

Analysis of the third Zr-H peak may have granted further insights into the solvation environments of the 4- systems. However, the 4- systems that contained 10, 20 and 40 explicit solvent molecules all have a third Zr-H RDF peak that extends from about 4.0 to 6.0 Å. Conversely, the other two 4- systems have a third and fourth Zr-H RDF peak in this same interval. Given this inconsistency, we could not be sure that we had an adequate number of explicit solvent molecules to describe this feature of the solvent environment. Therefore, we make no attempt at analyzing the Zr-H RDF beyond the second peak.

4. Conclusions

In this study, we modelled the solvent environment around aqua and hydroxo zirconium complexes. Specifically, we used density functional theory based MD calculations to calculate Zr-O and Zr-H RDF for each of these complexes. Using these RDF we concluded that, in addition to aqua or hydroxo ligands, at least approximately 30 explicit solvent molecules are needed to describe the first two solvation spheres of a zirconium ion. Furthermore, we also predicted that aqua zirconium exists as an eight coordinate complex, while hydroxo zirconate exists as a seven-coordinate complex.

It is our intention that this work will serve as a foothold for future computational chemistry studies into the mechanisms that underpin the solvent extraction of zirconium. For instance, attempts at modelling other aqueous zirconium species, such as dimers or tetramers, should include at least approximately 30 explicit solvent molecules. Attempts at modelling the attack of extractants on zirconium species in aqueous solvents might also need at least 30 explicit solvent molecules.

References

[1] Ohanian, H C, *Modern Physics*. Prentice-Hall: 1995
[2] Blumenthal, W B, *The chemical behavior of zirconium*. Van Nostrand: 1958
[3] Wang, L Y; Lee, M S, A review on the aqueous chemistry of Zr(IV) and Hf(IV) and their separation by solvent extraction. *Journal of Industrial and Engineering Chemistry* 2016, **39**, 1-9
[5] Hu, Y J; Knope, K E; Skanthakumar, S; Kanatzidis, M G; Mitchell, J F; Soderholm, L, Understanding the role of aqueous solution speciation and its application to the directed syntheses of complex oxidic zr chlorides and sulfates. *Journal of the American Chemical Society* 2013, **135** (38), 14240-48
[6] Cramer, J C, *Essentials of Computational Chemistry*. Wiley: 2005
[7] Delley, B, An all-electron numerical method for solving the local density functional for polyatomic molecules. *The Journal of Chemical Physics* 1990, **92** (1), 508-17
[8] Delley, B, From molecules to solids with the DMol3 approach. *Journal of Chemical Physics* 2000, **113** (18), 7756-64
[9] Perdew, J P; Burke, K; Ernzerhof, M, Generalized gradient approximation made simple. *Physical Review Letters* 1996, **77** (18), 3865-68
[10] Grimme, S, Semiempirical GGA-type density functional constructed with a long-range dispersion correction. *Journal of Computational Chemistry* 2006, **27** (15), 1787-99
[11] Klamt, A; Schüürmann, G, COSMO: A new approach to dielectric screening in solvents with explicit expressions for the screening energy and its gradient. *Journal of the Chemical Society, Perkin Transactions 2* 1993, **5**, 799-805
[12] Nosé, S, A unified formulation of the constant temperature molecular dynamics methods. *The Journal of Chemical Physics* 1984, **81** (1), 511-19
[13] Hoover, W G, Canonical dynamics: Equilibrium phase-space distributions. *Physical Review A* 1985, **31** (3), 1695-97

IOP Conf. Series: Materials Science and Engineering **430** (2018) 012018 doi:10.1088/1757-899X/430/1/012018

Thermodynamic stability of doped FeAl-X (X = Pd, Ag, Pt and Ru) systems

C S Mkhonto[1*], H R Chauke[1] and P E Ngoepe[1]

[1]Materials Modelling Centre, School of Physical and Mineral Sciences, University of Limpopo, Private Bag X1106, Sovenga, 0727, South Africa.

*chresimkhonto@gmail.com

Abstract. Iron-aluminides based systems play a significant role in many industrial applications due to their excellent resistance to oxidation at high temperatures. In particular, these alloys are applicable as thermocouples. In this paper, we used first-principles density functional theory to investigate the stability of FeAl and ternary FeAl-X alloys. Their equilibrium lattice parameters are in better agreement to within 2% with the experimental data. We employed virtual crystal approximation and supercell approaches to model various atomic compositions at $0 \leq x \leq 1$ for Ag, Pt, Pd and Ru. Their thermodynamic, electronic and mechanical properties were deduced from their equilibrium lattice constants. It was found that the addition of Ag, Pt, Pd and Ru enhances the stability at lower atomic percentage composition. It was found that Ag and Pd stabilizes the FeAl-X to lower percentage compositions. Furthermore, their heats of formation, density of states and elastic constants gave essential agreement to describe the structural, thermodynamic and mechanical stability of these systems.

1. Introduction

Iron-aluminides based alloys have created a great amount of interest due to their potential as candidates for applications in the energy sector, such as boilers and pipes. In order to improve the strength and ductility of the Fe-Al alloys various steps are considered, which included the addition of third metals such as Pt, Ru, Pd, Ag [1]. Previous reports showed that the presence of these elements increases the hardness of the alloys slightly [2]. Furthermore, addition of small amounts increased the corrosion resistance, though these alloys suffer limited ductility room temperature at 873 K [3]. Sharma *et al* demonstrated that the addition of Pt, Ru, Ag and Pd increases the resistance to tarnishing or the formation of silver sulphide [4].

Other work investigated the Fe40-Al alloy with addition of Ag, Ru, Pt and Pd for small concentrations of either 0.2 at. % or 0.5 at. %. On the other hand, studies had illustrated the hot corrosion behavior of Fe40-Al intermetallic alloyed with 2.5 at. % Ag at 723 K. Addition of Ag on Fe40-Al alloy enhances the corrosion behavior, particularly when the alloy was exposed to molten salt mixture at 943 K [5]. The binary Fe40-Al alloy exhibit a more active corrosion potential but at lower corrosion rate. The ternary FeAl-Ag alloy exhibited a more noble corrosion potential but a bigger corrosion rate. This behavior is due to the formation of a layer of corrosion products composed mainly by a mixture of Al and Fe protective oxides and other compounds such as metals chlorides [5] [6].

In this paper we employed density functional theory (DFT) to investigate the mechanical, thermodynamic and electronic properties of Pd, Pt, Ag and Ru doped on B2 FeAl intermetallic. A combination of ab-initio and molecular dynamics approach [4] were used to investigate the influence of ternary addition. The Heats of formation revealed stability for Pd and Ru addition at 0.2 at. %; while Ag and Pt at 0.5 at. % composition in agreement with the experimental data. The elastic constants were also calculated to describe the mechanical stability for the FeAl-X systems for small concentrations ($0 \leq x \leq 1$), and were found to be in accordance with the experimental findings [7]. The temperature dependence showed that Ru and Ag are stable over a wide range of temperature below 1000 K.

2. Methodology

Density functional theory (DFT) [8] within the generalized gradient approximation (GGA) by Perdew, Burke and Ernzerhof (PBE) [9] was used to study the FeAl-X alloys. We employed the plane wave pseudopotential method as implemented in CASTEP code [10, 11]. An energy cut-off of 500 eV was used, as it was sufficient to converge the total energy of the B2 FeAl phase. The Brillouin zone integrations were performed for suitably large sets of k-points of 10 X 10 X 10 according to Monkhorst and Pack [12] before and after doping.

The minimum and maximum Guassian smearing width were respectively set at 0.4 and 0.1 eV for the FeAl-Ag/Pd structures. For the exchange-correlation functional, the GGA-PBE was chosen. In the calculation of elastic constants and density of states, sufficient number of k-points were used. Optimization of structural parameters (atomic positions and lattice parameters) was achieved by minimization of forces and stress tensors. Initially, the optimised binary FeAl structures gave equilibrium lattice parameter with better agreement to the experimental findings. The incorporation of Pt, Pd, Ru and Ag as dopants in FeAl was achieved using virtual crystal approximation (VCA) [13]. The Dmol3 code was used to perform the temperature dependence calculations. A time step of 16ps and NVE ensemble was used to determine the transition temperature of FeAl-X systems.

3. Results and discussion

3.1. Thermodynamic stability

The heats of formation (ΔH_f) of the $Fe_{50\text{-}x}Al_{50}$-X is calculated as:

$$\Delta H_f = \frac{1}{N}\,[E^{FeAl} - [(1\text{-}x)E^{Fe} + xE^{Al}]]$$

Compositions	Dopant (at. %)	Lattice parameter (Å)	Volume (Å³/atom)	ΔH_f(eV/atom)
$Fe_{50}Al_{50}$	-	2.852	23.192	-0.5680 (-0.605) [14]
$Fe_{49.80}Al_{50}Pd_{0.2}$	0.2	2.276	9.077	-0.243
$Fe_{49.50}Al_{50}Ag_{0.5}$	0.5	2.264	8.931	-0.244

Table 1. Heats of formation (ΔH_f) of FeAl and doped FeAl-Pd/Ag structures.

Figure 1 shows the heats of formation (ΔH_f) against atomic composition for FeAl-X structures doped with Pd, Ag, Ru and Pt atoms, ΔH_f are listed in Table 1. The solid common-tangent line was constructed to illustrate the ground state structures of their different phase composition. Among the structures considered, the most favourable intermetallic phases was found to be $Fe_{49.80}Al_{50}Ru_{0.2}$, $Fe_{49.50}Al_{50}Ag_{0.5}$ and $Fe_{49.80}Al_{50}Pd_{0.2}$ with the lowest heats of formation [7]. Figure 1 (d) shows that at 0.35, 0.38, 0.45 and 0.48 at. % Pt may be considered meta-stable. However, we also note that figure 1 (b) points that 0.3, 0.4, 0.6 and 0.7 at. % may be considered as meta-stable phases since they fall above the stability tangent line. Theses phases have not been reported in literature. The dotted straight line show that the ΔH_f increases as the Ag concentration is increased, implying that the stability decreases. However, it is clear that at 0.5 at. % Ag, the structure becomes the most stable in agreement with experimental findings [7].

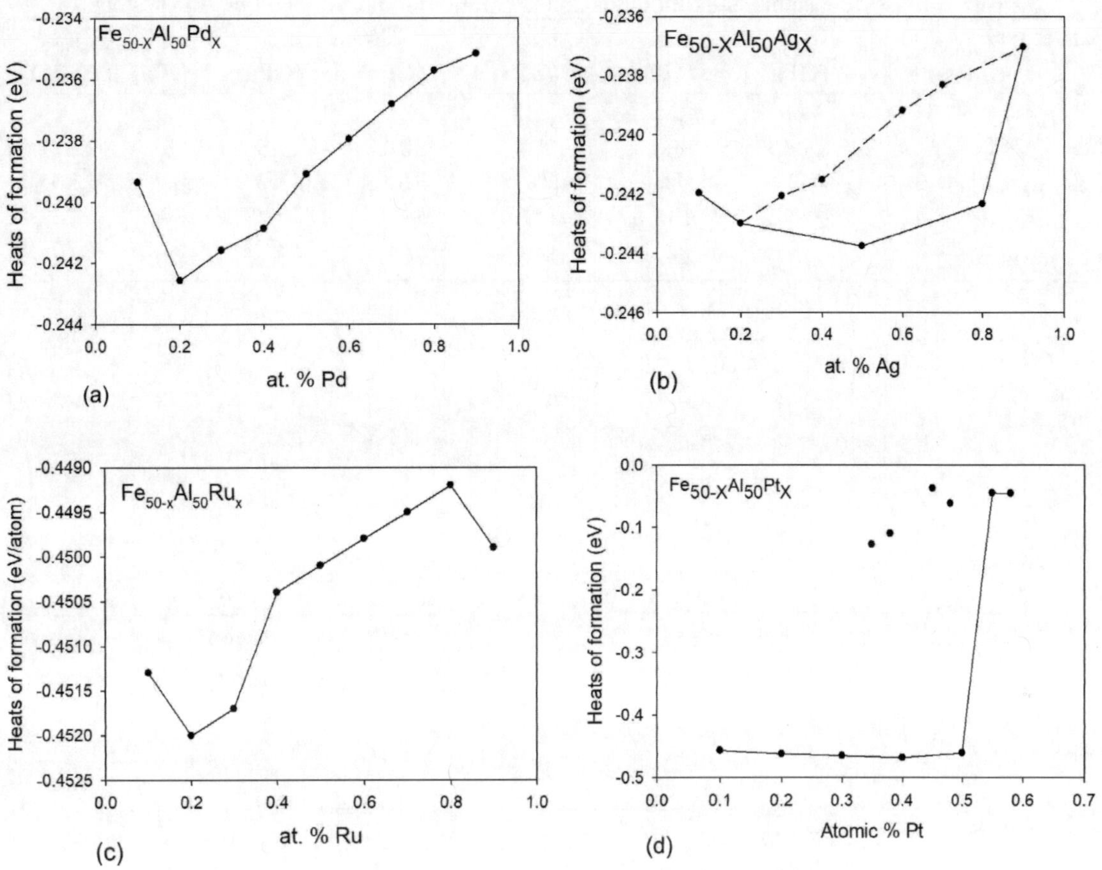

Figure 1. The graphs of heats of formation for the (a) $Fe_{50-X}Al_{50}Pd_X$, (b) $Fe_{50-X}Al_{50}Ag_X$, (c) $Fe_{50-X}Al_{50}Pt_X$, and (d) $Fe_{50-X}Al_{50}Ru_X$. In figure 1(b), a straight line is drawn to guide the eye on stability and meta-stability phases.

3.2. Mechanical Properties of Fe50-XAl50X (X= Pt, Pd, Ru, Ag)

In order to investigate the elastic stability of the ternary systems, we calculated the elastic constants (C_{ij}) for $Fe_{49.80}Al_{50}Pd_{0.2}$, $Fe_{49.50}Al_{50}Ag_{0.5}$, $Fe_{49.50}Al_{50}Pt_{0.5}$ and $Fe_{49.80}Al_{50}Ru_{0.2}$ systems at different concentrations as shown in table 2. Note that the symmetry was unchanged during doping, and thus three independent elastic constants (C_{11}, C_{12}, and C_{44}) have been found for the cubic lattice. The stability conditions for cubic lattice are:

$$C_{11} > 0,\ C_{11} - C_{12} > 0;\ C_{11} + 2C_{12} > 0;\ C_{44} > 0$$

We observed that all independent elastic constants are positive and satisfy the stability conditions for Ru doped. Furthermore, the calculated shear moduli is positive ($C' > 0$) indicating that the structure is elastically stable (see figure 2 (a)). The elastic constants in figure 2(b) confirmed instability at 0.3 at. % Pt, while Figure 2 (c) shows that C' is positive below 0.5 at. % Pd (condition of stability). However, the structure shows negative behavior (instability) above 0.5 at. % Pd. The elastic stability criteria for ductility and brittleness as proposed by Pugh was used (where B/G < 1.75 is for brittleness or B/G > 1.75 is for ductility) [15]. In this case, the predicted B/G ratio for Pd and Ag are greater than 1.75 (indicating ductility), while those for Ru and Pt are less than 1.75 (brittleness).

Table 2. Elastic constants, bulk modulus and B/G ratio (GPa) of doped $Fe_{50}Al_{50}$ systems

Structures	**C_{11} (GPa)**	**C_{12} (GPa)**	**C_{44} (GPa)**	**C′ (GPa)**	**B (GPa)**	**G (GPa)**	**B/G (GPa)**
$Fe_{49.80}Al_{50}Pd_{0.2}$	149.6	89.5	16.4	30.1	18.5	7.1	2.57
$Fe_{49.50}Al_{50}Ag_{0.5}$	-1.3	-15.4	-16.6	7.1	48.6	10.7	4.50
$Fe_{49.80}Al_{50}Ru_{0.2}$	279.4	140.6	147.4	69.4	186.9	108.9	1.72
$Fe_{49.50}Al_{50}Pt_{0.5}$	278.9	139.9	147.5	208.9	186.2	101.8	1.707

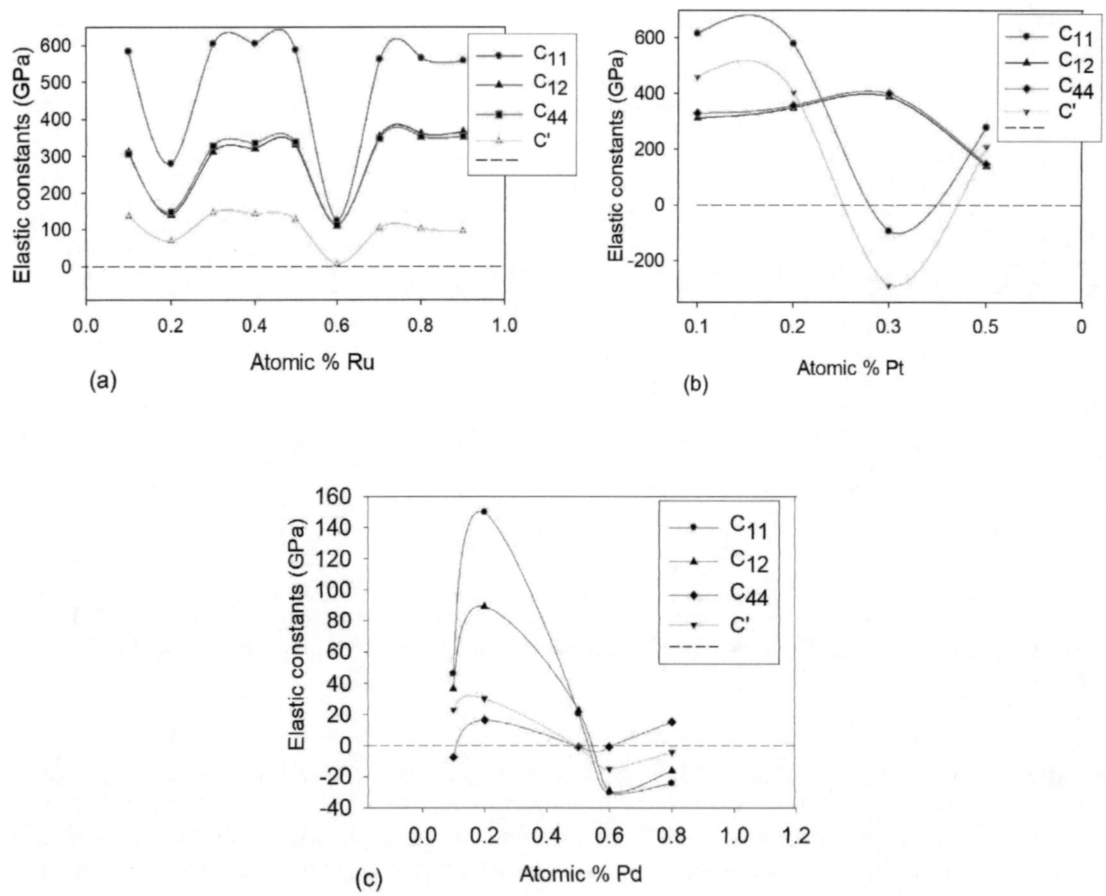

Figure 2. The elastic constants (GPa) for the (a) $Fe_{50-X}Al_{50}Pt_X$, (b) $Fe_{50-X}Al_{50}Ru_X$ and (c) $Fe_{50-X}Al_{50}Pd_X$ structures.

3.3. Density of states

Figure 3 shows the partial and total density of states of Pd and Ag doped FeAl systems. These systems display similar total DOS characterized by a strong *d*-orbital peak at E_f ($E-E_f = 0$). We see that the

Fermi level hits the top of the *d*- peak in both cases, while to that of Pd doped DOS is broader (figure 3(a)). The peak is due to the strong contribution of the 4*d*- orbital with less contribution of the *s*- and *p*- orbital. This signifies electronic stability which is in good agreement with the predicted heats of formation. The small sharp peak at about~0.5 eV (figure 3 (b)) emanates from hybridization of the 4*d*- orbitals (Fe, Ag) and 2*p*- orbital.

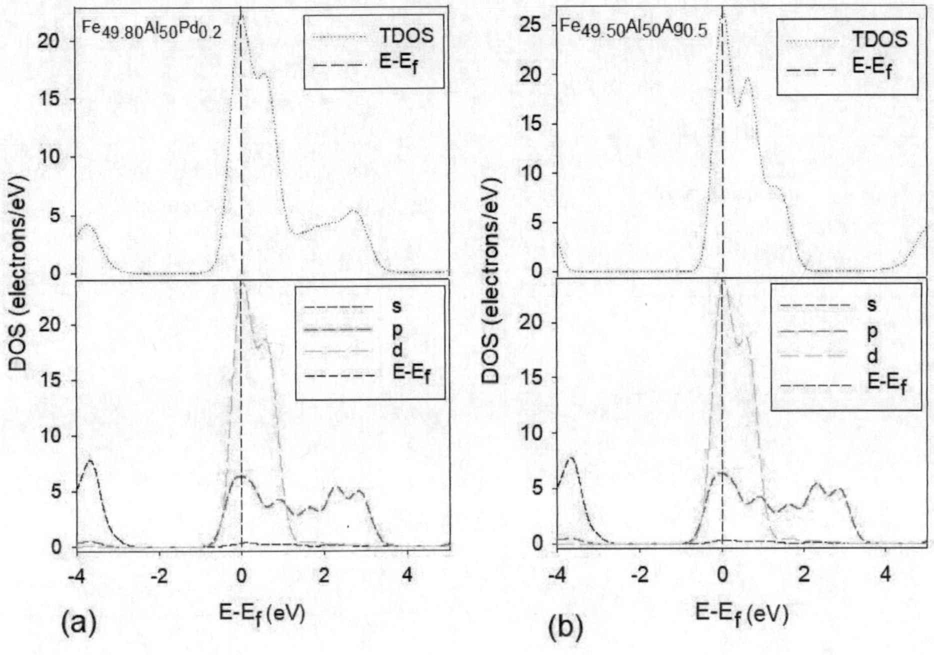

Figure 3. Total and partial density of states of ternary systems (a) $Fe_{49.80}Al_{50}Pd_{0.2}$ and (b) $Fe_{49.50}Al_{50}Ag_{0.5}$, with the Fermi energy taken as energy zero ($E-E_f= 0$).

3.4. *Effect of temperature on FeAl and FeAl-Pd/Ru/Ag.*

The results obtained showed preferential doping on the Fe sub-lattice with Pt and Ru being the most favourable dopant than Ag and Pd. According to literature, the binary FeAl have limited room temperature above 873 K [16]. In this study, we note from the binding energy curves that the binary $Fe_{50}Al_{50}$ phase is stable up to 900 K and becomes less stable above this temperature being in agreement with the experimental. However, the addition of 0.2 at. % Ru showed higher transition temperature of 1000 K, slightly lower for Ag (900 K), as compared to the experimental value of 1273 K and 1073 K [17, 18] and shows to be worse for Pd (400 K). The Ru and Ag doped systems shows a sudden jump at 1000 K and 900 K, respectively, which implies that the system is unstable above this temperature range.

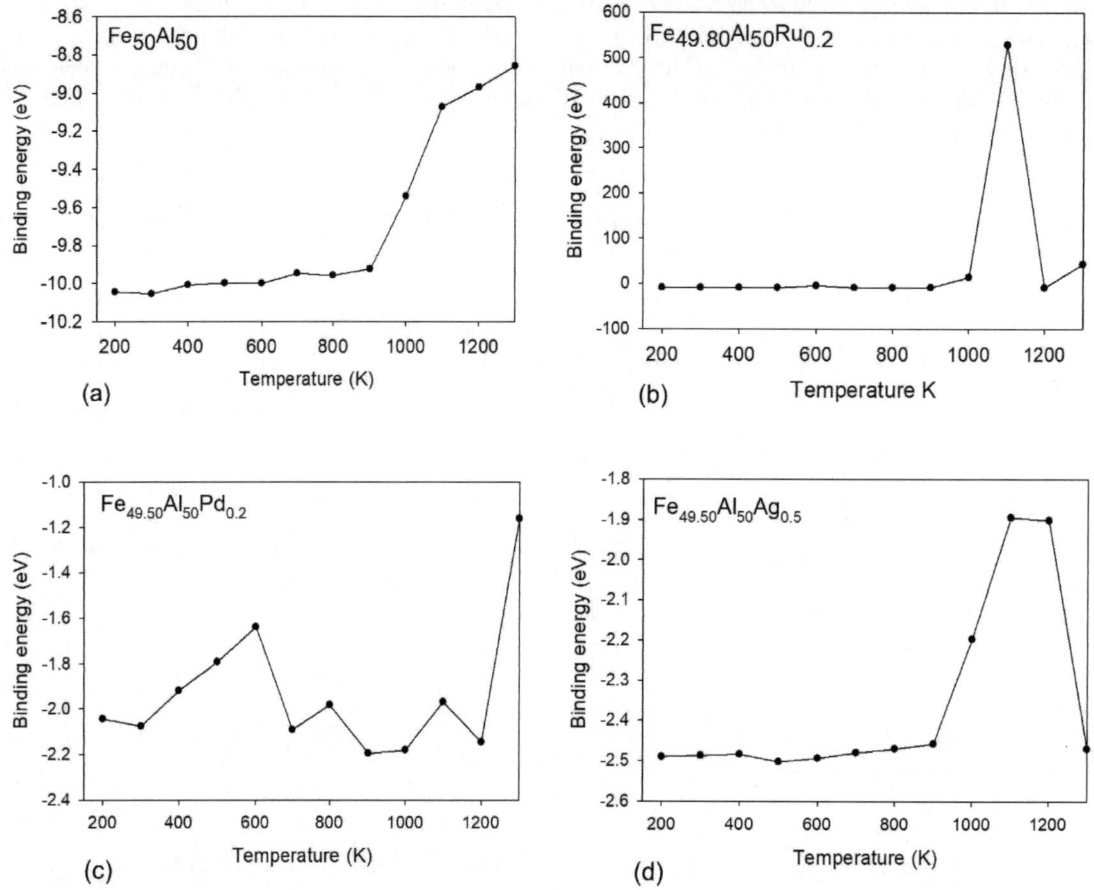

Figure 4. The graphs of temperature dependence for the (a) $Fe_{50}Al_{50}$, (b) $Fe_{49.80}Al_{50}Ru_{0.2}$, (c) $Fe_{49.80}Al_{50}Pd_{0.2}$ and (d) $Fe_{49.50}Al_{50}Ag_{0.5}$ systems.

4. Summary and conclusion

The DFT results were in good agreement with the experimental findings [7]. It was found that the $Fe_{50}Al_{50}$ structure was more stable (lowest heats of formation) and VCA showed preference of doping on Fe than Al sublattices: $Fe_{48.50}Al_{50}Ag_{0.5}$, $Fe_{48.80}Al_{50}Pd_{0.2}$, $Fe_{48.50}Al_{50}Pt_{0.5}$ and $Fe_{49.80}Al_{50}Ru_{0.2}$. The equilibrium heats of formation, lattice parameters, elastic properties, density of states and temperature effect on the B2 FeAl phase were determined using *ab initio* calculations. Interestingly, the B2 $Fe_{50}Al_{50}$ was found to be energetically and mechanically stable over the other phases at different compositions. These phases exist in the range between 50 and 100 atomic percent Al of the experimental phase diagram [19]. The shear modulus (C′) of the B2 $Fe_{50}Al_{50}$ and $Fe_{49.80}Al_{50}Ru_{0.2}$ phases are found to be positive, satisfying the condition of stability, whereas for $Fe_{48.50}Al_{50}Pt_{0.5}$, the C′ showed condition of instability at 0.3 at. %.

Acknowledgements

The work was carried at the Materials Modelling Centre, University of Limpopo. The support of the South African Research Chair Initiative of the Department of Science and Technology and the National Research Foundation is highly recognised.

References

[1] Joska L, Poddana M and Leitner J 2008 Dent Mater **24** 1009-16.
[2] Allisona E G and Bond G C 2006 *Catal. Rev.* **7** 233-89.
[3] Feng D and Taskinen P 2014 *J. Mater. Sci.* **49** 5790-8.
[4] Sharma B, Yadav H M and Kim J 2017 *IEEE Sens. Appl. Symp. (SAS)* **10** 1-5.
[5] Rodríguez-Diaz R A, Uruchurtu-chavarín J, Molina-Ocampo A, Porcayo-Calderon J, Mendoza M E, Valdez S and Juárez-Islas J 2011 *Int. J. Electrochem. Sci.* **8** 11877-95.
[6] Villagomez-Galindo M, Carbajal-De la Torre G, Romo-Castañeda J, Bedolla-Jacuinde A, González-Rojas H and Espinosa-Medina M 2016 *J. Mater. Res.* **31** 2473-81.
[7] Couperthwaite R A, Cornish L A and Mwamba I A 2016 *J. S. Afr. Inst. Min. Metall.* **116** 927-34.
[8] Kohn W and Sham L J 1965 *Phys. Rev.* **140** 1133-38.
[9] Perdew J P, Burke K and Ernzerhof M 1996 *Phys. Rev. Lett.* **77** 3865-8.
[10] Milman V, Winkler B, White J A, Pickard C J, Payne M C, Akhmatskaya E V and Nobes R H 1999 *Int. J. Quantum Chem.* **77** 895-910.
[11] Hedin L and Lundqvist B I 1971 *J. Phys. C: Solid State Phys.* **4** 2064-82.
[12] Monkhorst H J and Pack J D 1976 *Phys. Rev. B.* **13** 5188−92.
[13] Nordheim L 1931 *Ann. Phys.* **9** 607-40.
[14] Breur J, Gru A, Sommer F and Mittmeijer E J 2001 *Metall. Mater. Trans. B* **32** 913-8.
[15] Pugh S F 1954 *J. Sci.* **45** 823-43.
[16] X. Li 2016.
[17] Palm M, Inden G, Thomas N.1995 *J. Phase Equilib.* **16** 209-22.
[18] Palm M, Lacaze J. 2006 *Intermetallics* **16** 1291-1303.
[19] Atabaki M M, Nikodinovski M, Chenier P, Ma J, Harooni M and Kovacevic R 2014 J. *Manuf. Sci. Prod.* **14** 2191-4184.

Mechanical properties and stability of TiPt-M (M= Ru, Co, Cu, Zr and Hf) for high temperature applications

M P Mashamaite[1]*, H R Chauke[1] and P E Ngoepe[1]

[1]Materials Modelling Centre, School of Physical and Mineral Sciences, University of Limpopo, Private Bag X1106, Sovenga 0727, South Africa

*mordecai.mashamaite@ul.ac.za

Abstract. Shape memory alloys are smart materials that can remember their shape after being deformed. This is due to their unique properties, shape memory effect and super-elasticity, which occurs because of phase transformation. High temperature shape memory alloys are being developed in order to increase the application area of shape memory alloys. There is an increasing demand of shape memory alloys, which can be used in high temperatures environments such as in aerospace, actuators, sensors in the automotive industry. The first-principle approach was employed to investigate the effect of ternary alloying with Ru, Co, Cu, Zr and Hf on the equi-atomic B19 TiPt alloy. The supercell approach based on the B19 TiPt structure was used to construct various compositions containing a third element with reduced Pt content, and evaluate their thermodynamic and mechanical stability. The Ru, Co and Cu calculations were performed to compare their effects on the TiPt relative to Zr and Hf addition. It was found that partial substitution with Ru was the most stable structure as compared to the Hf, since it displayed the lowest heats of formation. Their martensitic transformation temperature increased due to their lowest C′ shear modulus. However, the addition of Cu reduced the transformation temperature of the TiPt. All structures were found to be completely anisotropic.

1. Introduction

Shape memory alloys (SMAs) are used in a wide range of industrial applications, such as medical, actuators, robotics, energy and aerospace, due to their unique properties, shape memory effect (SME) and pseudo-elasticity which occurs as a result of phase transformation [1]. Recently, high temperature shape memory alloys (HTSMAs) have been investigated in order to increase the application areas of shape memory alloys. A number of alloying elements such as Pt, Pd, Hf and Zr have been studied in the attempt to increase the transformation temperature of NiTi (373 K). The NiTi-Pt and NiTi-Pd was reported to be effective in increasing the transformation temperature of the NiTi binary [2, 3]. However, the martensitic transformation temperature of these alloys remains below 830 K. TiPd and TiPt at equi-atomic have a high temperature phase identified as the cubic B2 austenite and low temperature phase as orthorhombic B19 martensite. They undergo phase transformation from B2 phase to B19 phase at 850 K and 1300 K [4]. TiPd have poor shape recovery of ~10% after deformation at 773 K, due to low strength at high temperature causing plastic deformation during stress application [5]. TiPt based alloys exhibit very low SME due to low critical stress for slip deformation compared to the stress required for martensitic transformation (MT) [6]. Interestingly the alloys have the potential to be used for much higher-temperature shape memory applications due to their high martensitic transformation Tm if their shape memory properties can be enhanced [7, 8].

Numerous attempts to enhance the mechanical properties of TiPt have been reported by adding elements such as Ir, Co, Ru, Zr and Hf [9, 10]. The shape recovery ratio of TiPt-Ir remained lower around 10% [11, 12]. Co, Ru, Zr and Hf addition improved the shape recovery ratio to between 40% and 60% [13]. Among these alloying metals, Zr and Hf were found to be most effective to enhance the shape memory recovery ratio. In this work, density functional theory (DFT) based technique was used to study the effect of the third element Ru, Co, Cu, Zr and Hf on the B19 TiPt-M system (shown in Figure 1). The stabilities of the $Ti_{50}Pt_{50-x}M_x$ for x=5 were investigated by studying the heats of formation and elastic properties of the structures. The calculations revealed Ru is the most thermodynamically stable addition as compared to the other structures. It will be shown that the Ru addition is more preferential, to contribute as HTSMAs.

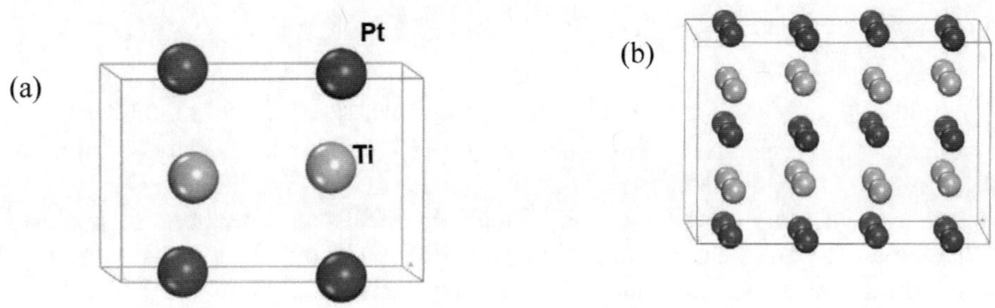

Figure 1. The atomic arrangement (a) Unit cell of B19 TiPt system with a space group Pmma. (b) Supercell for B19 TiPt-M structure.

2. Methodology

The calculations were carried out using the ab initio DFT in the Vienna ab initio Simulation Package (VASP) code [14, 15] with the projector augmented wave (PAW) [16]. An energy cutoff of 500 eV was used, to achieve a good convergence of the parameters. We have used the generalized gradient approximation (GGA) of the Perdew-Burke-Ernzerhof (PBE) exchange correlation functional [17]. The k-spacing of 0.18 for B19 TiPt was used according to Monkhorst and Pack [18].

3. Results and Discussion

3.1. Structural and Thermodynamic Properties

In Table1, the calculated equilibrium lattice parameters of the B19 $Ti_{50}Pt_{45}M_5$ are shown where M= Ru, Co, Cu, Zr, and Hf. The calculated results show that Ru, Co, and Cu decrease the equilibrium lattice parameter of the system, due to the atomic radii of Cu (135 pm) and Co (135 pm) being equal to that of Pt (135 pm), Ru being less than both Ti and Pt. The lattice parameters of both Zr and Hf increased minimally; this might be due to the larger atomic radii of Zr (155 pm) and Hf (155 pm) relative to Ti (140 pm) and Pt (135pm), respectively. Ru has a lower density of 12.37 g/cc than Pt (21.45 g/cc), the atomic radii of Ru is also lower (130 pm) than both Ti and Pt, but high in melting point of 2607 K. This attributed to Ru having lower lattice parameters. Similarly, the partial substitution of Co on Pt sub-lattice also reduced the lattice parameters of TiPt [19]. Therefore, the lattice parameter depends on the type of the alloying or third element [20]. The densities of both Zr and Hf are higher than that of Ti, which means that partial substitution of Ti with Zr or Hf will increase the density of the equi-atomic TiPt shape memory alloys.

Table 1. The equilibrium lattice parameters of B19 TiPt and $Ti_{50}Pt_{45}M_5$ (M= Ru, Co, Cu, Zr and Hf) ternaries.

Structures	$Ti_{50}Pt_{50}$	$Ti_{50}Pt_{45}M_5$ (M= Ru, Co, Cu, Zr and Hf)				
	-	Ru	Co	Cu	Zr	Hf
Lattice Parameters (Å)	a=4.618 (4.650) [4]	4.579	4.567	4.611	4.660	4.653
	b=2.857 (2.828)	2.805	2.778	2.768	2.817	2.819
	c=4.891 (4.940)	4.852	4.860	4.862	4.849	4.846
ΔH_f (eV/atom)	-0.932 (-0.962)	-0.864	-0.851	-0.834	-0.807	-0.804

The heats of formation (ΔH_f), of the intermetallic phase is calculated according to the relation [21],

$$\Delta H_f = E^{TiPt} - [(1-x)E^{Ti}_{solid} + xE^{Pt}_{solid}], \quad (1)$$

where E^{TiPt}, E^{Ti}_{solid} and E^{Pt}_{solid} are the total energies of the alloy, elemental Ti and Pt in their respective ground-state crystal structure; respectively whereas x and $1-x$ refers to the fractional concentrations of the constituent elements.

In Figure 2, we present plots of the predicted $Ti_{50}Pt_{45}M_5$ heats of formation for analysis. Ru substitution has the lowest value of heats of formation compared to Co and Cu dopants with Zr and Hf having the largest values. Co and Cu are more stable in comparison to Zr and Hf. Therefore, Ru is considered to be the most stable substitution with the value -0.8640 eV/atom.

Figure 2. Heats of formation of $Ti_{50}Pt_{45}M_5$ (M= Ru, Co, Cu, Zr and Hf) alloys.

3.2. Elastic Properties

The accurate calculation of elasticity is crucial to gain insight into the mechanical stability and elastic properties of compounds. Their elastic constants depend on the type of lattice i.e for an orthorhombic crystals; there are nine (C_{11}, C_{22}, C_{33}, C_{12}, C_{13}, C_{23}, C_{44}, C_{55}, C_{66},) independent elastic constants, respectively. This method has been successfully used to study the elastic properties of a range of materials including metallic systems [22].

For orthorhombic crystal, the mechanical stability conditions are given by [23].

$$C_{11} + C_{12} + C_{33} + 2C_{12} + 2C_{13} + 2C_{23} > 0,\ C_{22} + C_{33} - 2C_{13} > 0,\ C_{11} > 0,$$

$C_{22} > 0, C_{33} > 0, C_{44} > 0, C_{55} > 0, C_{66} > 0.$ (2)

The elastic constants and moduli for doped B19 $Ti_{50}Pt_{50-x}Mx$ structures are listed in Tables 3. The elastic constants were evaluated to observe the effect of ternary addition for Ru, Co, Cu, Zr and Hf concentrations of 5 atomic percent. The shear modulus (C′) for all the ternaries was positive, with Zr having the lowest value and Ru the largest, which translates to higher transformation temperature. The positive shear suggests that all the ternaries satisfy all conditions of mechanical stability of an orthorhombic crystal. Furthermore, anisotropic ratio (less micro cracks in the material), approaches unity (A=1) for all structures which indicates a strong correlation between C_{44} and C′ shown in figure 3. Considering that the shear modulus represent the resistance to plastic deformation and the bulk modulus represent the resistance to fracture [23], the Pugh ratio (B/G) was calculated and evaluated to determine the ductility and brittleness of the structures. The structures were observed to be ductile since all the structures were greater than the critical value of 1.75 [24]. The calculated young (E) moduli suggest that the Ru is least stiff material than Cu, which correlates greatly with C′ values. Therefore, Ru was found to be the most stable substitution.

Table 2. The elastic properties C_{ij} (GPa) of the $Ti_{50}Pt_{45}M_5$ (M= Ru, Co, Cu, Zr and Hf) ternaries and their anisotropy (A).

		$Ti_{50}Pt_{45}M_5$ (M= Ru, Co, Cu, Zr and Hf)				
Elastic Constant C_{ij} (GPa)	TiPt [8]	Ru	Co	Cu	Zr	Hf
C_{11}	319	322	312	312	289	290
C_{12}	133	138	143	135	133	130
C_{13}	152	153	145	139	138	137
C_{22}	361	346	356	357	322	306
C_{23}	112	115	109	112	105	104
C_{33}	317	310	309	315	296	297
C_{44}	-60	49	58	61	59	59
C_{55}	57	51	55	59	57	56
C_{66}	65	60	66	64	67	67
$C'=(C_{11}-C_{12})/2$	93	92	85	89	78	80
$A_1=2C_{66}/(C_{11}-C_{12})$	0.7	0.7	0.8	0.7	0.9	0.8
$A_2=2C_{44}/(C_{11}+C_{33}-2C_{13})$	0.7	0.3	0.4	0.4	0.4	0.4
$A_3=C_{44}/C_{66}$	0.9	0.8	0.9	1.0	0.9	0.9
Shear (G)		67	72	74	70	70
Bulk (B)		199	197	195	184	181
B/G		2.97	2.74	2.64	2.63	2.59
Young (E)		180	192	198	187	186

The thermal coefficient of linear expansion (α) with temperature is evaluated in Figure 4. The Ru is observed to have decreased bond length corresponding to lower value of α followed by Co, which indicate that the ternaries have higher melting points. On the contrary the higher values of α indicates that the structures expanded more, which was observed for both Zr and Hf, respectively. It can also been seen from the plot that the higher the value of α, implies reduced the transformation temperature.

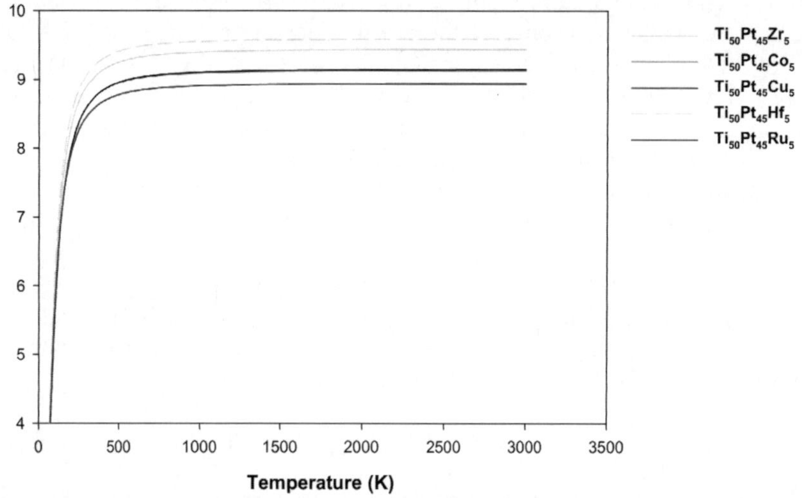

Figure 3. The graph of thermal coefficient of linear expansion against temperature for $Ti_{50}Pt_{45}M_5$ (M=Ru, Co, Cu, Zr and Hf) alloys.

4. Conclusion

First principle approach was used to study the mechanical stability of the B19 $Ti_{50}Pt_{45}M_5$ (M= Ru, Co, Cu, Zr and Hf) atomic composition. The mechanical properties of the ternary structures were investigated using the supercell approach. The thermodynamic stability for the ternary system was investigated by partially substituting some of the Pt with M, M= Ru, Co, Cu, Zr and Hf. It was observed that $Ti_{50}Pt_{45}Ru_5$ is the most stable structure, with the expansion of $Ti_{50}Pt_{45}Zr_5$ and $Ti_{50}Pt_{45}Hf_5$ being the least stable structures. We found the correlation with regard to mechanical stability of the ternaries (C' being positive), the competition between $Ti_{50}Pt_{45}Ru_5$ and $Ti_{50}Pt_{45}Co_5$ gave rise to Co substitution being the most mechanically stable ternary structure. It was also observed from the thermal coefficient of linear expansion that $Ti_{50}Pt_{45}Zr_5$ and $Ti_{50}Pt_{45}Hf_5$ reduced transformation temperature more than $Ti_{50}Pt_{45}Ru_5$ and $Ti_{50}Pt_{45}Co_5$.

Acknowledgements

The calculations were carried out using computer resources at the Materials Modelling Centre (MMC) at the University of Limpopo and the Centre for High Performance Computing (CHPC) in Cape Town. The support of the South African Research Chair Initiative of the Department of Science and Technology and the National Research Foundation is highly recognized.

References

[1] Suzuki Y, Xu Y, Morito S, Otsuka K and K Mitose 1998 Effects of boron addition on microstructure and mechanical properties of Ti–Td–Ni high-temperature shape memory alloys *Mater. Lett.* **36** 85

[2] Xu Y, Shimizu S, Suzuki Y, Otsuka K, Ueki T and Mitose K 1997 Recovery and recrystallization process in ti-pd-ni high-temperature shape memory alloys *Acta Mater.* **45** 1503

[3] Ma J, Karaman I and Noebe R D 2010 High temperature shape memory alloy *Int. Mater. Rev.* **55** 257

[4] Donkersloot H C and van Vucht J H N 1970 Martensitic transformations in gold-titanium, palladium-titanium and platinum-titanium alloys near the equiatomic composition *J. Less-Common Metals.* **20** 83

[5] Otsuka K, Oda K, Ueno Y, Piao M, Ueki T and Horikawa H 1993 The shape memory effect in a

$Ti_{50}Pd_{50}$ alloy *Scripta Metall. Mater.* **29** 1355

[6] Yamabe-Mitarai Y, Hara T, Miura S and Hosoda H 2010 Shape memory effect and pseudoelasticity of TiPt *Intermetallics* **18** 2275

[7] Biggs T, Cortie M B, Witcomb M J and Cornish L A 2001 Martensitic transformation, microstructure, and mechanical workability of TiPt *Metall. Mater.Trans.* **32A** 1881

[8] Mahlangu R, Phasha M J, Chauke H R, Ngoepe P E 2013 Structural, elastic and electronic properties of equiatomic PtTi as potential high-temperature shape memory alloy *Intermetallics* **33** 27

[9] Wadood A and Yamabe-Mitarai Y 2014 TiPt–Co and TiPt–Ru high temperature shape memory alloys *Mater. Sci. Eng.* **A601** 106

[10] Wadood A, Takahashi M, Takahashi S, Hosoda H and Yamabe-Mitarai Y 2013 High-temperature mechanical and shape memory properties of TiPt-Zr and TiPt-Ru alloys *Mater. Sci. Eng.* **A564** 34

[11] Yamabe-Mitarai Y, Hara T, Miura S and Hosoda H 2006 Mechanical properties of Ti–50(Pt,Ir) high-temperature shape memory alloys *Materi. Trans.* **47** 650

[12] Yamabe-Mitarai Y, Hara T, Miura S and Hosoda H 2012 Phase transformation and shape memory effect of Ti(Pt, Ir) *Mater. Trans. A* **43** 2901

[13] Yamabe-Mitarai Y, Arockiakumar R, Wadood A, Suresh K S, Kitashima T, Hara T, Shimojo M, Tasaki W, Takahashi M, Takahashi S and Hosoda H 2015 Ti(Pt, Pd, Au) based high temperature shape memory alloys *in Materials Today: Proceedings* **2** S517-S522

[14] Hohenberg P and Kohn W 1964 Inhomogeneous electron gas *Phys. Rev. B* **136** 864

[15] Kohn W and Sham L J 1965 Self-consistent equations including exchange and correlation effects *Phys. Rev* **140** A1133

[16] Blöchl P E 1994 Projector augmented-plane wave method Phys. Rev. B. **50** 17953

[17] Perdew J P, Burke K and Ernzerhof M 1996 *Phys. Rev. Lett.* **77** 3865

[18] Monkhorst H J and Pack J D 1976 Special points for Brillouin-zone integrations Phys. Rev. B. **13** 5188

[19] Mahlangu R 2015 Computational modelling studies of $Ti_{50}Pt_{50-x}M_x$ (M: Co, Ni, Pd, Ir) shape memory alloys. *University of Limpopo: Thesis*

[20] Chauke H R, Mashamaite M P, Modiba R and Ngoepe P E 2018 Advances in Ti-based systems as high temperature shape memory alloys *Key Eng. Mater.* **770** 230

[21] Fernando G W, Watson R E and Weinert M 1992 Heats of formation of transition-metal alloys: Full-potential approach and the Pt-Ti system *Phys. Rev. B* **45** 15 8233

[22] Zhang J M and Guo G Y 1997 Microscopic theory of the shape memory effect in TiNi *Phys. Rev. Letters* **78** 4789

[23] Mayer B, Anton H, Bott E, Methfessel M, Sticht J and Schmidt P C 2003 Ab-Initio calculation of the elastic constants and thermal expansion coefficients of laves phases *Intermetallics* **11** 23

[24] Pugh S F 1954 Relations between elastic moduli and plastic properties of polycrystalline pure metals *Philos. Mag* **45** 843

Computational study of binary titanium-based potential shape memory alloys

M E Baloyi[1*], R Modiba[2], H R Chauke[1] and P E Ngoepe[1]

[1]Materials Modelling Centre, School of Physical and Mineral Sciences, University of Limpopo, Private Bag X1106, Sovenga, 0727, South Africa
[2]Council for Scientific and Industrial Research, CSIR, Materials Science and Manufacturing, Lynnwood Manor, P.O. Box 395, Pretoria 0001, South Africa

*mphamela.baloyi@ul.ac.za

Abstract: First-principles density functional theory approach was employed to investigate the martensitic temperature of the equiatomic TiPt, TiZr and TiTa shape memory alloys. Firstly, their structural, thermodynamic, elastic and phonon dispersion have been deduced to mimic the relative stabilities. These alloys have potential to operate as high-temperature shape memory alloys since they possess a reversible B2 to B19 martensitic transformation. It was found that the lattice parameters are in good agreement with the experimental results to within 5%. Furthermore, the TiPt structure is thermodynamically stable displaying lowest heats of formation, while TiZr and TiTa are not. Our results show that the B2 phases of these binary alloys are vibrational unstable due to the presence of soft modes along high symmetry direction on the phonon dispersion curves. TiTa alloy is mechanically stable ($C' > 0$), which may be used to develop future high-temperature alloys.

1. Introduction

Shape memory alloys (SMAs) are classified as a group of metals that can recall their pre-deformed shape upon temperature increase and this is due to their two unique properties, shape memory effect (SME) and superelasticity [1]. These materials exhibit martensitic transformation between high-temperature austenite (B2) to low-temperature martensite phases (B19/19′) [2]. Ti-based SMAs have been used for various applications in the medical and engineering industry [3, 4]. Amongst the binary alloys, TiNi is the most studied and commercialized SMA, but its applications are limited by low martensitic transformation temperatures (Tm) of 373 K [5]. Thus, the development of SMAs that can operate at high temperature is ongoing, in particular to increase the application areas and shape memory properties [6]. Ti-based SMAs such as TiPd, TiTa and TiPt have been studied in the experimental and theoretical researches owing to their high transformation temperature which render them appropriate for potential high-temperature applications [7]. Amongst these alloys, TiTa was found attractive for advanced actuators applications due to their shape memory behaviour at relatively high temperatures [8]. Furthermore, TiPt is the most attractive for the development of high-temperature shape memory alloys (HTSMAs) since it has high Tm of about 1300 K [9]. This alloy has been reported to undergo a reversible martensitic transformation from B2 to B19/B19′ at this temperature.

Recently, many theoretical and experimental studies are being carried to explore the SMAs properties, particularly features pertaining transformation temperature and stability; this phenomenon is still not well understood [10]. In this paper, we investigate the stability of TiM (M = Pt, Ta, Zr) SMAs at 50:50 composition using the heats of formation, elastic properties and phonon dispersion curves of B2 structures. It was found that TiPt and TiZr structures are unstable at 0K, in agreement with experimental findings [10, 11]. Our results also showed that all the B2 structures are vibrational unstable displaying since soft modes are observed along high symmetry direction in the phonon spectra. However, TiTa has shown potential for new alloy development, being elastically stable at 0K.

2. Methodology

First-principles density functional theory (DFT) [12] approach with the projector augmented wave [13] was employed to investigate the stability of the equiatomic titanium-based SMAs using VASP code [14]. An energy cutoff of 500 eV was used, as it was adequate to converge the total energy of the systems. For exchange-correlation functional, the generalized gradient approximation of Perdew, Burke and Enzerhof (GGA-PBE) [15] was chosen. The suitable k-points mesh according to Monkhorst and Parck [16] of 8×8×8 was used. The phonon dispersion spectra were evaluated using PHONON code [17] as implemented in Materials Design within MedeA software of VASP code [14].

3. Results and Discussion

3.1. Structural properties

The calculated equilibrium lattice parameters and heats of formation of B2 Ti-M (M=Pt, Ta, Zr) alloys are shown in Table 1. It is found reasonable that equilibrium lattice parameters are in good agreement with experimental to within 5 %. TiPt lattice parameter is 3.179 Å compared to 3.192 Å [18]. In addition, TiZr lattice parameter agrees with a value of 3.418 Å (3.040) [19], while TiTa is well predicted as 3.269 Å (3.350) [20].

The heat of formation is used to predict the stability of alloys and may be calculated as,

$$\Delta H_f = E^{TiPt} - [(1-x)E^{Ti}_{solid} + xE^{Pt}_{solid}].$$

where E^{TiPt}, E^{Ti}_{solid} and E^{Pt}_{solid} are the total energies of an intermetallic alloy, and elemental Ti and Pt in their respective ground-state crystal structures, while *x* and *1-x* refer to the fractional concentrations of the constituent elements. The lowest heats of formation value indicate stability whereas highest indicates instability of the material. It is clear that TiPt is found to be more stable with the lower heats of formation of -0.821 eV/atom while that of TiZr and TiTa were found to be the least stable with the highest heats of formation of 0.065 eV/atom and 0.005 eV/atom, respectively.

Table 1. Equilibrium lattice parameters (a) and heats of formation (ΔH_f) of B2 Ti-M (M=Pt, Zr, Ta) binary alloys.

	a (Å)		
Structures	Calculated	Experimental	ΔH_f (eV/atom)
TiPt	3.179	3.192 [18]	-0.861
TiZr	3.418	3.040 [19]	0.065
TiTa	3.269	3.350 [20]	0.005

3.2. Elastic constants

The elastic constants (C_{ij}), anisotropy (A) and Pugh's ratio (B/G) of Ti-M (M=Pt, Zr, Ta) alloys were calculated as shown in Table 2. The elastic properties investigation is considered in this study to understand martensitic transformation behaviour of these alloys [11]. There are three independent elastic constants (C_{11}, C_{12} and C_{44}) for a cubic crystal structure, shear modulus C′ and A to evaluate the stability and transformation behaviour. In order to gain a deep understanding of the mechanical durability of these alloys, we calculated the anisotropic factor A,

$$A = \frac{C_{44}}{C'} . \qquad (2)$$

The stability criterion for the cubic system are as follows;

$$C_{44} > 0, C_{11} + 2C_{12} > 0, \text{ and } C_{11} > |C_{12}| . \qquad (3)$$

Note that the crystal structure is considered to be mechanically stable when C′ is positive, otherwise unstable. In this case, the calculated elastic constant show that $C_{11}<C_{12}$ for a cubic TiPt and TiZr structures leading to negative shear modulus C′ (C′<0). This indicate that the structures are mechanically unstable. It was reported that cubic TiPt is unstable at low temperature with C′ of -32 GPa. Our calculated elastic constants values are all positive, it is an indication of mechanical stability of a cubic TiTa at low temperature. In addition, the lower shear modulus C′ of TiZr (-25.49 GPa) suggesting the higher B19 transformation temperature. Higher A is necessary for B2-B19 phase transformation. Thus, there is a stronger correlation between C_{44} and shear modulus C′ of TiTa and is considered anisotropic (A≈1). Calculated A is smaller with value of 0.95 suggesting the B2 to B19′ phase transformation. To investigate the ductility of these alloys, bulk (B) and shear (G) moduli are considered. The Pugh's ratio clearly indicates that all binaries in this study are ductile with B/G>1.75 [21].

Table 2. Elastic constants C_{ij} (GPa), anisotropy (A) and Pugh's ratio (B/G) of B2 Ti-M (M=Pt, Zr, Ta) binary alloys.

Structures	C_{11}	C_{12}	C_{44}	C′	A	B/G
TiPt [7]	145	210	45	-32		2.28
TiZr	59.90	110.88	17.74	-25.49		3.38
TiTa	204.25	121.26	39.46	41.50	0.95	3.70

3.3. Phonon dispersion

Phonon dispersion and phonon density of states (DOS) were also calculated to investigate the vibrational properties of TiM (M=Pt, Zr, Ta) and are displayed in figure 3. There are soft modes observed on the curves signifying the instability of the phases with reverence to the vibrations of the atoms. TiPt, TiZr and TiTa are vibrational unstable due to the existence of soft modes observed in the negative frequency. In particular, along the M and R high symmetry directions and this could be attributed to high vibration of Pt atoms in the TiPt, Zr atoms in TiZr and Ta atoms in TiTa as depicted on the phonon DOS. The phonon dispersion curves of the binary B2 phase for TiPt and TiZr are mostly characterized by the negative value of shear modulus C′ which is related to elastic behaviour. These soft modes have the lower frequency at approximately -3.0 THz due to high vibration of Pt, Ta and Zr atoms.

IOP Conf. Series: Materials Science and Engineering **430** (2018) 012020 doi:10.1088/1757-899X/430/1/012020

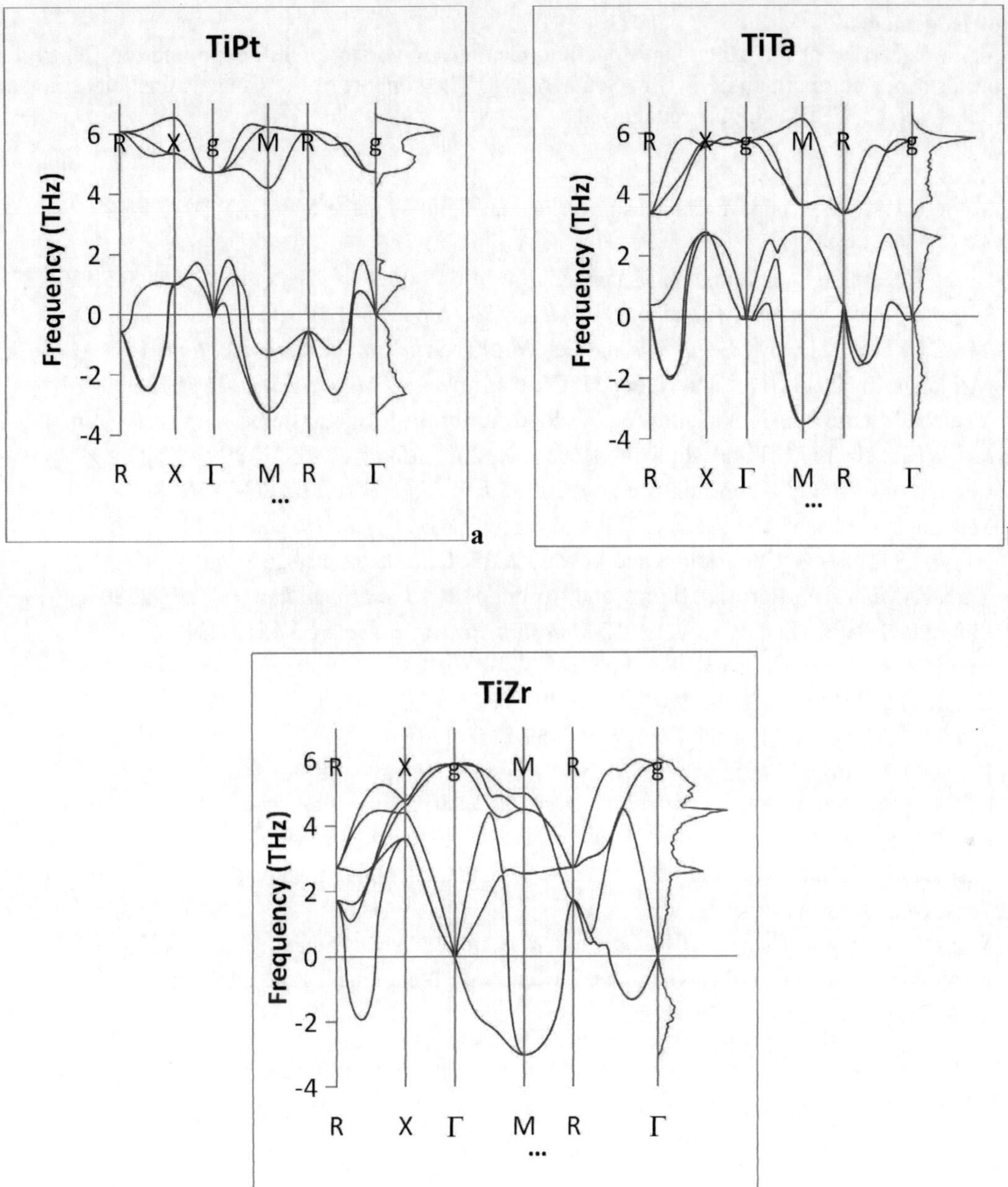

Figure 1. Phonon dispersion curves of (a) TiPt, (b) TiTa and (c) TiZr binary alloys.

4. Conclusion

Our DFT results displayed reasonable structural properties, elastic properties and phonon dispersion of TiM (M=Pt, Zr, Ta) alloys. The calculated equilibrium lattice parameters are in accord with the experimental findings to within 5 %. It is interesting to note that the predicted heats of formation confirmed the stability of TiPt. However, Pugh's ratio clearly indicates that all the structures in this study are ductile. Elastic constants show that TiTa is mechanically stable satisfying the cubic stability criterion. Furthermore, phonon dispersion spectra showed that TiPt and TiZr are vibrationally unstable, due to the presence of soft modes observed along high symmetry directions in the Brillouin zone , which is in agreement with the predicted elastic constants.

IOP Conf. Series: Materials Science and Engineering **430** (2018) 012020 doi:10.1088/1757-899X/430/1/012020

Acknowledgements
The work was carried at the Materials Modelling Centre, University of Limpopo and the Centre for High performance computing (CHPC) in South Africa. The support of the Council for Scientific and Industrial Research, CSIR is highly recognised.

References
[1] Wadood4 A, Takahashi M, Takahashi S, Hosoda H and Yamabe-Mitarai Y 2013 *Mater. Sci. Eng. A* **564** 34-41
[2] Biggs T, Cornish L A, Witcomb M J and Cortie M B 2004 *J. of Alloys and Comp.* **375** 120-127
[3] Wadood A and Yamabe-Mitarai Y 2014 *Mater. Sci. Eng. A* **601** 106-110
[4] Tan C L, Tian X H, Ji G J, Gui T L and Cai W 2008 *Solid State Communications* **147** 8-10
[5] Mahlangu R, Chauke H R and Ngoepe P E 2014 *Advanced Material Science* **1019** 385-390
[6] Yamabe-Mitarai Y, Arockiakumar R, Wadood A, Suresh K S, Kitashima T, Hara T, Shimojo M, Tasaki W, Takalashi M, Takalashi S and Hosoda H, 2015 *Materials Today* **2S** 517-522
[7] Chang-Long T, Wei C and Jing-Chuan Z 2006 *Chin. Phys. Lett.* **23** 2863-2866
[8] Mahlangu R, Phasha M J, Chauke H R and Ngoepe P E 2013 *Intermetallics* **33** 27-32
[9] Barzilai S, Toher C, Curtarlolo S and Levy O 2016 *Acta Mater.* **120** 255-263
[10] Yamabe-Mitarai Y, Hara T, Miura S and Hosoda H 2010 *intermetallics* **18** 2275-2280
[11] Plimpton S J and Thompson A P 2012 *Materials Research Society* **37** 513-521
[12] Kohn W and Sham L J 1965 *Phys. Rev.* **140** 33-38
[13] Kresse G and Hafner J 1994 *Phys. Rev. B* **49** 251-269
[14] Kresse G and Joubert D 1999 *Phys. Rev. B* **59** 1758-1775
[15] Perdew J P, Burke K and Enzerhof M 1996 *Phys. Rev. Lett.* **77** 88-92
[16] Monkhorst H J and Pack J D 1976 *Phys. Rev. B* **13** 88-92
[17] Parlinski K, Li Z Q and Kawazoe Y 1997 *Phys. Rev. Lett.* **78** 4063-4066
[18] Nishida M, Matsuda M, Yasumoto Y, Yano S, Yamabe-Mitarai Y and Hara T 2013 *Mater. Sci. Tech.* **24** 884-889
[19] Wang B, Li W and Zhang P 2012 *Condensed Matter Materials Science* **3** 1-9
[20] Dobromyslov A V, Dolgikkh G V, Dutkiewicz J and Trenogina T L 2006 *Archives of Metallurgy and Materials* **51** 547-550
[21] Pugh S F 1954 *J. Sci.* **45** 823-843

Computational modelling of $Ti_{50}Pt_{50-X}V_x$ potential shape memory alloys

R Modiba[1,2,*], E Baloyi[2], S Chikosha[1], H R Chauke[2] and P E Ngoepe[2]

[1] Materials Science and Manufacturing, CSIR, PO Box 395, Pretoria, 0001, South Africa
[2] Materials Modelling Centre, University of Limpopo, Private Bag X1106, Sovenga, South Africa

*Rmahlangu1@csir.co.za

Abstract. The demand for shape memory alloys for high temperatures have increased in recent years and TiPt alloys are promising candidates. The martensitic transformation temperature of TiPt is higher with the value of approximately 1300 K as compared to the commercialised TiNi which transforms at 300 K. Titanium-based SMAs have been widely used in the fields of engineering and medicine due to their shape memory effect (SME) and super-plasticity which are displayed in martensitic transformations. Some of the applications include actuators and medical stents. Previous studies showed that the alloy is mechanically unstable with the negative C′ modulus at 0 K. In this work, supercell approach is used to substitute Pt with vanadium on the B2 TiPt structure and evaluate the mechanical stability of the structures. The equilibrium lattice parameters, elastic properties and the phonon dispersions were calculated to investigate the stability of the $Ti_{50}Pt_{50-X}V_X$. Interestingly the vanadium addition was found to be stabilising the TiPt with all the C_{ij}'s being positive.

1. Introduction

Shape memory alloys (SMAs) are alloys that are capable of shape recovery when exposed to a certain critical temperature after deformation [1] . This is due to a reversible thermoelastic martensitic transformation [2], which leads to the properties of shape memory effect (SME) and superelasticity (SE). These properties make SMAs useful in applications such as sensors, actuators, microcontrollers, and as biomedical devices [1] [3] [4] [5] [6]. $Ti_{50}Ni_{50}$ is the most commercially successful SMA with a B2-B19' martensitic transformation. The alloys' transformation temperature (TT) can be varied between 100-500 K through careful selection of alloying element(s), often this is also accompanied by changes in crystal structure of the martensite phase [7]. It has long been established experimentally [8] and later theoretically [9] that this composition dependence of TT and martensite crystalline structure can be directly attributed to elastic properties, i.e. the softening and coupling of C' and C_{44} shear modes in the B2 cubic phase during cooling towards TT. First-principles approach, based on elastic constants and electronic structure, has been successfully used to explain the experimentally observed relationships between TT, martensite phase stability, and chemical composition for a number of $TiNi_{50-x}$-M_x alloys (M = Cu, Fe, Pt, Pd; $0 \leq x \leq 25$ at.%) [10] [11].

There has been growing interest to develop solid state actuators for applications in aeronautic and aerospace propulsion technologies with transformation temperatures in the range of 500-1300 K [12], this is beyond the capability TiNi-based alloys. Several binaries such as TiPt, TiIr, TaRu and NbRu

[12] [13] [14] are good candidates. In the current work, TiPt binary, with a B2-B19 martensitic TT of 1273 K [12] has been considered. However, the alloy exhibits poor shape recovery due to low strength of the B2 austenite phase as well as the absence of two-stage yield behaviour in stress-strain tests conducted below martensite finish temperature (M_f) [15]. There is a need to improve TiPt mechanical properties while maintaining high TT. Solid solution strengthening has been considered as one possible strengthening mechanism, previous work has shown that partial substitution of Pt with Ru and Co, and to a lesser extent Ir, in equiatomic TiPt improves strain recovery while maintaining high TT (M_s > 1000 K) [16].

Previous study on binary TiPt employing density functional theory (DFT) showed B2 phase to be unstable at 0 K with a negative *C′* modulus [17]. In a process analogous to the one followed for TiNi-based alloys, the current work uses first-principles to investigate the elastic constants and electronic structure of B2 phase in relation to martensitic TT and phase stability in TiPt-based alloys. We present results on the effect of partial substitution of Pt with V on the equiatomic cubic TiPt system using the supercell approach. Stability of the structures of B2 $Ti_{50}Pt_{50-x}V_x$ (x=6.25, 18.75 and 25 at.%) is determined with respect to their equilibrium lattice parameters, elastic constants and phonon dispersions using the DFT approach.

2. Methodology

The calculations were carried out using *ab initio* density functional theory (DFT) [18] [19] formalism as implemented in the Vienna *ab initio* simulation package VASP [20] with the projector augmented wave (PAW) [21]. An energy cut-off of 500 eV was used, as it was sufficient to converge the total energy of the cubic B2 TiPt alloys. For the exchange-correlation functional, the generalized gradient approximation of Perdew and Wang (GGA-PBE) [22] was chosen. The Brillouin zone integrations were performed for suitably large sets of *k*-points according to Monkhorst and Pack [23]. A 2x2x2 supercell of cubic B2 TiPt with 16 atoms was used. The substitutional search tool embedded in VASP was used to substitute Pt with V. The phonon dispersion spectra were evaluated using PHONON code [24] as implemented by Materials Design within their MedeA software. All the calculations were done at 0 K.

3. Results and discussion

3.1 Elastic properties

In Table 1 we show the calculated lattice parameters and elastic constants of the $Ti_{50}Pt_{50-x}V_x$ (x = 6.25, 18.75, 25). The calculated results show that as we increase the V concentration in the system the lattice parameter decreases. This is understood since the atomic radius of V (1.71 Å) is less than that of Pt (1.77 Å).

Table 1. Lattice parameters (a) Å, elastic properties C_{ij} (GPa) and anisotropy *A* of the $Ti_{50}Pt_{50-x}V_x$ ternaries.

Structures	a	C_{11}	C_{12}	C_{44}	C'	*A*
TiPt [17]	3.19	145	210	45	-32	
$Ti_{50}Pt_{43.75}V_{6.25}$	3.156	231	166	49	32	1.52
$Ti_{50}Pt_{31.25}V_{18.75}$	3.123	304	120	48	92	0.52
$Ti_{50}Pt_{25}V_{25}$	3.121	255	131	78	62	1.26

The accurate calculation of elasticity is important in determining the mechanical stability and elastic properties of any system. There are three (C_{11}, C_{12}, C_{44}) independent elastic constants for cubic structures. The mechanical stability criteria of cubic system as outlined elsewhere [17] [25] is given as follows:

$$C_{44} > 0,\ C_{11} > |C_{12}| \text{ and } C_{11} + 2C_{12} > 0 \tag{1}$$

Wherein the anisotropy is given by

$$A = \frac{C_{44}}{C'} \tag{2}$$

In order for the structure to be considered stable, the stability criterion for the elastic constants should be satisfied. In principle, positive C' indicates a mechanically stable crystal, otherwise it is unstable. It is interesting to note that all the independent elastic constants C_{11}, C_{12} and C_{44} are positive in the entire range of V compositions. Previous work on pure structures of TiPt at 0 K showed that the B2 structure was unstable with C' shear modulus of -32 GPa, while the C_{44} non-basal shear modulus is positive [17]. This implies that the C' modulus went through zero, transitioning from positive to negative at the transformation temperature, whereas the C_{44} remained positive throughout the phase transition. It follows from the above that the anisotropy ratio (A) must have increased towards a very large number at the transformation temperature, an indication of uncoupling between the C' and C_{44} shear. Upon addition of 6.25 at.% of V, the C_{11} becomes greater than C_{12}, resulting in a positive C'. This suggests mechanical stability of the B2 $Ti_{50}Pt_{43.75}V_{6.25}$ phase in accordance with the stability criterion of the elastic constants [17]. Similar trend is observed for 18.75 and 25 at.% V additions on the TiPt with all the C_{ij}'s being positive.

The anisotropy (A) of β-type martensitic alloys is typically in the range of 3 or greater [26]. Such alloys tend to form martensite characterised by basal plane stacking order structure as is the case with B19 martensite. The anisotropy calculated in the current work for 6.25, 18.75 and 25 at.% V alloys is considerably smaller, suggesting strong coupling between C' and C_{44} shear moduli, and thus a tendency of the B2 phase to transform into B19′ martensite as proposed by Ren and Otsuka [8]. It has been established that the C_{44} modulus controls B19′ TT in compositions with low-value anisotropy. From table 1, it can be observed that there is an overall hardening of C_{44} modulus with increasing V content. This indicates reduction in transformation temperature with V addition.

3.2 Phonon dispersions

Phonon dispersion curves and phonon density of states (DOS) for the $Ti_{50}Pt_{50-x}V_x$ (x = 6.25, 18.75 and 25) were calculated and are shown in Figure 1. The dispersion relations exhibit two types of phonons namely the optical and acoustic modes corresponding to the upper and lower sets of curves in the diagram, respectively. Upon analysis of the phonon dispersion spectra, the structure is considered stable if all the phonon frequencies are positive. However, if some of the phonon frequencies are imaginary, the system is considered unstable [17]. The most negative mode is called the soft mode. Previously, the phonon dispersion curves of the binary B2 phase were reported where the phonon spectra were mostly characterized by the soft modes attributed to the negative value of shear modulus C′ [17]. The phonon dispersion curves of the $Ti_{50}Pt_{43.75}V_{6.25}$ are unstable, displaying soft modes along the high symmetry directions. These soft modes have the lowest frequency at approximately -2.5 THz mainly due to Pt contributions. Interestingly, at higher content of vanadium the structures becomes vibrationally stable with no soft modes observed on the phonon curves below the negative frequency. This findings suggest that V content greater than 18.75 vibrationally stabilises the B2 TiPt phase in agreement with the elastic constants wherein the values of A for 18.75 and 25 at.% are slightly smaller compared to the $Ti_{50}Pt_{43.75}V_{6.25}$ structure. Stabilization of the B2 phase is associated with reduction in martensitic transformation temperature as previously indicated by analysis of elastic constants.

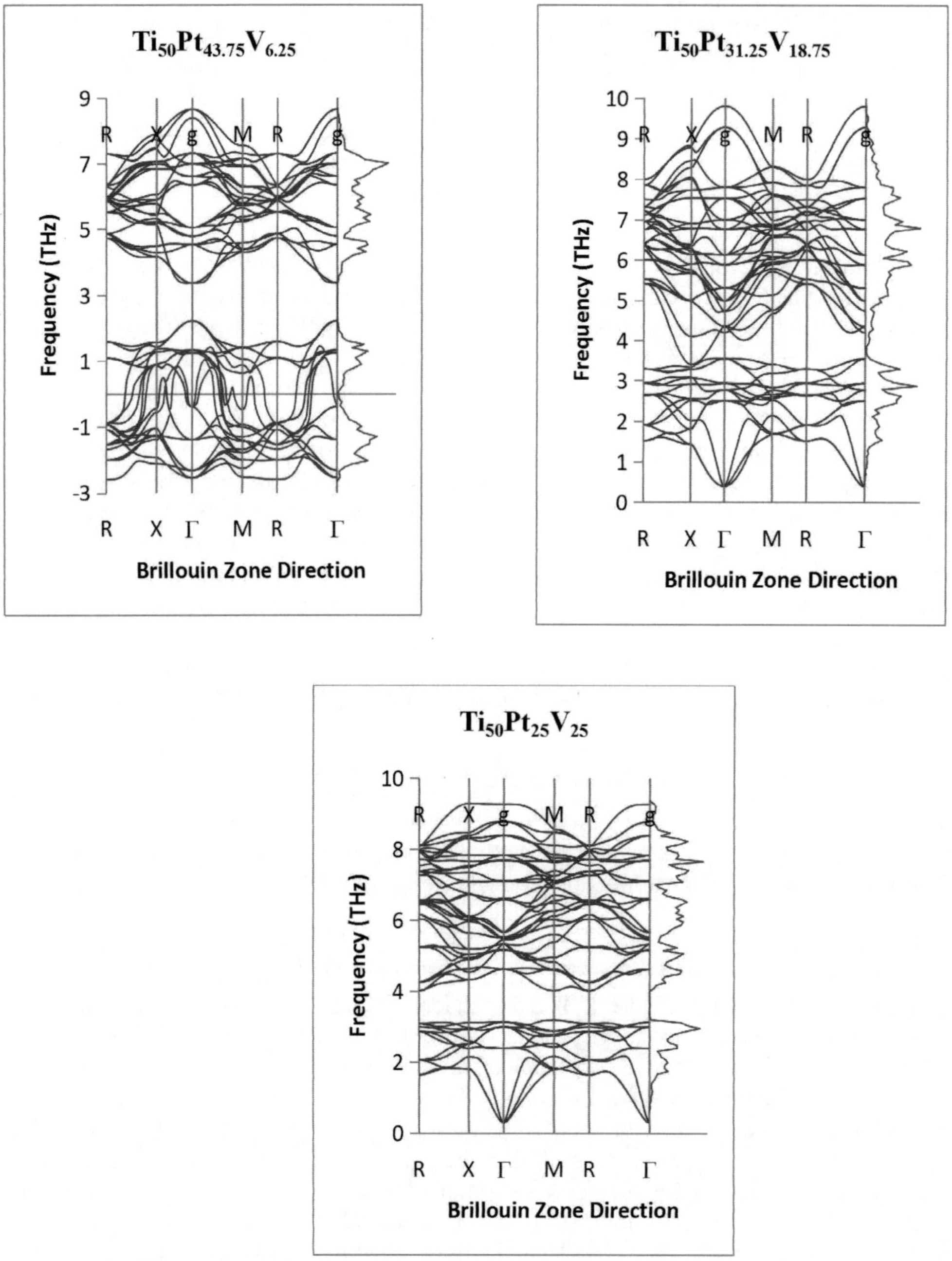

Figure 1: *Phonon dispersion curves for the $Ti_{50}Pt_{43.75}V_{6.25}$, $Ti_{50}Pt_{31.25}V_{18.75}$ and $Ti_{50}Pt_{25}V_{25}$.*

4. Conclusions

Computational modelling study on the elastic properties and phonon dispersion curves of the $Ti_{50}Pt_{50-x}V_x$ (6.25, 18.75, 25) were performed. The addition of V content in TiPt cubic phase was found to be increasing the C' moduli of the TiPt phase leading to all the C_{ij}'s being positive. The elastic constants were found to be stable with all the moduli obeying the elastic stability criterion. Interestingly the calculated anisotropy A of the structures is found to be less than 2 suggesting a B2 to B19′

transformation. The phonon dispersion of the 18.75 and 25 at.% of V were found to be vibrationally stable with no soft modes observed below the negative frequency. The investigations suggest a B2 to B19′ transformation with the TiPt martensitic transformation temperature being lowered.

Acknowledgements

The authors acknowledge the CSIR and NRF for their financial support. The support of the South African Research Chair Initiative of the Department of Science and Technology and the National Research Foundation is highly recognised. The calculations were carried out using computer resources at the Materials Modelling Centre (MMC) at the University of Limpopo and the Centre for High Performance Computing (CHPC) in Cape Town.

References

[1] Jani J M, Leary M, Subic A, Gibson M A 2014 *Mater Design* **56** 1078-1113
[2] Kubo H, Hamabe A and Shimizi K 1975 *Scr. Metall.* **9** 1083-1087
[3] Wu M H and Schetky L M, California: Pacific Grove, 2000 *SMST* 171-182
[4] Otsuka K and Kakeshita T 2002 *MRS Bull.* **27** 91-100
[5] Van Humbeeck J 1999 *Mater. Sci. Eng.* A **273–275** 134-148
[6] Duerig T, Pelton A and Stockel D 1999 *Mater. Sci. Eng.* A **273–275** 149-160
[7] K. Otsuka and X. Ren 2005 *Prog. Mater. Sci.* **50**, 511-678
[8] Ren X, Miura N, Taniwaki L, Otsuka K, Suzuki T, Tanaka K, Chumlyakov Y I and Asai M 1999 *Mater. Sci. Eng.* **A 273-275** 190-194
[9] Wagner M F X and W. Widl 2009 *Scripta Mater* **60** 207-210
[10] Tan C L, Tian X H, Ji G J, Gui T L and Cai W 2008 *Solid. state Comm.***147** 8-10
[11] Cai W, Tan C, Shen T and Tian X 2007 *J. Alloy Compd* **438** 30-33
[12] Hosoda H, Tsuji M, Mimura M, Takahashi Y, Wakashima K and Yamabe-Mitarai Y 2003 *MRS Bull.* **753** BB5-51-1–BB5-51-6
[13] Takahashi Y, Inamura T, Sakurai J, Hosoda H, Wakashima K and Miyazaki S 2004, *Trans. MRS-J.* **29** 3005–3008
[14] Fonda R, Jones H N and Vandermeer R A1998, *Scripta Materialia* **39** 1031-1037
[15] Yamabe-Mitarai Y, Hara T, Miura S, Hosoda H 2010, *Intermetallics* **18** 2275-2280
[16] Yamabe-Mitarai Y, Arockiakumar R, Wadood A, Suresha K S, Kitashima T, Hara T, Shimojo M, Tasaki W, Takahashi M, Takahashi S and Hosoda H 2015 *Materials Today: Proceedings* **2S** S517-S522
[17] Mahlangu R, Phasha M J, Chauke H R, Ngoepe P E 2013 *Intermetallics* **33** 27-32
[18] Hohenberg P and Kohn W 1964 *Phys. Rev.* B **136** 864-871
[19] Kohn W and Sham L J 1965 *Phys. Rev.* A **140** 1133-1138
[20] Kresse G and Hafner J 1993 *Phys. Rev.* B **47** 558-561
[21] Blöchl P E 1994 *Phys. Rev.* B **50** 17953-17979
[22] Perdew J P and Wang Y 1992 *Phys. Rev. B* **45** 13244-13249
[23] Monkhorst H J and Pack J D 1976 *Phys. Rev.* B **13** 5188-5192
[24] Parlinski K, Li Z Q and Kawazoe Y 1997 *Phys. Rev. Lett.* **78** 4063-4066
[25] Mayer B, Anton H, Bott E, Methfessel M, Sticht J, and Schmidt P C 2003 *Intermetallics,* **11** 23-32
[26] Wollmershauser J A, Kabra S, Agnew S R 2009, *Acta Mater* **57** 213-223

IOP Conf. Series: Materials Science and Engineering **430** (2018) 012022 doi:10.1088/1757-899X/430/1/012022

The effect of vanadium on structure and martensitic transformation temperature of TiPt alloy

S Chikosha[1,*], M L Mahlatji[1], R Modiba[1] and H K Chikwanda[1]

[1] Light Metals, Materials Science & Manufacturing, Council for Scientific and Industrial Research (CSIR), Meiring Naudé, South Africa

*SChikosha@csir.co.za

Abstract Shape memory alloys find application as sensors, microcontrollers or actuators. The TiPt binary system, a potential high temperature shape memory alloy due to its high transformation temperature of approximately 1000°C, displays negligible work output. This makes it unsuitable for actuator applications where work output is a prerequisite. A necessary condition for sufficient work output is for the alloy to display a critical stress for slip in austenite phase greater than detwinning stress of martensite phase. Solid solution strengthening, work hardening and precipitation hardening are possible mechanisms to improve work output. An indirect consequence of solid solution strengthening is altering of the crystal structure and the transformation temperature. This study evaluates the effect of partial substitution of Pt with V to form ternary $Ti_{50}Pt_{43.75}V_{6.25}$ (at.%) on the above-mentioned properties. Phase analysis and identification was done using scanning electron microscopy with energy-dispersive X-Ray spectroscopy and X-Ray diffraction. The transformation temperatures were studied by differential scanning calorimetry. Results show a three-phase microstructure with B19 $Ti_{50}(PtV)_{50}$ martensite as the major phase. The transformation temperature reduced with addition of vanadium in agreement with the predicted results from first-principles approach. However, the experimentally observed B19 structure did not agree with first-principles studies at 0K where B19' was the predicted stable phase.

1. Introduction

Shape memory alloys (SMA) have found use in applications such as sensors, actuators, microcontrollers, and as biomedical devices [1]. Several binary systems such as TiPt, TiIr, TaRu and NbRu [2-4] display exceptionally high transformation temperatures and are considered as candidates for high temperature shape memory applications. TiPt in particular, with a composition range of 45-56 at.% Pt [5], transforms between a high temperature B2 austenite and a low temperature B19 martensite phase. The transformation temperature exhibits a slight composition dependence and is maximum at the equiatomic composition [6]. Transformation temperatures for equiatomic TiPt measured with in-situ X-Ray diffraction (XRD) were reported as M_s = 1070, M_f = 1020, A_s = 1040, and A_f = 1085 °C [2]. Subsequent studies using DTA reported slightly lower transformation temperatures; M_s = 989, M_f = 963, A_s = 1000, A_f = 1057 °C from [3], and M_s = 1030, A_s = 1050 °C from [5]. The B2-B19 transformation enthalpies for the forward and reverse transformation were reported as -23.4 J/g and 28.7 J/g respectively from differential scanning calorimetry (DSC) analysis [7]. The martensitic transformation takes place thermoelastically and is thus reversible [8], giving TiPt alloys the related properties of shape memory effect (SME) and superelasticity (SE). The lattice parameters determined by XRD at room temperature are a_{B19} = 0.455, b_{B19} = 0.273, and c_{B19} = 0.479 nm; and from *in-situ* XRD at 1100 °C a_{B2} = 0.3192 nm [2]. Others reported similar values using *in-situ* XRD [9] and selected area electron diffraction (SAED) [7]. Optical microscopy on near equiatomic B19 TiPt showed a single phase fully martensitic microstructure with broad laths and a well-developed mid-rib [5]. Further analysis by transmission electron microscopy (TEM) revealed a heavily twinned martensite [3] as well as long-period stacking variants co-existing with the (2H) B19 stacking order [10,11].

The TiPt binary system, however, displays very negligible SME under load, only achieving 0-11 % shape recovery [3,9]. In actuator applications, SMAs are expected to perform work against an externally applied load. The TiPt binary system, displays negligible work output, which makes it unsuitable for actuator applications where work output is a prerequisite. This was attributed to its low critical stress for slip deformation in both martensite and austenite phases in comparison to the stress required for the martensitic transformation. A necessary condition for an alloy to be suitable actuator application is for it to display double yield phenomenon and have critical stress for slip in austenite phase greater than detwinning stress of martensite phase [3,12]. Double yield behavior enables SMAs to accommodate deformation strain elastically without suffering permanent slip and consequently optimizes shape recovery and SME. Solid solution strengthening, work hardening and precipitation hardening are possible mechanisms to improve the mechanical properties of TiPt. Various ternary TiPt-X solid solution strengthened ternaries have been reported on, some of which are discussed below [12,13]. In this work only stoichiometric ternaries with partial substitution for Pt were considered. $Ti_{50}(Pt,Ni)$ [12,13] with up to 30 at.% Ni substitution displayed improved work output, but at the cost of a greatly reduced transformation temperatures (A_f < 500 °C). Partial substitution of Pt with Zr, Ru, Co and Ir to form $Ti_{50}(Pt,X)$ (X = Zr, Co, Ru, Ir) ternaries maintained the high transformation temperatures (M_s > 900 °C), marginally improved martensite and austenite strength, and some of the compositions produced double yield phenomenon and SME in compression stress-strain tests conducted below M_f [3,14-16]. Substitutional elements differing the most in atomic radius from Pt were the most effective in improving mechanical properties [17]. Computational studies on TiPt-X ternaries [18] showed that V has no particular preference for Ti or Pt sites and would occupy sites on either sublattice with equal probability.

This study evaluates the effect of partial substitution of Pt with V to form ternary $Ti_{50}Pt_{43.75}V_{6.25}$ (at.%) on the transformation temperature and crystal structure.

2. Experimental Procedure

The materials used in this study were titanium hydride dehydride (TiHDH -45 μm) of 99.5 % purity supplied by Baoji, Lihua Non-Ferrous Metals Co. Ltd and platinum sponge (Pt -250 μm) of 99.99 % purity supplied by Anglo-Platinum. Three gram (3 g) powders were mixed in their respective compositions and pressed at 1055 MPa using a 13 mm tool steel die on Enerpac 100 tonne press. The $Ti_{50}Pt_{50}$ and $Ti_{50}Pt_{43.75}V_{6.25}$ (at.%) ingots were prepared by arc melting (in a water cooled Cu hearth) the 3 g compacts, samples were re-melted three times to ensure chemical homogeneity. Phase analysis on the as cast samples was conducted by SEM (Jeol JSM-6510) in backscattered mode with EDS. XRD (PANalytical X'Pert PRO) operated at a voltage and current of 45 kV and 40 mA respectively was used for structural analysis, using Ni-filtered Cu-K_α radiation. Lattice parameters were determined using Rietveld method using the X'Pert HighScore Plus software. Transformation temperatures and transformation enthalpies were measured with a Netzsch Jupiter 449 DSC. Measurement was conducted at a ramp rate of 20 °C/min under a flow of Ar gas in the temperature range of 100-1500 °C. Samples were subjected to two temperature cycles; only results from the second cycle are reported.

3. Results and Discussion

3.1. Phase identification

3.1.1 XRD

XRD analysis on as cast TiPt identified the structure as B19 martensite as shown in figure 1. The lattice parameters following Rietveld refinement were $a_{B19} = 0.459$, $b_{B19} = 0.278$ and $c_{B19} = 0.485$ nm ($R_{wp} = 5.748$ %, sigma = 1.184). Partial substitution of Pt with 6.25 at.% V did not significantly change the lattice parameters, estimated to be: $a_{B19} = 0.459$, $b_{B19} = 0.279$ and $c_{B19} = 0.482$ nm. The most notable change is shifting of the (101) reflection peak ($2\theta = 26.8$) to slightly higher angle, this corresponds to a reduction of the unit cell along the c-axis. Several unidentified peaks were observed on the $Ti_{50}Pt_{43.75}V_{6.25}$ scan, the most prominent appeared at 2θ angles of 21.6 and 29.5 degrees. Although the Ti_3Pt phase was readily observed in the vanadium sample during subsequent SEM analysis (see next sub-section), its cubic symmetry reflection peaks were not detected by XRD analysis, most likely due to the small volume fraction of the phase. First-principle calculations on the effect of partial substitution of Pt with 6.25 at.% V predicted B19' to be the stable ground state at 0 K [19]. It is not yet clear why experimental results on the as cast sample at a finite temperature show B19 to be the stable phase.

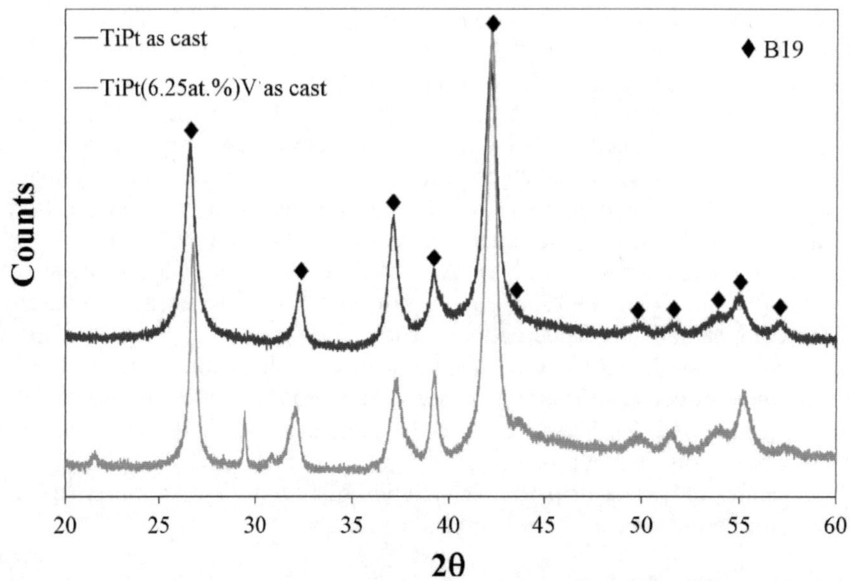

Figure 1. XRD of as cast TiPt and $Ti_{50}Pt_{43.75}V_{6.25}$ using Cu-K_{α} radiation. B19 martensite peaks are indicated.

3.1.2 Microscopy

SEM-EDS analysis on as cast TiPt showed a predominantly bright matrix dotted with grey and dark contrast regions. The microstructure was essentially single phase, with the bright, grey and dark contrasts corresponding to the slightly Pt-rich and Ti-rich variants of the TiPt phase respectively, figure 2(a-b) and table 1. This is expected since TiPt is not a line compound, but rather has a stability range of 45-56 at.% Pt [5]. The $Ti_{50}Pt_{43.75}V_{6.25}$ sample was less straight forward and showed three phases, see figure 2(c-d) and table 1. The bright matrix phase was identified as being isostructural with equiatomic TiPt phase with average compositions as shown in Table 1 and by points (1, 2) in figure 2d. It should be noted that points (1, 2) had a considerably reduced vanadium content in comparison with the intended composition of 6.25 at.%. The effect of a reduced vanadium content was not considered during first-principles computational studies [19], and along with temperature effects, could be the reason for stabilization of the B19 phase instead of the predicted B19' phase. Nevertheless, this is in agreement with the XRD results, where the major phase in the vanadium sample was identified as B19. Additionally, a dark contrast phase was observed and identified as being isostructural with Ti_3Pt, with composition approximated by points (3, 4) and average composition shown in table 1. Even though the bulk stoichiometric composition had vanadium partially substituting for Pt only, the as cast condition evidently has two phases where vanadium partially substitutes for titanium and platinum respectively. This is in line with the predictions of Bozzolo *et al.* [18] where it was determined that vanadium has a tendency to substitute for both platinum and titanium with equal probability.

A grey contrast phase with corresponding points (5, 6) was also detected. This phase was an intermediate between the $Ti_{50}(Pt_{47}V_3)$ and the $(Ti,V)_3Pt$ phases, "enclosing" all $(Ti,V)_3Pt$ regions. This composition, which can be approximated as $(Ti,V)_{57}Pt_{43}/(Ti,V)_4Pt_3$, as shown in Table 1 could not be readily identified by XRD. This is because the phase has been observed by other researchers but its crystal structure has not been resolved yet [20,21].

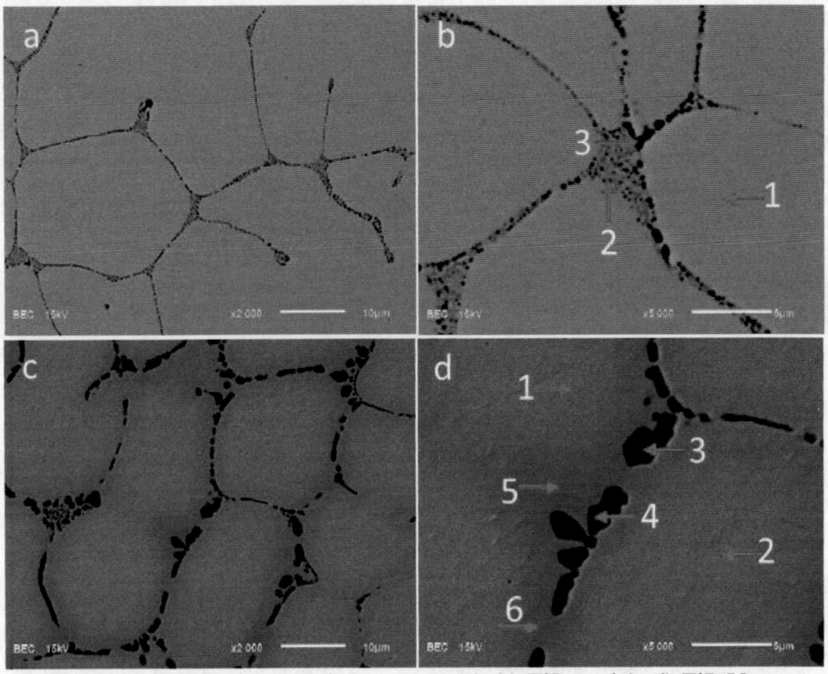

Figure 2. SEM micrographs on as cast (a-b) TiPt and (c-d) TiPtV.

Table 1: SEM-EDS analysis of as cast TiPt and $Ti_{50}Pt_{43.75}V_{6.25}$ samples.

	Ti-K	V-K	Pt-M	Phase Identification	
Figure 2b arrow labels/composition					
1	49	-	51	TiPt	bright
2	51	-	49	TiPt	grey
3	54	-	46	TiPt	dark
Figure 2d arrow labels/composition					
1	49	5	46	Ti(Pt,V)	bright
2	50	3	47	Ti(Pt,V)	bright
3	65	9	26	$(Ti,V)_3Pt$	dark
4	66	9	25	$(Ti,V)_3Pt$	dark
5	48	11	41	$(Ti,V)_4Pt_3$	grey
6	48	9	43	$(Ti,V)_4Pt_3$	grey

3.2 Transformation behaviour

Figure 3 shows DSC analysis results of as cast TiPt and $Ti_{50}Pt_{43.75}V_{6.25}$. While the transformation temperatures of the heating reactions are comparable for the two compositions, the cooling transformation of the ternary is suppressed by almost 100 °C compared to the binary. This reduction in transformation temperature is in general agreement with prediction of first-principles calculations where it was noted that 6.25 at.% V addition slightly increased the B2 C' shear modulus, indicating increased B2 stability [19]. Consequently, the transformation thermal hysteresis ($\Delta T = A_f - M_s$) is narrower for TiPt (ΔT = 101 °C) than for the ternary (ΔT = 163 °C). For actuator applications, materials with a narrow thermal hysteresis are preferred in order to achieve fast response times. The transformation enthalpy of ~20 J/g is slightly lower than previously reported for binary TiPt [7].

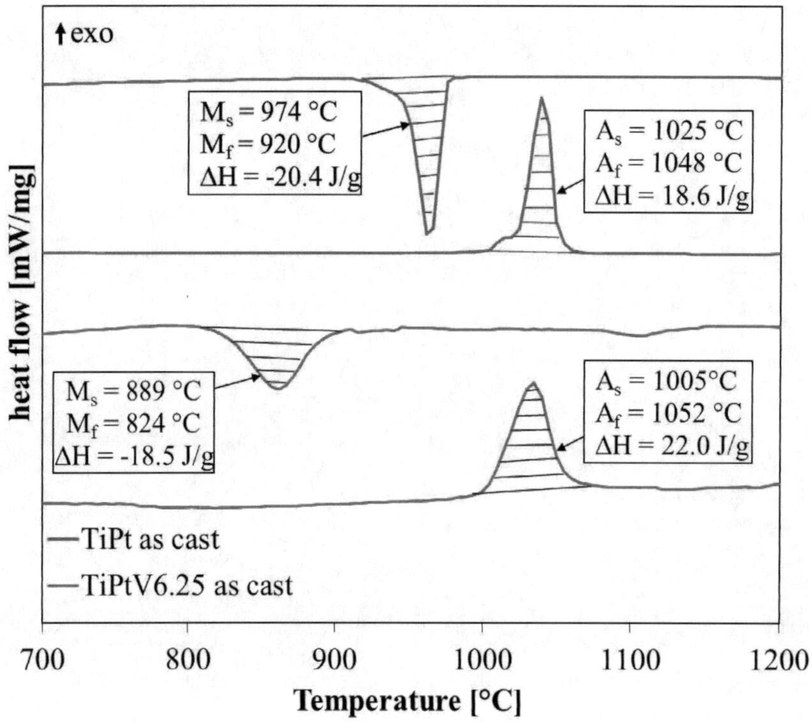

Figure 3. DSC analysis for TiPt and TiPtV from 2nd cycle at a ramp rate 20 °C/min in an Ar atmosphere.

4 Conclusions

Results show partial substitution of Pt with 6.25 at.% V reduced the martensitic transformation temperature of the alloy as predicted by first-principles studies. Vanadium has negligible effect on the martensite-to-austenite transformation temperatures, but considerably suppresses the austenite-martensite transformation temperature during cooling. Structural analysis indicated that the ternary structure remained orthorhombic B19, with vanadium addition only slightly reducing the c-axis lattice parameter while the *a* and *b* lattice parameters remained relatively unchanged. Microscopy revealed that the as cast ternary has a matrix with composition $Ti_{50}(PtV)_{50}$ and two precipitate phases, one that is isostructural with Ti_3Pt and a second Ti- and V-rich phase with an unknown structure. The experimentally observed B19 $Ti_{50}(PtV)_{50}$ structure did not agree with first-principles studies at 0 K where B19' was the predicted stable phase.

Acknowledgements

The Department of Science and Technology, CSIR and AFOSR are acknowledged for providing financial support for the work. Anglo American Platinum provided the Pt sponge used for sample preparation.

References

[1] J. Mohd Jani, M. Leary, A. Subic, and M. A. Gibson, A review of shape memory alloy research, applications and opportunities, *Mater. Des.,* vol. 56, pp. 1078–1113, 2014.

[2] H. C. Donkersloot and J. H. N. Van Vucht, Martensitic transformations in gold-titanium, palladium-titanium and platinum-titanium alloys near the equiatomic composition, J. Less-Common Met., vol. 20, no. 2, pp. 83–91, 1970.

[3] Y. Yamabe-Mitarai, T. Hara, S. Miura, and H. Hosoda, Mechanical properties of Ti-50(Pt,Ir) high-temperature shape memory alloys, *Mater. Trans.,* vol. 47, no. 3, pp. 650–657, 2006.

[4] R. W. Fonda, H. N. Jones, and R. A. Vandermeer, The shape memory effect in equiatomic TaRu and NbRu alloys, *Scr. Mater.,* vol. 39, no. 8, pp. 1031–1037, 1998.

[5] T. Biggs, M. B. Cortie, M. J. Witcomb, and L. A. Cornish, Martensitic transformations, microstructure, and mechanical workability of TiPt, *Metall. Mater. Trans. A Phys. Metall. Mater. Sci.,* vol. 32, no. 8, pp. 1881–1886, 2001.

[6] T. Biggs, L. A. Cornish, M. J. Witcomb, and M. B. Cortie, Revised phase diagram for the Pt-Ti system from 30 to 60 at.% platinum, *J. Alloys Compd.,* vol. 375, no. 1–2, pp. 120–127, 2004.

[7] G. M. Rotaru, W. Tirry, P. Sittner, J. Van Humbeeck, and D. Schryvers, Microstructural study of equiatomic PtTi martensite and the discovery of a new long-period structure, Acta Mater., vol. 55, no. 13, pp. 4447–4454, 2007.

[8] K. Kubo, H; Hamabe, A.; Shimizu, Shape memory effect in a Ag-Al alloy, *Yamadakami , Suita , Osaka 565 , Japan," Scr. Metall.,* vol. 9, pp. 1083–1087, 1975.

[9] Y. Yamabe-Mitarai, T. Hara, S. Miura, and H. Hosoda, Shape memory effect and pseudoelasticity of TiPt, *Intermetallics,* vol. 18, no. 12, pp. 2275–2280, 2010.

[10] M. Nishida, M. Matsuda, Y. Yasumoto, S. Yano, Y. Yamabe-Mitarai, and T. Hara, Crystallography and morphology of twins in equiatomic TiPt martensite, *Mater. Sci. Technol.,* vol. 24, no. 8, pp. 884–889, 2008.

[11] G. M. Rotaru and D. Schryvers, New (33) long-period microtwin variant in the martensitic phase of the PtTi alloy, *Mater. Sci. Eng. A,* vol. 481–482, no. 1–2 C, pp. 437–441, 2008.

[12] S. Padula, G. Bigelow, R. Noebe, D. Gaydosh, and A. Garg, Challenges and progress in the development of high-temperature shape memory alloys based on nitix compositions for high-force actuator applications, *SMST-2006 - Proc. Int. Conf. Shape Mem. Superelastic Technol.,* pp. 787–802, 2008.

[13] O. Benafan, D. J. Gaydosh, R. D. Noebe, S. Qiu, and R. Vaidyanathan, In Situ Neutron Diffraction Study of NiTi–21Pt High-Temperature Shape Memory Alloys, *Shape Mem. Superelasticity,* vol. 2, no. 4, pp. 337–346, 2016.

[14] Y. Yamabe-Mitarai, T. Hara, S. Miura, and H. Hosoda, Phase transformation and shape memory effect of Ti(Pt, Ir), *Metall. Mater. Trans. A Phys. Metall. Mater. Sci.,* vol. 43, no. 8, pp. 2901–2911, 2012.

[15] A. Wadood, M. Takahashi, S. Takahashi, H. Hosoda, and Y. Yamabe-Mitarai, High-temperature mechanical and shape memory properties of TiPt-Zr and TiPt-Ru alloys, *Mater. Sci. Eng. A,* vol. 564, pp. 34–41, 2013.

[16] A. Wadood, H. Hosoda, and Y. Yamabe-Mitarai, Phase transformation, oxidation and shape memory properties of Ti-50Au-10Zr alloy for high temperature applications, *J. Alloys Compd.,* vol. 595, pp. 200–205, 2014.

[17] Y. Yamabe-Mitarai, R. Arockiakumar, A. Wadood, K. S. Suresh, T. Kitashima, T. Hara, M. Shimojo, W. Tasaki, M. Takahashi, S. Takahashi, and H. Hosoda, Ti(Pt, Pd, Au) based high temperature shape memory alloys, *Mater. Today Proc.,* vol. 2, pp. S517–S522, 2015.

[18] G. Bozzolo, H. O. Mosca, and M. F. del Grosso, Energy of formation, lattice parameter and

bulk modulus of (Ni,X)Ti alloys with X = Fe, Pd, Pt, Au, Al, Cu, Zr, Hf, Intermetallics, vol. 16, no. 5, pp. 668–675, 2008.

[19] R. Modiba, E. Baloyi, S. Chikosha, H.R. Chauke, P.E. Ngoepe, Computational Modelling of $Ti_{50}Pt_{50-x}V_x$ Potential Shape Memory Alloys, *paper submitted, CoSAAMI 2018 conference proceeding, IOP Materials Science and Engineering 2018.*

IOP Conf. Series: Materials Science and Engineering **430** (2018) 012023 doi:10.1088/1757-899X/430/1/012023

Phase analyses of the Co-Fe-Pd ternary alloys

K Dyal Ukabhai[1, 2,*], I A Mwamba[2, 3] and L A Cornish[1, 2]

[1] School of Chemical and Metallurgical Engineering, University of the Witwatersrand, Johannesburg, South Africa
[2] DST-NRF Centre of Excellence in Strong Materials, hosted by University of the Witwatersrand, Johannesburg, South Africa
[3] Advanced Materials Division, Mintek, Randburg, South Africa

*Kiyaasha.dyalukabhai@students.wits.ac.za

Abstract. Samples of different compositions of Co-Fe-Pd were made in a button arc furnace, under an argon atmosphere on a water-cooled copper hearth. They were sectioned and a half was kept in the as-cast state. The samples were analysed in a scanning electron microscope with energy dispersive X-ray spectroscopy, and in an X-ray diffractometer to derive the microstructures and to determine the overall and phase compositions. Coring occurred in many samples, which was expected due to the wide composition range of the main solid solution in the system. The microstructures were not similar to any of the work of sixty years ago when the Co-Fe-Pd phase diagram was studied. Some of the phases were too fine to be analysed with EDS in the as-cast samples.

1. Introduction

Magnetic materials are essential in many industries, and there is a need to replace expensive rare earth metals with cheaper and more readily available metals. Historically, patents held by American and Japanese companies have been a barrier to entry into the magnet market. However, China's establishment of research institutes in the 1950s and 1960s, and the expiry of the American and Japanese patents in the 1980s has enabled China to enter and dominate the magnet market. Currently, China has a monopoly on the magnet value chain, producing 90% of the world's permanent magnets, and China also has most of the world's rare earth resources.

There is potential to use Co-Fe-Pd rare earth-doped alloys for magnets, because FeCo is a soft magnetic material with high saturation magnetism and a high Curie temperature, due to the cobalt [1]. Palladium nano-particles have localized paramagnetism, with the potential to become magnetic, and so Pd can improve the magneto-crystalline anisotropy of FeCo. Thus, the Co-Fe-Pd system offers possible alternatives to the heavy and expensive rare-earth magnets, which are in current use.

The Fe-Co-Pd alloy has potential applications in biotechnology, electronics and green technologies. Magnetic thin films and nanostructures, which are magnetized perpendicular to their surfaces, are essential to many developing technologies, including spintronics devices [2] and patterned media [3, 4]. This maintained the thermal stability even as the device dimensions became increasingly nanoscale. Multilayers or superlattice structures consisting of alternating ferromagnetic and non-magnetic layers are highly suitable for these applications, due to their tunable perpendicular magnetic anisotropy (PMA) and saturation magnetization [5, 6].

By combining the soft magnetic properties of iron and the hard magnetic properties of cobalt and palladium, an optimal magnetic material could be developed. Fe-Co alloys have a B2 structure which makes them extremely brittle at room temperature [1]. The workability of these alloys can be improved by adding nickel or palladium, and palladium especially leads to a higher tensile strength and elongation at room temperature [1].

Apart from the mechanical properties [1] of the Fe-Co-Pd alloys and research on the magnetic shape memory properties [7, 8, 9], little work has been done on the ternary phase diagram [10-13]. The published phase diagram work is from four papers by a group in Russia [10-13] and these papers are contradictory, even sometimes within the same paper. Ragavhan [14] attempted to assess the system, and was unable to satisfactorily interpret it due to the contradictory data, and admitted that his interpretation was very speculative. His isothermal section at 25 °C [14] is given in figure 1, and is much simpler than that of Grigorev and Kuprina [13].

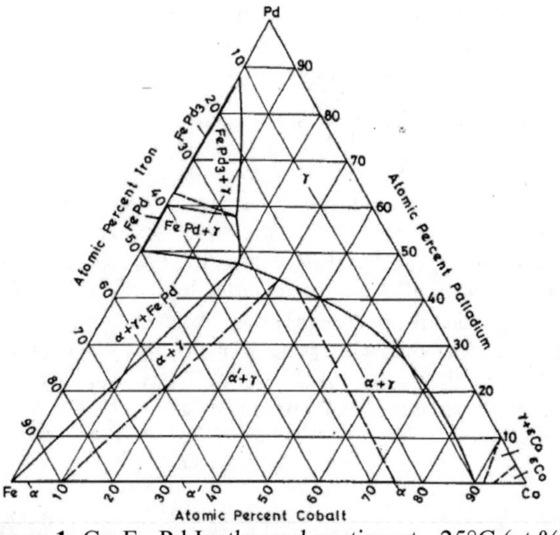

Figure 1. Co-Fe-Pd Isothermal section at ~25°C (at.%) [14].

Table 1. Nominal composition of samples.

Alloy	Fe (at.%)	Co (at.%)	Pd (at.%)
1	60	30	10
2	15	10	75
3	30	30	40

Since Co-Fe-Pd alloys are being studied to improve their magnetism [15], it is necessary to know the extent of the promising Co-Fe phases into the ternary system. There are problems with the phase diagram [14], and so this study was to identify the extents of those phases.

2. Experimental procedure

The composition of the samples shown in table 1 were determined by choosing three-phase regions from the 25°C isothermal section of Raghavan [14], as well as compositions of interest for magnetic properties [15]. The alloys were prepared from 99.99% pure elemental powders of Fe, Co and Pd. The powders were weighed, mixed and cold compacted (to hold the components together prior to melting) and then melted in a button arc-furnace (BAF), according to standard preparation method PMD-SOP-O37. An argon atmosphere and a water-cooled copper hearth were used, together with a titanium oxygen-getter, which was melted before making each sample. The samples were cut using a Struers automatic cutting machine with an Accutom cutting wheel, mounted in a mounting press, ground and polished. Grinding and polishing were done using a Saphire 520 automatic machine. One half of each sample was studied in the as-cast condition. Once the samples were polished, they were removed from the polyfast resin and X-ray diffraction (XRD) was conducted using a Bruker D2 Phaser and the Diffrac Plus EVA program with the 2018 database to analyse the data.

Due to the wide range of compositions of the samples (from 10-75 at.% Pd), different etching techniques were tried, and the best results for each sample was used. The first etch tried was using a steel wire cathode and a solution of 90ml water and 10ml hydrochloric acid [16]. The second etch was immersion in a solution of distilled water (50 ml), hydrochloric acid (SG 1.19, 100 ml) and nitric acid (SG 1.4, 10ml) [16]. Once the samples were etched, they were analyzed on a Sigma Zeiss scanning electron microscope (SEM) using backscattered (BSE) and secondary electron (SE) modes. Once the microstructures were visible, EDX analyses were done to determine the composition of the different phases and the overall compositions of each sample.

3. Results

For Alloy 1 (Fe_{60}:Co_{30}:Pd_{10} (at.%)), the low Pd content allowed a less aggressive etchant to be used: electric etching with a steel wire cathode in a solution of water and hydrochloric acid. The dendrites in figure 2a were due to coring from the fast cooling from arc-melting, given the lack of a definite interface between the dendritic and interdendritic regions, and they also had similar compositions, table 1. There was also a finer, darker phase visible, which was too small (<2 μm) to analyse accurately [17], without collecting X-rays from the surrounding phase. The XRD results (figure 3) showed that the major phase was γ (fcc), and the minor phase was α (bcc).

Alloy 2 (Fe15:Co10:Pd75 (at.%)) needed the most aggressive etchant method, 2, because of its high Pd content. The microstructure in figure 2b shows dendrites, and two-phase interdendritic regions, where the second phase was darker. Once again, the smallest phases were too fine to analyse accurately [17], resulting in similar compositions of the dendritic and interdendritic areas, table 2. The XRD pattern (figure 4) indicated FePd3 and FePd.

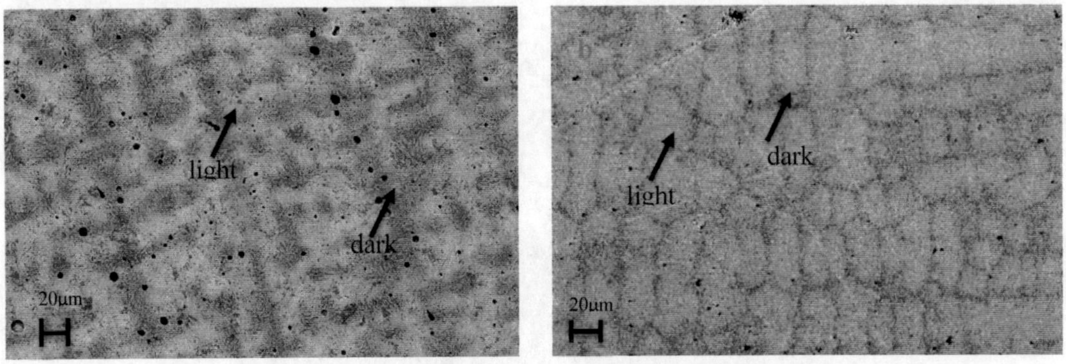

Figure 2. SEM-BSE images of as-cast a) Alloy1 Fe_{60}:Co_{30}:Pd_{10} (at.%) showing dendrites (darker) and b) Alloy 2 Fe_{15}:Co_{10}:Pd_{75} (at.%) showing gamma dendrites (lighter) and two phase interdendritic area (dark).

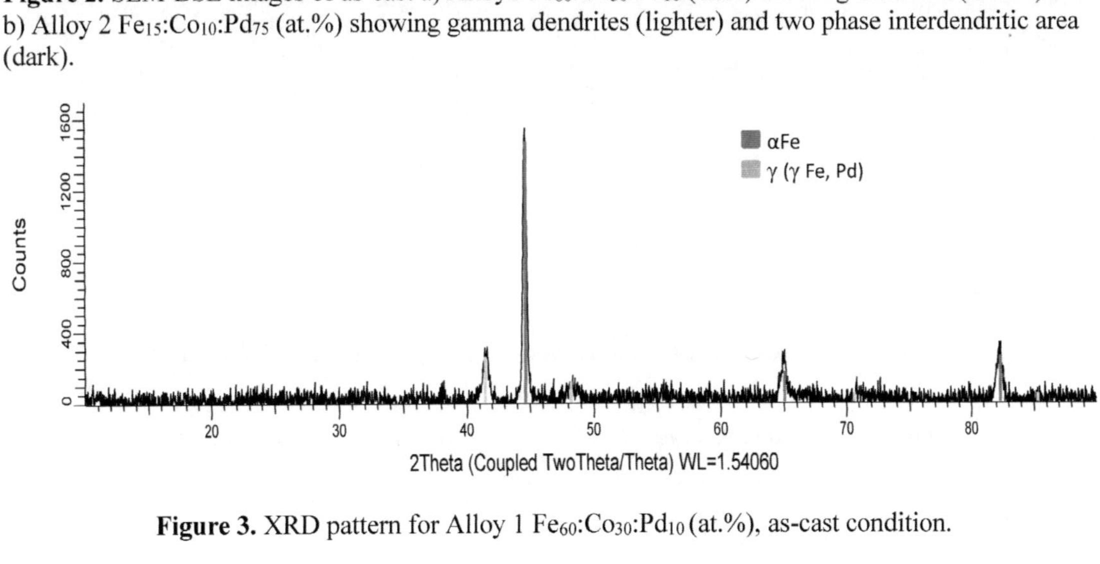

Figure 3. XRD pattern for Alloy 1 Fe_{60}:Co_{30}:Pd_{10} (at.%), as-cast condition.

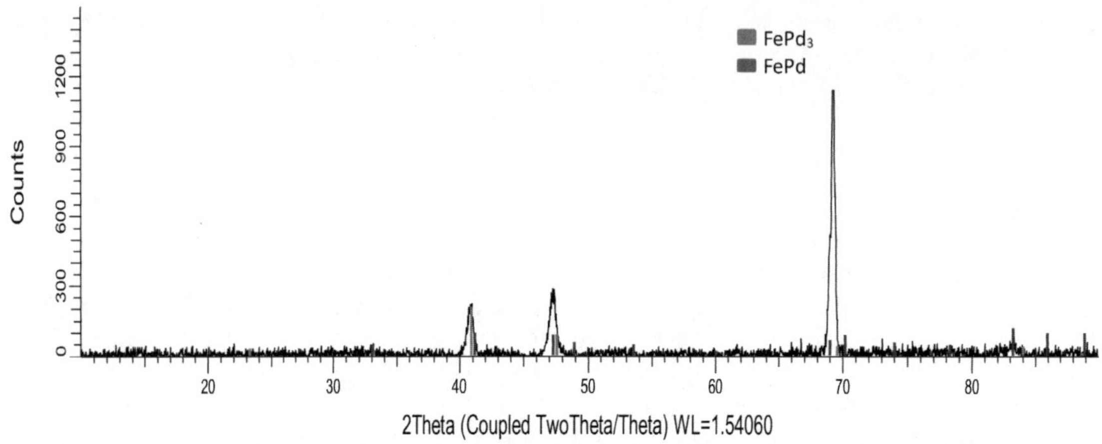

Figure 4. XRD pattern for nominal Alloy 2 Fe_{15}:Co_{10}:Pd_{75} (at.%), as-cast condition.

Figure 5 of Alloy 3 (Fe30:Co30:Pd40 (at.%)) shows a single phase, with pores filled with debris from grinding. The XRD pattern in figure 6 shows one major phase which matched two different compositions of γ, with wide peaks, indicating a cored phase.

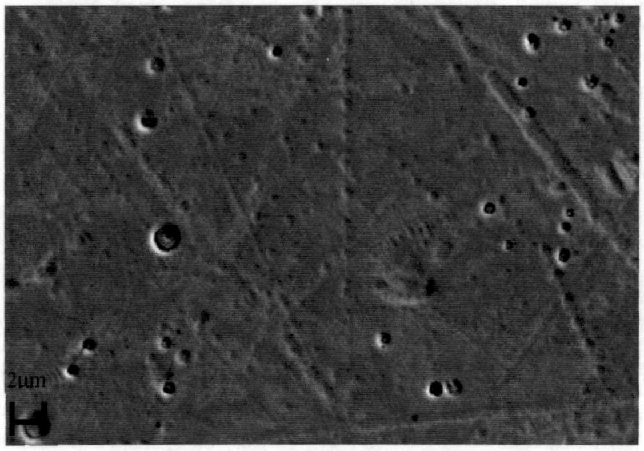

Figure 5. SEM-BSE image of as-cast Alloy 3 Fe_{30}:Co_{30}:Pd_{40} (at.%) showing a single phase with pores.

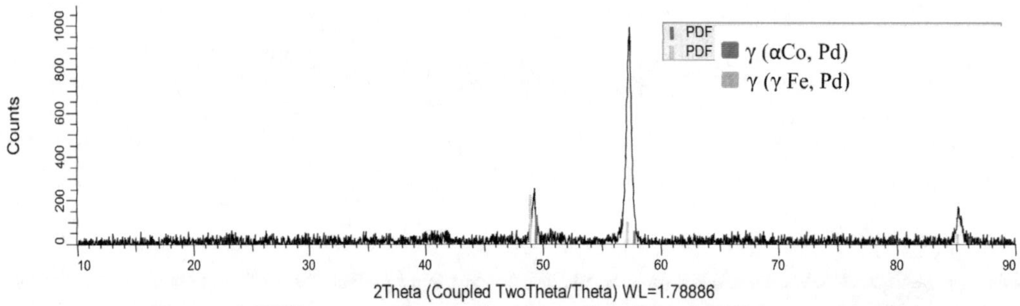

Figure 6. XRD pattern for nominal Fe_{30}:Co_{30}:Pd_{40} (at.%), as-cast condition.

Table 2. SEM-EDS analyses of the sample overalls and the phases of different contrasts.

Alloy	Overall	Dark: γ	Light: γ	Light: FePd	Dark: $FePd_3$
	Fe 59.6	59.1±0.1	54.8±1.2		
1	Co 29.1	28.5±0.3	25.1±1.3	--	--
	Pd 11.3	12.4±0.3	20.1±2.5		
	Fe 13.6			13.4±0.4	14.5±0.1
2	Co 9.0	--	--	8.2±0.4	11.1±1.3
	Pd 77.4			78.4±0.5	74.4±1.2
	Fe 28.2	28.7±0.1	29.2±0.1		
3	Co 28.3	28.3±0.1	27.7±0.1	--	--
	Pd 43.5	43.0±0.1	43.1±0.1		

4. Discussion

Due to the samples containing Pd, a high acceleration voltage (20kV) had to be used on the SEM in order to excite the Pd X-rays for EDS. Using a high kV causes the electron beam to spread out further [17], making the area for accurate EDS analysis larger (typically ≥2 microns), hence the samples with fine microstructures could not be accurately analysed in the SEM.

Given the very wide γ liquidus of the system and the few solid phases [10-14], it is not surprising that the alloys were cored. None of the microstructures found to date was similar to those of the previous workers [10-14], although the decomposition of Alloy 1 would be expected from the ~25°C isothermal section of Raghavan [14] because the alloy composition lies in a two-phase region. However, the phases found here were γ and α, whereas Raghavan [14] showed that the phases should be γ and α′ (Figure 1). Thus, Alloy 1 suggests a wider γ + α phase field than Raghavan [14]. Alloy 2 shows that the γ phase was more narrow than Raghavan [14] indicated (Figure 1). Conversely, Alloy 3, is a single-phase solid solution of γ, showed that this phase had a greater extent at this alloy composition than in Raghavan's isothermal section [14], Figure 1.

There was a better agreement of the as-cast samples with the 25°C isothermal section of Raghavan [14] than with that of Kuprina and Grigorev [11]. Strictly, the as-cast samples of this work should not have been compared to the ~25°C isothermal sections of Kuprina and Grigorev [11] and Rhagavan [14], but there was no other information available for comparison.

The coring in Alloy 1 was used to assess the validity of the two liquidus surfaces [11, 14] since the dendrites would form first, and the interdendritic area would form next, and the direction from the former to the latter should follow the slope of the liquidus surface. The deduced liquidus slope agreed with both of the liquidus surfaces [11, 14], so it was not possible to choose a better one. The XRD patterns showed very few peaks, making identification of the phases difficult.

5. Conclusions

Three alloys of different compositions were manufactured by arc melting and studied in the as-cast condition. Alloy 1 comprised γ and α, suggesting a wider γ + α phase field than Raghavan [14]. Alloy 2 showed that the γ phase was more narrow than Raghavan [14], whereas Alloy 3, a single-phase solid solution of γ, showed a greater extent than Raghavan's isothermal section [14]. The results were in better agreement with Raghavan's [14] isothermal section at ~25°C The compositions of the cored microstructures of Alloy 1 agreed with both published liquidus surfaces.

References

[1] Matsuda M, Sago R, Akamine K, Tsurekawa S, Takashima K and Nishida M 2016 Enhancement of Ductility in Fe-Co Based Alloys by Substitution of Pd (in English) *J. Alloys*

Compd. **682** 124-131
[2] Ravelosona M S, Katine D, Carey J A, Terris M J and Fullerton E E 2006 Current-induced magnetization reversal in nano-pillars with perpendicular anisotropy *Nat. Mater.* **5** 210–215
[3] Todorovic M, Schultz S, Wong J and Scherer A 1999 Writing and reading of single magnetic domain per bit perpendicular patterned media *Appl. Phys. Lett.* **74** 2516–2518
[4] Ding J and Adeyeye A O 2013 Binary ferromagnetic nano-structures: fabrication, static and dynamic properties *Adv. Funct. Mater.* **23** 1684–1691
[5] Terris B D and Thomson T 2005 Nano fabricated and self-assembled magnetic structures as data storage media *J.Phys.* D **38** 199–222
[6] Rippard W H, Deac A M, Pufall M R, Shaw J M, Keller M W, Russek S E, Bauerand G E and Serpico C 2010 Spin-transfer dynamics in spin valves with out-of-plane magnetized CoNi free layers *Phys. Rev.* B **81**
[7] Koeda J, Nakamura Y, Fukuda T, Kakeshita T, Takeuchi T and Kishio K 2001 *Trans. Mater. Res. Soc.* (Japan) **26** 215-217 as quoted by [18]
[8] Oshima R 1981 *Scripta Metall.* **15**, 829-833 as quoted by [18]
[9] Kauffmann-Weiss S, Hamann S, Gruner M E, Buschbeck J, Ludwig A, Schultz L and Fähler S 2012 Understanding the Magnetic Shape Memory System Fe-Pd-X by Thin Film Experiments and First Principle Calculations *Adv. Eng. Mater.* **14** (8) 724-749
[10] Grigorey A T and Kuprina V V 1961 Order-Disorder Transformatiom in Alloys of Iron with Cobalt and Palladium (in Russian) *Zh. Neorg. Khim.* **6** (8) 1891-1898, English translation: *J. Inorg. Chem.* **6**(8) 966-970
[11] Kuprina V V and Grigorev A T 1958 Investigation of Alloys of Fe-Co-Pd System. I. Fusibility Diagram of the Fe-Co-Pd System (in Russian) *Zh. Neorg. Khim.* **3** (12) 2736-2739
[12] Kuprina, V V and Grigorev AT 1959 Solid-State Transformations in Alloys of Iron with Cobalt and Palladium (in Russian) *Zh. Neorg. Khim.* **4** (7) 1606-1610 English translation: *J. Inorg. Chem.* **4** (7) 724-727
[13] Kuprina V V and Grigorev A T 1961 Polymorphic Transformation in Alloys of Iron with Cobalt and Palladium (in Russian) (Izv. Vyss. Uchebn. Zaved. Khim., Khim. Tekhnol) **4** (1) 7-10
[14] Raghavan V 1992 The Co-Fe-Pd (Cobalt-Iron-Palladium) System (in English) Phase Diagrams of Ternary Iron Alloys (Indian Institut of Metals, Calcutta) **6** A, 623-627 .
[15] Manyti H 2017 Personal communication
[16] Wing G 1996 A lexicon of metallographic etchants, MSc Thesis (University of the Witwatersrand South Africa)
[17] Goldstein G I, Newberry D E, Echlin P, Joy D C, Fiori C and Lifshin E 1981 *Scanning Electron Microscopy and X-Ray Microanalysis* (New York: Plenum Press) 56-59
[18] Tsuchiya K, Nojiri T, Ohtsuka H and Umemoto M 2003 Effect of Co and Ni on Martensitic Transformation and Magnetic Properties in Fe-Pd Ferromagnetic Shape Memory Alloys *Mater. Trans.* **44** (12) 2499-2502

Finite-element analysis of the effect of sheath-gas composition in an inductively-coupled plasma

N J M Grobler[1,*], H Bissett[2], G J Puts[1], P L Crouse[1]

[1] Department of Chemical Engineering, University of Pretoria, Private Bag X20, Hatfield 0028, South Africa
[2] Division of Applied Chemistry, Nuclear Energy Corporation of South Africa, R104 Pelindaba, Madibeng 0240, South Africa

*njmgrobler@gmail.com

Abstract. The sheath gas plays an important role in inductively-coupled plasmas (ICPs) in preventing thermal damage to the side wall of the torch. The sheath gas (hydrogen in our case) is more difficult to ionise than argon (the main plasma gas) due to its lower electrical conductivity at the working temperature, preventing plasma formation and concomitant high temperatures in the immediate vicinity of the torch inside wall. The sheath gas also has a higher flow rate, reducing the time for heat transfer from the plasma to the inside wall. Numerous models to simulate an ICP reactor have been published in the open literature, none of which take into account the effect of characteristics of the sheath gas on the performance of an ICP on the heat transfer from the plasma to the sidewall of the reactor. The H_2:Ar gas ratio can have severely detrimental effects on the efficiency of the plasma because of the higher ionisation potential of H_2. Excess hydrogen in the sheath gas may also constitute material wastefulness. Both of these factors have an implication on the economics of the plasma process. This research aims at finding the optimum sheath-gas flow for the Necsa torch, used for spheroidisation. The work was done on the commercial finite-element programme COMSOL Multiphysics.

1. Introduction

Inductively coupled plasma (ICP) torches find application in a number of industrial settings, including the preparation of metal powders for additive manufacturing [1]. Spheroidised metal particles exhibit a uniform particle morphology and this uniformity imparts superior processing characteristics to the metal powders. The Nuclear Energy Corporation of South Africa (Necsa) has embarked on a program of spheroidisation technology development, with the ultimate aim of creating a commercial plasma-based metals processing facility. The developmental research is being conducted on a Tekna TEKSPHERO-15 [2] small-scale spheroidisation system using argon as operating gas. This system (and all other ICP plasma systems) suffers from energy efficiency issues. The plasma torch loses large amounts of energy to the water cooling in the torch sidewall. Successful commercialization of the technology requires mitigation of this energy loss.

The energy lost through the sidewall can be reduced by moving the plasma heated zone away from the sidewall. This may be effected by isolating the sidewall with a sheath gas that is more difficult to ionize than the operating gas. Hydrogen is more difficult to ionize than argon, therefore introducing hydrogen into the sheath gas moves the heated zone away from the torch sidewalls. Excess hydrogen can, however, be detrimental to the plasma, extinguish the plasma, and constitutes material wastefulness.

Unfortunately, the extremely high temperatures reached inside the plasma makes it difficult to perform simple measurements (such as temperature and flow rate) inside the torch. Computational models offer a facile method for investigating the physical and chemical properties of an ICP and present useful tools for diagnosing behavioural problems in industrial systems, without having to resort to experimental measurements. Obviously, models have to be verified experimentally; they are nevertheless essential for interpreting any measurement made.

There are several existing computational models in the literature. Boulos [3] was the first to develop a computational model for an ICP. Several elementary models followed, focusing on investigating the temperature and flow fields [4-6]. Contemporary models use finite element/volume method software packages and these software suites come with built-in facilities for performing temperature and flow field calculations. Examples of such packages include COMSOL [7] Multiphysics and ANSYS Fluent [8]. For example, Ikhlef *et al.* [9] developed a FEM COMSOL model for an ICP at atmospheric pressure with argon gas using a resistive magnetohydrodynamic model. There are several modern models using FEM or FVM to study different cases or uses of the ICP [10-13]. To the best of our knowledge, none of these takes into account the effect of sheath gas composition.

The aim of this research is the determination of the optimum sheath gas composition such that the energy loss through the sidewall is kept to a minimum. The modelling and simulation was done in COMSOL Multiphysics, using the built-in computational modules. No additional code was written and no experimental measurements are reported here.

2. Computational model

2.1 *Constitutive equations*

Boulos *et al.* [14] developed a mathematical model for the heat of powders in an ICP under dense loading conditions. This model is employed by the COMSOL package. The model is described in the subsequent paragraphs.

The laminar compressible flow in the system is described by the continuity equation and the momentum equation, shown in equation 1 and equation 2, respectively.

$$\nabla \cdot (\rho \boldsymbol{u}) = 0 \tag{1}$$

$$\rho(\boldsymbol{u} \cdot \nabla)\boldsymbol{u} = \nabla \cdot \left[-p\mathbf{I} + \mu\left(\nabla \boldsymbol{u} + (\nabla \boldsymbol{u})^T\right) - \frac{2}{3}\mu(\nabla \cdot \boldsymbol{u})\mathbf{I} \right] + \boldsymbol{F} \tag{2}$$

Here ρ and $\boldsymbol{u}$ are the density and the velocity field, respectively. $\boldsymbol{F}$ is the body force vector, which accounts for all the forces in the volume, inclusive of gravity, electric- and magnetic fields. T, μ, and p are the temperature, viscosity, and pressure, respectively.

The total heat transfer is described by equation 3 and the conductive heat flux described by equation 4.

$$\rho C_p \frac{\partial T}{\partial t} + \rho C_p \boldsymbol{u} \cdot \nabla T + \nabla \cdot \boldsymbol{q} = Q + Q_p + Q_{vd} \tag{3}$$

$$\boldsymbol{q} = - k \nabla T \tag{4}$$

Here Q, Q_p, Q_{vd}, $\boldsymbol{q}$, C_p, and k are the plasma heat source, the pressure work, the viscous dissipation, the conductive heat flux, the specific heat of the gas, and the thermal conductivity of the gas [7], respectively. The Maxwell equations are given in equation 5 to equation 8.

$$\nabla \times \boldsymbol{H} = \boldsymbol{J} \tag{5}$$

$$\boldsymbol{B} = \nabla \times \boldsymbol{A} \tag{6}$$

$$\boldsymbol{J} = \sigma \boldsymbol{E} + j\omega \boldsymbol{D} + \sigma \boldsymbol{v} \times \boldsymbol{B} + \boldsymbol{J}_e \tag{7}$$

$$\boldsymbol{E} = - j\omega \boldsymbol{A} \tag{8}$$

IOP Conf. Series: Materials Science and Engineering **430** (2018) 012024 doi:10.1088/1757-899X/430/1/012024

Here ***H***, ***J***, ***B***, ***A***, ***E***, ω, ***D***, ***J****e*, σ, and ***v*** stand for the magnetic field intensity, current density, magnetic flux density, magnetic vector potential, electric field intensity, frequency, electric displacement, externally generated current density, electrical conductivity, and velocity of the geometry relative to the reference system, respectively.

The plasma heat source *Q*, used in equation 3, is defined by equation 9.

$$Q = \frac{\partial}{\partial T}\left(\frac{5k_BT}{2q}\right)(\nabla T \cdot \boldsymbol{J}) + (\boldsymbol{E} \cdot \boldsymbol{J}) + Q_{rad} \tag{9}$$

Here k_B, q, and Q_{rad} are the Boltzmann constant, the unit charge, and the total volumetric emission loss, respectively. The first term on the right-hand side represents the enthalpy transport. The second term accounts for Joule heating.

2.2 *Computational Modelling*

The mathematical model, presented in equations 1 to 9, was used to simulate the Tekna torch system using COMSOL Multiphysics. The model was implemented *via* three physics nodes *viz.* fluid dynamics, heat transfer and the electromagnetic field. The magnetic field node solves the Maxwell equations for the electromagnetic fields produced due to an alternating current in the coil. The COMSOL plasma package was used to interface the magnetic field node and the heat transfer node, providing the plasma heat source. Laminar flow, optically thin plasma, and local thermodynamic equilibrium was assumed.

The model geometry is presented in figure 2. This geometry is a practical simplification of the technical schematic. Since the torch exhibits cylindrical symmetry, the geometry was simplified to permit 2D axisymmetric simulation, significantly cutting down on required computational resources and decreasing the simulation time.

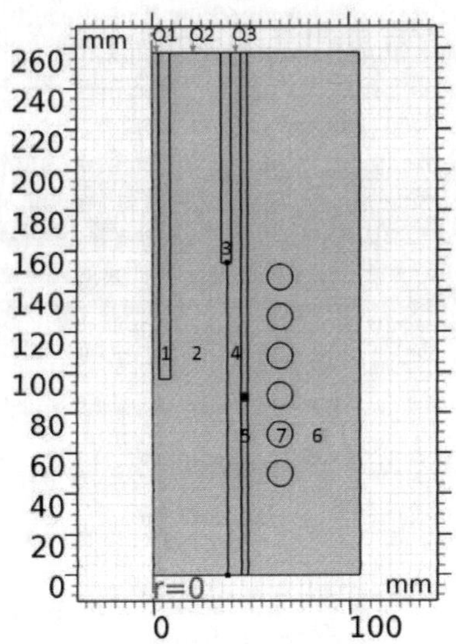

Figure 1. Simulated geometry indicating different domains and inlets.

Domain 1 (in figure 2) is the wall of the carrier gas probe and Q1 is the carrier gas inlet. Domain 2 is the central plasma gas, domain 3 the quartz tube separating the plasma gas and the sheath gas. Domain 4 is the sheath gas and domain 5 the torch wall. The fluid dynamic equations were only solved over domain 2 and 4 with a no-slip boundary condition on the torch sidewall (the boundary between Domain 4 and 5). The boundary between Domain 2 and 4 is open with no physical effects on the flow and heat transfer.

COMSOL does not properly account for the mixing of gases in a plasma. Modelling a system with hydrogen in the sheath gas and pure argon in the main plasma gas requires one to trick the software with a *faux* partition to produce two separate regions with dissimilar composition, while still

IOP Conf. Series: Materials Science and Engineering **430** (2018) 012024 doi:10.1088/1757-899X/430/1/012024

permitting the unified treatment of heat transfer and flow. Domain 4 is, therefore, a homogenous mixture of hydrogen and argon and the properties of this mixture programmed manually.

The heat transfer was determined in Domains 2, 4, and 5. The heat transfer in the rest of the geometry was regarded as of no concern since the heat is lost through the cooling water which is included in the model through a fixed temperature boundary condition of 300 K on the outside wall. The initial temperature was set to 300 K for the entire geometry and the initial velocity field was set to 0 m/s.

The system was studied in a frequency-transient domain for 30 s of operating time. 5 simulations were run at varying amounts of H_2 in the sheath gas (0 vol%, 5 vol%, 7.5 vol%, 10 vol% and 50 vol%). Table 1 shows the operating conditions and initial values kept constant for all the simulations.

Table 1. Operating conditions and initial values.

Property	Value	Unit
Power	11	kW
Frequency	3	MHz
Inlet 1 (Q1)	2	slpm
Inlet 2 (Q2)	10	slpm
Inlet 3 (Q3)	40+*x*	slpm
Initial temperature	300	K
Initial flow field	0	m/s
Initial magnetic vector potential	0	Wb/m
Pressure	85	kPaa

3. Results and discussion

The main results are provided graphically in figure 4 to figure 8. Figure 3 graphs the electrical conductivity of hydrogen and argon as a function of temperature, as used by COMSOL. In the temperature range of interest, *viz.*, 5 000 K to ~ 20 000K, the electrical conductivity of argon is higher than hydrogen and will, therefore, ionize more easily and at lower temperatures than hydrogen. This is due to the fact that in the case of hydrogen energy goes into both bond breaking and ionization. Figure 4 shows the temperature at a fixed point in the centre of the sidewall. This point was set at the height where the maximum temperature was achieved in each simulation. Figure 5 shows the temperature 1 mm away from the torch sidewall. Figure 6 shows the total normal heat flux across the boundary on the outer sidewall. Figure 7 shows the temperature profile of the geometry at 0 vol%, 10 vol% and 50 vol% H_2. Figure 8 shows the radial temperature profile at the maximum temperature for each simulation case.

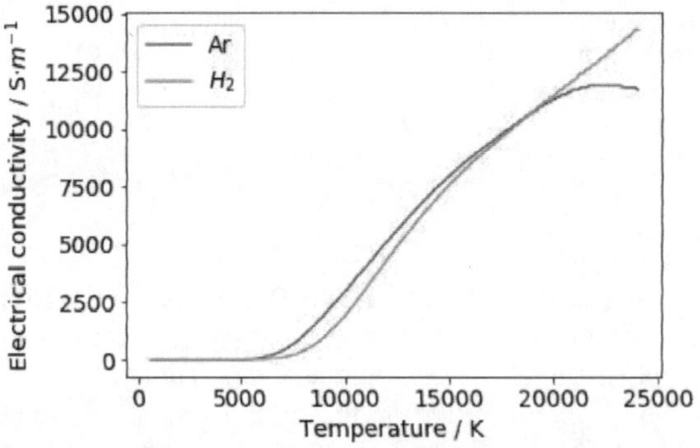

Figure 2. Electrical conductivity of argon and hydrogen gas as a function of temperature [14].

The time-dependent temperature (figure 4) calculated for the sidewall appears to reach a steady-state after about 20 s. Therefore all simulations were run for a total of 30 s. The maximum Reynolds number over the whole geometry was found to be 18.97, indicating the assumption of laminar flow to be correct.

It should be noted that the 0.0 vol% H_2 and 7.5 vol% H_2 lines overlap. The presence of hydrogen moves the heated zone to the left and reduces plasma formation near the torch wall (as shown in figure 6 and figure 7); however, it also increases heat transfer to the torch sidewall. This is expected as hydrogen exhibits a greater thermal conductivity than argon. The overlap implies that the mitigation of heat loss to the reactor walls essentially becomes a trade-off between the effects of improved heat transfer to the wall, the increasing distance of the hot zone from the wall due to reduced plasma formation and the energy loss to H-H bond dissociation.

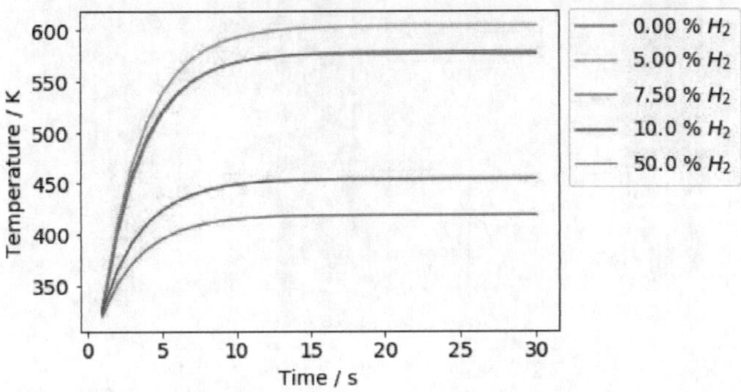

Figure 3. Temperature in the centre of the torch sidewall at the height of the maximum

Figure 5 shows the temperature 1 mm away from the torch sidewall over the full torch length for different concentrations H_2 in the sheath gas at the final time step. Above 5 vol% H_2 in the sheath gas, the maximum temperature is below 5000 K. This is the temperature at which argon begins to ionize. The implication is that no plasma formation occurs at the torch sidewall. Figure 6 shows the total normal heat flux across the boundary on the outer sidewall for the different concentrations of H_2 gas. Table 2 shows the total power lost through the torch side wall for each case.

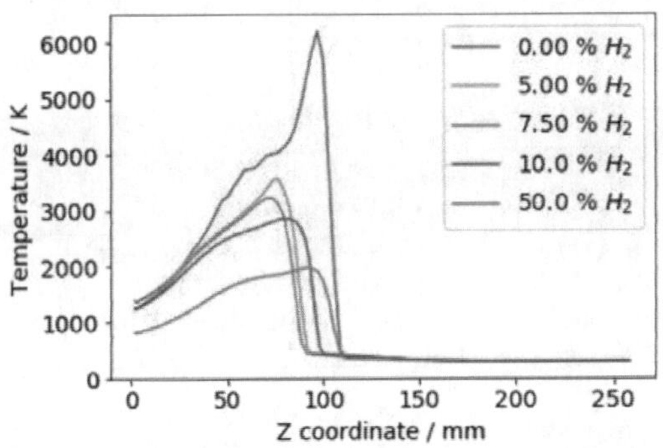

Figure 4. Temperature 1 mm from the torch sidewall for different concentrations of H_2 in the sheath gas.

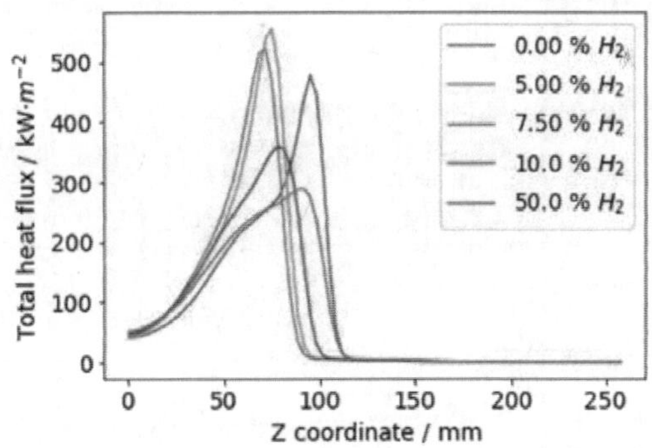

Figure 5. Total heat flux across the boundary at the outer sidewall of the torch at different concentrations of H_2 gas.

Table 2. Total power lost through torch outside wall

Percentage H_2 in sheath gas	Total heat lost
0	6.37 kW
5	6.08 kW
7.5	5.91 kW
10	5.64 kW
50	5.73 kW

IOP Conf. Series: Materials Science and Engineering **430** (2018) 012024 doi:10.1088/1757-899X/430/1/012024

Figure 7 shows the temperature profile of the geometry at 0 vol%, 10 vol% and 50 vol% H_2 to indicate how the heated zone has moved to the left with the increased H_2 gas in the sheath.

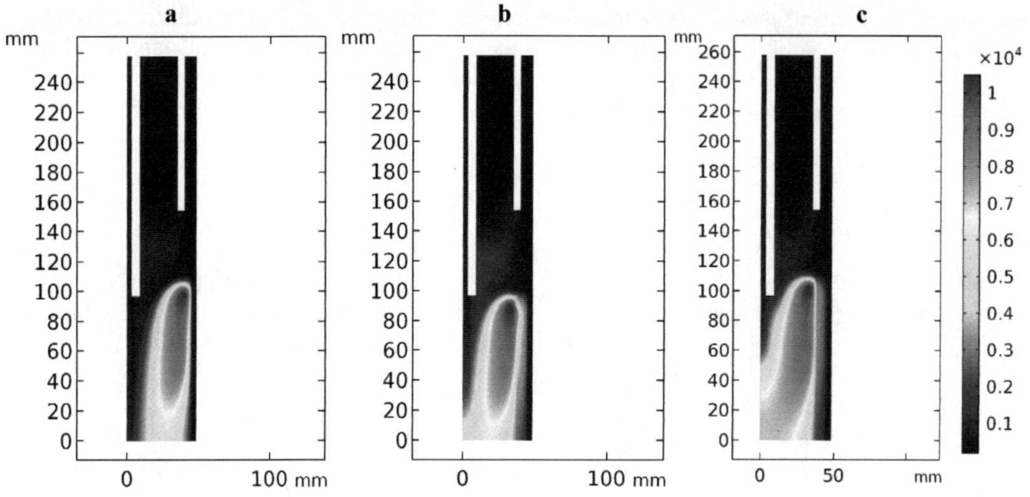

Figure 6. Temperature profile with: a) 0 vol%, b) 10 vol% and c) 50 vol% H2 in the sheath gas.

The absence of data in Figure 8 in the 0 vol% and 50 vol% H_2 line is due to the heated zone being higher in the geometry so the radial profile passes through the carrier gas probe where the heat transfer was not solved.

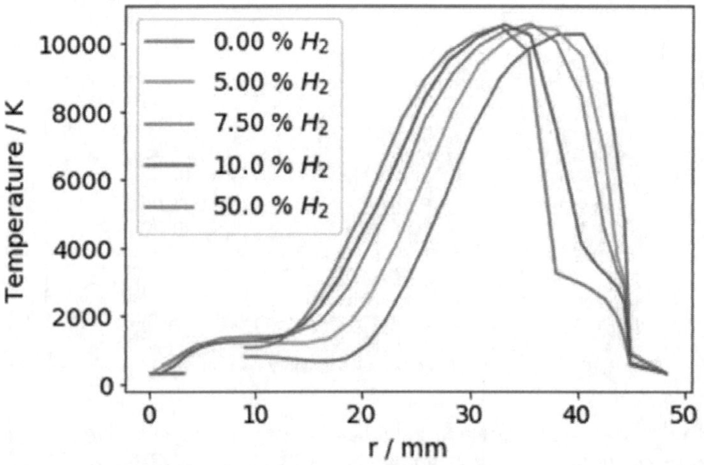

Figure 7. Radial temperature profile at the maximum temperature.

Hydrogen in the sheath gas clearly reduces the temperature at the torch sidewall and results in less heat lost through the sidewall. The temperature on the inside wall reaches a maximum of 914 K with 7.5 vol% H2 in the sheath gas. This is a low enough temperature to prevent thermal damage to the torch and still have a minimal effect on the plasma and energy costs. Less than 7.5 vol% H2 causes an increased temperature in the wall due to the increased heat transfer capability of hydrogen. The total energy loss appears to reach a minimum at 10 vol% H2. The sheath gas composition for minimum energy loss seems to be between 7.5 vol% and 10 vol% H2.

4. Conclusion

The computational model indicates that isolating the side wall of an ICP torch from the plasma with a hydrogen/argon sheath gas reduces the energy lost through the wall. Hydrogen increases the sheath gas heat transfer capability but shifts the plasma away from the side wall. These two competing mechanisms result in a minimum energy loss when the sheath gas hydrogen content ranges between 7.5 and 10 vol%. If the hydrogen content is below 7.5 vol%, more energy is lost to the side wall than with a pure argon sheath gas. The more computational effort is required to determine the exact composition of the sheath gas needed to achieve minimum energy loss.

References

[1] Murr LE, Martinez E, Amato KN, Gaytan SM, Hernandez J, Ramirez DA, Shindo PW, Medina F, and Wicker RB 2012 Fabrication of Metal and Alloy Components by Additive Manufacturing: Examples of 3D Materials Science *J. Mater. Res. Technol.* **1** 42

[2] Tekna 2016 Tekna Spheroidization system Teksohero-15 system http://www.tekna.com/landing-page/teksphero-15 [16 July 2018]

[3] Boulos MI 1978 Heating of Powders in The Fire Ball of an Induction Plasma *IEEE Trans. Plasma Sci.* **6** 93

[4] Proulx P, Mostaghimi J and Boulos MI 1983 *Int. J. Heat Mass Transfer.* **28** 1327

[5] Mostaghimi J, Proulx P and Boulos MI 1985 Computer modeling of the emission patterns for a spectrochemical ICP *Numer. Heat Tran* **8** 187

[6] Yang P, Barnes RM, Mostaghimi J and Boulos MI 1989 Application of a two-dimensional model in the simulation of an analytical inductively coupled plasma discharge *Spectrochim. Acta, Part B* **44** 657

[7] COMSOL 2018 COMSOL Multiphysics 5.3a https://www.comsol.com/ [16 July 2018]

[8] Ansys 2018 Ansys Fluent https://www.ansys.com/products/fluids/ansys-fluent [16 July 2018]

[9] Ikhlef N, Hacibl T, Leroy O, and Mékiddèche MR 2012 Nonlinear compressible magnetohydrodynamic flows modeling of a process ICP torch *Eur. Phys. J. Appl. Phys.* **58** 10804

[10] Aghaei M and Bogaerts A 2016 Particle transport through an inductively coupled plasma torch: elemental droplet evaporation *J. Anal. At. Spectrom.* **31** 631

[11] Bolot R, Coddet C, Schreuders C, Leparaoux M and Siegmann S 2007 Modeling of an Inductively Coupled Plasma for the Synthesis of Nanoparticles *J. Therm Spray Technol.* **16** 690

[12] Bogaerts A and Aghaei M 2017 Inductively coupled plasma-mass spectrometry: insights through computer modeling *J. Anal. At. Spectrom.*, **32**, 203

[13] Colombo V, Ghedini E and Mostaghimi J 2008 Three-Dimensional Modeling of an Inductively Coupled Plasma Torch for Spectroscopic Analysis *IEEE Trans. Plasma Sci.* **36** 1040

[14] Boulos MI, Fauchais P and Pfender E 1994 Plasma tables *Thermal Plasmas: Fundamentals and Applications* (New York: Plenum Press) Appendix A.3 pp 388-447

IOP Conf. Series: Materials Science and Engineering **430** (2018) 012025 doi:10.1088/1757-899X/430/1/012025

Numerical simulation of plastic deformation and mechanical response of strip rolled aluminium alloys

G J Jansen van Rensburg[1,2,*] and A E J Bogaers[1,2]

[1] Modelling and Digital Science, Council for Scientific and Industrial Research, Pretoria, South Africa
[2] Computer Science and Applied Mathematics, University of the Witwatersrand, Johannesburg, South Africa

*jjvrensburg@csir.co.za

Abstract. During the optimal design and simulation of machining or forming processes, detailed simulation of structural response is typically required for use in Finite Element Analysis (FEA). In this study, the bulk temperature and rate dependent resistance to deformation of strip formed aluminium alloys is modelled using the Mechanical Threshold Stress model. The model is characterised to AA5182 alloy data and used in the FEA simulation of a strip rolling process. The effects of element choice, dynamic or quasi-static simulations and the use of particular modelling algorithms on the computational cost and overall accuracy associated with each simulation have to be considered. Given the correctly implemented material tangent, an implicit analysis is illustrated to allow larger stable time-steps. Dynamic and quasi-static solutions are very similar for the simulated process meaning inertia effects are negligible. It is further demonstrated that great care be given to the maximum allowable time step size in order to capture the expected interaction between temporal and spatial discretisation in a fully Lagrangian FEA simulation.

1. Introduction

During the optimal design and simulation of machining or forming processes, a better understanding and detailed simulation of alloy deformation mechanisms or structural response are typically required for use in detailed Finite Element Analysis (FEA). If the effect on energy input and reaction forces or moments are required given a specific processing rate and temperature, a continuum model that models the correct bulk temperature and strain rate dependent mechanical response is needed [1,2].

The purpose of this study is to inspect the development of steady state deformation so that the resulting forces and moments involved in the continuous roll may be assessed as an analogue for required energy input. The accurate modelling of bulk temperature and rate dependent resistance to deformation of strip formed aluminium alloys is investigated using the dislocation density based Mechanical Threshold Stress (MTS) material model [3]. The model is first implemented into the Abaqus FEA package [4] and then characterised to experimental temperature and rate dependent stress-strain response data of an AA5182 aluminium alloy. Given a model that captures the correct bulk mechanical response, considerations for use in a Lagrangian FEA simulation of strip formed bulk material is then investigated. The effects of dynamic or quasi-static simulations and the use of various modelling or contact algorithms and friction considerations have a further effect on the computational cost and overall accuracy associated with each simulation. These are important aspects to consider if the intent is a study on the effects of, or optimising for, specific processing parameters.

2. The Mechanical Threshold Stress Model

The Mechanical Threshold Stress (MTS) model is a continuum plasticity model that captures the dominant bulk temperature (T) and strain rate ($\dot{\alpha}$) dependent post-yielding behaviour of metals. Dislocation density (ρ) is introduced as a microstructural variable of state through its relationship to the stress-like variable ($2\widehat{\sigma} = M\mu_r b\sqrt{\rho}$) using the average Taylor factor measure of polycristalinity (M), a reference elastic shear modulus (μ_r) and length of the Burger's vector (b) [3]. In the MTS model, a material has a theoretical maximum flow stress (the mechanical threshold) $\widehat{\sigma}$ at 0 K.

Any yield stress value σ_y is a scaled version of the mechanical threshold to accommodate for rate and temperature dependence. The effective yield stress may be calculated as a function of various threshold contributions. The standard MTS formulation calculates a yield stress value using three contributions [3]:

$$\sigma_y = \widehat{\sigma}_a + S_i(\dot{\alpha},T)\widehat{\sigma}_i + S_\varepsilon(\dot{\alpha},T)\widehat{\sigma_\varepsilon} \tag{1}$$

where $\widehat{\sigma}_a$ is an athermal, constant contribution to the yield stress with $\widehat{\sigma}_i$ a constant scalable initial threshold value and $\widehat{\sigma}_\varepsilon$ an evolving scalable contribution to the yield stress. The scaling function $S_i(\dot{\alpha},T)$ takes the form [5,6]

$$S_i(\dot{\alpha},T) = \frac{\mu(T)}{\mu_r}\left[1 - \left(\frac{T}{\gamma_{0i}\,\mu(T)}\ln\frac{\dot{\alpha}_{0i}}{\dot{\alpha}}\right)^{1/q_i}\right]^{1/p_i} \tag{2}$$

with $\dot{\alpha}_{0i}$ a reference maximum strain rate while p_i and q_i are statistical constants ($0 \leq p_i \leq 1$, and $1 \leq q_i \leq 2$) that characterise the shape of obstacles in the path of mobile dislocations. γ_{0i} is a convenient grouping of physical lattice parameters and normalised activation energy into a single material parameter value [5,6]. $\mu(T)$ is a temperature dependent shear modulus with μ_r the maximum reference value.

The parameters are chosen so that the scaling function is between 0 and 1 for any expected value of strain rate ($\dot{\alpha}$) and temperature (T). The other scaling function $S_i(\dot{\alpha},T)$ is similarly calculated using $\gamma_{0\varepsilon}$, $\dot{\alpha}_{0\varepsilon}$, q_ε and p_ε a values instead so that the evolving mechanical threshold component is allowed a different scale compared to that of the initial value. The temperature dependent shear modulus is modelled using [7]

$$\mu(T) = \mu_r - \frac{D_r}{\exp\left({}^{T_r}/_{T}\right) - 1} \tag{3}$$

in which T_r and D_r are further reference parameter constants.

The evolution of $\widehat{\sigma}_\varepsilon$ is given in rate form using the modified Voce hardening law [8]

$$\theta = \frac{d}{d\alpha}\widehat{\sigma}_\varepsilon = \theta_0\left(1 - \frac{\widehat{\sigma}_\varepsilon}{\widehat{\sigma}_{\varepsilon S}}\right)^{\kappa} \tag{4}$$

where θ_0 is the constant hardening due to dislocation accumulation while $\widehat{\sigma}_{\varepsilon S}$ is the saturation stress. A power law exponent κ chosen as unity exhibits saturating behaviour while $\kappa > 1$ mimics a non-saturating final hardening rate.

The saturation threshold stress $\widehat{\sigma}_{\varepsilon S}$ is a function of both strain rate and temperature, through [3]

$$\widehat{\sigma}_{\varepsilon S}(\dot{\alpha},T) = \widehat{\sigma}_{0\varepsilon S}\exp\left(\frac{T}{\gamma_{0\varepsilon S}\,\mu(T)}\ln\frac{\dot{\alpha}}{\dot{\alpha}_{0\varepsilon S}}\right) \tag{5}$$

where $\widehat{\sigma}_{0\varepsilon S}$, $\gamma_{0\varepsilon S}$ and $\dot{\alpha}_{0\varepsilon S}$ are further empirical material constants.

The MTS material model was built around physically observed deformation mechanisms in fcc metals. A desirable benefit to using this physically motivated material model is that it can therefore reasonably interpolate and extrapolate material response for temperatures and strain rates slightly outside of the experimental data used to calibrate it.

2.1. Material parameter values for AA5182

A point integration implementation of the MTS model was coded and characterised to temperature and rate dependent stress-strain data of an AA5182 alloy. The response data and certain MTS parameter values were extracted from the work by Chen et al. [8].

Of the 18 total MTS material parameter values, $\dot{\alpha}_{0i}$, p_i, q_i, $\dot{\alpha}_{0\varepsilon}$, p_ε, q_ε, $\dot{\alpha}_{0\varepsilon S}$, κ, μ_r, T_r and D_r were fixed while the remaining seven parameters are fine-tuned so that the model best replicates the experimentally observed response. Numerical optimisation is used to fine tune the material parameter values as follows:

Given a candidate set of parameter values $\boldsymbol{x} = \{\hat{\sigma}_a, \hat{\sigma}_i, \gamma_{0i}, \gamma_{0\varepsilon}, \gamma_{0\varepsilon S}, \hat{\sigma}_{0\varepsilon S}, \theta_0\}$, an objective function is constructed by first evaluating the model predicted values $\sigma^{(k)}_{\text{model}}(\boldsymbol{x})$ corresponding to each of the (k) experimental data points $\sigma^{(k)}_{\text{experiment}}$. The smallest objective function value

$$f_{\text{obj}}(\boldsymbol{x}) = \sum_{(k)} \left| \frac{\sigma^{(k)}_{\text{model}}(\boldsymbol{x})}{\sigma^{(k)}_{\text{experiment}}} - 1 \right| \tag{6}$$

is then sought using the downhill simplex algorithm [9]. The optimal material parameter values are reported in table 1 with a visual comparison of experimental and model predicted response at selected temperatures and rate of deformation is presented in Figure 1.

Table 1. Mechanical Threshold Stress parameter values used to model the temperature and rate dependent response of AA5182.

Equation	Parameter	Value	Unit	Equation	Parameter	Value	Unit
(1) : σ_y	$\hat{\sigma}_a$	4.86	MPa	(4) : θ	θ_0	4993.8	-
	$\hat{\sigma}_i$	354.6	MPa		κ	2	-
(2) : S_i	γ_{0i}	0.757	K/MPa	(2) : S_ε	$\gamma_{0\varepsilon}$	3.083	K/MPa
	$\dot{\alpha}_{0i}$	10^7	s^{-1}		$\dot{\alpha}_{0\varepsilon}$	10^7	s^{-1}
	q_i	1	-		q_ε	1	-
	p_i	1	-		p_ε	1	-
(3) : μ	μ_r	28815	MPa	(5) : $\hat{\sigma}_{\varepsilon S}$	$\hat{\sigma}_{0\varepsilon S}$	730.6	MPa
	D_r	3440	MPa		$\gamma_{0\varepsilon S}$	0.243	K/MPa
	T_r	215	K		$\dot{\alpha}_{0\varepsilon S}$	10^7	s^{-1}

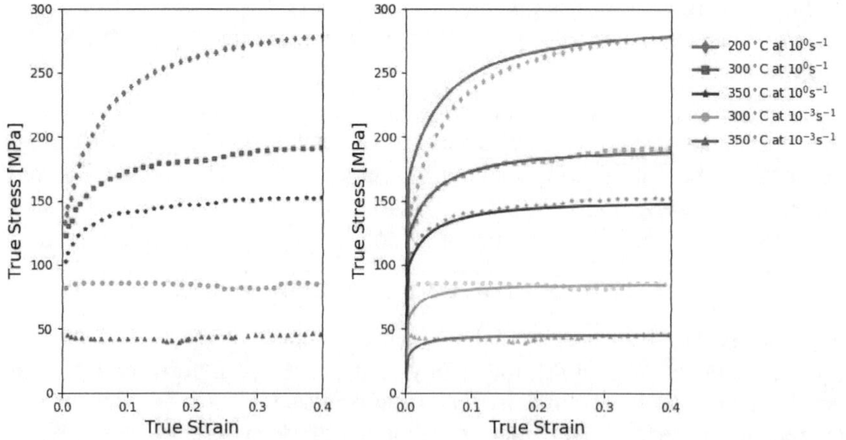

Figure 1. AA5182 experimental stress-strain data (markers / left) and MTS response (solid lines / right) using parameter values in table 1.

3. Simulating strip form reduction

The characterised MTS model is now used to investigate different considerations in simulating a roll pass reduction of AA5182. A 300mm thick slab at 300°C is rolled to a true (logarithmic) reduction of 10%. The rollers are 1.2m in diameter with a peripheral velocity of 1m/s ($\omega = 1.6667\text{s}^{-1}$). Given symmetry in the roll process, only half of the slab and a single roller is modelled.

Only 1m length of the slab is modelled so that there is enough material to move past any initial edge effects. The half slab is initially pushed into the roller at 0.3m/s. Once sufficient contact is made with the roller at a constant $\omega = 1.6667\text{s}^{-1}$, the half slab is pulled through as a result of friction between the two surfaces assuming a friction coefficient of 0.3. The simulations are done in Abaqus Standard and Explicit [10] using user coded Fortran subroutines for the MTS model.

The different simulation considerations investigated included the use of reduced integration quadratic plane strain elements (CPE8R) and reduced integration linear plane strain elements (CPE4R) with enhanced hourglass stiffness treatment. The reduced integration elements are considered to relax the pressure variational and prevent volumetric locking as a result of an overly stiff element during plastic deformation. The other big consideration is the use of implicit compared to explicit solvers as well as including or excluding inertia effects (i.e. dynamic versus quasi-static FEA simulations).

The simulation using quadratic elements with inertia effects in an implicit solver with small time steps results in likely the most accurate simulation. This is expected given that (1) higher order accuracy elements are used, (2) all of the physics is included and (3) solutions, states and variables are solved simultaneously instead of only being a function of the previous time step values. Some resulting contours for this baseline roll reduction simulation using a maximum allowable step size of 0.0001 seconds is given in Figure 2.

In Figure 2(a) the von Mises equivalent stress contours illustrate a maximum stress concentrated just below the surface. The steady state residual stresses downstream from the roll (at the far right of the figure) indicate that the highest residuals are below the surface with a slightly higher residual in the middle of the slab than a third into its reduced thickness. The roll process forces material through a complex shear deformation history as illustrated by the contours in Figure 2(b). In 2(c) the importance of a rate dependent plastic deformation material model is illustrated given a highly non-linear plastic strain rate distribution. The rates of deformation vary from above $3\,\text{s}^{-1}$ just in front of the roll to around 1s^{-1} at the centre and orders of magnitude variation around the deformation zone.

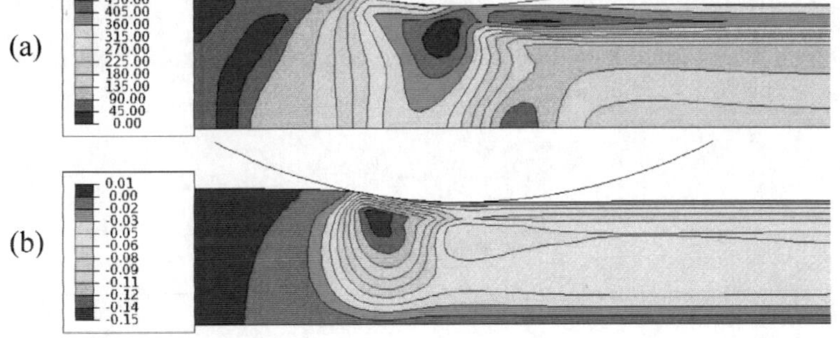

Figure 2. Results of a Finite Element Analysis to simulate AA5182 roll pass reduction at 300°C using CPE8R elements in a dynamic implicit analysis. Contours show the steady state developed and through thickness variation of internal material state. (a) von Mises equivalent stress σ_{vM}

IOP Conf. Series: Materials Science and Engineering **430** (2018) 012025 doi:10.1088/1757-899X/430/1/012025

(c)

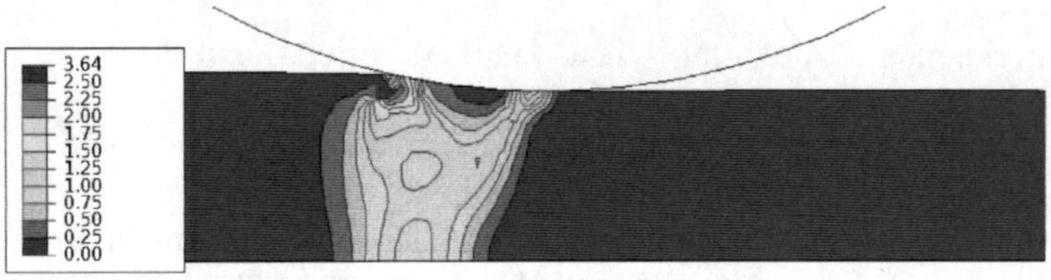

in MPa; (b) Logarithmic shear strain and (c) Equivalent plastic strain rate $\dot{\alpha}$ in s^{-1}.

In Figure 3, the evolving internal state variable $\hat{\sigma}_\varepsilon$ contours for the same simulation (apart from maximum allowable step size) is illustrated. Figure 3(a) shows the evolving threshold stress $\hat{\sigma}_\varepsilon$ associated with the same analysis in figure 2 using a maximum allowable step size of 0.0001 seconds. The contours in Figure 3(b) are as a result of the same simulation using a maximum allowable automatic time step size of 0.011s. If the slab and each element of length 0.01m moves at the roll peripheral velocity of 1m/s, a 0.011 second time step results in information skipping over the entire element length. In this case spurious cyclic behaviour is transferred into the material state evolution.

(a)

(b)

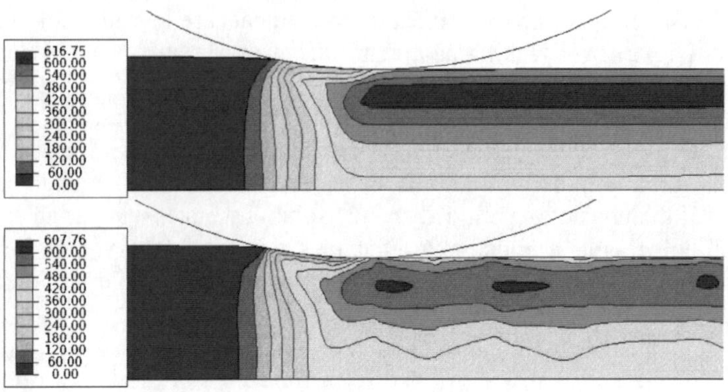

Figure 3. Comparison of CPE8R dynamic implicit internal state variable for the evolving threshold stress $\hat{\sigma}_\varepsilon$ in MPa using a maximum allowable automatic time step of (a) 0.0001 and (b) 0.011 seconds. While implicit simulation means the results are stable, greater cyclic behaviour is transferred into the material state evolution using too large a step size.

The reaction forces and moments at the centre of the roller are extracted for each simulation for comparison in Figure 4. From the baseline dynamic implicit result using quadratic CPE8R elements, the reaction forces exhibit a reasonably steady oscillating behaviour around 0.5 seconds after initial contact. The detail reaction moment and force plots in Figure 4 show the results extracted from five different FEA simulations: (1) the baseline dynamic implicit simulation using quadratic CPE8R elements, (2) a dynamic implicit simulation using linear CPE4R elements, (3) a dynamic explicit simulation using CPE4R, (4) a quasi-static implicit simulation using CPE4R and (5) a dynamic implicit CPE4R simulation with a larger allowable automatic time step size. In the first four simulations a maximum allowable automatic time step size of 0.0001 seconds is prescribed while in simulation (5) the allowable step size is 0.011.

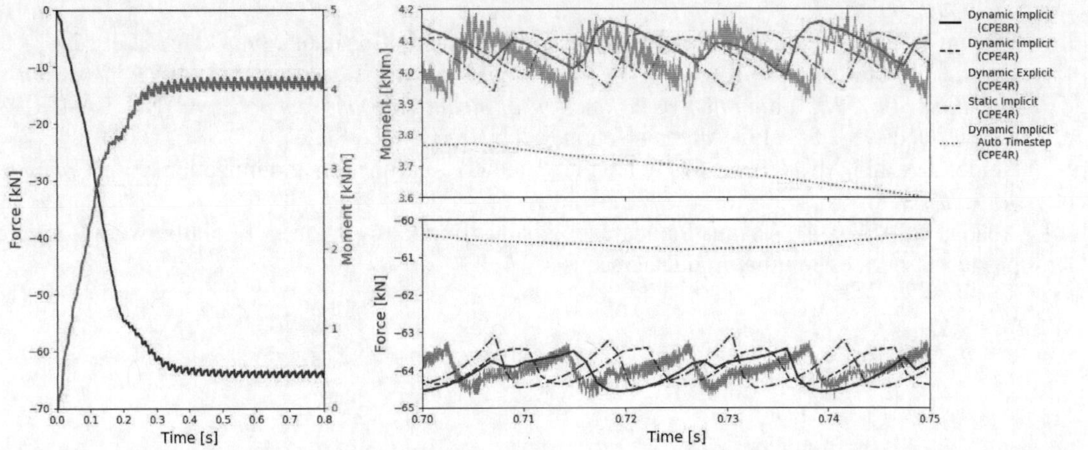

Figure 4. Reaction forces and moments experienced at the roll centre for different modelling considerations. Left: Time history using a reduced integration quadratic element to simulate full physics (dynamic) using the implicit solver. Right: Detail of cyclic response in the steady state forming process as a result of the interaction between temporal and spatial discretisation.

4. Discussion of results and conclusion

Given the Lagrangian FEA results, the benefit of an implicit rather than explicit analysis is evident in the number of time steps required. The implicit analysis makes use of solutions and sensitivities at the time step of interest to allow larger stable step sizes while the explicit analysis required close to two orders of magnitude more time steps.

The use of automatic time stepping with a large allowable step size gives incorrect estimates on the bulk reactions. If a step size is so large that an element does not go through the expected strain history, the through thickness material state becomes increasingly cyclic. A maximum step size should be chosen so that the effects of a non-linear deformation history does not skip through any entire element.

There is very little difference observed between the bulk response in the dynamic and quasi-static simulations. This means that though the quasi-static case simulates less physics, the inertia effects are so limited that it could be neglected altogether without any major implication on the accuracy of the solution.

References

[1] Jansen van Rensburg G J, Kok S and Wilke D N. 2017. Steel Alloy Hot Roll Simulations and Through-Thickness Variation Using Dislocation Density-Based Modeling. *Metall and Materi Trans B* **48** 2631.

[2] Jansen van Rensburg G J. 2016. *Development and implementation of state variable based user materials in computational plasticity*. PhD thesis. Univeristy of Pretoria

[3] Follansbee P S and Kocks U F. 1988. A Constitutive Description of Copper Based on the Use of the Mechanical Threshold Stress as an Internal State Variable. *Acta Metall* **36** 81.

[4] Jansen van Rensburg G J and Kok S. 2012. Tutorial on state variable based plasticity: an Abaqus UHARD subroutine. *In: 8th South African Conference on Computational and Applied Mechanics (SACAM 2012)*. 3-5 September 2012. Johannesburg. South Africa.

[5] Jansen van Rensburg G J, Kok S, Wilke D N. 2016. Cyclic effects and recrystallisation in temperature and rate dependent state variable based plasticity. *In: 10th South African Conference on Computational and Applied Mechanics (SACAM 2016)*. 3–5 October 2016. Potchefstroom. South Africa.

[6] Jansen van Rensburg G J and Kok S. 2018. Modelling multiple cycles of static and dynamic recrystallisation using a fully implicit isotropic material model based on dislocation density.

Comput Mech
[7] Varshni Y P. 1970. Temperature Dependence of the Elastic Constants. *Phys Rev B* **2** 3952.
[8] Chen S R, Stout M G, Kocks U F, MacEwen S R and Beaudoin A J. 1998. *Constitutive modeling of a 5182 aluminum as a function of strain rate and temperature* (No. LA-UR-98-2731; CONF-981054-) Los Alamos National Lab NM (United States)
[9] Nelder J A and Mead R. 1965. A simplex method for function minimization. *The Computer Journal* **7** 308.
[10] Abaqus 6.14 (2014) Abaqus/Standard – Abaqus/Explicit version 6.14. http://www.3ds.com/ products- services/simulia/products/abaqus

IOP Conf. Series: Materials Science and Engineering **430** (2018) 012026 doi:10.1088/1757-899X/430/1/012026

Determination of roll compaction parameters required for high green density, defect free Ti-6Al-4V strips

N S Muchavi[1,2], S Chikosha[1*], H K Chikwanda[1] and E M Makhatha[2]

[1] Light Metals, Materials Science and Manufacturing, Council for Scientific and Industrial Research, P.O. Box 395, Pretoria 0001, South Africa
[2] Department of Engineering Metallurgy, University of Johannesburg, South Africa

*SChikosha@csir.co.za

Abstract: Direct powder rolling (DPR) is an unconventional powder metallurgy technique for fabricating flat mill products. It offers significant cost reduction by using low processing temperatures and reduced fabrication steps compared to ingot metallurgy. Roll compaction is the first stage in the DPR process, hence proper selection of roll compaction parameters is vital for achieving high density defect free (no visible cracks, alligatoring or centre split) green compacts. Unsuitable roll compaction parameters can result in fragile low green density strips which could be difficult to handle during subsequent processing or high green density strips with defects. The focus of this study was therefore, to determine the range of roll compaction parameters required in producing high density, defect free strips. The Ti-6Al-4V strips studied were produced by mixing 150 μm titanium hydride de-hydride powder with a 45 μm 60Al40V master alloy. The powder mix was roll compacted at varying roll gaps and rolling speeds. The compacted strips were evaluated by measuring the green strip density and thickness. It was concluded that for the Ti-6Al-4V powder studied in this work, the preferred roll compaction parameters for achieving high density, defect-free green strips are 0.3 mm to 0.5 mm roll gap with roll speeds from 3 rpm to 10 rpm.

1. Introduction

Direct powder rolling (DPR) of metal powders is a process that has been around for many years [1] but is however still not yet well understood [2] as there are multiple factors (machine design parameters, process parameters as well as properties of the feed material) affecting the roll compaction stage of the process [3,4]. The DPR process typically consists of roll compaction, sintering and/or hot rolling, mechanical working and/or heat treatment [5–7]. Roll compaction is a critical stage of the DPR process in achieving dimensional stability and full density of the compact after subsequent processing [8]. The DPR process entails using powder (pure powders, blended elemental powders or pre-alloyed powders) as feed material for a roll compacting mill. The roll compacting mill consolidates the feed into a green compact which is then sintered to produce the final product [9–11].

The key process variables in roll compaction include powder properties [9,10], compaction pressure, roll gap, rolling speed [10,12–14], roll surface finish [12,13,15] and roll diameter [16]. Roll gap is the most critical parameter of compaction as it has a significant effect on the resulting green density [12]. Decreasing the roll gap has been found to yield an increase in green density of roll compacted powder materials [10,17]. Roll speed determines the dwell time for particle compaction which in turn also affects density. Rolling at high speeds can result in inadequate compaction of material passing through the rolls [12] which can result in formation of a non-continuous strip [16] or no compaction taking place at all; contrary, rolling at low speeds could result in over-compaction [18]. Roll compaction parameter selection plays an important role in the consolidation process as it has a marked effect on the compaction behaviour of powders. Unsuitable roll compaction parameters will result in a green strip with side cracking, alligator cracking, centre splitting, non-uniform thickness distribution etc. [19,20]. Defects obtained in the roll compaction stage may be difficult to rectify or eliminate with subsequent processing. This in turn could result in the green strip being discarded thereby leading to loss in material. It is, consequently, of utmost importance to define roll compaction processing parameters for a given powder particle size and morphology [9]. In the attempts to evaluate the potential application of DPR on commercially pure titanium powder, Park *et al.* [9] were able to show side cracking and the alligatoring phenomenon by rolling -100 mesh Ti powder. Joo *et al.* [21] used direct reduced iron (DRI) powder to study the maximum allowable strip thickness before alligatoring can occur. Park *et al.* [9] and Joo *et al.* [21] successfully demonstrated roll compaction defects associated with unsuitable roll compaction process parameters.

In their work, Hong *et al.* [22] used -200 mesh and -400 mesh Ti powder to study the effect of particle size and distribution, roll gap and roll speed on densification. Their results concluded that density increases with a decrease in roll gap and that roll speed has no significant effect on density and thickness. The results obtained from the work of Freeman *et al.* [17] showed a general trend that increasing the roll gap decreases density. Chikosha *et al.* [10] studied the effect of particle size and morphology on roll compaction of Ti based powders. In their work, they concluded that density and strip thickness are dependent on the roll gap. Their results showed a general trend where decreasing the roll gap resulted in an increase in density and a decrease in strip thickness.

In this paper, we study the effects of roll gap and roll speed to determine the range at which high density; defect-free green strips are produced. During the roll compaction of 150 μm Ti-6Al-4V powder mix, all other parameters such as powder properties, compaction pressure, diameter roll compaction mill were kept constant.

2. Experimental methods

2.1. Materials

The starting powders used in this study to produce a Ti-6Al-4V powder mix were titanium hydride-dehydride TiHDH supplied by Baoji Lihua Non-Ferrous Metals Co., Ltd and 60Al40V (60 wt.%Al and 40 wt.%V) master alloy supplied by Reading Alloys. The TiHDH powder was reported by the supplier to have a particle size distribution of 150 μm whereas the 60Al40V master alloy powder was reported by the supplier to have a particle size distribution of 45 μm. The chemical compositions of the powders according to the suppliers are listed on Table 1.

Table 1: Chemical composition (wt.%) of starting powders according to supplier.

Material ID	Fe	C	N	H	O	Cl	Al	V	Ti	Other
150 μm TiHDH	0.018	0.002	0.017	0.002	0.10	0.020			Bal.	0.40
45μm Master alloy	0.500	0.100	0.050	0.015	0.10		54-60	40-45		1.4

2.2. Powder characterisation and mixing

The powders were characterised for morphology, particle size distribution and oxygen content. Powder morphology analysis was done using the Jeol JSM-6510 Scanning Electron Microscope (SEM). Particle size distribution was analysed using the Microtrac Bluewave X3500 particle size analyser and the oxygen analysis was achieved using ELTRA ONH 2000 analyser. Powder feed preparation was achieved by mixing the Ti and master alloy powders for 30 minutes on a Diex Tubular mixer operating at a speed of 96 rpm. The powder feed formulation used in the study consisted of 90 wt. % titanium and 10 wt. % master alloy mix.

IOP Conf. Series: Materials Science and Engineering **430** (2018) 012026 doi:10.1088/1757-899X/430/1/012026

2.3. Roll compaction

Roll compaction of the powder was carried out on a reversible two roll mill supplied by Chico precision machine Co. (China). The diameter of the rolls was 400 mm with the roll face width being 30 mm. The process parameters used in the study were roll speeds of 3, 5, 7, and 10 rpm; roll gaps of 0.2, 0.3 and 0.5 mm

2.4. Density and thickness analysis

Bulk densities of the roll compacted strips were measured using volumetric density which uses the mass, length, width and thickness of the strips to calculate the density. The relative densities of the strips were then calculated as the ratio of the bulk density to the theoretical density (4.42 g/cm^3). The mass of the strips were measured using the OHAUS Voyager Pro balance and the length and width were measured using a digital calliper and the thickness using a TA micrometer screw gauge.

3. Results and Discussions

3.1. Powder characterisation

The 150 μm TiHDH, 45 μm 60Al40V (60 wt.%Al and 40 wt.%V) master alloy powder and the mixed Ti-60Al-40V alloy powder was analysed for powder morphology, particle size distribution and oxygen content. Figure 1(a,b,c) shows particle size distributions and morphologies of the powders whereas Figure 1(d) shows how the powder deforms and interlocks to form a strip after roll compaction. All powders possessed an angular morphology. The particle size distribution of the powders showed a narrow distribution; this is also confirmed by the D_{10}, D_{50} and D_{90} values on Table 2. The 150 μm TiHDH and the 150 μm Ti-60Al-40V both show a somewhat bimodal distribution with peaks at around 45 μm and 150 μm. The measured oxygen content of the powders were higher than those indicated by the supplier; consequently, the resulting 150 μm Ti-60Al-40V mixture has higher oxygen content as compared to the ASTM B265-15 [23] standard for Ti-6Al-4V alloy.

(a)

(b)

(c) **(d)**

Figure 1. Particle morphology of the powders: (a) 150 μm TiHDH, (b) 45 μm master alloy, (c) 150 μm Ti-60Al-40V, (d) roll compacted powder forming a strip

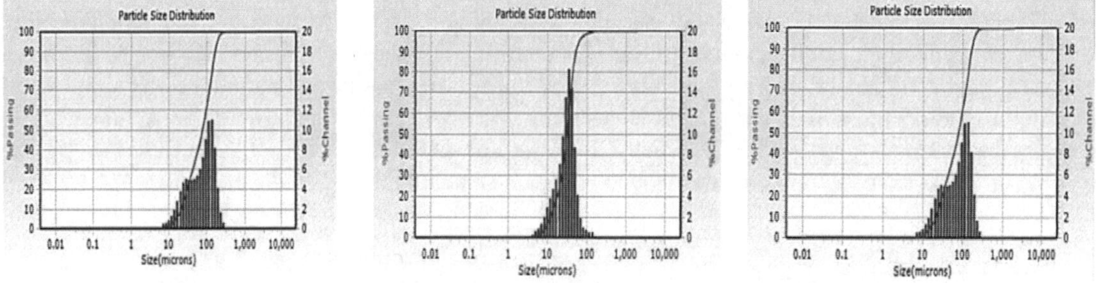

(a) **(b)** **(c)**

Figure 2. Particle size distribution of the powders: (a) 150 μm TiHDH, (b) 45 μm master alloy, (c) 150 μm Ti-60Al-40V

Table 2. Particle size distribution and oxygen analysis of the starting powders.

Material ID	**Particle size distribution**			**Oxygen content**
	D_{10} (μm)	D_{50} (μm)	D_{90} (μm)	(%)
150 μm TiHDH	35.4	125.1	160.1	0.27
45 μm Master Alloy	22.2	45.7	150.7	0.29
150 μm Ti-6Al-4V	28.1	92.1	142.7	0.27

3.2. Roll compaction

Table 3 shows the rolling parameters tested in this study. During roll compaction at a roll gap of 0.2 mm, the roll compacted strips resulted in side cracking and centre splitting. At 0.3 mm and 0.5 mm roll gap, the powder roll compacted to varying densities with no defects. Figure 3(a) shows a typical defective strip containing side cracks whilst Figure 3(a) shows a typical defect-free strip produced from the roll compaction process.

Table 3. Rolling parameters and resulting green density and thickness of strips used in the study.

Roll gap (mm)	Roll Speed (rpm)	Defects	Density (%)	Thickness (mm)
0.2	3	Yes	As a result of cracking, density and thickness of strips rolled at 0.2 mm roll gap were not measured.	
	5	Yes		
	7	Yes		
	10	Yes		
0.3	3	No	84	0.20
	5	No	87	0.21
	7	No	85	0.21
	10	No	86	0.21
0.5	3	No	83	0.21
	5	No	79	0.22
	7	No	52	0.22
	10	No	80	0.22

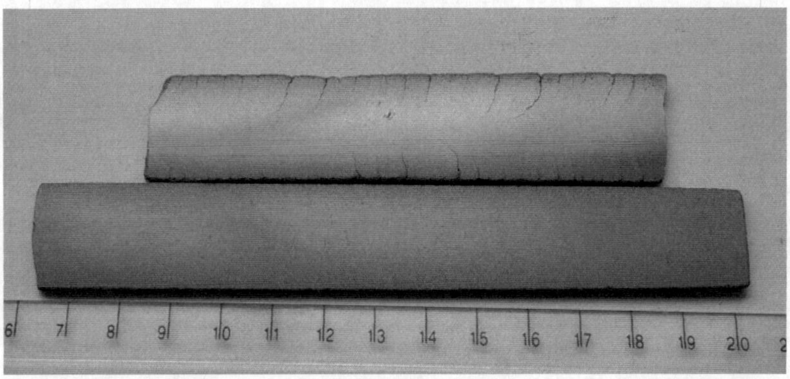

Figure 3. Roll compacted strips (a) defective strip showing side cracking and (b) defect free strip

3.3. Density, and thickness characterisation

The cracked strips rolled at a roll gap of 0.2 mm were not characterised for density and thickness as the cracking could make the results unreliable. Seeing as to how the strips cracked in different ways, the types and causes of cracking will be studied later to understand the relationship between density, roll gap and types of cracking.

The density results on Figure 4 are represented with the standard error of the mean (SEM). The relative green density results (Figure 4 (a)) show a general trend that a smaller roll gap gives a higher relative green density regardless of the rolling speed employed. This behaviour is expected and has been observed in the work of Freeman *et al.* [17]. However, the strips rolled at a 0.3 mm roll gap and a rolling speed of 3 rpm yielded a significantly low green density compared to those rolled at higher roll speeds at the same roll gap. This behaviour is uncommon and unexpected as literature states that rolling at low speeds implies more dwell time for particle compaction [12] which could even result in over-compaction of the powder [18]. This behaviour could be attributed to particle packing of the powder feed prior to roll compaction whereby the fine powders could have agglomerated resulting in inhomogeneous densification after roll compaction. Apart from this unexpected result, the remaining results indicate that there is no significant change in density with rolling speed. This result is contrary to literature since rolling at high speeds is said to result in inadequate compaction of material passing through the rolls [12] hence low densities in compact rolled at higher speeds. However, Hong *et al.* [22] also found that there is no significant effect on density when rolling at high speeds.

Figure 4 (b) shows the effect of roll gap on strip thickness. The results show that increasing the roll gap results in an increase in the strip thickness. This behaviour is expected as increasing the roll gap means more material is passing through the roll gap during roll compaction and has been documented in the work of Chikosha *et al.* [10] and Hong *et al.* [22]. Varying the roll speed has no distinct effect on the strip thickness. This behaviour is also expected and has been documented in a study conducted by Hong *et al.* [22]. The strip thicknesses for all parameters were found to be thicker than the set roll gap. This behaviour has also been previously reported on by Zhang [16] who attributed to the roll spring back effect during roll compaction that occurs as a result of the elastic strain recovery after roll compaction. Overall, green density and strip thickness are strongly dependant on the roll gap where green density is inversely proportional to the roll gap with thickness being contrary. Both green density and strip thickness were found to be independent of the rolling speed. This behaviour was expected from the strip thickness and not from the density as rolling at high speeds results in inadequate compaction of material passing through the rolls.

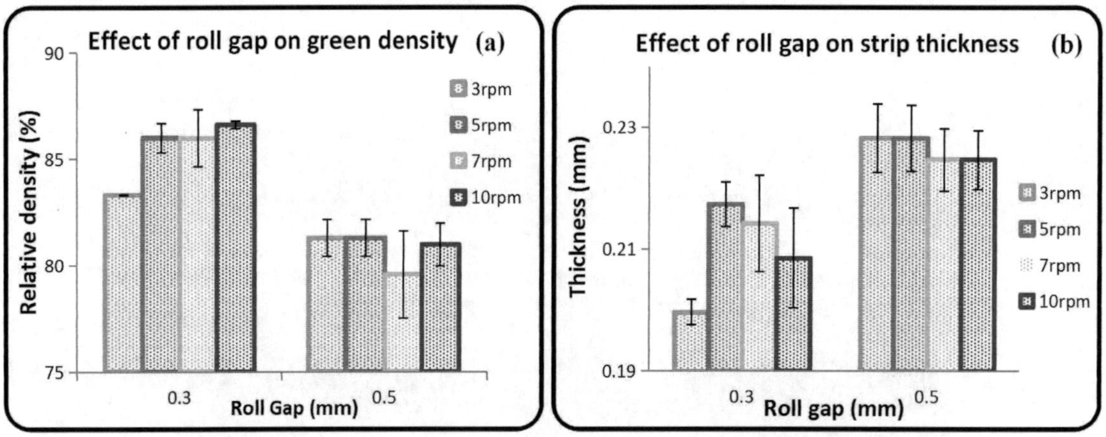

Figure 4. Effect of roll gap on (a) green density and (b) strip thickness.

4. Conclusions

High density and defect-free strips could be achieved at roll gap ranges of 0.3 mm to 0.5 mm and roll speed of 3 rpm to 10 rpm specifically:

- Rolling 150 µm Ti-60Al-40V powder mix is unsuitable at a 0.2 mm roll gap as the powder over compacts resulting in defects.
- Rolling between 0.3 mm and 0.5 mm roll gaps would be suitable for achieving green densities of greater than 80%.
- Increasing the roll gap from 0.3 mm to 0.5 mm resulted in a decrease in the green density.
- Reducing roll gap from 0.5 mm to 0.3 mm increases the green density by 4.9% while reducing the strip thickness by 1.4%.
- Increasing the roll gap from 0.3 mm to 0.5 mm resulted in an increase in strip thickness.
- Green density and thickness are independent of the rolling speed.

Acknowledgements

The authors would like to acknowledge the Council for Scientific and Industrial Research (CSIR), the Department of Science and Technology (DST) and the National Research Foundation (NRF) for the funding used to carry out the work.

References

[1] Esawi A M K and El Borady M A 2008 Carbon nanotube-reinforced aluminium strips *Compos. Sci. Technol.* **68** 486–92
[2] Cantin, G M D&Gibson M 2015 Titanium sheet fabrication from powder *Titanium Powder Metallurgy: Science, Technology and Applications* ed M Froes, Francis. H. Sam & Qian (Butterworth-Heinemann) pp 383–403
[3] Cunningham J C 2005 *Experimental studies and modeling of the roller compaction of pharmaceutical powders*
[4] Yu S 2012 Roll compaction of pharmaceutical excipients (University of Birmingham)
[5] Muchavi N S, Bam L, De Beer F C, Chikosha S and Machaka R 2016 X-ray computed microtomography studies of MIM and DPR parts *J. South. African Inst. Min. Metall.* **116** 973–80
[6] Cantin G M D, Kean P L, Stone N A, Wilson R, Gibson M A, Yousuff M, Ritchie D and Rajakumar R 2011 Innovative consolidation of titanium and titanium alloy powders by direct rolling *Powder Metall.* **54** 188–92
[7] Ro D H, Toaz M W and Moxson V S 1983 The Direct Powder-Rolling Process for Producing Thin Metal Strip *JOM J. Miner. Met. Mater. Soc.* **35** 34–9
[8] Mothosi K L, Chikosha S, Madyira D M and Chikwanda H K 2017 Determination of Residual Stresses in Roll Compacted Titanium Strips *Procedia Manuf.* **7** 309–15
[9] Park N K, Lee C H, Kim J H and Hong J K 2012 Characteristics of powder-rolled and sintered sheets made from HDH Ti powders *Key Eng. Mater.* **520** 281–8
[10] Chikosha S, Shabalala T C and Chikwanda H K 2014 Effect of particle morphology and size on roll compaction of Ti-based powders *Powder Technol.* **264** 310–9
[11] Cantin D, Kean P, Stone N A, Wilson R, Gibson M A, Ritchie D and Rajakumar R 2010 *Direct Powder Rolling (DPR) of Titanium. Light Metals Flagship*
[12] Srikant P, Akash J, Mahesh D and Astik S 2015 Roller compaction design and critical parameters in drug formulation and development: Review *Int. J. PharmTech Res.* **7** 90–8
[13] Hadzovic E 2008 *Roller Compaction of Theophylline*
[14] Reynolds G, Ingale R, Roberts R, Kothari S and Gururajan B 2010 Practical application of roller compaction process modeling *Comput. Chem. Eng.* **34** 1049–57
[15] Guigon P and Simon O 2003 Roll press design - Influence of force feed systems on compaction *Powder Technol.* **130** 41–8
[16] Zhang Y 2015 *A study of direct powder rolling route for CP-Titanium* (University of Cape Town)
[17] Freeman T, Vom Bey H, Hanish M, Brockbank K and Armstrong B 2016 The influence of roller compaction processing variables on the rheological properties of granules *Asian J. Pharm. Sci.* **11** 516–27
[18] Des E and Carmaux M D A- 2003 *Numerical Methods for Predicting Numerical Methods for Predicting*
[19] Sharma P . 1996 A textbook of production technology *(Manufacturing Processes)* (New Dehli: S. Chand & company pvt. Ltd.)
[20] Youssef A H, El-Hofy A H and Ahmed H M 2012 M*anufacturing technology Materials, Processes, and Equipment* (Florida: CRC Press Taylor & Francis Group)
[21] Joo S H, Chang H J, Bang W H, Han H N and Oh K H 2005 Analysis of Alligatoring Behavior during Roll Pressing of DRI Powder with Flat Roller and Indentation-Type Roller *Mater. Sci. Forum* **475–479** 3223–6
[22] Hong J-K, Lee C-H, Kim J-H, Yeom J-T and Park T-K 2010 Ti strip properties fabricated by powder rolling method *Surf. Rev. Lett.* **17** 229–34
[23] ASTM International ASTM B265-15, Standard Specification for Titanium and Titanium Alloy Strip, Sheet, and Plate *ASTM Stand.*

IOP Conf. Series: Materials Science and Engineering **430** (2018) 012027 doi:10.1088/1757-899X/430/1/012027

Characterization of A356 investment cast component produced using controlled liquid metal cooling technique

TF Sishuba [1*], P Rossouw [1], U Curle [1], K Mutombo [1]

[1]Light Metals, Materials Science & Manufacturing, Council for Scientific and Industrial Research, Meiring Naudé Road, Brummeria, Pretoria 0185, South Africa

[*]fsishuba@csir.co.za

Abstract. In this work Controlled Liquid Metal Cooling (CLMC) integrated in gravity investment casting process is evaluated in order to minimize the internal defects and subsequently improve the casting yield and mechanical properties. A356 was cast and cooled using conventional air cooling and controlled liquid metal cooling technique. Optical light stereo-microscopy, optical light microscopy, scanning electron microscopy equipped with energy dispersive x-ray scope and tensile tests were performed to investigate internal defects, microstructural features, grain sizes, shapes and performance. Microstructures of CLMC sample showed modified eutectic structure with spherical, dendritic primary alpha grains and reduced shrinkage porosity. Mechanical properties of CLMC are competitive to conventional air cooled samples.

1. Introduction

Shrinkage porosity is major concern in conventional investment casting in still air. This is due to the fact that aluminum tends to shrink more as it cools from liquid to solid. The problem is even aggravated by the radial heat rejection from the solidifying casting primary aluminum dendrites that grow perpendicular to the direction of molten metal flow as runners system feeds the cavity. Growth of such dendrite interrupts the flow of molten metal to interdendritic regions during solidification and this promotes micro porosities in those regions since now last metal to solidify has no molten aluminum to feed. To eliminate this problem it will be necessary to control the rejection of heat from the mold, particularly the direction of heat flow as casting is cooling down. This can be achieved by promoting directional solidification and axial heat flow from the casting. Controlled Liquid metal cooling (CLMC) has been used in investment casting for over 40 years. The potential for this technique was evaluated around 1970s in attempt to eliminate the poor quality experience by the Bridgman process during investment casting of large industrial gas turbine blades with super alloys. The laboratory scale development test was carried out around 1976 [1]. The benefits of the CLMC technique is the provision of high thermal gradients at the liquid-solid interface as the alloy solidifies compared to the convectional Bridgman process. This high thermal gradient leads to homogenous and refined microstructures [2,3]. Therefore, the objective of this work is to evaluate the effect of CLMC technique on the production of near net shape complex aluminum thin walled components with minimum shrinkage porosity.

2. Methods and material

The alloy A356, with the chemical composition given in Table 1, was used for this investigation. Two investment molds were prepared for liquid metal cooling and still air cooling. Mold temperature kept the same for two cooling processes at 850°C. Melting was carried with an electrical resistance furnace equipped with a thermocouple. The molten aluminum was poured at 665°C. During solidification one mold was cooled using CLMC technique [4] and the other mold was conventionally cooled in still air. A356 plates of 3 mm thick were cast (Figure 1a). The cast plates were cut as show in the (Figure1b). The tensile properties of both cooling techniques were evaluated post T6 heat treatment.

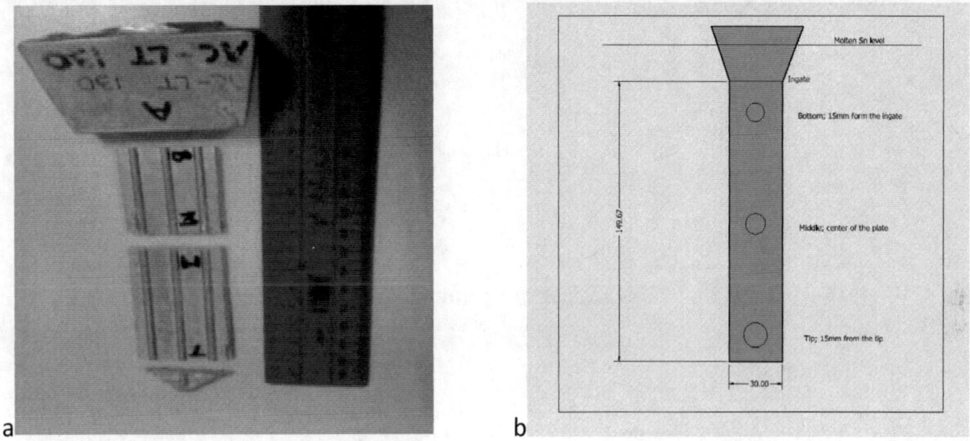

Figure 1. a) The A356 plate and (b) schematic of the sampled regions

Table 1. A 356 alloy composition

Elements	Si	Cu	Mg	Mn	Fe	Zn	Ni	Ti	Al
Wt. %	7.2	0.2	0.29	0.01	0.18	0.01	0.02	0.11	balance

3. Results and discussions

The microstructures of the conventional air cooled and CLMC as cast plates are shown in Figure 2a & 2b respectively. Air cooled sample showed coarser plate like eutectic precipitates (Figure 2a). This as cast CLMC microstructure promises better alloy ductility and low strength due to more dendrites formations towards cast in gates. A eutectic microstructure and refined Si-Al phases are observed in the as cast CLMC plate (Figure 2b). These finely distributed Si-Al phases can be attributed to high thermal gradients of the controlled liquid metal cooling technique. Such refined microstructures in particular of precipitates will be easily dissolved during heat treatment and more of solute content will provide solute solution strengthen after heat treatment.

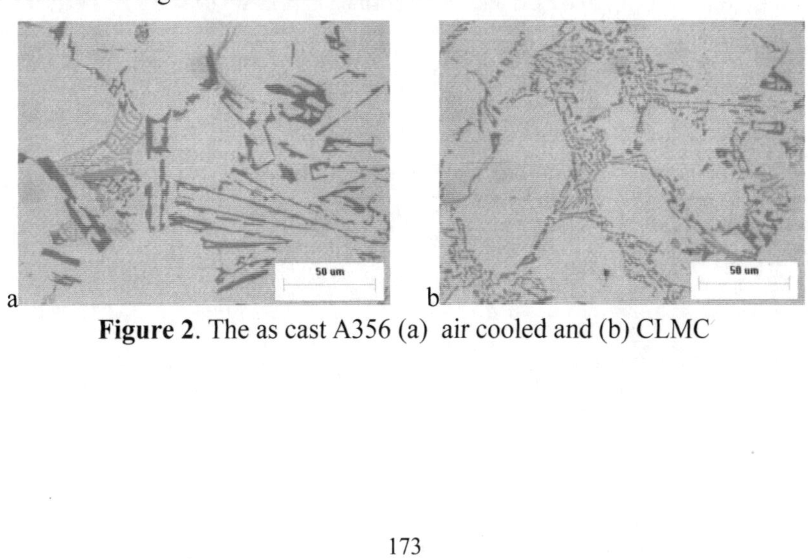

Figure 2. The as cast A356 (a) air cooled and (b) CLMC

Micrographs below displays the representing microstructural features of both CLMC and air cooled samples. Focusing particularly to primary α-morphology, at the tip of both samples (Figure3a & 3b) we observe elongated primary α grains. At the casting in gates we observe coarser elongated primary α grains coupled with primary α dendrites in both samples (figure4a & 4b). Very long and more primary-α dendrites are observed in the CLMC at in gates (figure4b).

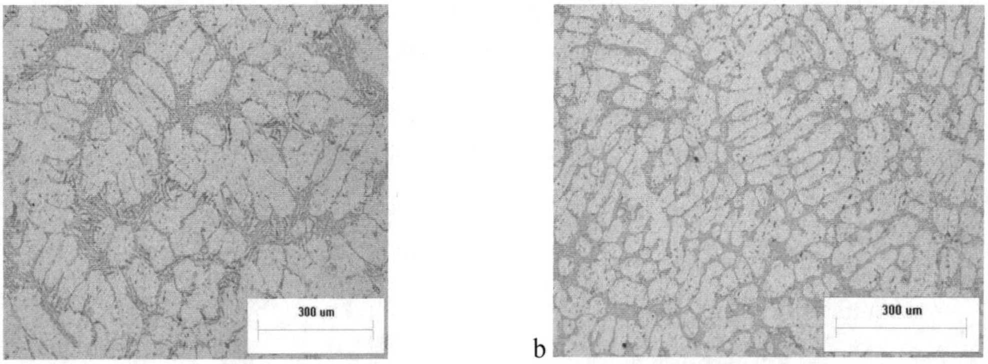

a b

Figure 3. Elongated **α**-primary grains in (a) conventionally air cooled and (b) CLMC as cast.

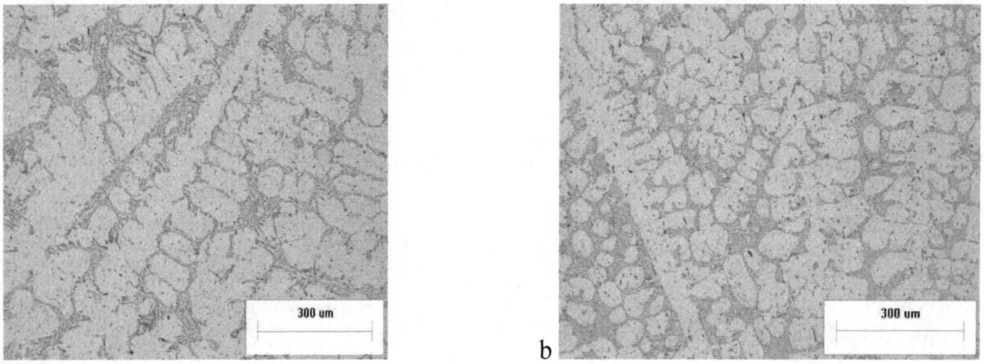

a b

Figure 4. Dendrites and coarser elongated α grains at in-gates of the (a) conventionally air cooled and (b) CLMC plate.

The formation of more long primary α dendrites in the controlled liquid metal cooling process (Figure 4b) is due to the drop in the under cooling as compared to the tip region .This also takes place in the air cooled process, but to a lesser extent.

Table 2 As cast - Composition of primary α

Element % at	CLMC	Air-cooled
Si	1.06	0.77
Mg	0.15	0.37
Al	balance	balance

The composition of primary alpha can be said to be homogeneous generally. Al-Si-Mg and Si phases identified around the matrix as shown in (Figure 5a & 5b).

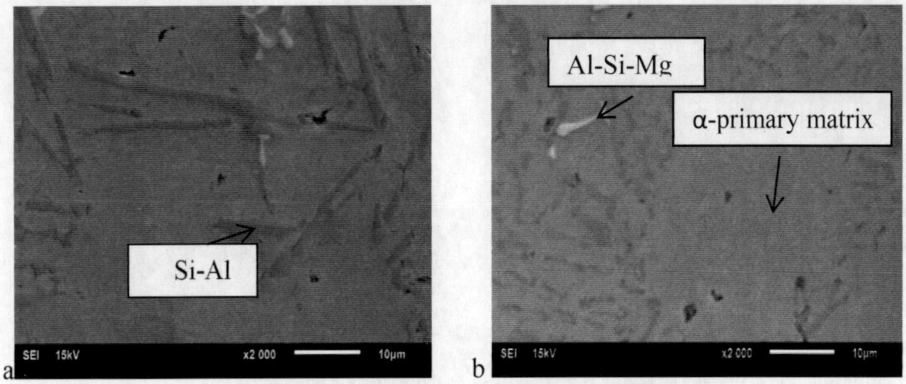

Figure 5. Air cooled (a) and (b) CLMC

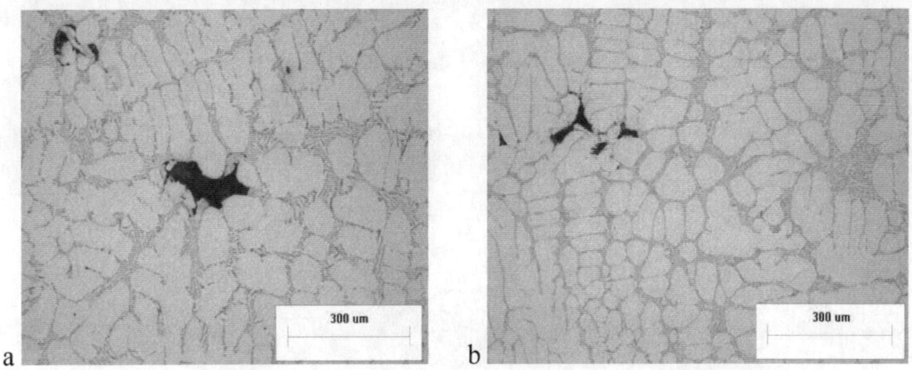

Figure 6. Air cooled (a) and CLMC (b).

These optical micrographs show 2-D cross-sections of the samples Figure 6. Notice reduction in shrinkage pores size in Controlled Liquid Metal Cooling process (Figure 6b).This will be beneficial in terms of improving fatigue life and mechanical properties of the casting.[5,6]. The ultimate, yield strength and the percentage elongation of the conventionally air cooled and CLMC plates after T6 heat treatment are shown in Figure 7, 8 and 9 respectively. A slight increase in ultimate and yield strength was observed in CLMC samples, however, there was drop in percentage elongation Figure 9.

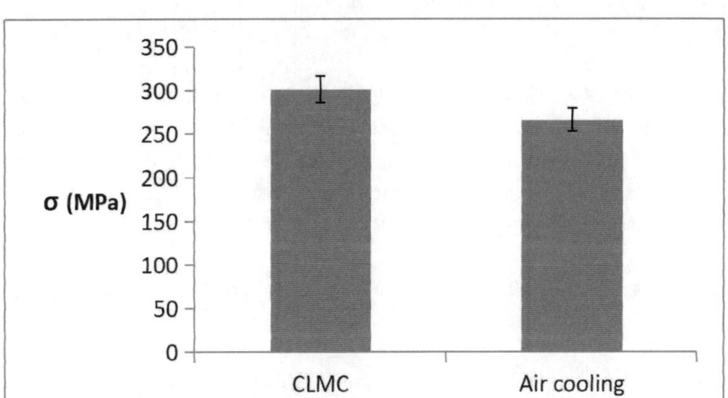

Figure 7. Ultimate strength post T6 heat treatment.

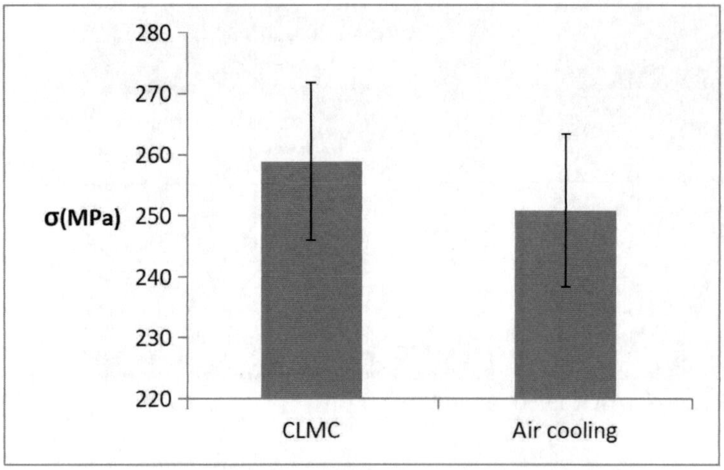

Figure 8. Yield strength post T6 heat treatment.

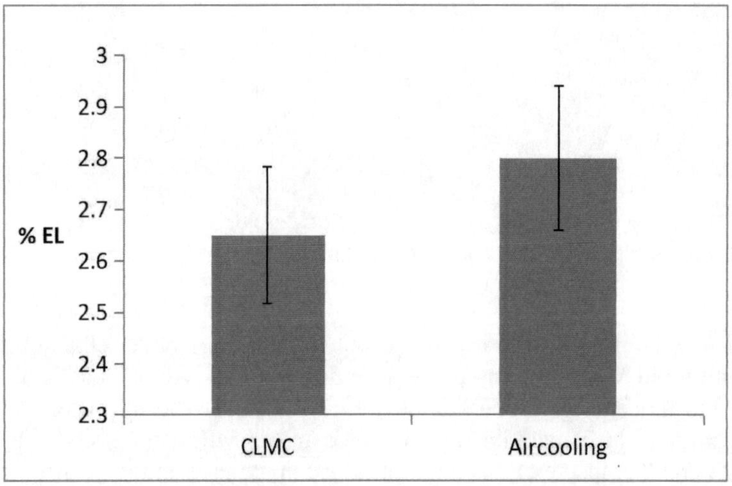

Figure 9. Elongation post T6 heat treatment

4. Conclusions

The Controlled Liquid Metal Cooling (CLMC) technique integrated in gravity investment casting process was investigated on the production of A356 Al casting, the following conclusions were drawn:

1. Controlled Liquid Metal Cooling technique refines the eutectic microstructure and reduces shrinkage pore sizes of A356 durring gravity investment casting.
2. Controlled liquid metal cooling technique improved tensile strength of A356 after T6 heat treatment.
3. Yield strength of A356 is increased by CLMC as compared to air cooling after T6 heat treatment.

4. There is small drop of elongation after T6 heat treatment by CLMC.

Acknowledgements

Authors would like to thank Department of Science and Technology (DST) and Council for Scientific and Industrial Research (CSIR) for funding this work.

IOP Conf. Series: Materials Science and Engineering **430** (2018) 012027 doi:10.1088/1757-899X/430/1/012027

References

[1] Giamei A F and Tschinkel J G 1976 Liquid metal cooling: A new solidification technique *Metall. Trans. A* **7** 1427–34
[2] Lohmtillert A, Eerz W, Gromann J, Hbrdler M, Preuhs J and Singer R F Improved qijality and economics of investment castings by liquid metal cooling - TJ3E Selection of cooling media
[3] Elliott A J, Karney G B, Gigliotti M F X and Pollock T M 2004 Issues in processing by the liquid-sn assisted directional solidification technique
[4] US6276433
[5] Nguyen H 2005 The Effects of Solidification Rates on Porosity Formation and Cast Microstructure of Aluminum Alloy A356
[6] Deegan P T, Stephens R I, Hardin R a and Beckermann C 2007 The Effect of Porosity on the Fatigue Life of 8630 Cast Steel *Trans. Am. Foundry Soc.* **115** 1–17

IOP Conf. Series: Materials Science and Engineering **430** (2018) 012028 doi:10.1088/1757-899X/430/1/012028

The effect of selective laser melting build orientation on the mechanical properties of AlSi10Mg parts

B J Mfusi[1,2,*], L C Tshabalala[2], A P I Popoola[1] and N R Mathe[2,*]

[1] Tshwane University of Technology, Department of Mechanical and Metallurgical Engineering, Staatsartillerie Rd, Pretoria West, Pretoria, 0183, South Africa
[2] National Laser Centre, Council for Scientific and Industrial Research, Meiring Naudé Road, Brummeria, Pretoria, 0185, South Africa

*mfusibusisiwe08@gmail.com, nmathe@csir.co.za

Abstract. Additive manufacturing is currently applied across a broad material selection and it is growing in production volume and economic impact. Aluminium is one of the materials that play a significant role in the aerospace industry because of its good strength-to-weight ratio. However, comparatively lower interest is shown in the additive manufacturing of aluminium alloys compared to titanium alloys, therefore there is much scope for data gathering and development of this material. In this investigation, the microstructure and mechanical properties of AlSi10Mg that were built using selective laser melting at different orientations are explored. The specimen were built in the XY, 45° and Z orientations. The microstructures of the obtained samples showed typical scan patterns with the density and hardness values similar to literature values. The XY built samples showed the lowest level of porosity and also possessed the lowest ultimate tensile strength and elongation. In contrast, the 45° built samples showed the highest ductility with the Z build samples showing the highest ultimate tensile strength. This behaviour is anisotropic where different properties are observed for different build orientations, thus the build orientation should be taken into consideration during the optimization of the laser processing parameters.

1. Introduction

Selective laser melting (SLM) is a process that uses a laser beam to build up a part by melting metal powder layer on top of a layer from computer-aided design data [1]. The core parameters that affect the built part includes; laser power, powder layer thickness, layer overlap and the scanning speed [2]. Various materials are used for additive manufacturing (AM) and aluminium (Al) alloys are increasingly becoming desirable in the aircraft and railway transportation industry according to Li et al., [3], which could increase the applications of SLM Al produced parts. In the limited amount of literature reviewed, there are numerous difficulties encountered regarding the success of SLM processing of aluminium based alloys, which are the high thermal conductivity and reflectivity. These properties lead to a need for intensified laser power essential for melting with the drawback being the oxidation of the alloys which leads to entrapment of oxide inclusions causing weak spots inside the parts [1]. Therefore, it is important to optimize the SLM processing parameters in order to obtain the optimum mechanical properties and porosity levels of less than 0.05% for as-built parts as required in the aerospace industry [4]. The other aspect of SLM that is deem crucial is the build orientation. It has been determined that the build orientation affects the various properties such as porosity and mechanical properties as different microstructures are obtained based on the selected build orientation.

Therefore, this study aims to investigate the effect of build orientation on the mechanical properties of SLM produced AlSi10Mg before heat treatment or stress relieving. The samples were built in three different orientation of XY horizontal, 45° diagonal and Z vertical. These samples will be examined for various properties such as hardness, part density, porosity, microstructural changes and mechanical properties in order to determine the orientation that can be better modified and give ideal properties that are desirable for industrial use. All the samples were produced using fixed laser processing parameters.

2. Experimental Procedure

The tensile samples presented in figure 1 were built using the SLM Solutions M280, which is a commercial system with fixed parameters of; 150W power, 1000 mm/s scan speed, 50 μm hatch spacing and 50 μm powder layer thickness. A set of four samples per orientation were produced, with tensile testing performed using 3 of the samples and the remaining sample was sectioned for microstructural characterization. The samples are denoted as: XY (batch A), 45° (batch B) and Z (batch C) as seen in figure 1.

Figure 1: SLM produced AlSi10Mg samples in different build orientations.

2.1. Characterization.

Density measurements were performed on the tensile specimens using the OHAUS EX244/AD densitometers applying the Archimedes method in ethanol. The samples were sectioned, mounted, polished and etched using Keller's reagent prior to microstructural characterisation with the Olympus BX51M light optical microscope. The mounted specimen were also characterized for hardness on the Zwick Micro/Macro Vickers hardness tester. The load used in this case was 300 g and 20 indents were taken and averaged per sample. The 20kN Zwick/ Roell Tensile tester was used applying standard ASTM E8/E8M-16a for tensile strength measurements.

3. Results

3.1. Density and porosity measurements

The density and porosity results of the SLM built AlSi10Mg parts using different build orientations are presented in table 1. The densities obtained for the different orientations were similar and in agreement with the literature value of 2.67 g/cm^3 obtained for commercially produced AlSi10Mg samples [5]. In these results, it was observed that for all the build orientations, batch A (XY orientation) showed less porosity than the rest of the orientations, thus having a higher relative density. Whereas the highest amount of porosity was observed on batch C (Z orientation). The porosity levels obtained in these samples was less than the 0.15% with a relative density of above 99.8% as required for as-built samples for applications in the aerospace industry [4, 6]. To improve the porosity levels post heat treatments are performed on the samples but this will not be addressed in this work. The hardness measurements were also similar and will be discussed further in section 3.3.

Table 1: Properties of SLM produced sample in different orientations

Build orientations	Density (g/cm^3)	%Porosity	Relative density	Hardness (Hv)
A (XY)	2.679 ± 0.0133	0.037 ± 0.00019	99.96 ± 0.50	127 ± 0.635
B (45°)	2.677 ± 0.0132	0.112 ± 0.00056	99.88 ± 0.51	128 ± 0.641
C (Z)	2.676 ± 0.0014	0.149 ± 0.00075	99.85 ± 0.50	126 ± 0.630

For clarity on the porosity results observed in table 1, the microstructures of the samples were analysed to determine the presence of pores on the internal surfaces of the samples.

3.2 Microstructural analysis

The scanning direction microstructures of the SLM samples produced in different orientations are presented in figure 2. The images of batch A at different magnifications (figure 2a&d), show a microstructure consistent with the laser scan pattern during SLM process. This scale-like morphology was also observed by Rosenthal et al, [1]. The scanning pattern sizes in figure 2a were measured using Image J and they were; 50 – 90 μm height and 120 – 330 μm width and at higher magnifications, figure 2d, shows the presence of cellular and dendrite growth within the grain. The same was observed by Brandl et al., [6], Rosenthal et al., [7], and Lam et al., [8], which they attributed to the eutectic silicon particles within the grain boundaries.

The pores (black spots on the microstructure) of this sample were scattered and they were not observed in all tracks. For their work on SLM processed AlSi10Mg samples, Kempen et al. [9], and Awd et al., [10], also observed different types of pores, which are irregular and located at the melt pool boundaries, resulting from entrapped gasses as oxides or evaporated powder.

The microstructures of batch B samples (figure 2b&d), also showed a laser scan pattern though in this case the grains were diagonal to the laser scan path. The scan pattern sizes were also measured as height; 28 – 156 μm, and width; 113 – 344 μm and there were pores observed on every track including grain boundaries. Figure 2b&e showed the growth of dendrites on the boundaries and cellular growth which is larger than the dendrite growth in the middle of the structural morphology. Lam et al., [8], mentioned that the dendritic growth was supressed as a result of experiencing high solidification rate. A scale pattern similar to that of batch A was also observed in figure 2c&f, batch C. The scan pattern sizes were measured as height; 50 – 135 μm and width; 140 – 248 μm. There were pores observed in this structure especially on figure 2c. These pores could be as a result of undissolved powder or insufficient overlapping between scan tracks and spherical pores are usually as a result of trapped gasses, which might be oxides or evaporated powder. The pores are located within the melt pools which was also observed by Kempen et al., [11]. The levels of porosity observed in the microstructures are in agreement with measured porosity from Archimedes method presented in table

1, where the Z orientation possessed the highest quantified porosity and visible pores on the microstructures.

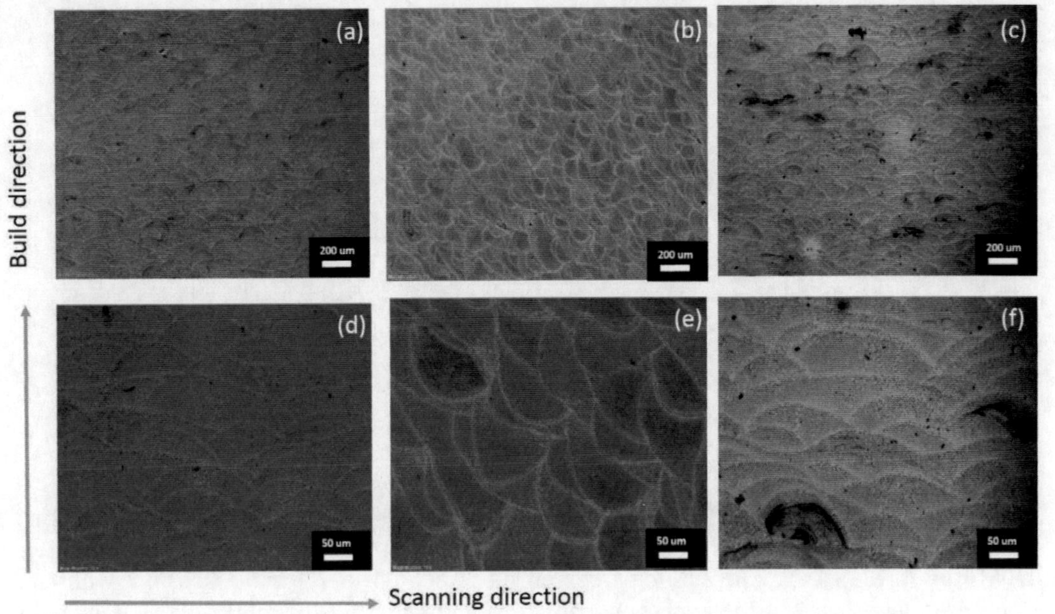

Figure 2. Microstructure of the SLM produced AlSi10Mg samples in various orientations, (a&d) XY, (b&e) 45°, and (c&f) Z orientations. (Images are in low and high magnification respectively.

3.3. Mechanical properties analysis

The hardness measurements are presented in table 1, where the values obtained varied from 126 – 128 Hv, which were relatively higher than values of 92 ± 5 Hv obtained by Rosenthal et al, [12], for SLM produced parts. The values were similar and showed little variations with the change in build orientation. Tensile tests were performed on the samples as shown in figure 3 in order to determine the effect of build orientation on the mechanical properties of the built parts. Typical stress-strain curves were obtained for the samples.

IOP Conf. Series: Materials Science and Engineering **430** (2018) 012028 doi:10.1088/1757-899X/430/1/012028

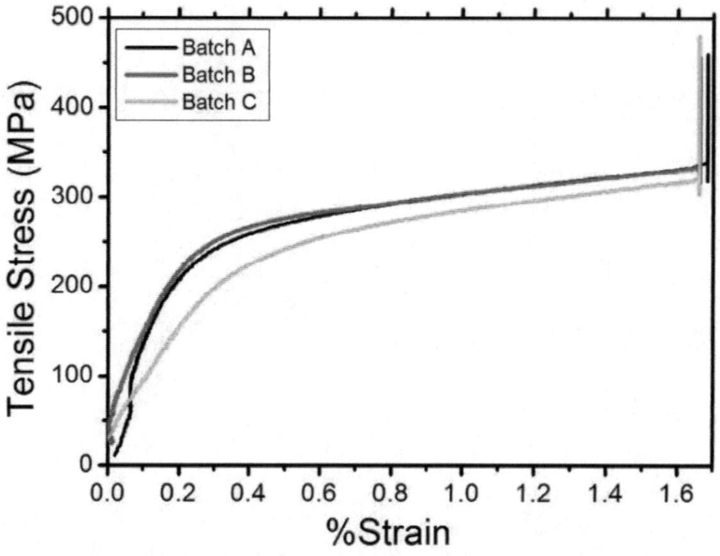

Figure 3. Stress-strain curves of the SLM produced AlSi10Mg samples with build orientation.

The modulus of elasticity for batch A was 83.5 MPa, which was above the expected elasticity of AlSi10Mg of 70 MPa [12]. The elasticity was only endured up to 2.4 mm extension; withstanding stress of up to less than 264 MPa and breaking at 5 mm at stress below 460 MPa. The modulus of elasticity of batch B was 82 MPa, which was also higher than obtained in literature. The elasticity in this case was endured up to 3.8 mm extension; withstanding stress of up to less than 268 MPa and breaking below 7 mm at stress below 462 MPa. This was due to the large number of pores as explained by Takata et al., [13], that SLM produced porous Al alloys exhibit an unstable compressive stress nature characterized by the peaks and troughs. The modulus of elasticity of batch C was 71 MPa, which was almost equal to the expected elasticity of AlSi10Mg of 70 MPa. The elasticity was only endured up to 2 mm extension; withstanding stress of up to less than 244 MPa and actually breaks below 5mm at stress below 473 MPa.

The average ultimate tensile strength (UTS) values of triplicate measurements derived from the stress-strain curves are presented in figure 4a plotted against the build orientation. The highest UTS was obtained for batch C at 473 ± 6 MPa, with batch A showing the lowest UTS value of 460 ± 4 MPa. The values obtained were in the same UTS range for selective laser sintering (460 MPa) [14], and higher than values obtained by Rosenthal et al., [12], for SLM produced AlSi10Mg after stress relieving (272 MPa), and Calagno et al.,[15], for casted A357 alloy (388 MPa). Figure 4b shows the plot of elongation against build orientation. Batch B possessed the highest elongation at 7.4%, and batch A showed the lowest elongation at 6.3%. However, these elongations are lower than the stress relieved AlSi10Mg SLM samples produced by Rosenthal et al.,(8.5%), [12], and they are comparable to the heat treated AlSi10Mg SLS data (6 – 9%), [15], and higher than the elongations for die-casted Aluminium A357 alloy (5.3%). The trend obtained for the samples in this case contradicts what was obtained in literature for the SLS where the Z-orientation samples showed lower elongation compared to XY-orientation and the SLM where the Z-orientation showed the highest elongation [5]. In this case, the highest ductility was obtained in batch B which is the 45° build orientation.

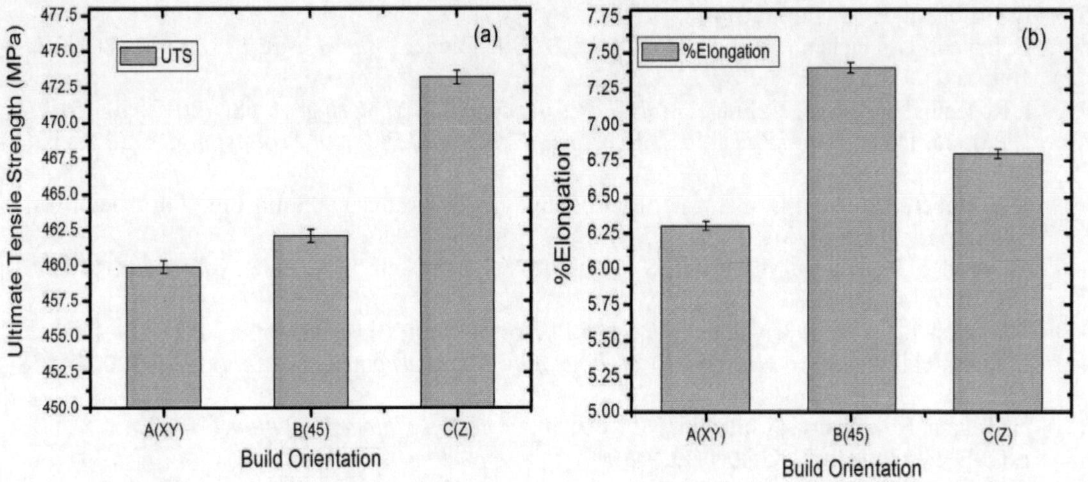

Figure 4. Tensile stress results of the SLM produced AlSi10Mg samples with build orientation and (a) ultimate tensile strength, and (b) %elongation.

4. Conclusions

The aim of this work was to demonstrate the effect of SLM build orientation on the microstructure, density, porosity and mechanical properties of AlSi10Mg samples. The results of this work substantiated anisotropy, which is the demonstration of different mechanical properties in different build orientations before thermal treatment. The samples varied in mechanical properties by built orientation with the Z orientations exhibiting the highest UTS though it also showed the highest levels of porosity from density measurements and microstructural analysis. Initially from density measurements, it was apparent that batch A was to be selected for further modification relatively to the rest of the samples, but with further tests for tensile strength, it was distinguished that batch B would be the best to be chosen practically to manufacture a part. Therefore, based on the obtained results, it can be concluded that the final application will determine the chosen build orientations as part of the process parameters in order to obtain the desired results. For instance, in applications that require ductile components, batch B will be ideal whereas for high strength, batch C is comparable to currently available literature values.

Acknowledgments

Metal Heart Additive Manufacturing, Tshwane University of Technology, Council for Scientific and Industrial Research (South Africa) is gratefully acknowledged. The authors would also like to thank National Laser Center (NLC) Metallurgical Laboratory and Material Science and Manufacturing Mechanical Testing Laboratory for sample preparation and characterization. The National Research Foundation Grant No. 114675 is acknowledged for conference funding. Acknowledgement also goes to the following individuals for their support, Mr. Thabo Lengopeng, Mr. Khoro Malabi, Mr Nana Arthur, Mr. Paul Lekoadi and Mr. Rendani Nemagovhani.

References

[1] I. Rosenthal, A. Stern, N. Frage, *Metallography, Microstructure, and Analysis*, **3** (2014) 448-453.
[2] P. Hanzl, M. Zetek, T. Bakša, T. Kroupa, *Procedia Engineering*, **100** (2015) 1405-1413.
[3] W. Li, S. Li, J. Liu, A. Zhang, Y. Zhou, Q. Wei, C. Yan, Y. Shi, *Materials Science and Engineering: A*, **663** (2016) 116-125.
[4] L. Portolés, O. Jordá, L. Jordá, A. Uriondo, M. Esperon-Miguez, S. Perinpanayagam, *Journal of Manufacturing Systems*, **41** (2016) 65-75.

[5] E. Gmbh, in: EOS (Ed.), 2014.
[6] E. Brandl, U. Heckenberger, V. Holzinger, D. Buchbinder, *Materials & Design*, **34** (2012) 159-169.
[7] I. Rosenthal, A. Stern, N. Frage, *Materials Science and Engineering: A*, **682** (2017) 509-517.
[8] L.P. Lam, D.Q. Zhang, Z.H. Liu, C.K. Chua, *Virtual and Physical Prototyping*, **10** (2015) 207-215.
[9] K. Kempen, L. Thijs, E. Yasa, M. Badrossamay, J.P. Kruth, in: Solid Free Form Fabrication Symposium, Texas, USA, 2011.
[10] M. Awd, J. Tenkamp, M. Hirtler, S. Siddique, M. Bambach, F. Walther, *Materials*, **11** (2017) 10.3390/ma11010017.
[11] K. Kempen, L. Thijs, J. Van Humbeeck, J.P. Kruth, *Physics Procedia*, **39** (2012) 439-446.
[12] I. Rosenthal, A. Stern, N. Frage, *Metallography Microstructure and Analysis*, **3** (2014) 448 - 453.
[13] N. Takata, H. Kodaira, A. Suzuki, M. Kobashi, *Materials Characterization*, (2017).
[14] E.G.-E. Systems, in: E. GmbH (Ed.), 2017.
[15] F. Trevisan, F. Calignano, M. Lorusso, J. Pakkanen, E. Ambrosio, M. Lombardi, M. Pavese, D. Manfredi, P. Fino, in: World Powder Metallurgy Congress 2016, Hamburg, Germany, 2016.

IOP Conf. Series: Materials Science and Engineering **430** (2018) 012029 doi:10.1088/1757-899X/430/1/012029

Development and characterization of nanocrystalline cobalt powder prepared via high energy ball milling process

A L Rominiyi[1]*, M B Shongwe[1] and B J Babalola[1]

[1]Institute of NanoEngineering Research, Department of Chemical, Metallurgical and Materials Engineering, Faculty of Engineering and Built environment, Tshwane University of Technology, Pretoria, South Africa.

*rominiyipedi@gmail.com

Abstract. This work investigated the influence of varied milling time and ball-to-powder weight ratio (BPR) at constant milling speed on micron-sized cobalt powder. Stainless steel balls of 2, 3 and 4 mm diameter were used to prevent cold welding of the charged powders and increase the collision energy available. Powder processing was performed in an argon filled chamber to avoid contamination of the powder. Detailed X-ray diffraction (XRD) studies were carried out to evaluate the lattice strain, crystallite size and the planes at different peaks using Williamson-Hall method. The morphologies of the as-received and milled samples were investigated with scanning electron microscope equipped with energy- dispersive spectroscope (SEM/EDS). The results indicated that the lowest crystallite size and highest lattice strain of 9.05 nm and 3.74 % respectively was achieved for powders subjected to 12 hrs milling at constant BPR of 20:1. A constant milling time of 2 hrs and varied BPR also yielded a crystallite size of about 13.73 nm and lattice strain of 0.79 % at BPR of 10:1. The XRD results show the influence of this process on the diffraction width and peaks. Hence, varying the BPR at constant milling time helps to save time, conserve energy and guarantee the economy of the operation.

1. Introduction

Cobalt is viewed as a key raw material and an important metal in areas where high temperature properties, energy storage, process efficiency and environmental benefits are essential requirements. It is used in the fabrication of materials required for diverse applications, ranging from production of magnets, hard metals, superalloys and gas turbine components to the manufacturing of lithium-ion batteries and industrial catalysts [1]. Concerted efforts have been made by researchers to obtain nanoparticles materials from micrometer sized particles through high energy ball milling process [2, 3]. Nanoparticle materials possess enhanced properties which improve the performance of the resultant components in service. During mechanical milling/alloying, ball-powder-ball collision results in cold welding and fracturing of particles, leading to a reduction in crystallite size and introduction of lattice defects [4]. High energy ball milling process can either be mechanical alloying (MA) or mechanical milling (MM). MA is a process which involves milling together of powders of different metals, alloys or compounds. It results in materials transfer and formation of homogeneous alloy whereas MM is conducted on pure metals or powders of uniform composition which does not require material transfer for homogenization to occur [2].

Micron-sized cobalt powder can be reduced to the nano-sized range by mechanical milling with close attention paid to some important process variables which include: milling speed, milling time, BPR, milling vial, milling medium, ball diameter and milling temperature. These parameters influence the powder quality [2, 5]. A nanostructured material is a material consisting of one or more crystallites size measuring in nanometer (10-9 m) range [6]. These materials are important as they exhibit fascinating and novel properties such as electrical, mechanical, optical and magnetic properties which are superior to those found in conventional bulk materials as a result of their quantum-size effect, small-size effect as well as large number of grain boundaries [7]. This made them the desirable materials for many biomedical, technological and industrial applications [3, 6]. In recent years, efforts have been geared towards studying the behaviour of mechanically milled cobalt powder under different operations [8-11]. However, extensive work has not been conducted on the effect of high energy planetary ball milling parameters on the crystallite size and lattice strain of cobalt powder. Therefore, this work investigates the influence of varying milling time and ball-to-powder weight ratio on the crystallite size and lattice strain of nanocrystalline cobalt powder developed through mechanical milling.

2. Experimental Procedure

Commercially available cobalt powder (FlomasterTM metal powder, F.J. Brodmann & Co., ltd., USA) with average particle size < 44 μm (99.9 % Co) was used as the starting material for the milling experiment. The experiment was conducted in two stages (mechanical milling and vacuum drying). Mechanical milling was carried out using the pulverisette 5 planetary ball milling machine. The milling media consisted of stainless steel lined vial and balls of diameter 2, 3 and 4mm. The experiment was carried out in two batches. BPR was kept constant at 20:1 for a milling duration up to 16 hrs at an interval of 4 hrs for the first batch whereas milling duration of 2 hrs was maintained over a varying BPR of 20:1, 15:1, 10:1 and 5:1 for the second batch. A process control agent (PCA), ethanol (3 wt.%) was added to prevent cold welding. Withdrawal of powder from the vial after milling for the desired duration was performed under argon atmosphere in a glove box. The milling speed was maintained at 350 rpm with allowed reverse rotation after 2 hrs and 10 min pause (to avoid overheating during milling) throughout the duration of the experiment. After the milling process, the milled powders were dried at 40 °C in Vacutech drier.

The unmilled and milled powders were subjected to X- ray diffraction studies with the aid of PW1710 Philips X-ray diffractometer utilizing Cu Kα radiation at 40 kV and 40 mA to determine the phases present in the milled powder. Diffraction patterns were obtained over a range of 2θ between 5 and 90°, with a step size of 0.02. A search-match routine was carried out to compare the data obtained with the data base of the equipment using X'Pert High Score Plus software for identification of phases. The crystallite size and lattice strain of the analysed samples were calculated from the XRD pattern by applying the Williamson-Hall (W-H) method. The applicable equation is as follows:

$$\beta \cos(\theta) = \frac{k\lambda}{D} + \eta \sin(\theta) \qquad (1)$$

where θ s the Bragg diffraction angle, k is the Scherrer constant (0.91), λ the radiation wavelength (0.15406 nm), D the crystallite size, β the diffraction peak at half maximum intensity and η is the lattice strain. By plotting β cos(θ) versus sin(θ), the slope of the straight line graph gives the estimated average lattice strain and the point at which sin(θ) = 0 gives the estimated average crystallite size of the analysed powder. The morphology of the ball milled powders was investigated using scanning electron microscope (JOEL FE-SEM JSM- 7600F). The energy-dispersive X-ray spectroscope (EDS) incorporated with SEM was employed for compositional analysis of the milled powder.

3. Result and discussion

The morphology of the unmilled cobalt powder is shown in Figure 1a. The SEM image revealed clusters of irregularly shaped particles. The particles flatten out into sheet-like shapes due to heavy deformation during cold milling operation as shown by the SEM images in Figures 1b-1d. It was observed from Figure 2 that milling over varied times reduced the crystallite size of the cobalt powder from 64.8 nm to 9.05 nm and lattice strain of 3.74 % which represented about 3.44 % increment over the unmilled (0.3 %) at 12 hrs was obtained. This can be attributed to the influence of lattice defects such as stacking faults, dislocation and other imperfections introduced as the powder particles undergone severe plastic deformation due to interaction of the powder particle with the stainless steel balls and the wall of the vial. [4, 12]. Continued milling (16 hrs), caused an increase in the crystallite size and reduction in lattice strain (Figure 2) due to cold welding and subsequent agglomeration of the particles (Figure 1d).

Milling at a constant time of 2 hrs and varied ball-to-powder weight ratio (BPR) also yielded a reduction in the crystallite size and an increase in the lattice strain of the cobalt powder. The lowest crystallite size of about 13.73 nm and highest lattice strain of 0.79 % was achieved for this set of powders at an optimum BPR of 10:1 after, which cold welding and agglomeration resulted in increasing crystallite size and reducing lattice strain as shown in Figure 3. Crystallite size reduction introduces strains into the lattice of the material as a result of an increase in the volume of particles per unit area. According to Bhatt et al. [6], introduction of lattice strain into nanostructured materials has been found to be a beneficial and economical method of altering the band structure and enhance the performance of energy devices.

The EDS analysis results (not shown here) indicated that there was no contamination of the powder either during handling or with any interstitial element as a result of milling operation.

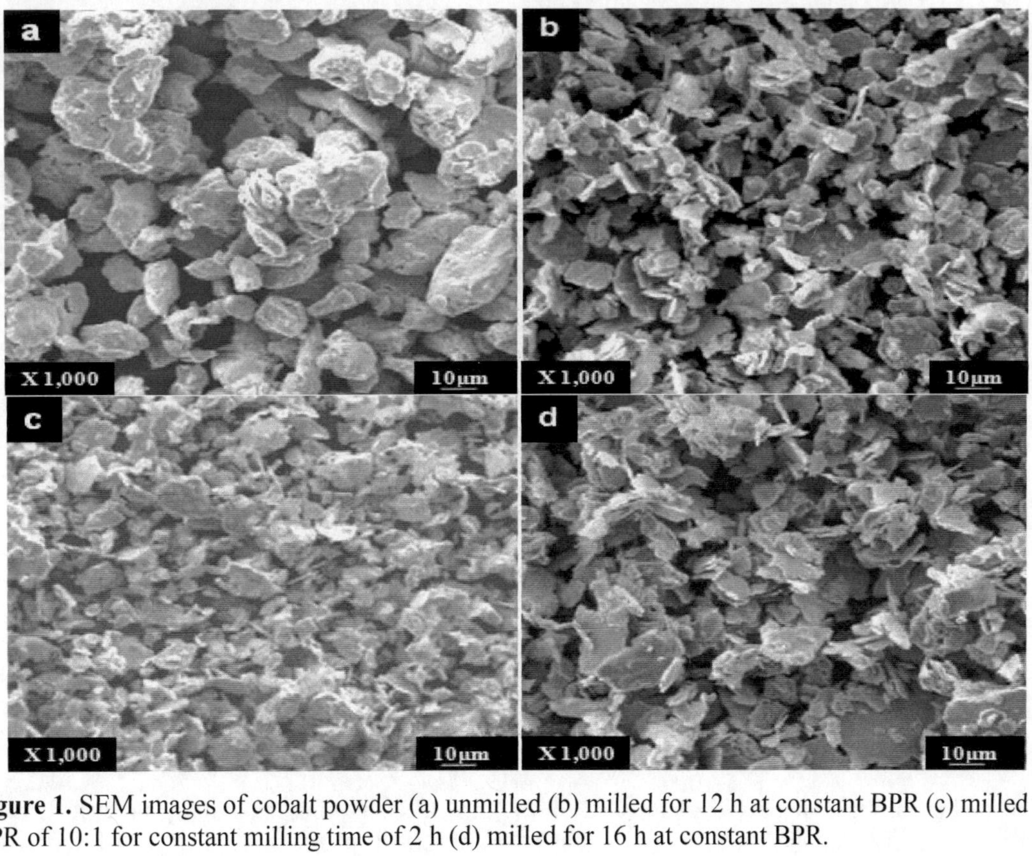

Figure 1. SEM images of cobalt powder (a) unmilled (b) milled for 12 h at constant BPR (c) milled at BPR of 10:1 for constant milling time of 2 h (d) milled for 16 h at constant BPR.

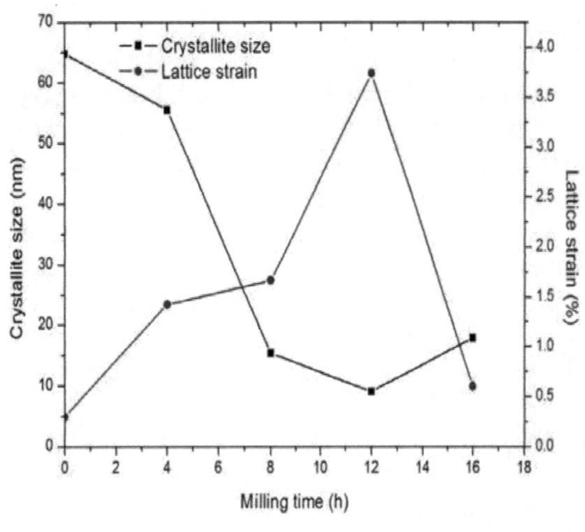

Figure 2. Variation of milling time with crystallite size and lattice strain of the milled cobalt powder.

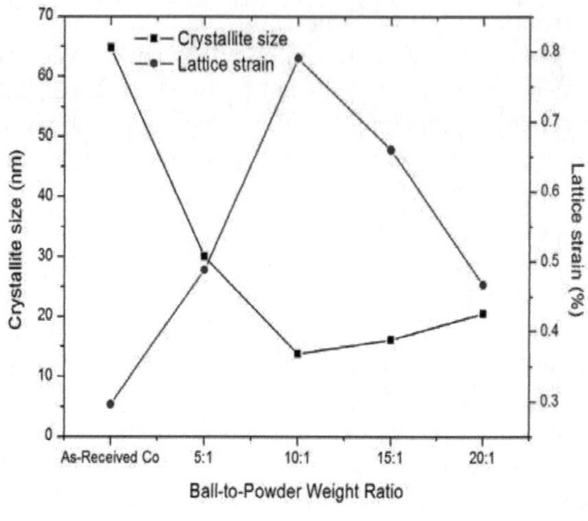

Figure 3. Variation of BPR with crystallite size and lattice strain of the milled cobalt powder.

XRD patterns shown in Figures 4 and 5 were obtained at different stages and processing parameters. In Figure 4, three pronounced peaks of (100), (002), (101) and one weak (110) planes were noticed in the unmilled Co powder. These diffraction peaks were suppressed leaving only the peak of (002) plane which was shifted towards left as the milling time increased. This can be attributed to the formation of amorphous phase in the milled cobalt powder [8, 9]. The intensity of the (002) plane does not really change because the crystals in this plane were not significantly affected by the milling operation [13]. During mechanical milling, fragmentation of the particles occurred, this leads to variation in strain from point to point of the irradiated area of the samples, thereby causing the line broadening observed in Figure 4. This observation is in agreement with the previous findings of Ravi et al.[5].

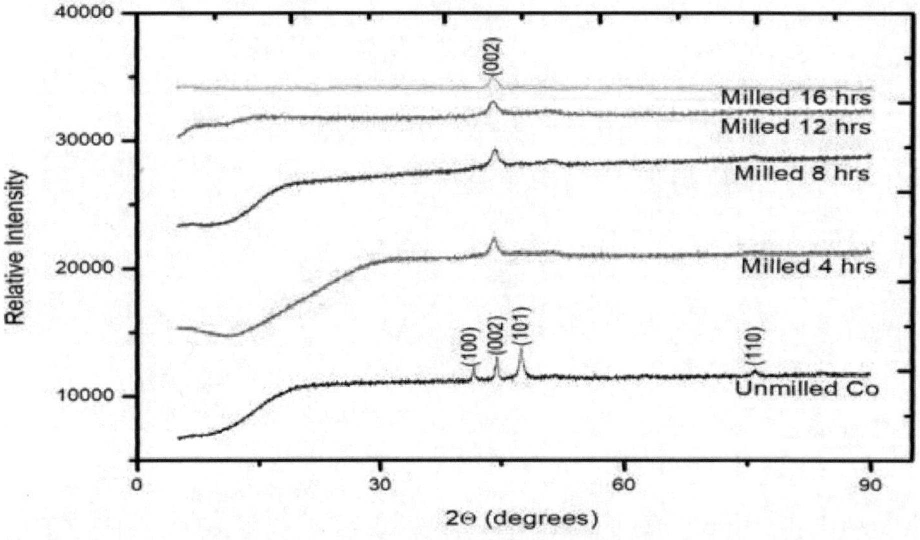

Figure 4. XRD patterns of cobalt powder milled over varied time.

Diffraction patterns of powders subjected to varying BPR and constant milling time of 2 hrs is presented in Figure 5. The least diffraction peak was observed in powders milled at BPR of 10:1.

Amorphization was not observed in the milled cobalt. This might be due to insufficient available milling energy which is a function of milling speed, milling time and ball-to-powder weight ratio [14]. Nonetheless, broadening in diffraction peaks and movement in position of the diffraction maxima due to varying BPR at constant milling time resulted in reduction of crystallite size and increase in the developed strain within the material. The observed peak at BPR of 10:1 correlated with the lowest crystallite size observed in Figure 3.

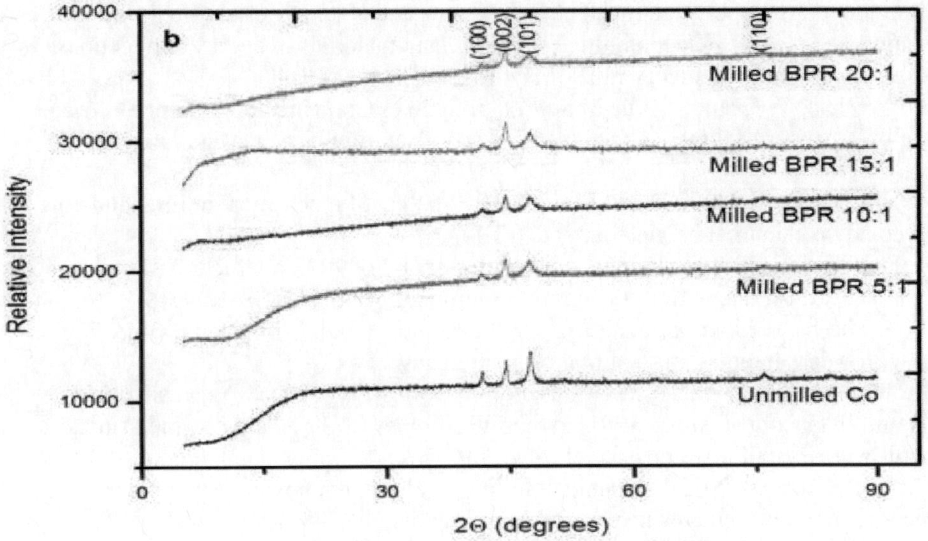

Figure 5. XRD patterns of milled cobalt powder with varied BPR.

4. Conclusions

Development and characterization of nanocrystalline cobalt powder prepared through high energy ball milling process was investigated. The important findings in this study are summarized as follows:

1. High energy ball milling process was found to reduce crystallite size and increased lattice strain until the right combination of parameters which yielded optimum result(s) were attained under the different conditions investigated.
2. A shorter milling time of 2 hrs at BPR of 10:1 yielded better results which could positively affect the efficiency of the operation since less energy was used.
3. High energy ball milling resulted in the development of nanocrystalline cobalt with increased lattice strain. Increasing the lattice strain of nanostructured materials is an economic way of enhancing their utilization in energy devices. Hence, this process aids in the production of nanostructured cobalt utilized in energy devices.

Acknowledgement

The authors acknowledge the Department of Chemical, Metallurgical and Materials engineering, Tshwane University of Technology and the management and staff of the Institute of NanoEngineering Research (INER), Pretoria Campus, South Africa for their technical support in the course of this research.

References

[1] Schmidt T, Buchert M and Schebek L 2016 Resources, Conservation and Recycling: 112 107-22
[2] Suryanarayana C 2001 Progress in Materials Science Mechanical alloying and milling 46 1-184
[3] Ansari S M, Bhor R D, Pai K R, Sen D, Mazumder S, Ghosh K, Kolekar Y D and Ramana C

V 2017 Applied Surface Science Cobalt nanoparticles for biomedical applications: Facile synthesis, physiochemical characterization, cytotoxicity behavior and biocompatibility 414 171-87

[4] Molladavoudi A, Amirkhanlou S, Shamanian M and Ashrafizadeh F 2012 The production of nanocrystalline cobalt titanide intermetallic compound via mechanical alloying Intermetallics 29 104-9

[5] Ravi Kiran U, Prem Kumar M, Sankaranarayana M, Singh A K, and Nandy T K 2015 High energy milling on tungsten powders Int. J. of Ref. Metals and Hard Materials 48 74-81

[6] Bhatt M D and Lee J S 2017 Effect of lattice strain on nanomaterials in energy applications: A perspective on experiment and theory Int.J.of Hydrogen Energy 42 16064-107

[7] Zeng H, Wu Y, Zhang J, Kuang C, Yue M and Zhou S, 2013 Grain size-dependent electrical resistivity of bulk nanocrystalline Gd metals Progress in Natural Science: Materials International 23 18-22

[8] Bolokang A, Phasha M, Motaung D and Bhero S 2012 Effect of mechanical milling and cold pressing on co powder Journal of Metallurgy 2012 1-7

[9] Maziarz W, Dutkiewicz J, Rogal L, Grzonka J and Cesari E 2009 Microstructure of ball milled and compacted Co-Ni-Al alloys from the β range Journal of Microscopy 236 143-8

[10] Bolokang A S, Phasha M J and Motaung D E 2013 Metallurgy Ball Milling and Annealing of Co-50 at% W Powders International Journal of Nonferrous 02 41-6

[11] Bolarín-Miró A M, Jesús F S-D, Torres-Villaseñor G, Cortés-Escobedo C A, Betancourt-Cantera J A and Betancourt-Reyes J I 2011 Amorphization of Co-base alloy by mechanical alloying J.l of Non-Crystalline Solids 357 1705-1709

[12] Ramezani M and Neitzert T 2012 Mechanical milling of aluminum powder using planetary ball milling process J. of Achievements in Mat and Man Eng 55 790-798

[13] Zhang S, Li L and Kumar A 2008 Materials Characterization Techniques 1st edition (Boca Raton, London: CRC Press, Taylor and Francis Group) 115

[14] Munir K S, Oldfield D T and Wen C 2016 Control Agent in the Synthesis of Multi-Walled Carbon Nanotubes Reinforced Titanium Metal Matrix Powder Mixture Advanced Engineering Materials 18 294-303

Effect of milling speed on the formation of Ti-6Al-4V via mechanical alloying

P Daswa[1], **Z Gxowa**[1], **MJI Monareng**[1] and **K Mutombo**[1]

[1]Light Metals, Materials Science & Manufacturing, Council for Scientific and Industrial Research, Meiring Naudé Road, Brummeria, Pretoria 0185, South Africa

E-mail: KMutombo@csir.co.za

Abstract. Ti-6Al-4V alloys are well known for good corrosion resistance, excellent surface oxide biocompatibility and high specific strength. This paper investigates the influence of milling parameters on the formation of Ti-6Al-4V alloy processed using mechanical alloying. A mixture of CP Ti and master alloy (60Al-40V) powder were milled using a high energy ball mill (HEBM) in order to produce a mechanically alloyed Ti-6Al-4V alloy powder. Milling was carried out at a constant milling time of 5 hours at varying speeds of 800, 900, 1000 and 1100 rpm in order to investigate the influence of speed on the formation of Ti-6Al-4V alloy. Scanning electron microscopy with energy dispersive analysis (EDX) and X-ray diffraction (XRD) were performed on the mechanically alloyed powder to investigate the chemical homogeneity and the formation of alpha (α) and beta (β) phase during the mechanical alloying of CP Ti and 60Al-40V.

1. Introduction

Titanium and its alloys have been widely used for aerospace and biomedical applications due to its low modulus of elasticity, excellent corrosion resistance, and high strength to weight ratio, good biocompatibility and attractive mechanical properties [1–6]. The most commonly used Ti alloy is the Ti-6Al-4V alloy which contains aluminium and vanadium as major alloying elements. They fall in the two phase (α + β) region where aluminium acts as an alpha (α) stabilizer and vanadium as a beta (β) stabilizer [1,3,6]. The formation of α and β phase leads to the attainment of a variety of microstructures and property combinations in the alloy.

Ti-6Al-4V alloy has been usually produced using the conventional melting and casting process routes, followed by thermomechanical processes such as forging and extrusion. These process routes are associated with problems such as high costs, high energy consumption and excessive loss of material [3–7]. Recently, it has been shown that the abovementioned problems can be overcome by using powder metallurgy and solid state production techniques [3]. Powder metallurgy methods such as mechanical alloying have become a viable and promising route for cost effective fabrication of Ti alloys. The mechanical alloying process involves the transfer of material by repeated deformation, cold welding, fracturing of blended powder particles in a high energy ball mill to produce an alloy through highly energetic ball collisions [3,8]. This process is capable of producing nanosized structures, metallic alloys (high temperature resistant, high entropy), composites (metals-polymers-oxides), nano-crystallines (alloys & composites), and high temperature alloys [3,8]. Therefore the current study aims to investigate the influence of milling parameters on the formation of the Ti-6Al-4V alloy powder produced via mechanical alloying.

2. Experimental Procedure

2.1. Starting materials
The starting materials were CP Ti powder (99.5 wt. %) and the master alloy containing (60 wt. % Al - 40 wt. % Al) according to suppliers. The titanium powder was ASTM Ti Grade 2 quality with particles of 100 mesh (-150 μm) particle size range. The master alloy powder had a particle size of -45 μm.

2.2. Method
CP Ti and master alloy (60Al-40V) powders were mixed in the proportions of 90% CP Ti and 10% master alloy. The mixture was milled using the Simoloyer high energy ball mill (HEBM), using a high strength stainless steel jar in order to produce a mechanically alloyed Ti-6Al-4V alloy powder. The mill charge was composed of 100g powder blend and 2000 g 5 mm diameter 100Cr6 stainless steel milling balls giving a ball to powder ratio of 20:1. Milling was carried out at speeds of 800, 900, 1000 and 1100 rpm with constant milling duration of 5 hours in order to investigate the effect of milling speeds on the formation of the Ti-6Al-4V alloy. Scanning electron microscopy (SEM) with energy dispersive analysis (EDS) was used to study the morphology and chemical analysis using a JEOL® JSM-6510 instrument with a tungsten electron gun. X-Ray powder diffraction analysis was carried out in a PAN analytical X-Pert X-Ray diffractometer with Cu Kα radiation (λ = 1.5421 Å). Several powder samples were analysed over the 2θ range of 5 – 90°, with a scanning step size of 2θ = 0.026°, and a scan rate of 58 seconds/step. The resulting diffraction patterns were analysed using X'Pert-HighScore® analysis software package.

3. Results and Discussion
Figure 1 shows micrographs of CP Ti and 60Al40V master alloy powder in the as- received condition after mixing (Figure 1). Individual powders had fine and coarse particles.

(a) (b) (c)

Figure 1. Scanning electron microscopy (SEM) micrograph of as-received (a) CP Ti, (b) 60Al-40V master alloy and (c) mixture of CP Ti and 60Al-40V powders.

Figure 2 shows micrographs of the mechanical alloyed powder at different speeds of 800, 900 and 1000 and 1100 rpm using a constant milling duration of 5 hours. It is worth noting in figure 2 that with varying speeds there were no significant changes taking place in the powder so this might mean that the change of speed had no effect on the formation of the alloy. The Ti-6Al-4V alloy powder was already formed at a speed of 800 rpm with a milling duration of 5 hours and this was confirmed by the EDS results in Figure 3 and in Table 1. EDS analysis was done on a few powder particles of different milling speeds in order to confirm the formation of Ti-6Al-4V alloy. Table 1 shows the EDS results of the powder particles at different speeds. It can be seen that with all the different milling speeds, there was no significant change in the chemical composition of the powder and Ti-6Al-4V alloy had formed.

Figure 2. SEM micrographs after milling Ti and master alloy at a constant ball to powder ratio (20:1), milling duration (5 hours) and different speeds of (a) 600rpm, (b) 800rpm, (c) 1000 rpm and (d) 1100rpm.

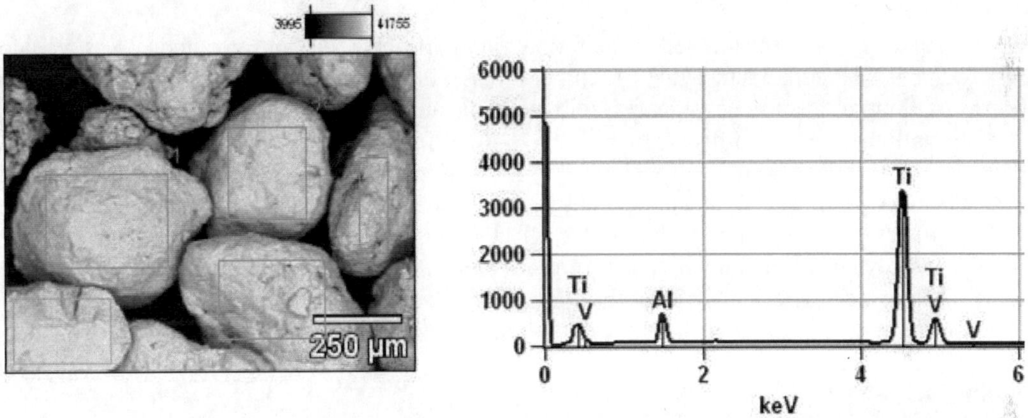

Figure 3. EDX analysis of the typical Ti6Al4V milled at 5 hours milling time.

Table 1. EDS analysis results in wt.% of the as -mixed and mechanically alloyed powders of Ti-6Al-4V for 5 hours at varying speeds of 800, 900, 1000, 1100 rpm.

Elements	As-mixed	800 rpm	900 rpm	1000 rpm	1100 rpm
Ti (wt. %)	90	90.84	89.4	91.25	89.03
Al (wt. %)	6	5.73	6.70	5.47	6.32
V (wt. %)	4	3.43	3.90	3.28	4.65

The X-Ray diffraction (XRD) analysis of the milled CP Ti and 60Al-40V master alloy is shown in Figure 4. XRD patterns of Ti-6Al-4V before mechanical alloying, standard Ti-6Al-4V alloy and after mechanical alloying for 5 hours using different milling speeds are revealed. Mechanically alloyed samples' XRD results were compared with standard Ti6Al4V XRD to confirm the formation of the alloy produced using mechanical alloying. It is worth noting that after 5 hours of milling, Ti peaks had broadened whereas Al and V peaks had disappeared with all different milling speeds.

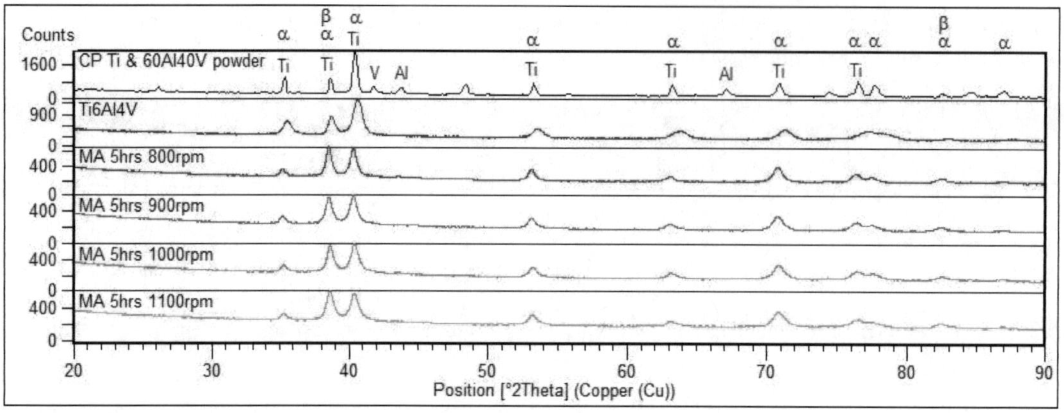

Figure 4. XRD results of the (a) CP Ti and 60Al40V powder, (b) standard Ti6Al4Valloy and mechanically alloyed CP Ti and master alloy for 5hours at different speeds of (c) 800 rpm, (d) 900 rpm, (e) 1000 rpm and (f) 1100 rpm.

Milling CP Ti and 60Al-40V master led to the formation of Al rich phase and V rich phase as revealed by the EDX analysis (Figure 3 and Table 1). This is a result of the chemical segregation caused by the high energy milling process. The repeated cold welding and fracturing of particles and increased density of crystalline defects and dislocations leads to the diffusion of Al and V which also led to the broadening of the mechanical alloyed XRD peaks [3]. Aluminium and vanadium dissolve in the Ti and forms a homogeneous Ti (Al, V) solid solution or phase depending on Al or V content.

The Al and V rich phases are revealed by the XRD analysis (Figure 3) to be α-phase and β-phase respectively. Increasing the rotational speed proportionally increases the kinetic energy input in the ball milling, which was expected to affect the formation of α and β-phase. However, varying milling speeds within the range of 800 to 1100 rpm, evidently is not affecting the formation of α and β-phase.

4. Conclusions

A mixture of CP Ti and 60Al-40V powder were milled using a high energy ball mill at speed range of 800 to 1100 rpm and at constant time of 5 hours. The following conclusions were drawn:

- The high energy milling process may lead to the formation of α-phase (hcp) and β-phase (bcc) which are stabilized aluminium and vanadium respectively.
- The formation of Ti6Al4V is not affected by the milling speed within the rotation range of 800 to 1100 rpm
- Ti-6Al-4V alloy can be produced using high energy ball milling process.

References

[1] Pederson R, "Microstructure and Phase Transformation of Ti-6Al-4V," pp. 27–30, 2002.

[2] Wojtaszek M, Sleboda T, Czulak A, Weber G, and Hufenbach W A, "Quasi-Static and Dynamic Tensile Properties of Ti-6Al-4V Alloy," *Arch. Metall. Mater.*, vol. 58, no. 4, 2013.

[3] Soufiani A M, Karimzadeh F, and Enayati M H, "Formation mechanism and characterization of nanostructured Ti6Al4V alloy prepared by mechanical alloying," *Mater. Des.*, vol. 37, pp. 152–160, 2012.

[4] Yang J, Yu H, Yin J, Gao M, Wang Z, and Zeng X, "Formation and control of martensite in Ti-6Al-4V alloy produced by selective laser melting," *Mater. Des.*, vol. 108, pp. 308–318, 2016.

[5] Chao Q, Hodgson P D, and Beladi H, "Ultrafine grain formation in a Ti-6Al-4V alloy by thermomechanical processing of a martensitic microstructure," *Metall. Mater. Trans. A Phys. Metall. Mater. Sci.*, vol. 45, no. 5, pp. 2659–2671, 2014.

[6] Semenova I P, Smyslova M K, Selivanov K S, Valiev R R, and Modina Y M, "Enhanced

fatigue properties of Ti-6Al-4V alloy turbine blades via formation of ultra-fine grained structure and ion implantation of surface," *IOP Conf. Ser. Mater. Sci. Eng.*, vol. 194, no. 1, 2017.
[7] Ter Haar G M and Becker T H, "Selective laser melting produced Ti-6Al-4V: Post-process heat treatments to achieve superior tensile properties," *Materials (Basel).*, vol. 11, no. 1, 2018.
[8] Suryanarayana C, "Mechanical Alloying and Milling Mechanical Engineering," *Prog. Mater. Sci.*, vol. 46, pp. 1–184, 2001.

IOP Conf. Series: Materials Science and Engineering **430** (2018) 012031 doi:10.1088/1757-899X/430/1/012031

A comparative study of oxygen pick-up of TiHDH powder during press and sinter and loose sintering processing

T M Motsai[1,2], S Chikosha[1,*], C Machio[1] and M E Makhatha[2]

[1] Light Metals, Materials Science & Manufacturing, Council for Scientific and Industrial Research, Meiring Naudé Road, Brummeria, Pretoria 0185, South Africa
[2] Department of Metallurgy, University of Johannesburg, PO Box 524, Auckland park, 2006, South Africa

*SChikosha@csir.co.za

Abstract: Press and sinter is considered to be the most cost effective powder metallurgy process for producing parts. However, loose powder sintering shows to be a more promising cost effective powder metallurgy process as it entails pouring powder into a mould followed by sintering. The differences in their behaviour with respect to densification, maintaining dimensional stability and their oxidation behaviours determine the choice of their industrial applicability. Titanium has a high affinity for oxygen which in turn makes it difficult to process components from powder. It also affects the mechanical properties of products significantly; particularly in applications where ductility is imperative. The focus of this study was therefore to evaluate the oxygen pick-up of the two cost effective powder metallurgy processes (press and sinter and loose sintering). A 100 Mesh TiHDH powder was used for sintering. Sintering was performed at 1500 °C for 4 hours. The oxygen contents of the green and sintered compacts were then compared. High oxygen contents were observed in tap density powder and pressed samples. The loosely sintered components showed high oxygen pick-up after sintering while oxygen pick-up decreased with increasing pressure in pressed samples. These results show that press and sinter is advantageous over loose sintering where oxygen control is critical.

1. Introduction

Titanium (Ti) is a structural metal that is used mostly for its light weight and corrosion resistant characteristics [1]. Powder metallurgy (PM) of titanium alloys was established in the 1980s [2]. It is a particularly attractive manufacturing process for titanium components [3]. This is due to its potential for true net shape capability combined with mechanical properties that are equivalent to or exceed cast and wrought products [4]. PM is a near-shape and solid state method which significantly limits the need for machining [5]. The main driving force behind PM processing is cost reduction [2]. So far, the most cost effective powder metallurgy process (for parts production) is press and sinter [6], [7]. This process involves compaction of powders and sintering stages [8]. However, loose powder sintering of Ti shows to be a more promising innovative cost effective process of producing powder metallurgy parts as it eliminates the compaction stage. Loose powder sintering is a process which entails pouring or mechanically vibrating powder into a mould of a desired shape followed by sintering [9].

One of the critical parameters in powder metallurgy of titanium is the selected process' ability to possess minimal oxidation behaviour. The oxidation process occurs through oxygen adsorption, oxide nucleation, lateral growth of the oxide and compact oxide scale formation [10]. Oxygen adsorption involves the attachment of oxygen on the metal surface through chemisorption or physical adsorption. After adsorption, the oxygen molecules dissociate and get adsorbed as atoms. During the oxide nucleation process, oxide nuclei form after the metal surface is saturated with adsorbed oxygen. Lastly, the lateral nuclei growth result in the formation of a compact oxide scale that covers the metal surface.

Small amounts of oxygen, nitrogen and carbon can significantly alter the mechanical properties of Ti, particularly its ductility [5]. To prevent or minimise oxidation [11], sintering of titanium is usually carried out in high-vacuum [12] or flowing argon gas [13]. The authors could not find literature on work done to compare the amount of oxygen pick up during sintering of loosely packed and tightly packed titanium powder. This work therefore aims at studying and comparing the oxidation behaviour of TiHDH 100 mesh during press and sinter and loose sintering processes.

2. Experimental procedure

TiHDH 100 mesh (-150 μm) powders were supplied by Baoji Lihua Non- ferrous Metals Co., Ltd in China. The powder and pressed compacts were analysed for oxygen content using the ELTRA ONH 2000 Oxygen Nitrogen Hydrogen Determinator PC controlled machine. The powder was pressed using a 17.5 mm high-speed tool steel die at three different nominal pressures of 120, 440 and 1170 MPa. The mass of powder per sample was approximately 7 g. Cold compaction was carried out using an Enerpac 100 T press. Densities were determined from the mass and dimensions of the samples. The masses were measured using an Ohaus explorer mass balance while dimensions were measured using the Vanier calliper. The theoretical densities (%TD) were determined using Equation 1.

$$Theoretical\ density\ (\%TD) = \left(\frac{\mathrm{m}}{\pi \mathrm{h}(\frac{\mathrm{d}}{2})^2} / 4.51 \right) * 100 \qquad (1)$$

where *m* is the mass (g), *h* is the height (cm) and *d* is the diameter (cm) of the compact.

The loose sintering samples were achieved by sintering powders in their aerated (untapped) and tapped states using alumina crucibles. Each powder sample was poured into the crucibles and left untapped or tapped before sintering. The tapped state was obtained after 60 taps. Each tapping consisted of lifting the crucible containing the powder to a height of approximately 10 mm from the working surface. Equation 1 was used to calculate both bulk densities.

The sintering of the powders (aerated and tapped) and green compacts was carried out at 1500°C for 4 hours. Ceramic reef discs were used to close the tops of the alumina crucibles. Sintering was done in a Carbolite tube furnace using a heating rate of 10°C/min in an inert argon atmosphere at a flow rate of 1 L/min and the experiment was terminated by a furnace cool. The ends of the work tube were fitted with gas-tight end seals to keep out air from the atmosphere.

The densities of the sintered samples were determined as for the green compacts, from their masses and dimensions. For overall oxygen content analysis, and in order to accommodate location differences, approximately 100 mg sections of the sintered compacts were cut from the top, middle and bottom of the compacts using a BRILLANT 221 abrasive cutter. The sections were analysed for oxygen content using the ELTRA ONH 2000 and the overall oxygen pick-up (from the top, middle and bottom sections of the samples) was reported as an average. The ELTRA ONH 2000 was calibrated using standards. The level of deviation in the oxygen content of the three standards tested was 0.00168/1.18775 %.

3. Results and discussion

3.1 Initial oxygen content

Table 1 gives the oxygen contents of the powders used in this work. The tested oxygen contents were found to be higher than those given by the suppliers on the supplier certificates. The tested oxygen content of TiHDH -150 μm was found to match that of the ASTM Ti grade 2 [4].

Table 1. The starting oxygen contents of TiHDH with varying starting densities.

TiHDH 100 mesh		wt. % O_2
Powder	Supplier	0.10

	Tested	0.19

3.2 Visual examination

Photograph in figure 1 shows a change in colour of the sintered compacts with increasing density from the aerated state (1) to a pressure of 1170 MPa (2). Compacts sintered in the aerated state were gold in colour as compared to those sintered after pressing at 1170 MPa which were silver in colour. The colour change observed is due to oxidation and nitridation which is commonly observed during sintering of titanium powders [10]. Another interstitial that can be picked up, besides oxygen and nitrogen, is carbon [14]. However, it should be noted that the focus of this study is on oxygen pick-up of titanium during sintering. The source of oxygen and nitrogen could have been the oxygen on the particle surface, air in the pores of the loose powder "compacts", air that leaked into the furnace or the argon gas. Loosely packed powders have greater porosity, and therefore, more entrapped air, as compared to tightly packed or compacted powders. This resulted in high discolouration in the aerated sample (1) with the highest porosity as compared to samples pressed at the highest nominal pressure of 1170 MPa (2), which had the least porosity. Oxygen and nitrogen studies were carried out using the ELTRA ONH 2000.

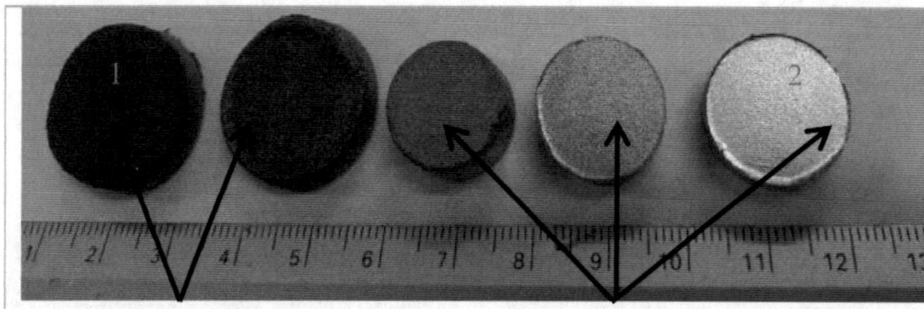

Figure 1. Photograph of samples sintered 1500 °C for 4 hours.

3.3 Density and oxygen results

The initial (a) densities and (b) oxygen contents are given as grey bars and lines in figure 2 respectively; the grey bars represent the initial densities while the grey lines represent the starting oxygen contents. Tapping led to a statistically significant increase in the bulk density from 38 % TD in the aerated state to 44 % TD for the tapped state. This is in agreement with previous study where tapping led to increased contact area between the particles, resulting in increased bulk density [15].

With the pressed samples, the green density also showed an increase with increasing pressure, increasing from 62 % TD when pressing at 120 MPa to 90 % TD when pressing at 1170 MPa. This was expected as the volume of pores in commercially pure (CP) Ti decreases with the increase in pressure, resulting in the increase of sample theoretical density as observed in other titanium powders [16].

The bulk densities of the loose powders were lower than the densities of the compacted powders, tentatively because of the application of a load during compaction causing the powder particles to move closer and decrease the volume of pores, hence the increased density [16].

The oxygen content of the powder that was loose sintered increased with tapping from 0.19 to 0.23%. This was unexpected as the aerated powder is expected to have more open pores due to increased distance between particles as compared to the tapped and pressed compacts. Oxygen content increases with increase in porosity [17]. There is no significant change in oxygen content of the pressed compacts with the increase in pressure. This is in agreement with the study done by Baril *et al.* where oxygen content does not change with pressing [18].

The lowest oxygen content of 0.19 % was observed in the aerated powder. The highest oxygen content of 0.23 % was achieved with the tapped powder and compacts pressed at 120 and 1170 MPa.

The reason for the increased oxygen content with decreased porosity from aerated to both the tapped powder and cold compacted samples are unknown and still under investigation.

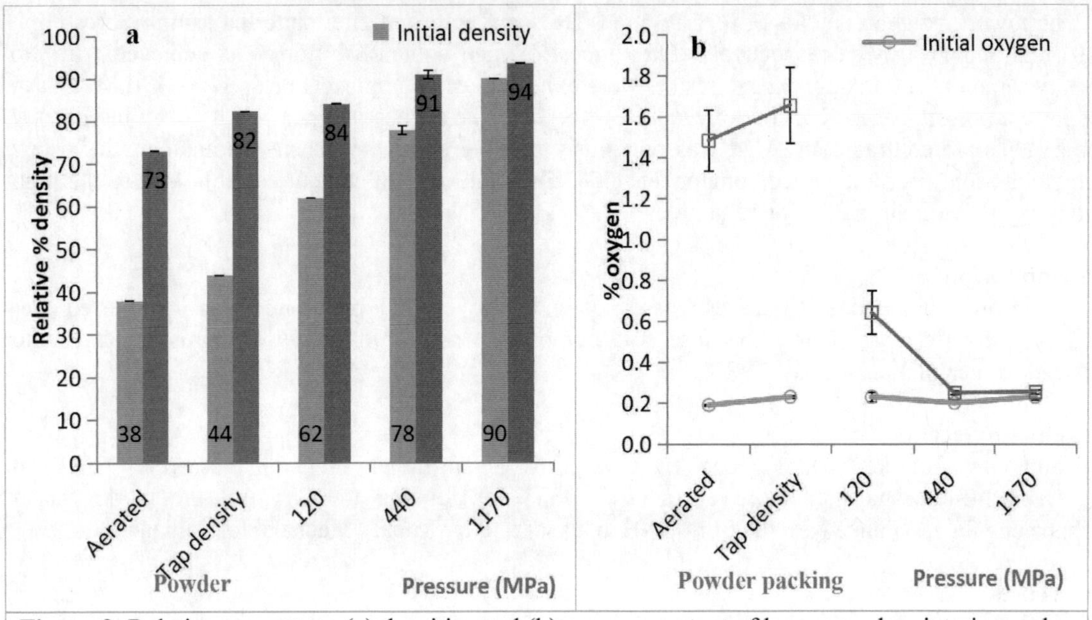

Figure 2. Relative percentage (a) densities and (b) oxygen content of loose powder sintering and press and sinter at 1500 °C for 4 hours.

The final (a) densities and (b) oxygen contents of compacts sintered at 1500 °C for 4 hours are also given in figure 2. These are represented by the blue and red bars, and the blue and red lines respectively. There is an increase in density of loose sintered samples from 73 % in the untapped state to 82 % in the tapped state. This could be due to the higher starting density in the tapped state as compared to the untapped state. The increase in density with tapping after loose sintering is attributed to the increased contact area between the particles, resulting in increased bulk density, hence high sintered density [15].

The brown density of the compacted samples also increases with the increase in pressure. The increase is significant in samples pressed at 120 and 440 MPa, and insignificant between samples pressed at 440 and 1170 MPa. The increase in brown density with increasing pressure could be due to decreased porosity as pressure increases, leading to increased sintered density [16].

The brown densities of the loose powder sintering process (blue bars) are lower than those of the press and sinter samples (red bars). This could be due to the high starting densities in press and sinter as compared to loose sintering as high starting densities lead to high final densities [15].

The final oxygen content of loose sintered samples showed an insignificant difference with the increase in packing; see figure 2 (b), blue line. This behaviour was unexpected as tapped powder has lower porosity, and therefore, less entrapped air compared to untapped powder. Increased porosity/ entrapped air will lead to high oxidation [17]. The reason for the insignificant increase in oxygen with the increase in packing (decreased porosity) in loose sintered samples could be due to the high starting oxygen contents of 0.23 % in the tapped powder compared to 0.19 % in the untapped powder.

There is a decrease in oxygen content with the increase in pressure from 120 to 1170 MPa in the press and sinter compacts. This is supported by previous work of others where low porosity samples had low oxygen content as compared to high porosity samples [17].Looking at the final oxygen content, blue and red lines, there is a significant difference in the amount of oxygen in the two processes. The press and sinter samples picked up less oxygen as compared to the loose sintered samples. The significant difference is observed between the tap density samples and the sampled pressed at 120 MPa. The difference in the oxygen pick-up of loose sintering and press and sinter of TiHDH 100 mesh was expected as press and sinter samples had less initial porosity/ high starting density as compared to loose sintered samples [17].

The lowest oxygen contents of 0.25 % and 0.26 % are achieved after sintering samples pressed at 440 MPa and 1170 MPa respectively. The highest oxygen content of 1.66 % is achieved with tap density samples. The lowest oxygen results were expected as the compacts pressed at 1170 MPa have the lowest porosity, therefore low oxygen pick-up [17]. The highest oxygen results from tap density components are contradicting what was previously discovered as tap density components have low porosity as compared to aerated components [17]. The high oxygen content could be due to the high initial oxygen content in the tapped samples.

4. Conclusions

It is therefore concluded that, in cases where oxygen control is vital, press and sinter is preferred over loose powder sintering. This is because cold compaction results in decreased porosity, leading to reduced oxygen pick up.

Acknowledgements

The authors would like to acknowledge the Council for Scientific and Industrial Research (CSIR) and the Department of Science and Technology (DST) for funding to carry out the work; Suzan Nchabeleng for assistance with the Eltra ONH analyser and Dr Ronald Machaka for guidance.

References

[1] S. Petersen, *Investigation of the hot deformation sintered titanium compacts produced from direct reduction powder*, 2010.

[2] F. H. Froes, A historic perspective of titanium powder metallurgy, in *Titanium Powder Metallurgy: Science, Technology and Applications*, F. H. Qian, Ma & Froes, Ed. Butterworth-Heinemann, 2015, pp. 1–19.

[3] M. Qian, G. B. Schaffer, and C. J. Bettles, *Sintering of Advanced Materials*. Elsevier, 2010.

[4] Donachie MJ, Titanium - A Technical Guide, *Mater. Inf. Soc.*, 2000.

[5] L. Bolzoni, P. G. Esteban, E. M. Ruiz-Navas, and E. Gordo, Mechanical behaviour of pressed and sintered titanium alloys obtained from master alloy addition powder, *J. Mech. Behav. Biomed. Mater.*, vol. 15, pp. 33–45, Nov. 2012.

[6] E. J, Kováčik ;Š, Sintering of HDH Ti powder," *Sci. Proc.*, pp. 50–55, 2014.

[7] J. Lou, B. Gabbitas, and D. Zhang, Improving the uniformity in mechanical properties of a sintered Ti compact using a trace amount of internal lubricant, *J. Mater. Process. Technol.*, vol. 214, no. 9, pp. 1798–1805, Sep. 2014.

[8] Robertson, I. M. Robertson, and G. B. Schaffer, Review of densification of titanium based powder systems in press and sinter processing, *Powder Metall.*, vol. 53, no. 2, pp. 146–162, 2010.

[9] Y. Torres, S. Lascano, J. Bris, J. Pavn, and J. A. Rodriguez, Development of porous titanium for biomedical applications: A comparison between loose sintering and space-holder techniques, *Mater. Sci. Eng. C*, vol. 37, no. 1, pp. 148–155, 2014.

[10] B. Sefer, Oxidation and Alpha–Case Phenomena in Titanium Alloys used in Aerospace Industry: Ti–6Al–2Sn–4Zr–2Mo and Ti–6Al–4V, Luleå University of Technology, 2014.

[11] I. Skotnicova, K; Kursa, M & Szurman, *Powder metallurgy (University textbook)*. 2014.

[12] W. H. Fang Z.Z, Sun, P., Hydrogen Sintering of Titanium to Produce High Density Fine Grain Titanium Alloys, *Adv. Eng. Mater.*, vol. 14, no. 6, pp. 383–387, 2012.

[13] M. Qian, G. B. Schaffer, and C. J. Bettles, 13 – Sintering of titanium and its alloys, *Sinter. Adv. Mater. - Fundam. Process.*, pp. 324–355, 2010.

[14] V. A. R. Henriques, S. L. G. Petroni, M. S. M. Paula, C. A. A. Cairo, and E. T. Galvani, Interstitial Control in Titanium Alloys Produced by Powder Metallurgy, *Mater. Sci. Forum*, vol.

660–661, pp. 3–10, 2010.
[15] E. Abdullah and D. Geldart, The use of bulk density measurments as a flowability indicators, *Powder Technol.*, vol. 102, no. March 1998, pp. 151–165, 1999.
[16] W. Chen, Y. Yamamoto, and W. H. Peter, Investigation of Pressing and Sintering Processes of CP-Ti Powder Made by Armstrong Process, *Key Engineering Materials*. 2010.
[17] D. Y. Chen, Z. C. Li, W. Miao, and Z. J. Zhang, Effects of Porosity and Temperature on Oxidation Behavior in Air of Selected Nuclear Graphites, *Mater. Trans.*, vol. 53, no. 6, pp. 1159–1163, 2012.
[18] E. Baril, L. P. Lefebvre, and Y. Thomas, Interstitial elements in titanium powder metallurgy: sources and control, *Powder Metall.*, vol. 54, no. 3, pp. 183–186, 2011.

Exploring Micro-focus X-ray computed tomography for metal injection moulded green parts

M Seerane[1,2,3]

[1] Light Metals, Materials Science & Manufacturing, Council for Scientific and Industrial Research, Meiring Naudé Road, Brummeria, Pretoria 0185, South Africa
[2] Titanium Centre of Competence, Materials Science & Manufacturing, Council for Scientific and Industrial Research, Meiring Naudé Road, Brummeria, Pretoria 0185, South Africa
[3]School of Chemical and Metallurgical Engineering, University of the Witwatersrand, Johannesburg, South Africa

*mseerane@csir.co.za

Abstract: Metal injection moulding (MIM) involves mass production of simple to intricate near-net shape parts at a relatively low cost by blending metal powders of defined characteristics with a polymeric binder system. The blend, popularly known as a feedstock, is injected into a mould cavity of a desired shape to form a green part. Most defects such as cracks, sink marks and voids manifest during the injection stage of the process. Internal cracks and voids are impossible to visualise after injection moulding. Micro-focus X-ray computed tomography (µXCT) is one of the techniques that can be used to evaluate and characterise such defective features. µXCT is a non-destructive method that uses penetrating X-ray radiation to probe objects or samples and create a complete virtual 3D representation of the object to visualise the internal structures and morphologies such as porosity, voids, cracks and inclusions as well as their location and their distribution in 3D. This work is intended to explore the µXCT approach MIM green parts and characterise the internal defects. The samples were scanned at 135 keV and 100 µA for X-ray penetration and contrast while placed onto a rotating sample manipulator which facilitated scanning through 360°. The samples were then reconstructed and further analysed for volume (3D) graphics.

1. Introduction

Metal Injection Moulding (MIM) is a relatively new technology that produces near net shaped medium-to-complex metal components at low cost [1]. This process involves blending of a metallic powder with a polymeric binder to produce a feedstock for forming a part of desired shape. MIM consists of four basic steps: 1) blending 2) injection moulding 3) debinding 4) sintering [2]. Injection moulding forms part of the most critical steps in MIM and affects the subsequent steps in the process [3].

Defects such as weld lines, jetting, incomplete filling, air traps, gate and ejection pin marks are common during injection moulding. These defects are generated during mould filling and part ejection as a result of the process settings and the quality of the feedstock. It is, therefore, important that this step is fully understood and that the filling pattern of the feedstock into the mould cavity is well monitored. The process settings are studied and optimised in different ways. Some of the common methods of studying and obtaining the optimum process settings for injection moulding is Factorial design and Taguchi approach [4], [5]. The filling pattern of the feedstock into the mould cavity and the evolution of defects is often difficult to trace and detect *in-situ*. Rheological models [6] and computer aided engineering (CAE) analysis [7] are normally adopted to predict the complex flow behaviour of the feedstock into the mould cavity.

For defect analysis, microscopic examination and visual inspection of the green parts have been useful methods to identify sample defects [8]. These examinations can be conducted destructively and non-destructively, respectively. Destructive inspection include breaking or slicing of the green part. Non-destruction examination includes penetrant inspection (PI), magnetic particle inspection (MPI) and eddy current testing (ECT). These inspections are chosen on the basis of the nature and size of the defect, or whether the defects are systematic or random. Defects can be located either on the surface of the green specimen or internally. Surface defects are easily identified and can be located visually or through PI, MPI and ECT. However, internal defects are more difficult to detect using the latter traditional techniques unless the suspected internal defect generates an external morphology. In addition, MPI is only limited to ferromagnetic materials such as Fe, Ni, Co and their alloys; whereas ECT requires conductive materials and PI is suitable for materials with relatively non-porous surface. Scanning electron microscope (SEM) and optical microscope (OM) are suitable tools for analysing defects from sintered components; also, these techniques are limited to surface analysis.

It is therefore important to adopt methods that can resolve and characterise internal porosity and defects (large or small) of a sample. Large internal defects can be detected using radiography or ultrasonic echography. The biggest challenge is encountered when the defects are less than 1 mm in dimension. Micro-focus X-ray computed tomography (μXCT) is becoming an attractive technique to resolve and characterise defects as small as 1 mm. μXCT is a non-destructive radiation based technique for high quality micron-level information of the interior as well as composition of samples [9]. This technique uses an X-ray beam to probe objects or samples to create a complete virtual 3D volume that reveals internal structures such as defects, porosity and inclusions at high spatial resolution better than 1 μm [10], [11]. However, the big limitation in to this system is that the focal spot is relatively large (3 mm) and as a result the spatial resolution obtained, even with small sample, is in the order of 0.08 mm. Also, metals are often a challenge due to the relatively higher absorption of the X-rays. This is normally improved by increasing the X-ray energy. The technique is commonly applied in geoscience, archaeological science and medical field.

μXCT is gaining a lot of attention in MIM for quality inspection and process optimisation. Muchavi et al. [12] adopted this technique to explore and characterise the size, concentration and distribution of pores in MIM green and sintered parts. It was claimed that the technique can be used as a reliable tool to reveal detailed 3D internal structure of MIM parts qualitatively and quantitatively. Mannschatz et al. [13] and Yang et al. [14] used this method to analyse MIM parts and optimise the process. However, the authors did not give a detailed information on how the technique works. Hoffman and De Beer [9] and Bam et al. [10] outlined the characteristics of μXCT facility at Necsa and how it can be applied in geological science. This work is aimed at exploring and understanding the capabilities of μXCT technique in MIM research, especially for MIM practitioners who are beginning to adopt the technique.

2. Materials and experimental approach

The MIM green components were produced from titanium alloy powder and wax-based binder using a 40 tonne ARBURG Allrounder 400-70 injection moulding machine. Injection moulding trials were done through design of experiments (DoE) approach and each trial process parameters were recorded. The injection moulding parameters used are shown in Table 1. A total number of trials for each combination of the parameter values was reduced to 9 using Minitab software as a DoE tool. The specimens made are tensile test specimens with dimensions 89 mm (overall length), 40 mm (gauge length), 14.8 mm (grip section diameter) and 5 mm (gauge section diameter) (See figure 1).

Table 1. Injection moulding parameters

	Factor	Level		
		1	2	3
A	Injection temperature [°C]	130	135	140
B	Injection pressure [Bar]	600	700	800
C	Volumetric Injection speed [$cm^3.s^{-1}$]	30	40	50
D	Mould temperature [°C]	30	40	50

The samples were taken for tomography scans. The scanning unit is a Nikon XTH 225L µXCT (Nikon Metrology, Leuven, Belgium) based at the MIXRAD laboratory at the South African Nuclear Energy Corporation (NECSA, Pelindanba). Important scanning parameters include X-ray energy, exposure time, beam current and number of projections. A combination of these parameters give different image contrasts and quality for optimal data acquisition. The MIM specimens were mounted on a polystyrene mould and placed onto a 360° rotating sample manipulator for scanning process. After scanning, the samples were then reconstructed using Nikon CTPro software (Nikon Metrology, Leuven, Belgium) to transform 2D projections into virtual 3D volumes. The reconstruction parameters are optimised automatically for quality scanning to avoid error of measurements by the operator. The samples were further analysed using VGStudioMax V2.2 (Volume Graphics GmbH, Heidelberg, Germany) for material-volume isolation and defect/pore/inclusion identification.

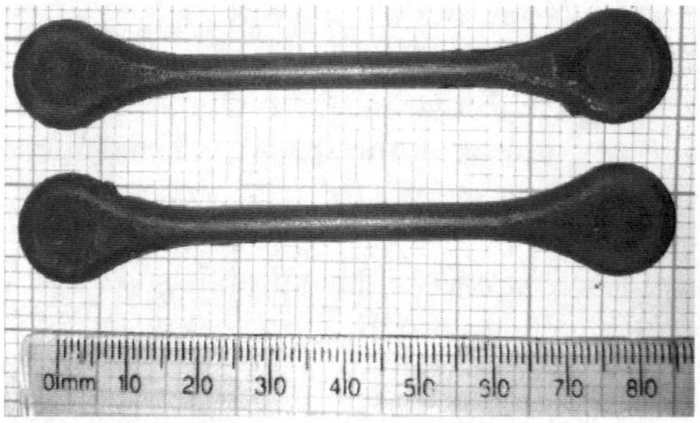

Figure 1. MIM green parts as tensile test specimens

Figure 2 shows a micro-focus XCT process from beam release to part reconstruction and analysis. An X-ray beam is released from the source to interact with the sample. A filter, normally copper, is placed in front of the specimen to remove low/weak energy X-rays to avoid beam hardening. Another common artefact is cone beam artefact which result from the positioning of the specimen within the beam. High energy X-rays interact with the sample and are attenuated. The attenuated beam is collected by the detector. The detector projections are reconstructed within a field of view at a full rotation of 360°. This stage allows optimum image acquisition and the reduction of artefacts such as misalignment artefacts, ring artefacts and line artefacts [10]. The acquired image is transformed from 2D (pixels) to 3D (voxels) volume. This volume is then analysed for internal features.

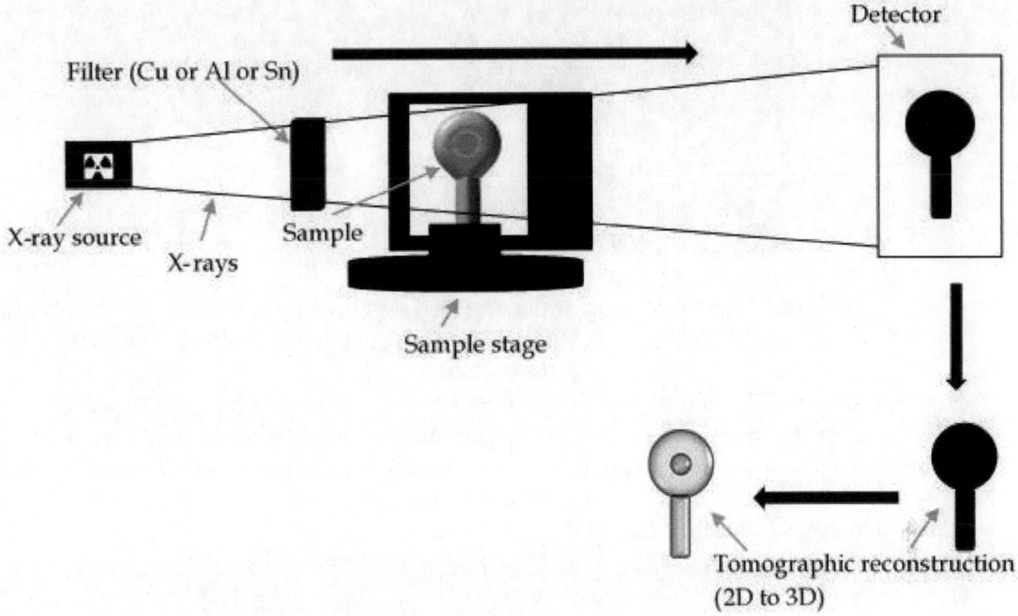

Figure 2. A schematic representation μXCT process

3. Results and discussion

Figure 3(a-d) shows a sequential process of the reconstructed MIM test piece from material isolation to a 3D volume using a VGStudioMax. In Figure 3(b) the volume is isolated from the material to reveal the surface and internal features of the sample. Blisters and bubble-like defects can be seen. Further analysis show finely distributed pores and cracks on the sample volume (Fig. 3c and d). These defects resulted from a set of conditions as follows: injection temperature – 140 °C, injection pressure – 800 bar, injection speed – 40 $cm^3.s^{-1}$ and mould temperature – 30 °C. The order of these defects and pores were quantified in the range of 0.01 to 2.03% combined, using VGStudioMax. This type of defects affect MIM subsequent processes such as debinding and sintering since they act as crack initiation sites and weak areas during mechanical testing, and hence part failure. This technique has proved to be a reliable source to reveal and quantify surface and internal features of MIM samples without any destruction, whereas other NDT test methods (ECT, MPI and PI) can only analyse surface defects and that the defects cannot be easily quantified.

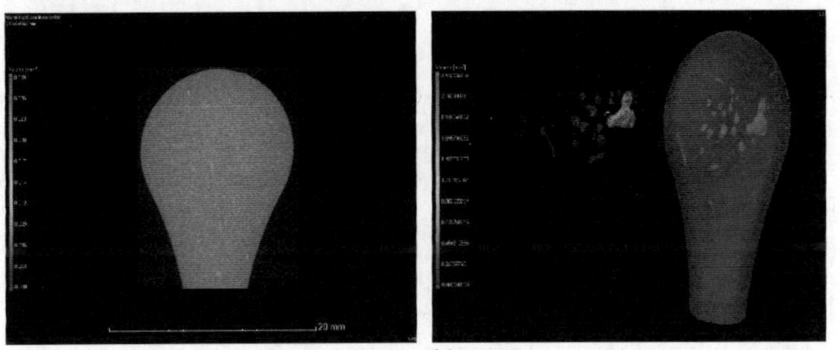

a) b)

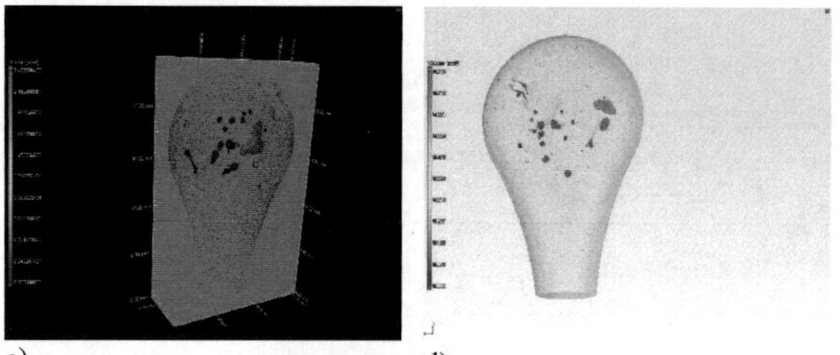

c) d)

Figure 3. Transformation of the acquired sample from pixels (2D) to voxels (3D): (a) reconstructed 2D image, (b) voxel isolation (c) voxel analysis (d) completed 3D volume

Figure 4 shows a 3D volume of MIM sample produced from optimum MIM process and it can be clearly seen that the sample does not contain any defects, both internally and externally. The defect free optimum process parameters are as follows: injection temperature – 130 °C, injection pressure – 800 bar, injection speed – 40 $cm^3.s^{-1}$ and mould temperature – 40 °C. The porosity level was found to be 0.01% and this is acceptable for MIM green parts.

Figure 4. A test piece volume produced from MIM optimised process parameters

4. Conclusions

The μXCT technique was explored using MIM test piece samples produced at the CSIR. The technique proved to be a quick, reliable and most effective source to analyse not only geoscience, medical and archaeological samples but also MIM samples with no significant limitation in terms of material type and sample size. Also, the technique was able to reveal internal features of the part unlike traditional NDT methods such as ECT, MPI and PI. Set of parameters that led to noticeable defects during injection moulding were found to be as follows: injection temperature – 140 °C, injection pressure – 800 bar, injection speed – 40 $cm^3.s^{-1}$ and mould temperature – 30 °C. However, the optimum set of parameters that showed no signs of defects were 130 °C for injection temperature, 800 bar for injection pressure, 40 $cm^3.s^{-1}$ for injection speed and 40 °C for mould temperature.

Acknowledgements

The author would like acknowledge the Council of Science and Industrial Research (CSIR) and Department of Science and Technology (DST) for funding this work. The author also extends acknowledgements to the South African Nuclear Energy Corporation (NECSA), MIXRAD laboratory for providing a μXCT facility and lessons, in particular Mr. L. Bam and Mr. FC De Beer.

References

[1] Li S, Huang B, Li Y, Qu X, Liu S and Fan J 2003 *J. Mater. Process. Techno.* **137** 70–73.
[2] Moballegh L, Morshedian J and Esfandeh M 2005 *Mater. Lett.* **59** 2832–2837.
[3] Fang W, He X, Zhang R, Yang S and Qu X 2014 *Powder Technol.* **256** 367–376.

IOP Conf. Series: Materials Science and Engineering **430** (2018) 012032 doi:10.1088/1757-899X/430/1/012032

[4] Kamaruddin S, Khan Z A and Foong S H 2010 *Int. J. Eng. Technol.* **2**
[5] Packianather M, Chan F, Griffiths C, Dimov S and Pham D T 2013 *Procedia CIRP* 12 300–5.
[6] Machaka R, Ndlangamandla P and Seerane M 2018 *Powder Technol.* 326 37–43.
[7] Shin K, Heo Y, Park H, Chang S and Rhee B 2013 *Materials (Basel)* 6 5878–92.
[8] Schlieper G, Dowson G and Williams B 2013 *Metal injection moulding*, 3rd ed. Shrewsbury (European Powder Metallurgy Association).
[9] Hoffman J W and de Beer F C 2012 *18th World Conf. Nondestruct. Test.* 1–12.
[10] Bam L C, Miller J A, Becker M, de Beer F C and Basson I 2016 *The Third Ausimm International Geometallurgy Conference* 15–6.
[11] Landis E N and Keane D T 2010 *Mater. Charact.* **61** 1305–16.
[12] Muchavi N S, Bam L, de Beer F C, Chikosha S and Machaka R 2016 *J. South. African Inst. Min. Metall.* **116** 973–80.
[13] Mannschatz A, Müller A and Moritz T 2011 *J. Eur. Ceram. Soc.* **31** 2551–2558.
[14] Yang S, Zhang R and Qu X 2015 *Mater. Charact.* **104** 107–115.

Metal injection moulding of a 17-4 PH stainless steel: a comparative study of mechanical properties

Ronald Machaka[*,1]

[1] Light Metals, Materials Science & Manufacturing, Council for Scientific and Industrial Research, Meiring Naudé Road, Brummeria, Pretoria 0185, South Africa

[*] rmachaka@csir.co.za

Abstract. While the metal injection moulding (MIM) of the precipitation-hardenable 17-4 PH stainless steel has found wide-ranging applications in a number of industries and popularity as a workhorse alloy, data on the as-sintered properties 'that can be expected' is limited and scattered in the literature. A quantitative comparison of the mechanical properties of as-sintered 17-4 PH materials prepared via MIM is currently not available in the literature. -5 μm, -15 μm, and −45 μm starting 17-4 PH stainless steel powder materials and their bimodal mixtures were used to formulate with feedstocks investigated in this work. This paper compares the as-sintered mechanical properties resulting from the measured tensile tests against those properties reported in the literature, those specified in the ASTM and MPIF standard minimum specification for MIM materials, those properties specified in the material suppliers' datasheets and the mechanical properties of 17-4 PH wrought material – for completeness. While the as-sintered properties of the CSIR 17-4 PH material are generally lower than expected, the comparison shows that the measured properties are within the industry standard specifications. Findings from this study are a testament that a range of mechanical properties is feasible from the 17-4 PH alloy.

1. Introduction

The precipitation-hardenable 17-4 PH stainless steel has a remarkable combination of superior properties. It is the most common and most used type of the precipitation-hardenable stainless steels [1] and it suitably finds wide-ranging applications in a number of industries [2,3].

Many researchers have reported attempts to fabricate 17-4 PH components via numerous techniques alike. The challenges associated with fabricating of complex, near-net shaped 17-4 PH stainless steel components via the conventional press and sinter (P&S) and additive manufacturing (AM) techniques are well documented. These challenges include low powder compressibility for P&S [4] and the undue residual stress accumulation from martensitic transformation during a deposition for AM [1,5,6]. More recent reports have however widely reported on the attractiveness of metal injection moulding (MIM) in the fabrication of complex, near-net shaped 17-4 PH components, for example, Wu et al have reported MIM 17-4 PH stainless steel as a promising material for the industry because of its superior mechanical and corrosion performances [2]. The MIM technique is suitable for the cost-effective manufacturing of high volume production of intricately shaped parts where high dimensional accuracy is required – further details are discussed elsewhere, see Ref.s [7–11].

The current article builds on some of our prior reported preliminary findings [12–14] from the studies underway at the CSIR on the development of versatile custom 17-4 PH MIM feedstocks. The

IOP Conf. Series: Materials Science and Engineering **430** (2018) 012033 doi:10.1088/1757-899X/430/1/012033

current article reports on a comprehensive comparison of the static properties of the 17-4 PH materials prepared at the CSIR and those 17-4 PH MIM material properties reported by others in literature, the materials standard specifications for as-MIM 17-4 PH [15,16], 17-4 PH material properties specified in the material (powder and feedstock) suppliers' datasheets.

To make the comparison more meaningful, only the as-sintered 17-4 PH MIM properties are considered from at least fifty prior studies. The material properties of wrought 17-4 PH are included for completeness of the manuscript. This current report neglects the influences of the powder and feedstock characteristics, injection moulding and sintering parameters on the static mechanical properties (tensile, hardness, density and shrinkage) of the metal injection moulded 17-4 PH stainless steel materials.

2. Materials and Experimental Methods

Three starting 17-4 PH stainless steel powder materials, the -45 µm Grade 630 (supplied by Praxair Surface Technologies), the -5µm and the -15µm (both supplied by Epson Atmix Corp) were used for this study. The morphologies of the starting metal powders are reported in Figure 1 while a summary of the powder particle size distributions and chemical composition have been reported elsewhere [12,14].

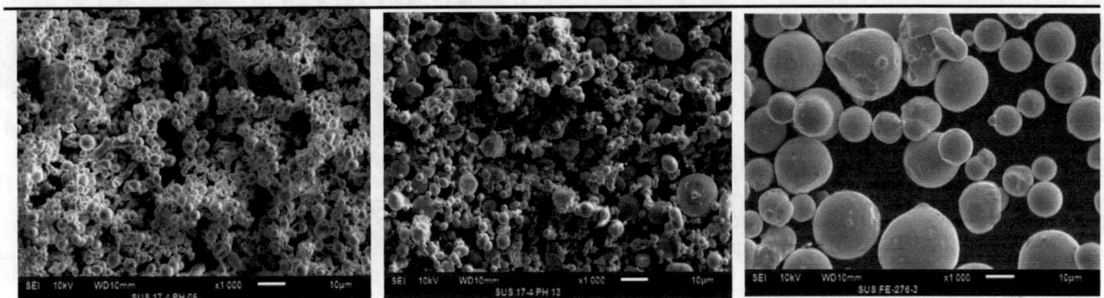

(a) (b) (c)

Figure 1. The SEM micrographs of the starting 17-4 PH stainless steel powder materials -5µm (a), -15µm (b) and -45µm (c) at 1000x magnification [12,14].

2.1. *Feedstock preparation and injection moulding*

The MIM feedstocks were prepared by mixing 17-4 PH stainless powder materials with a proprietary wax-polymer binder system developed at the CSIR as explained elsewhere, see Ref.s [7,8,12,14]. The binder system consists of paraffin wax, low-density polyethene, polypropylene and stearic acid proprietary grades. The solids loading (volume ratio of the powder to the total volume of the powder and binder) used for the study ranged from 55 to 65 vol%. The feedstocks were granulated and then injection moulded into the standard 'dog-bone' tensile test specimen according to ISO 2740 specifications (see Figure 2(a)) at 140°C using a 40-tonne ARBURG Allrounder 270U 400-70 injection moulding machine. The optimization of the injection moulding process parameters (injection temperature, speed and pressure and mould temperature) in our previous work [7,8].

2.2. *Debinding and sintering*

Debinding of the green components was performed in two steps; i.e. solvent debinding in *n*-heptane (Merck) at 60 °C for 24 hours followed by thermal debinding at 550°C for 1 hour in a Carbolite tube furnace under a controlled atmosphere of argon gas flowing at 1.0 L/min. The debinding process parameters have been reported on before, see Ref.s [7,8]. The sintering of the debound components was also performed in two steps; a pre-sintering stage at 1000 °C for 1 hour in the tube furnace (flowing Ar. gas) followed by final sintering in a high-vacuum furnace. The final sintering step was performed at 1300 °C for 3 hours. A Xerion cold-wall high-vacuum furnace with molybdenum shield packs and tungsten heating was used. The dimensions of the produced as-produced, solvent debound, and the as-sintered specimen are illustrated in Figure 2(b).

2.3. Post-sintering evaluation

The sintered samples were evaluated for density, microstructure, pore morphology, hardness and tensile properties. The density of the sintered samples was determined according to the immersion method (Archimedes' principle) as outlined in ASTM B311. The microstructure and pore distribution were determined using a Leica DMI500 M optical microscope. Apparent hardness measurements were carried using an automated Vickers microhardness tester (FM-700) as outlined in ASTM-E384. The tensile testing was conducted as per ISO 2740 standard on the INSTRON™ Servo Hydraulic 1342 test instrument at a rate of 0.5 mm/min.

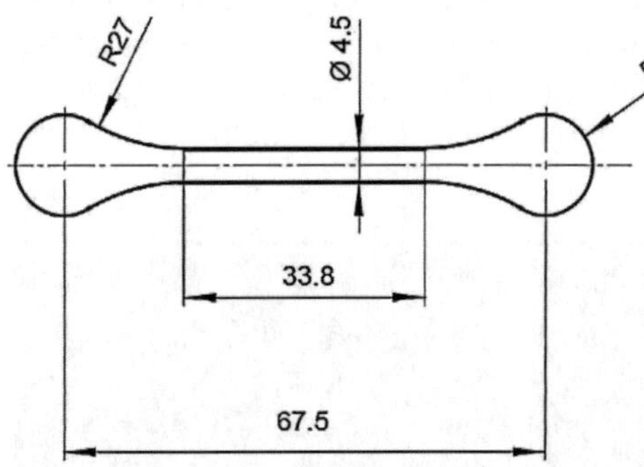

(a)

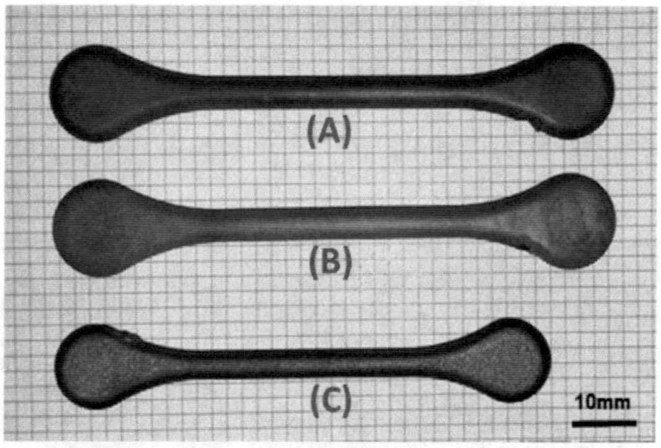

(b)

Figure 2. the (a) design drawing of the as-sintered 'dog-bone' test specimen as per ISO 2740 standard and (b) photographs of the produced as-moulded (A), solvent-debound (B) and sintered (C) tensile test specimen as reported previously in Ref. [12]. The dimensions shown are all in millimetres.

3. Results and Discussions

The processing of 17-4 PH via MIM is now fairly documented; however, a recent review published by German (during the preparation of this article) also submits that despite the widespread popularity of the 17-4 PH alloy, data on the sintered properties 'that can be expected' is still limited [3].

(a)

(b)

Figure 3. (a) the influence of feedstock solids loading and starting bimodal powder materials on the measured as-sintered density and (b) the typical optical micrograph of the as-sintered specimen

3.1. The microstructure and as-sintered densities of the sintered MIM 17-4 PH materials

The effects of the solids loading and using bimodal feedstocks on the sintered density is shown in Figure 3(a). The sintered density was shown in Figure 3(a) to increase with the increase in solids loading of the feedstock. At low feedstock solids loading, the starting bimodal powder materials appear to have adversely negative effects on the sintered density while the effect is significantly positive at higher feedstock solids loading. These observations together ratify foundational PM tenets that the sintered density can be enhanced through the formulation of either highly loaded feedstocks or bimodal feedstocks, without necessitating longer sintering times or at higher sintering temperatures.

Figures 3(b) is included only to exhibit the typical microstructure observed from the 17-4 PH MIM specimens. Tentatively, the microstructural constituents of the shown dual-phase as-sintered are the delta-ferrite (light grey) and the martensitic (dark grey) phases – the dark regions are oxide inclusions and spherical pores. The microstructure is consistent with previously reported findings [2,3,17,18].

3.2. *The mechanical properties of the as-sintered MIM 17-4 PH materials*

The experimentally measured 17-4 PH as-sintered MIM mechanical properties resulting from the tensile tests (*MIM CSIR*) are displayed in Figure 4 along with the comparative (i) material properties specified in the powder and feedstock material suppliers' datasheets (*MIM Pwd* and *MIM FS*), (ii) as-sintered MIM mechanical properties reported in the literature (*MIM Exp*), (iii) material properties reported from various minimum standard specifications (*MIM Std*), and (iv) the mechanical properties 17-4 PH wrought material reported also in the literature properties (*Wrought*). From the author's previous work low feedstock solids loadings (e.g. 55 vol. %) and coarse starting 17-4 PH stainless steel powder materials (e.g. -45 μm) tend to typically exhibit experimentally measured 17-4 PH mechanical properties [12] when compared to the material properties specified in the material suppliers' datasheets.

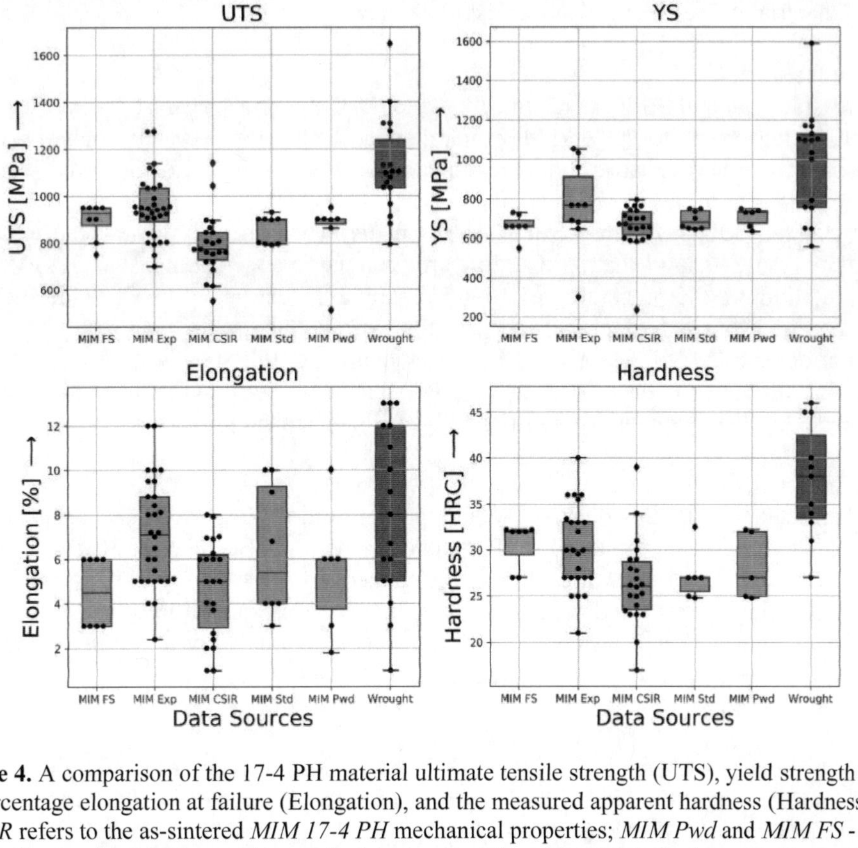

Figure 4. A comparison of the 17-4 PH material ultimate tensile strength (UTS), yield strength (YS), percentage elongation at failure (Elongation), and the measured apparent hardness (Hardness). *MIM CSIR* refers to the as-sintered *MIM 17-4 PH* mechanical properties; *MIM Pwd* and *MIM FS* - material properties specified in the powder and feedstock material suppliers' datasheets, respectively; *MIM Exp* - properties reported prior in the literature; *MIM Std* - properties reported from various minimum standard specifications; and *Wrought* - the literature reported mechanical properties 17-4 PH wrought material.

Available mechanical properties data for wrought 17-4 PH stainless steels materials is generally higher by up to 10 25 % compared to the 17-4 PH as-sintered MIM properties experimentally measured in this study. Reasons cited by Schade *et al.* [19] include the presence defects as a result of sintering (retained pores and grains, and residual carbon impurities), chemical composition deviations due to (Cr, Cu and Ni) elemental loss via evaporation during sintering and the inclusion of impurities (O, C, and N) [3], and but chiefly precipitation hardening during the fabrication of wrought 17-4 PH stainless steels materials typically increases the properties by as much as 15 to 20 %. While on the face of it is trivial, the fact that PM processes such as MIM seek to substitute wrought 17-4 PH stainless steels materials in various applications strongly necessitates this comparison.

The current article makes use of over thirty prior studies available in the literature reporting on the mechanical properties of the as-sintered MIM. The large scatter in the literature reported mechanical properties is understandably due to numerous reasons including a heterogeneity of the starting material sources, and differences in the processing conditions as lately pronounced by German [3]. Likewise, the large scatter in this study's experimentally measured mechanical properties can also be attributed to the numerous MIM feedstocks prepared and used. As clarified by Coleman *et al.* for MIM AISI 4605 materials [20], the huge variations depicted in Figure 4 from the tensile tests from the author's experimental work and properties reported in the literature is a testament that a wide range mechanical properties are feasible from the 17-4 PH alloy.

The experimentally measured as-sintered 17-4 PH MIM from this study are generally comparable to the minimum standard specifications for 17-4 PH MIM materials and the mechanical properties specified by various suppliers (powder and MIM feedstock) in data sheets, and the prior-reported MIM 17-4 PH material properties available in the literature.

4. Concluding Remarks

A comparative study of the CSIR metal injection moulded materials against the as-sintered MIM material properties according to the ASTM standard specification for MIM materials, prior-reported 17-4 PH material properties available in the literature, as well as 17-4 PH wrought and as-cast material properties were presented in this paper.

While the sintered density of the CSIR 17-4 PH material is generally lower than expected, the mechanical test results show that properties are comparable with industry ASTM standard specification. In addition, the processing of 17-4 PH stainless steel using the CSIR-developed MIM feedstocks and processing led to properties typically within those obtained when the parts were processed via prior-reported MIM studies. It is also evident from this study that low feedstock solids loadings (e.g. 55 vol. %) and coarse starting 17-4 PH stainless steel powder materials (e.g. -45 μm) tend to typically exhibit experimentally measured 17-4 PH mechanical properties.

Acknowledgements

The author dedicates this article to *Ntate* Sam Papo and Filipe Pererra on the occusion of their retirement – Thank you! The contributions of P. Ndlangamandla, M. Seerane, N. Muchavi, S. Papo, F. Pererra, M. William and P. Rossouw are fittingly acknowledged. This work was financially supported by the CSIR; the binder system and feedstocks development was funded through the CSIR Young Researcher Establishment Fund award. The author declares that Epson Atmix Corp. (Japan) generously provided the -5μm and -15μm powder used in this study.

References

[1] Murr L E, Martinez E, Hernandez J, Collins S, Amato K N, Gaytan S M and Shindo P W 2012 Microstructures and Properties of 17-4 PH Stainless Steel Fabricated by Selective Laser Melting *J. Mater. Res. Technol.* **1** 167–77

[2] Wu M-W, Huang Z-K, Tseng C-F and Hwang K 2015 Microstructures, mechanical properties, and fracture behaviors of metal-injection molded 17-4PH stainless steel *Met. Mater. Int.* **21** 531–7

[3] German R M 2018 MIM 17-4 PH Stainless Steel: Processing, properties and best practice *Powder Inject. Mould. Int.* **12** 49–76
[4] Reinshagen J H and Witsberger J C 1994 Properties of Precipitation Hardening Stainless Steel Processed by Conventional Powder Metallurgy Techniques ed C Lall and A J Neupaver *Adv. Powder Metall. Part. Mater.* **7** 313–23
[5] Murr L E, Gaytan S M, Ramirez D A, Martinez E, Hernandez J, Amato K N, Shindo P W, Medina F R and Wicker R B 2012 Metal Fabrication by Additive Manufacturing Using Laser and Electron Beam Melting Technologies *J. Mater. Sci. Technol.* **28** 1–14
[6] DebRoy T, Wei H L, Zuback J S, Mukherjee T, Elmer J W, Milewski J O, Beese A M, Wilson-Heid A, De A and Zhang W 2018 Additive manufacturing of metallic components – Process, structure and properties *Prog. Mater. Sci.* **92** 112–224
[7] Machaka R, Seerane M N and Chikwanda H K 2014 Binder development for metal injection moulding: A CSIR perspective *Advances in Powder Metallurgy and Particulate Materials - 2014, Proceedings of the 2014 World Congress on Powder Metallurgy and Particulate Materials, PM 2014* (Metal Powder Industries Federation) pp 443–58
[8] Machaka R and Chikwanda H K 2014 Kinetics of titanium metal injection moulding feedstock thermal debinding *Light Metals Technologies 2015* ed H K Chikwanda and S Chikosha (Port Elizabeth) pp 1–6
[9] Ferri O M, Ebel T and Bormann R 2010 Influence of surface quality and porosity on fatigue behaviour of Ti-6Al-4V components processed by MIM *Mater. Sci. Eng. A* **527** 1800–5
[10] Barriere T, Liu B and Gelin J C 2003 Determination of the optimal process parameters in metal injection molding from experiments and numerical modeling *J. Mater. Process. Technol.* **143–144** 636–44
[11] Pachauri P and Hamiuddin M 2015 Optimization of Injection Moulding Process Parameters in MIM for Impact Toughness of Sintered Parts *Cloud Publ. Int. J. Adv. Mater. Metall. Eng.* **1** 1–11
[12] Seerane M, Ndlangamandla P and Machaka R 2016 The influence of particle size distribution on the properties of metal-injection-moulded 17-4 PH stainless steel *J. South. African Inst. Min. Metall.* **116** 935–40
[13] Muchavi N S, Bam L, {de Beer} F C, Chikosha S and Machaka R 2016 X-ray computed microtomography studies of MIM and DPR parts *J. South. African Inst. Min. Metall.* **116** 973–80
[14] Machaka R, Ndlangamandla P and Seerane M 2018 Capillary rheological studies of 17-4 PH MIM feedstocks prepared using a custom CSIR binder system *Powder Technol.* **326** 37–43
[15] ASTM 1998 ASTM B883-98, Standard Specification for Metal Injection Molded (MIM) Materials
[16] MPIF 2016 MPIF Standards 35 - Materials Standards for Metal Injection Molded Parts
[17] Sung H, Kwon T, Ahn S and Won Y 2002 Powder injection molding of a 17-4 PH stainless steel and the effect of sintering temperature on its microstructure and mechanical properties *J. Mater. Process. Technol.* **131** 321–7
[18] Gülsoy H Ö, Özbek S and Baykara T 2007 Microstructural and mechanical properties of injection moulded gas and water atomised 17-4 PH stainless steel powder *Powder Metall.* **50** 120–6
[19] Schade C, Stears P, Lawley A and Doherty R 2006 Precipitation Hardening P/M Stainless Steels *Adv. Powder Metall. Part. Mater.* **1** 7
[20] Coleman A J, Murray K, Kearns M, Tingskog T A, Sanford B and Gonzalez E 2013 Processing and Properties of MIM AISI 4605 via Master Alloy Routes *2013 Int. Conf. Powder Metallurgy Particulate Material, Chicago* (Chicago: Metal Powder Industries Federation, Princeton, NJ) pp 412–9

IOP Conf. Series: Materials Science and Engineering **430** (2018) 012034 doi:10.1088/1757-899X/430/1/012034

Densification behavior of spark plasma sintered duplex stainless steel reinforced with TiN nanoparticles

S R Oke[1,2*], O O Ige[1,3], O E Falodun[1], M R. Mphahlele[1], PA Olubambi[1]

[1]Center for Nanoengineering and Tribocorrosion, University of Johannesburg, Johannesburg, South Africa
[2]Department of Metallurgical & Materials Engineering, Federal University of Technology Akure, Ondo State, Nigeria
[3]Department of Materials Science and Engineering, Obafemi Awolowo University, Ile-Ife, Osun State, Nigeria

*sroke99@outlook.com

Abstract. Duplex stainless steel (SAF 2205) reinforced with titanium nitride (TiN) nanoparticles ranging from 0 to 8 wt% were fabricated in vacuum via spark plasma sintering (SPS) using optimized SPS process parameter of 1150 °C, 10 min and 100 °C/min. The influence of TiN addition on the densification and shrinkage mechanism of the fabricated duplex stainless steel composite were evaluated. The results indicate even dispersion of the TiN nanoparticles in the steel matrix during tubular mixing. The displacement and shrinkage rates showed three densification stages relating to micro-nano particles rearrangement, plastic deformation of the particles and rapid densification of the composite. The steel composite samples displayed relatively high densities in the range of 96-99 % of the calculated theoretical density but were noted to decrease with TiN content.

1. Introduction

The development of nano-engineered materials has attracted a lot of interest from researchers in recent times. The concept requires that at least one of the components of a multiphased material be in the nano-size range (less than 100 nm) [1]. The potential of nanoparticle reinforced materials relies on their multifunctionality and their ability of achieving improved properties beyond levels obtainable with convectional materials [2]. The challenges of accomplishing homogeneous dispersion and prevention of grain growth of the nano-sized component, selection of appropriate fabrication route and understanding the role of densification in obtaining a less porous sintered product are tremendous [3-4].

This study presents an investigation into the mechanisms involved during the densification of SAF 2205 and TiN nanoparticles by SPS technique. SPS is a pressure assisted sintering technique and has been extensively used for the fabrication of different types of materials such as alloys and metals, ceramics, intermetallic and composite [5]. SPS has gained attention due to its advantages of lower sintering temperature and shorter holding time as compared with other conventional sintering processes [6]. The densification data obtained during SPS of powders can be used to have a thorough grasp of the underlying deformation/densification mechanism during SPS [7]. A previous study by Bernard-Granger and Guizard [7] formulated a hypothesis based on creep deformation to understand the underlying mechanisms that govern densification during the SPS process. The creep model has been employed to predict the densification phenomena of metal matrix composite during the SPS but the understanding of the fundamental microscopic mechanisms of densification is still largely unexplored. In another study by Oke et al. [8] on the optimization of SPS process parameters, the authors reported that densification during SPS is achieved in three distinctive stages of powder particles rearrangement, deformation at powder contact point, and bulk deformation of the sintered material.

An attempt is therefore made in this work to study the influence of TiN nanoparticle addition on the densification mechanism of SPS consolidated duplex stainless steel samples reinforced with TiN nanoparticles.

2. Materials and method

Starting materials used in this study were commercially available SAF 2205 (C-≤0.03, Si- ≤1.0, Mn-≤2.0, P-≤0.03, S-≤0.015, Cr-22, Ni-5, Mo-3.2, N-0.18, Fe-bal wt.%) as well as TiN nanopowders (C-0.03, Si-<0.003, Ni-5, N-<21.91, Ti-77.83, Fe-<0.001 wt. %). Blending of the powders were done for 6 h at 72 rpm using a Tubular Mixer T2F. The mixed stainless steel powders with varying amount of TiN ranging from 0 to 8 % were sintered using a SPS apparatus (model HHPD-25, FCT GmbH Germany) using optimized process parameters from our previous study [8]. On the completion of SPS, the sintered samples were removed from the die and sand blasted to get rid of graphite contaminations on the surface of the samples. The sintering data was obtained and analyzed.

The morphology of the mixed powders was assessed using scanning electron microscope (FESEM, JSM-7600F, JEOL, Japan). The density of the sintered samples was determined using Archimedes technique. Six repeated tests were ensured for each sample for reliability of result. The SPS data obtained during sintering were used to analyze the densification and shrinkage mechanisms.

3. Results and discussion

3.1 Microstructure of mixed powders

The efficiency of tubular mixing of the powder particles is judged by the degree of dispersion of the reinforcing nanoparticles in the duplex matrix. Figs 1(a) and (b) present SEM morphology of admixed SAF 2205 and TiN powders. An even dispersion of the nanoparticles around the each of the SAF 2205 powder was observed. This indicates that tubular mixing is an effective method for dispersing nanoceramics in stainless steels. Fig. 1c depicts the X-ray diffraction pattern of the tubular blended powders, only peaks corresponding ferrite, austenite and TiN phase were identified. The microstructure of the unreinforced stainless steel (Figure 1d) revealed only primarily phases of ferrites (α) and austenite (γ). Energy Dispersive Spectroscopy (EDS) analysis of the composite show Cr and Mo as dominant in the α phase while Ni is dominant in the γ phase as expected. Figure 1e shows the microstructure of the duplex steel samples with 6 wt% TiN contents containing α, γ and nitride phases (evident in the XRD presented in Figure 1f) at α. γ grain boundaries. During the sintering process, the applied electrical potential causes a current flow through powder particles and is followed by sparks formation at the contact points but in the case of TiN nanoceramics and conductive DSS, a potential difference across stainless steel particles separated by the nitride phases may develop.

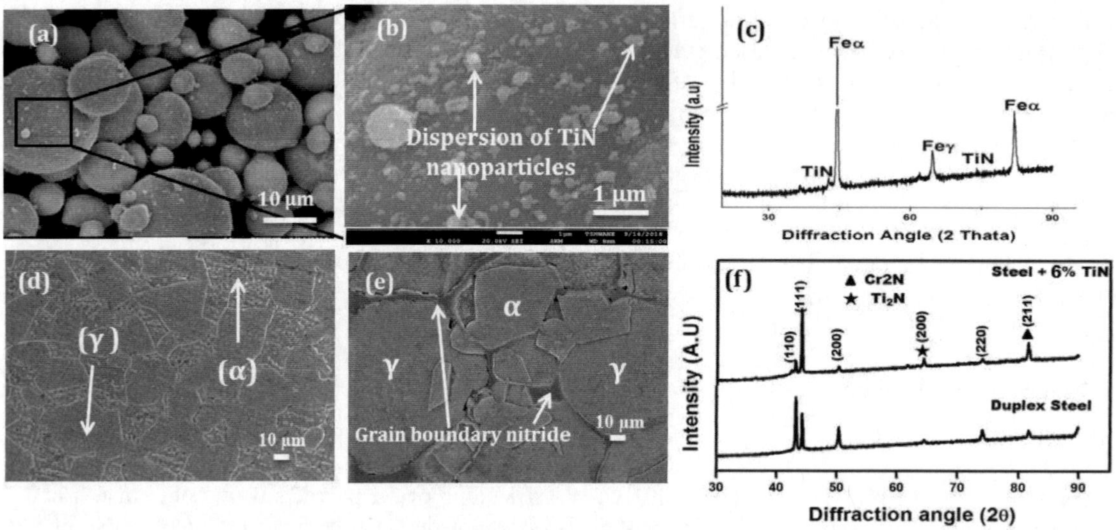

Figure 1. Micrographs showing (a) SEM of tubular mixed powders (b) High mag. Image showing dispersion of TiN nanoparticles. (c) XRD of tubular mixed powders (d) unreinforced duplex stainless steel (e) Steel composite with TiN contents (f) XRD of sintered composite showing evolution of nitride phases.

3.2. Displacement and shrinkage rates of the SAF 2205-TiN

Figures 2 and 3 present the displacement and shrinkage rates as a function of total sintering time during SPS at 1150 °C of the TiN nanoparticles reinforced stainless steel. The figures depict the densification phenomena occurring throughout the sintering process. It is noted from Figure 2 that the total sintering time increased with TiN nanoparticles. The rise in total sintering time could be due to decreased conductivity of the composite as a result of the addition of a nanoceramic phase (TiN). It is imperative to state that the SAF 2205 and TiN nanoparticles are conductive and non-conductive materials respectively, during sintering a constant pulse electric current is used to initiate a current flow through the contact point between conductive (SAF 2205) and non-conductive nanoceramics (TiN). Due to difference in thermal conductivity and potential between the SAF 2205 and TiN, an increased amount of energy is needed to raise the sintering temperature up to 1150 °C, thereby increasing the total sintering time. Similar trend in punch displacement behavior is observed for all the composite irrespective of the amount of reinforcement but the amount of total punch displacement increased with TiN content. The behavior can be attributed to increased energy devoted in consolidating the mixed powders from the initial height of powder bed to final sintered height 5 mm [9].

The shrinkage rate during SPS as a function of total sintering time for the unreinforced and 8 wt% TiN reinforced steel is shown in Figure 3. The composite displayed similar shrinkage behavior; three distinct shrinkage stages were identified. The first shrinkage stage (0 – 10 min) is characterized by initial rearrangement of the mixed powder particles, gas removal and creation of spark between SAF 2205 and TiN particles [8]. The curves revealed an increase in shrinkage rate with TiN nanoparticle additions. The enhanced shrinkage obtained for the unreinforced steel could be attributed to the low melting point of the steel as compared to the reinforced composite. Similar observation was reported by Ghasali et al., the authors argued that metal matrix with lower melting point tend to deform easily [10].

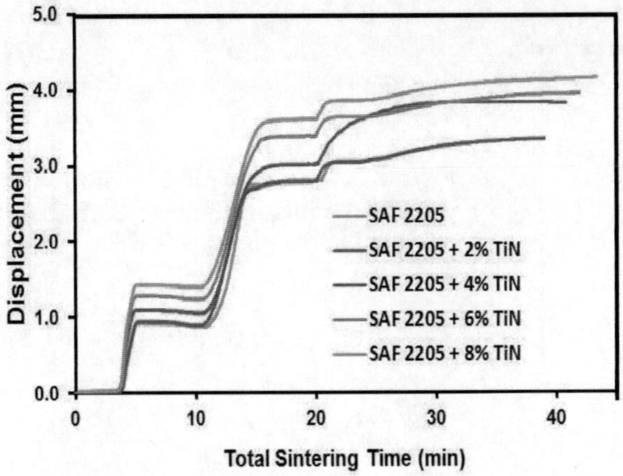

Figure 2. Variation of punch displacement with total sintering time for the SAF 2205 reinforced with TiN nanoparticles.

The second shrinkage stage which occurred between 10 – 20 min is attributed to Joule heating effect accompanied by localized plastic deformation of the particles at contact point. In this stage of SPS, powder surface activation and partial melting, formation of neck at contact points, atomic diffusion and plastic flow occurs [11]. The surface of the stainless steel and TiN nanoparticles was activated due to high pulse discharge plasma created at powder contacts. With increase in sintering temperature, sintering takes place between the powder particles due to softening effects at powder surface, thereby forming necks. At this point, joule heating becomes the dominant heating mode. Huge amount of Joule heating increases the temperature at particle surface resulting in melting, evaporation and growth of the sintering neck. Densification is enhanced at this stage and proceeds further via plastic deformation under the pressure. The third stage commenced after 20 min until the end of sintering operation, a drastic reduction in shrinkage rate was observed. This could be attributed to the decrease in temperature caused by cooling of the SPS chamber to room temperature. Densification is also completed by mass transportation at this stage [8,10.11].

3.3 Influence of TiN addition on density

The result of the relative densities of the stainless steel reinforced TiN nanoparticles is depicted in Figure 4. The steel composite samples displayed relatively high densities; the density was noted to be in the range of 96-99 % of the calculated theoretical density. This could be attributed to the positive effect of created plasma on the mobility and diffusion of atoms by vaporizing and melting of powder particle surface during SPS [12]. It was however noted that the relative density of composite decreased with an increase in TiN content of the duplex stainless steel matrix. The occurrence of the TiN reinforcing phase at boundaries tends to inhibit atomic diffusion and reduces the sintering rate during SPS (Figure 1). Owing to the complex phenomena occurring during SPS and the diffusion inhibition effect of the TiN phase, the sintered composite showed a diminishing densification with increasing TiN contents. Maja *et al.* [13] reported similar behavior with the addition of TiN to metal matrix. The authors advocated that TiN addition could decrease the contacting area between particles of the ductile metal matrix, impede their diffusion and decrease their deformation ability. These factors have an adverse influence on densification. The appropriate application of selected SPS process parameters as established in our previous study [8] resulted in near fully dense steel composite in short sintering times. In the SPS process of synthesizing SAF 2205 and TiN nanoparticles, a high temperature condition is created at the contact point of the stainless steel and TiN nanoparticles. The sparks which were generated due to applied pulsed currents causes evaporation and melting on the powder surface, thus helping in the fabrication of high-density sintered samples [14].

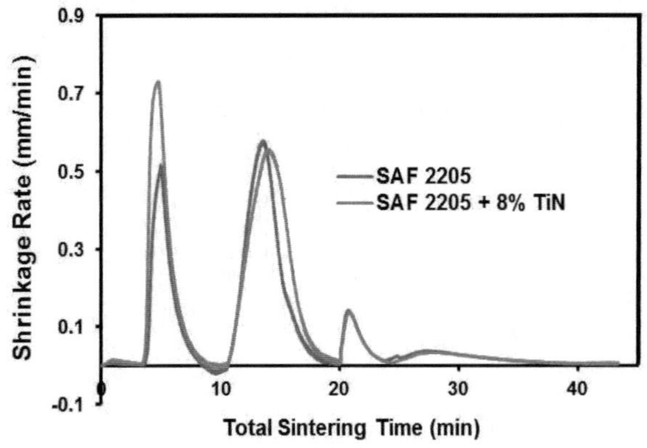

Figure 3. Variation of shrinkage rate with total sintering time of SAF 2205 reinforced with 8 wt% TiN nanoparticles.

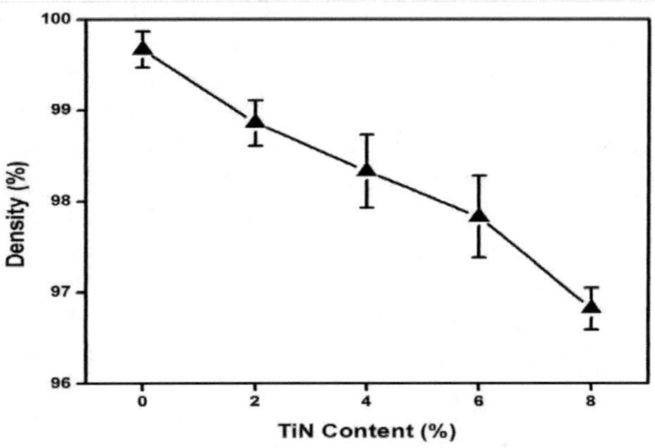

Figure 4. Variation showing decreasing trend of experimental density with the increased addition of TiN nanoparticles to duplex stainless steel.

4. Conclusions

This study fabricated duplex stainless steel reinforced with TiN nanoparticles via SPS using optimized parameters from our previous work [8]. The effects of TiN nanoparticles addition on displacement and shrinkage rate of the duplex stainless steels composite was investigated. The results are summarized as follows;

1. Similar shrinkage behavior was noted for all the composite irrespective of the amount of reinforcement but the total sintering time and punch displacement increased with increasing content of TiN nanoparticles.
2. Three densification stages characterized by powder particle rearrangement, localized plastic deformation and rapid densification by mass transport and cooling respectively were observed during SPS process.
3. TiN nano addition decreased the relative density of the stainless steel composite.

Acknowledgments

This work is supported by the National Research Foundation (NRF) of South Africa and the University of Johannesburg Global Excellence and Stature (GES).

References

[1] Ajayan P M, Schadler L S, Braun PV 2006 Nanocomposite science and technology *John Wiley & Sons.*

[2] Moya J S, Lopez-Esteban S, Pecharroman C 2007 The challenge of ceramic/metal microcomposites and nanocomposites *Prog. Mater. Sci.* **52** 1017-90.

[3] Tjong SC 2007 Novel Nanoparticle-reinforced metal matrix composites with enhanced mechanical properties *Adv. Eng. Mater.* **9** 639-52.

[4] Wang L, Zhang J, Jiang W 2013 Recent development in reactive synthesis of nanostructured bulk materials by spark plasma sintering *Int. J. Refract. Met. Hard Mater.* **39** 103-112.

[5] Oke SR, Ige OO, Falodun OE, Obadele BA, Mphahlele MR, Olubambi PA 2018 Influence of sintering process parameters on corrosion and wear behaviour of SAF 2205 reinforced with nano-sized TiN *Materials Chemistry and Physics* **206** 166-173.

[6] Diouf S, Molinari A 2012 Densification mechanisms in spark plasma sintering: Effect of particle size and pressure *Powder Technology* **221** 220 –227

[7] Bernard-Granger G, Guizard C 2007 Spark plasma sintering of a commercially available granulated zirconia powder: I. Sintering path and hypotheses about the mechanism (s) controlling densification *Acta Materialia* **55(10)** 3493-3504.

[8] Oke SR, Ige OO, Falodun OE, Obadele BA, Shongwe MB, Olubambi PA 2018 Optimization of process parameters for spark plasma sintering of nano structured SAF 2205 composite *Journal of Materials Research and Technology* **7(2)** 126-134.

[9] Cheng Y, Cui Z, Cheng L, Gong D, Wang W 2017 Effect of particle size on densification of pure magnesium during spark plasma sintering. *Advanced Powder Technology* **28(4)** 1129-1135.

[10] Ghasali E, Shirvanimoghaddam K, Pakseresht A.H, Alizadeh M, Ebadzadeh, T 2017 Evaluation of microstructure and mechanical properties of Al-TaC composites prepared by spark plasma sintering process *Journal of Alloys and Compounds* **705** 283-289.

[11] Bonifacio CS, Rufner JF, Holland TB, Van Benthem K 2012 In situ transmission electron microscopy study of dielectric breakdown of surface oxides during electric field-assisted sintering of nickel nanoparticles *Appl. Phys. Lett.* **101** 4583–4586.

[12] Song Y, Li Y, Zhou Z, Lai Y, Ye Y 2011 A multi-field coupled FEM model for one-step-forming process of spark plasma sintering considering local densification of powder material *J. Mater. Sci.* **46** 5645–5656

[13] Maja ME, Falodun OE, Obadele BA, Oke SR, Olubambi PA 2018 Nanoindentation studies on TiN nanoceramic reinforced Ti–6Al–4V matrix composite *Ceramics International* **44(4)** 4419-4425

[14] Oke SR, Ige OO, Falodun OE, Obadele BA, Mphahlele MR, Olubambi PA 2018 Dependence of wear and corrosion properties on holding time of spark plasma sintered SAF 2205 reinforced with TiN *nanoparticles InMechanical and Intelligent Manufacturing Technologies (ICMIMT) IEEE 9th International Conference* 80-84 IEEE.

Spark plasma sintered Ti6Al4V-ZrO2 bio-composites: optimization of ball milling and turbula mixing of powders and their subsequent sintering parameters

N A Moshokoa[1*], **T V Mukhudwana**[1] , **L M Raganya**[2*,] , **J I Lekgoathi** [1] , **B A Obadele**[1] and **P A Olubambi**[1]

[1]Centre for Nanoengineering and Tribocorrosion, School of Mining, Metallurgy and Chemical Engineering University of Johannesburg. Doornfontein Campus Johannesburg, South Africa
[2]Light Metals, Materials Science & Manufacturing, Council for Scientific and Industrial Research, Meiring Naudé Road, Brummeria, Pretoria 0185, South Africa

[*]LRaganya@csir.co.za

Abstract. The use of Ti6Al4V as a biomaterial has been limited due to the dissociation of V and Al ions into the surrounding environment which can cause inflammations, cell damage and toxicity. To prohibit these undesirable challenges, the addition of ZrO_2 into Ti6Al4V was proposed because ZrO2 is biocompatible, it has good wear and corrosion resistant properties, it has low affinity to bacterial colonization and it has high fracture toughness.The aim of this work is to determine the optimum powder synthesising parameters by Tubular mixer and ball mill for fabrication of sintered composites . Powder mixtures were prepared by 49 rpm,72 rpm and 74rpm for 2, 4 and 8 hours followed by spark plasma sintering at 1100 °C at heating rate of 100 °C/min for 10 min and pressure of 50 MPa.from the results obtained it was discovered that Tubular mixing of 90 Vol% Ti6Al4V and 10 Vol% ZrO_2 at 72 rpm and ball milling of 90 Vol% Ti6Al4V and 10 Vol% ZrO_2 at 74rpm speed have proved to produce homogeneous admixed powders, which can further produce fully dense SPS compacts with higher microhardness values.

1.Introduction

Ti6Al4V is a grade of titanium alloy which is considered suitable for the fabrication of implants. This alloy was developed to improve the properties of commercially pure titanium such as low strength and wear resistance for hard tissue replacements [1].Ti6Al4V properties range from good biocompatibility, good mechanical properties to good corrosion and wear resistance as such, it was considered a superior metallic biomaterial compared to cobalt-alloys and stainless [2]. However, it was discovered by Manivasagam [3] that under a biological environment such as the human body it releases toxic metal ions of vanadium and aluminium thus leading to diseases such as Alzheimer. This release of toxic ions occurs even though the metal is protected by the surface oxide layers from the environmental attack; as such this was a clear indication that the corrosion resistance of Ti6Al4V is not sufficient and may results in the implant failure. To hinder the release of these ions and the formation of debris which can go into the blood, the addition of ZrO_2, which possesses high fracture toughness, high wear resistance, high strength and excellent biocompatibility to Ti6Al4V was proposed [3]. The bio-composite was fabricated by Spark Plasma Sintering, a powder metallurgy technology based on simultaneous application of high temperature, high axial pressure, and low-mode (low voltage, high current) field (plasma) associated with an electric current on powders through a graphite die and plungers. In comparison with the conventional sintering methods, it offers products with fine microstructures and high density at relatively low sintering temperatures, very short sintering times and easy control of the sintering conditions [4]. Hence the work was aimed at studying the effect of the ad-mixed and milled powder metallurgy preparation methods on the densification and microhardness of Ti6Al4V-ZrO_2 bio-composites produced via SPS.

2. Materials and Methods

Ti6Al4V (average particle size of 18.94μm) and ZrO2 (65, 26μm) powders were utilized as the starting materials for this research. The two powders were mixed and milled in a dry environment using a Turbular shaker mixer (T2F) and planetary ball mill (RETSCH PM 400 MA-type) at a speed of 49 and 72/74rpm for 2, 4 and 8 hours with the powder ratio of 90Vol%Ti6Al4V and 10Vol%ZrO2. The morphology and powder particle size distribution (PSD) of the mixed and milled powders were examined using scanning electron microscope (SEM) and particle size analyser (Microtrac s3500). The amount of mixed powders required to produce a 40mm diameter and 5mm thickness of Ti6Al4V-ZrO2 Nano structured bio-composites were poured into graphite die and then sintered using a spark plasma-sintering machine (HHPD-25 FCT GmbH Germany).

The optimized powders were sintered at 1100°C at a heating rate of 100°C/min and 50Mpa pressure in vacuum condition. The relative densities of the samples were measured using an OHAUS density kit and by Archimedes principle of the staring powders constituents using the rule of mixtures. The samples were prepared for metallography following the metallographic standard procedure prescribed for griding and polishing for Ti6Al4V-ZrO2. The samples were etched using kroll etchant for 30 seconds and the microstructure was analysed using a (Tescan) scanning electron microscope. The Vickers micro hardness (Hv20) was measured by a micro-hardness tester machine (model falcon 500 series) whereby each sample was indented 10 times. The samples were labelled based on the ZrO2 ratio added to Ti6Al4V, the mixing/milling speed, the mixing/milling time and the type of mixing process. For example, 10-49-2T means that this sample has 10Vol% ZrO2 added, it was mixed at 49rpm for 2 hours in a Turbula mixer and 10-49-2B means the sample has 10Vol% ZrO2 at a speed of 49rpm for 2 hours in a ball mill.

3. Results and Discussion

3.1 SEM results

Figure 1: SEM Micrographs of as-received (a) Ti6Al4V and (b) ZrO_2 powders.
The SEM results of as-received powders are shown in figure 1, revealed that the Ti6Al4V powder is spherical and non- porous while the ZrO_2 powders are doughnut shaped, hollow and porous.

3.1.1 Turbular mixing.
Figure 2 illustrates SEM Micrographs of (a)10-49-2T ,(b) 10-49-4T, (c) 10-49-8T, (d)10-72-2T,(e)10-72-4T and (f)10-72-8T.In the obtained results, it is evident that the doughnut-like shape of ZrO2 particles was strongly broken down in all Turbula admix powders due to the combination of motion of rotation, translation and inversion as well as the impact of the balls that led to the effective reduction of the particle size of the product during mixing. The ZrO2 particles are homogeneously distributed in the Ti6Al4V particles. Furthermore, an interesting trend discovered in these SEM results was the powders became more homogenous with an increase in speed and time from 49-72rpm and 2-8 hours.

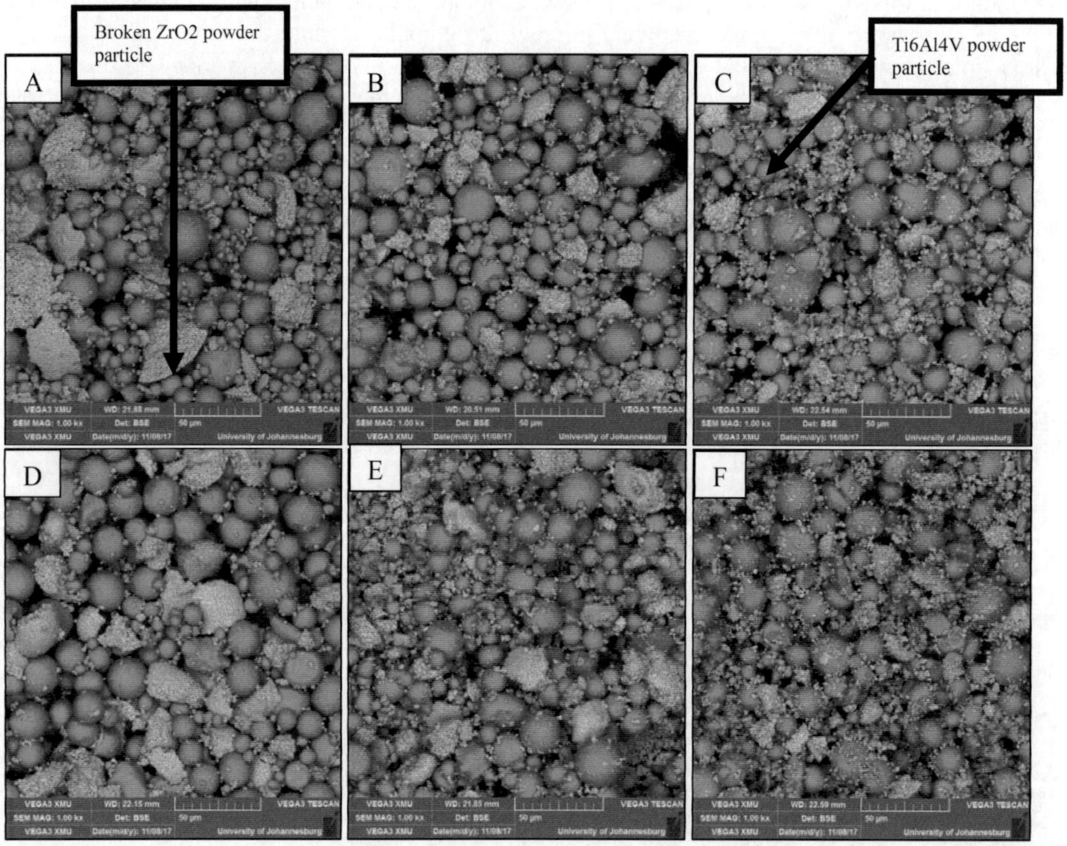

Figure 2: SEM Micrographs of (a)10-49-2T,(b)10-49-4T,(c)10-49-8T,(d)10-72-2T,(e)10-72-4Tand (f) 10-72-8T.

3.1.2 Ball milling
Figure 3 illustrate the SEM Micrographs of (a)10-49-2B,(b)10-49-4B,(c)10-49-8B,(d)10-74-2B,(e)10-74-2B and (f) 10-74-8B. An interesting trend discovered in these results was that satellites of Ti6Al4V have agglomerated in the unbroken hallow ZrO2 powder particles, this trend is more evidential in

(c)10-49-8B, additionally the results indicate that the ZrO_2 powder particles are gradually homogeneously distributed in the Ti6Al4V particles as the speed and time of milling increases. According to these results, it is evident that ZrO_2 powder particles are also broken down during ball milling but only at high speeds such as 74rpm.

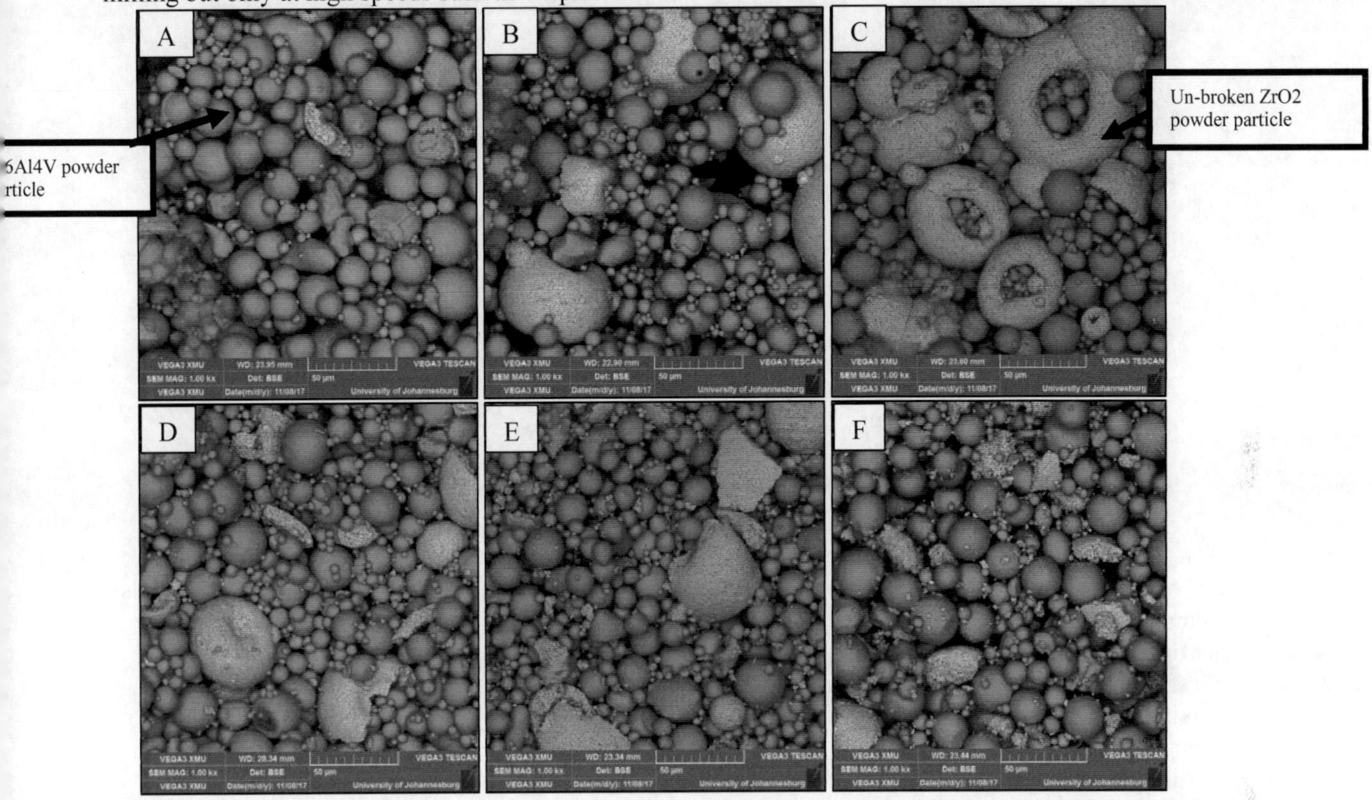

Figure 3: SEM Micrographs (a)10-49-2B,(b)10-49-4B,(c)10-49-8B,(d)10-74-2B,(e)10-74-2B and (f) 10-74-8B.

3.2 PSD results

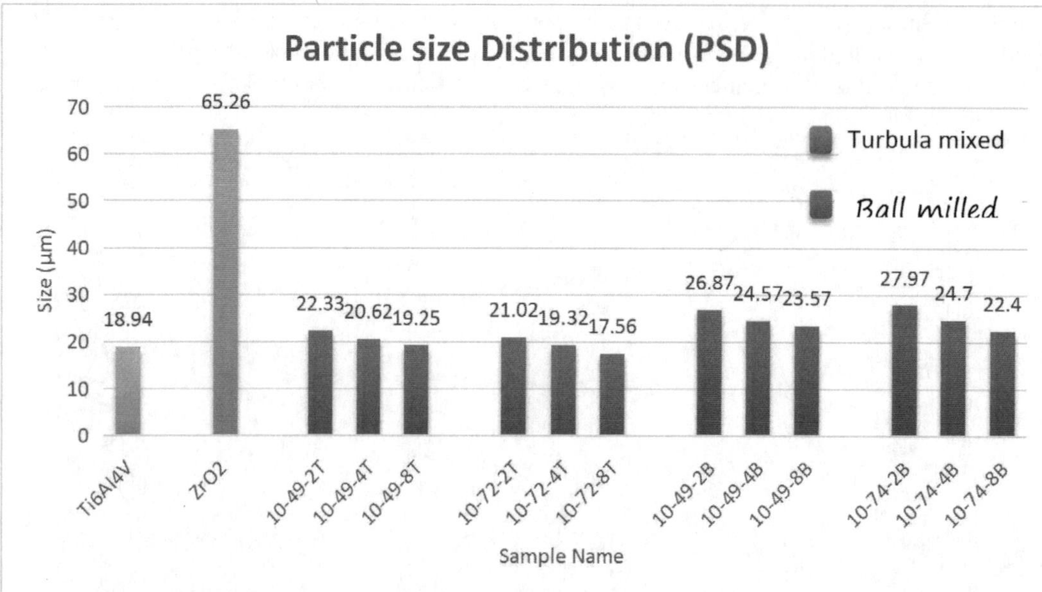

Figure 4: shows the particle size distribution of received powders and mixed powders

The obtained PSD results showed that the ZrO_2 powder has coarse particle size of 65 μm while Ti6Al4V has fine particle size of 19 μm as shown in figure 4. It was discovered that the PSD results compliments the SEM results whereby the Turbula mixed powders have the lowest values of average particle size than that of ball milled powders (illustrated in figure 4). A pattern was also discovered in these results where there is a continuous reduction in powder particle size reduction as we increase mixing/milling speed and time. Moreover, the SEM results alongside with the PSD results indicate that 72 rpm and 74rpm are the optimum mixing/milling speed since they produce powders with low particle size distribution than powders mixed/milled at 49rpm. The samples from the optimised speeds of 72rpm and 74rpm were considered to be the best choice to sinter Ti6Al4V-ZrO_2 bio-composites.

3.3 Sintered Composites Results

SEM Micrographs of sintered Ti6Al4V-ZrO2 composites, which were mixed and milled at the speed of 72rpm and 74rpm, are illustrated by figure 5. The microstructure of all sintered samples contains a pool of Ti with island of ZrO2 and they also contain micro pours that lead to crack propagation during metallographic preparation.

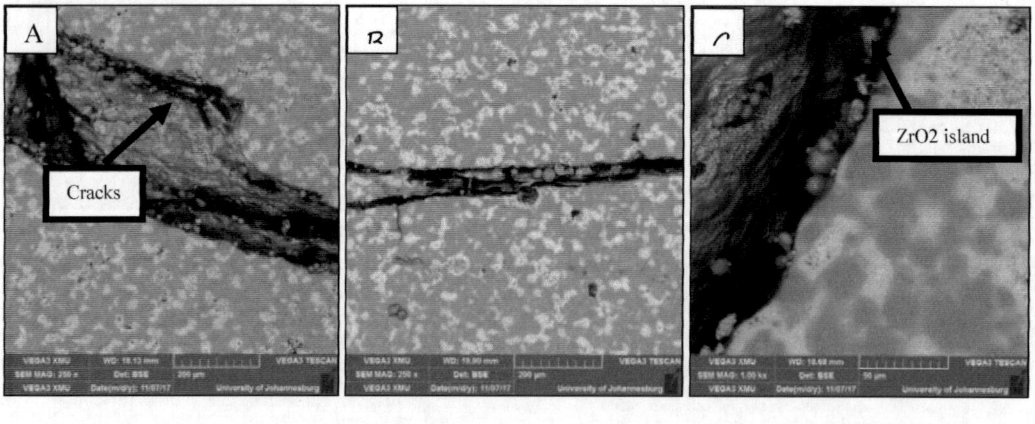

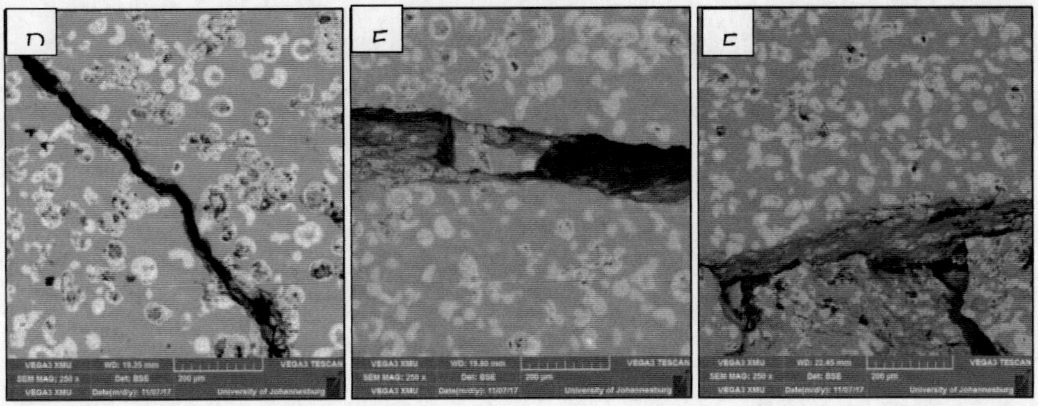

Figure 5: SEM Micrographs of sintered composites (a) 10-72-2T,(b)10-72-4T, (c)10-72-8T, (d)10-74-2B, (e)10-74-4B and (f) 10-74-8B.

3.4 Densification

Table 1 illustrate that density increase with increasing with sintering temperature, this is because porosity decreases when the atoms are bonded together during sintering. Considering the above statement, this could be one of the reasons that the Standard sample (Ti6Al4V) sintered at 1000 has low densification value of 99.6% and the Ti6Al4V- ZrO2 bio-composite samples sintered at 1100 have high densification values of 100%.

Table 1: Densities of Sintered samples.

Sample name	Holding Time, min	Pressure, MPa	Heating Rate, °C/min	Sintering temperature (°c)	Densification %
Standard sample (Ti6Al4V)	10	50	100	1000	99.6
10-72-2T	10	50	100	1100	100
10-72-4T	10	50	100	1100	100
10-72-8T	10	50	100	1100	100
10-74-2B	10	50	100	1100	100
10-74-4B	10	50	100	1100	100
10-74-8B	10	50	100	1100	100

3.5 Hardness

Figure 6 Is Microhardness of sintered composites (10-72-2T, 10-72-4T and 10-72-8T) and (10-74-2B, 10-74-4B and 10-74-8B) the results shows that mixed samples have the highest microhardness values than the ball milled samples. A trend was observed where microhardness values increase as the mixing and milling time increases from 2 to 8 hours. This could be due to high powder homogeneity as speed and time increased and thus producing a well dense sintered sample. Theoretically, sintering rates increases with increasing reduction in powder particle size, hence the increase in microhardness values as the mixing time increase.

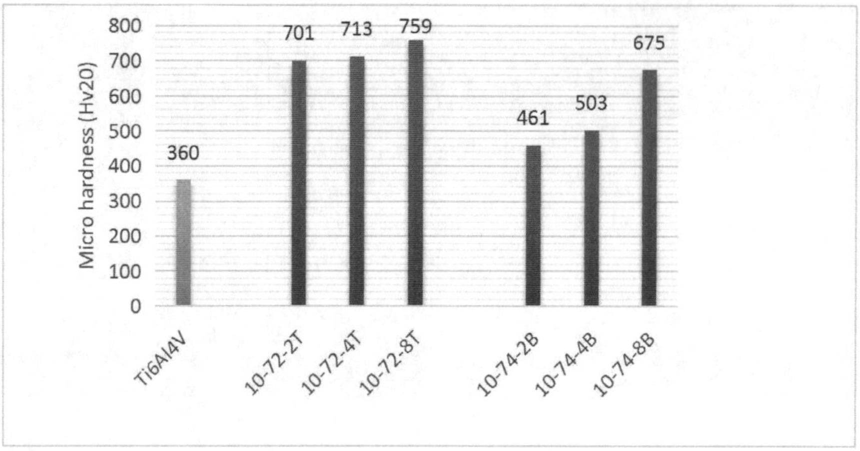

Figure 6: Hardness values of sinterd samples.

4. Conclusion

In conclusion Turbular mixing using balls has proved to be more effective than ball milling in producing homogeneously distributed powder mixtures with finer particles, which can be facilitated in the production of fully dense composites. Furthermore, the optimisation of mixing, milling and sintering parameters done in this paper could be utilised as a reference during further work based on Ti6Al4V- ZrO_2 bio-composites.

References

[1] Li Y, Yang C, Zhao H,Qu S,Li X and Li Y 2014 New Developments of Ti-Based Alloys for Biomedical Applications *Materials* 1709-1800.

[2] Ribeiro A L R, Fla R C J, Cardos F, Belon R and Vaz F F L G 2009 Mechanical, physical chemical characterization of Ti-35Nb-5Zr cast alloys, *J master. SCI mater*. 1629-1636.

[3] Manivasagam G, Dhinasekaran D and Rajamanickam A 2010 'Biomedical Implants: Corrosion and its Prevention - A Review', *Recent Patents on Corrosion Science*. **2** 40-54.

[4] Bauman I, Curi´c D and Boban M 2008 'Mixing of solids in different mixing devices, *S¯adhan¯a*. 721–731

[5] Tohgo K, Fujii T, Harada M, Isono H and Shimamura Y 2015 *Materials Science and Engineering* **621** 166-172

[6] Saheb N 2013 *Intl J of Min, Met and Materials* **20(2)** 152

IOP Conf. Series: Materials Science and Engineering **430** (2018) 012036 doi:10.1088/1757-899X/430/1/012036

Cytocompatibility evaluation of nano-sintered Ti-15Zr-4Nb-2Ta-0.2Pd alloy produced by spark plasma sintering technique

Annastasia Mantshiu[1], Wallace Matizamhuka[1,*], Akiko Yamamoto[2], Brendon Shongwe[3], Ronald Machaka[4]

[1] Department of Metallurgical Engineering, Vaal University of Technology (VUT), South Africa.
[2] Biomaterials Unit, International Center of Materials Nanoarchitectonics (WPI-MANA), National Institute for Material Science, Namiki 1-1, Tsukuba, Ibaraki 305-0044, Japan.
[3] Department of Metallurgical, Chemical Engineering and Materials Engineering, Tshwane University of Technology (TUT), South Africa
[4] Light Metals, Materials Science & Manufacturing, Council for Scientific and Industrial Research (CSIR), South Africa.

*wallacem@vut.ac.za

Abstract. Nano-sintered Ti-15Zr-4Nb-2Ta-0.2Pd alloy made up of non-toxic elements was produced by mechanical alloying using high energy ball mill and synthesized by spark plasma sintering (SPS) hot furnace with an objective to develop cytocompatible metallic implant. These were done to develop an alloy capable of avoiding adverse body reaction and increase the longevity of human life by replacing the lost or diseased organs to restore form and function. Microstructure obtained shows well defined grain boundary lines for sintered CP Ti with no surface porosity. Sintered Ti-6Al-4V microstructure showed well defined sinter-bonding of grain matrix with white rings on grain boundary lines and sintered Ti-15Zr-4Nb-2Ta-0.2Pd alloy showed no definite orientation of grains on the surface and had obvious visible grain boundary lines with a few observable surface pores across the area of the specimen. Cytocompatibility test was conducted by culturing CP Ti, Ti-6Al-4V and Ti-15Zr-4Nb2Ta-0.2Pd alloys separately with murine fibroblast L929 at a concentration of 70 cells/well, and incubated for 7 days at a temperature of 37 °C. Amongst CP Ti, Ti-6Al-4V and Ti-15Zr-4Nb-2Ta-0.2Pd specimen surfaces, the murine fibroblast L929 cells were well attached, grew and formed more colonies on CP Ti and Ti-15Zr-4Nb-2Ta-0.2Pd and least attached to the Ti-6Al-4V surface.

1. Introduction

The success of the biomaterials mainly depends on the reaction of the human body to the implant and this measures the biocompatibility and cytocompatibility of the material [1]. Cell attachment in cytotoxicity analysis is the approach which is used in the comparing of adhesive interactions of cells to the implant in the field of biomedical applications. The main source of the toxic effects that metallic implants have on cells is the metal ion release which is reduced due to high surface integrity as well as corrosion resistance of the implant. The healthy adhesion of cells to the implant surface, as well as their growth and proliferation, are the most important indicators of material cytocompatibility. Little attention has been given to the use of powder metallurgy techniques, mechanical alloying (MA) and

Spark Plasma Sintering (SPS) in the production of nano-grained materials made up of non-toxic and biocompatible elements for orthopaedic applications.

The influence of SPS consolidation on the microstructure of mechanically alloyed Ti-20-13Zr has been previously studied by Hussein and results obtained indicated a mixture of β-titanium in a partially amorphous phase that formed as a result of crystal defect evolution during the mechanical alloying process [7]. Furthermore, the microstructural examination of TNZ alloy displayed grain boundaries together with the secondary structure of dendritic structure. Apart from that, same year Hussein et al., 2015 further studied the effect of sintering parameters on microstructure, mechanical properties and electrochemical behaviour of Nb-Zr for biomedical applications. Results showed that the combination of mechanical alloying and spark plasma sintering could be a promising approach to successfully produce a nano/submicron grain structured Nb-Zr alloy [7]. Specifically, the use of SPS prevents undesirable grain growth within the nano/submicron-grain structured and resulted in a grain size between 100-300 nano-meters.

Surface characterization such as composition and topography exhibit a high influence on the success of implant treatment as it has a direct influence on cell proliferation and differentiation of bone cells, due to this reason the cell adhesion to bone apatite is consequently sacrificed. In o, an implant requires a surface that promotes osteogenic differentiation (i.e. includes human or mouse stem cells functionally defined by their capacity to self-renew and differentiate into multiple cell types such as osteocytes) [2]. Titanium and titanium alloys such as Ti-64 and TiNi have found extensive applications in dentistry and orthopaedics. When the titanium alloy is implanted into a human body, the release of metal ions from the implant may generate an adverse biological effect or allergic reactions. Furthermore, it was observed that both aluminium and vanadium making up the Ti-64 alloy are toxic and can cause mutagenic cytology in addition to triggering allergic reactions. Li et al., 2010 studied the cytotoxicity of titanium and titanium alloying elements such as Ta, Zr, Mo and Si in forms of bulk and powders and different degrees of cytotoxicity was observed. Amongst which Mo and Si exhibited strong cytotoxicity, meanwhile Ta, Zr and Sn exhibited good biocompatibility in both bulk and powder forms. These results indicated that Mo should be limited or rather not be used in designing and developing titanium alloys for dental and orthopaedic device applications [3]. The findings by Li et al., 2010 provided fundamental knowledge of alloying elements that can be used in the future as implants. In this work, cytocompatibilty evaluation of Ti-15420.2 in comparison with CP Ti and Ti-64 was assed using murine fibroblast L929 cells. The assessment was carried out by recording the number of colonies (i.e. a group of cells) attached to the sample surface.

2. Materials and methods

2.1. Materials

High purity elemental powders of titanium, aluminium and vanadium (of -44 microns and 99.0-99.5 % purity); zirconium (of -149 microns, 99.5% purity), niobium (of 1-5 microns, 99.8% purity), tantalum (of 2 microns, 99.9% purity) and palladium (of < 1 micron, 99.9% purity) supplied by Alfa Aesar and Sigma Aldrich were used as starting materials. The powders were mixed in weight percentage as shown in Table 1. The powders were prepared using MB 200B ECO glove box under very high vacuum or an inert atmosphere of argon and mixed for 30 minutes at 49 rpm using the T2F Tabular shaker mixer to obtain a homogeneous mixture. The powders were then mechanically alloyed (MA) for 5 hours at a speed of 500 rpm using a stainless steel pot with <10mm diameter stainless steel balls with an objective to reduce the powder particle size to <100 nanometers (nm) using a powder to ball ratio of 10:1 under argon atmosphere. Mechanically alloyed powders were sintered using the spark plasma sintering (SPS) machine (FCT system-model HHP D25, Germany). The powders were loaded into a 40mm diameter graphite die and punch where a thin graphite foil was placed between the powders and the die to reduce the friction between the die walls and the powders. The test was carried out in a vacuum at a pressure of 50 MPa, the sintering temperature of 1200 °C, heating rate and holding time used was 100°C/min and 10 minutes respectively [4]. The sintered sample of 40mm diameter and 2mm thickness were ground using grit 320 and 600 grinding paper and water as a lubricant. First and second stages of polishing were obtained by ultra Pad and Trident polishing cloths, polycrystal 9µm and polycrystal 3µm as lubricants. The metallographically prepared samples were further etched with a mixture of hydrofluoric acid, nitric acid and ultra-pure water prior to microscopic evaluation using JSM 7600F scanning electron microscope (SEM) equipped with energy-dispersive X-ray (EDX) and microscopic evaluation on cytocompatibility was done on Nikon Eclipse ME600L optical microscope (OM).

Table 1. Chemical composition (wt. %) of CP Ti, Ti-6Al-4V and Ti-15Zr-4Nb-2Ta-0.2Pd alloys.

Titanium alloyed powders	Weight percentages of the powder mix
CP Ti	100 %Ti
Ti-6Al-4V	6%Al, 4 %V, & 90%Ti
Ti-15Zr-4Nb-2Ta-0.2Pd	15%Zr, 4%Nb, 2%Ta, 0.2% Pd & 78.8%Ti

2.2. Cytocompatibility test

CP Ti, Ti-6Al-4V and Ti-15Zr-4Nb-2Ta-0.2Pd samples were ultrasonically cleaned using acetone, methanol and ethanol for 5 minutes each. Sterilization on as polished samples was done by autoclaving for two hours at a temperature of 120 °C in a dry condition in preparation for cytocompatibility test. The cytocompatibility evaluation was performed using the direct contact method, where cell attachment on each sample (three of each alloy) was evaluated by recording the number of colonies (i.e. a group of cells) attached to the samples. A 24 well microplates were used to locate samples separately and cultured with murine fibroblast L929 at a concentration of 70 cells/will in 1mL of the culture medium (E-MEM + FBS). The same number of cells were seeded into an empty well to serve as a control well and incubated for 7 days at 37°C in a 5% vol% C02 atmosphere with 95% relative humidity. After incubation, the cells were fixed and stained using 25% glutaraldehyde solution for 10 minutes and 10 vol% Giemsa's staining solution for 15 minutes, respectively. Fixing was done to kill the cells such that they do not move on the slide or dish and staining was done such that the cells and their organelles holds the stain and appear in purple colour. This helps the observer to see and count cells using their naked eye. Stained cells were observed and captured using a NIKON optical microscope. Results obtained from the counting were presented by calculating the relative plating efficiency (*PE*) using the equation 1:

$$\text{PE}\ (\%) = (Ns\ /\text{As})/(Nc\ /Ac\ ) \qquad (1)$$

where N_s and N_c indicate the number of colonies on the sample top surface and in control well, respectively. A_s indicates the sample top surface area and A_c indicates the well bottom area in cm^2. The data obtained was analysed statistically using student *t*-test [5].

3. Results and discussions

3.1. The microstructures of the sintered specimens

Figure 1 shows the microstructure of the sintered specimens obtained from scanning electron microscope (SEM) and elemental mapping of selected areas obtained through energy dispersive spectroscopy (EDS). The chemical analysis of all specimen consisted of only elements included in the compositional design: sintered CP Ti (Figure A & A.1), Ti-6Al-4V (Figure B & B.1) and Ti-15Zr-4Nb-2Ta-0.2Pd (C & C.1). The surface micrograph of CP Ti sintered specimen shows no indication of surface porosity which is associated with incomplete sintering in terms of neck formation. The grain boundaries are well defined with visible homogeneous grain boundary lines. From observation of SEM micrographs of sintered Ti-6Al-4V alloy, the surface has a clear surface topography with a white ring shaped compounds at the grain boundary sites of the matrix. The SEM micrographs of sintered Ti-15Zr-4Nb-2Ta-0.2Pd alloy show a high volume of grain boundary sites, with irregular orientation.

Regardless of the obvious of metallographic finishing on the specimen surface (Figure 1 C.2), it can be observed that the Ti-15Zr-4Nb-2Ta-0.2Pd alloy sintered at 1200oC consists of large surface pores across the area of the sample.

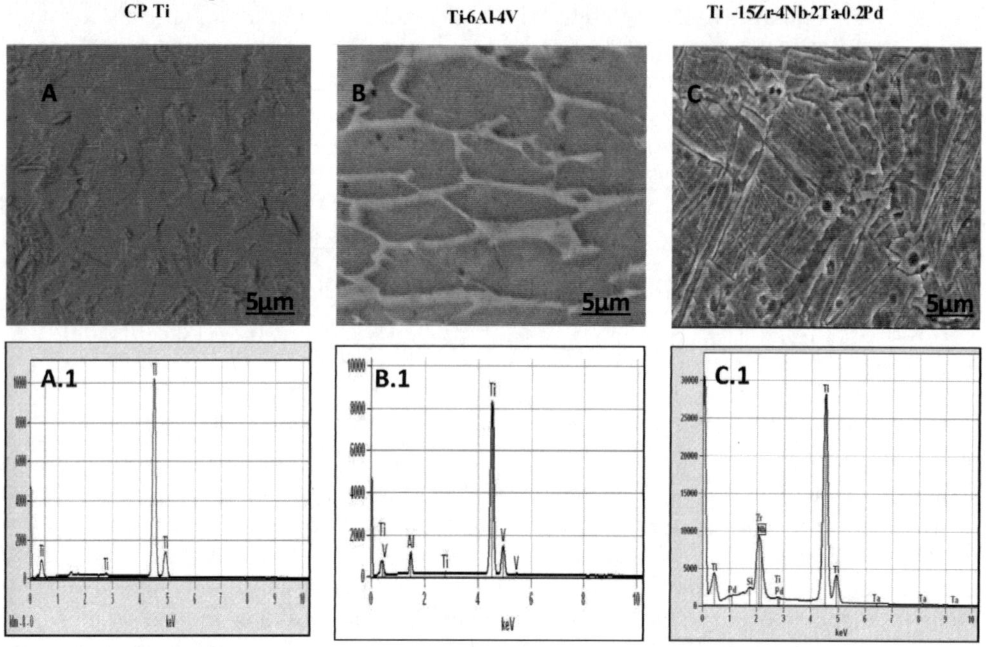

Figure 1. SEM surface micrographs and microstructure characteristics and EDS analysis of CP Ti (A & A.1), Ti-6Al-4V (B & B.1) and Ti-15Zr-4Nb-2Ta-0.2Pd (C & C.1) sintered at 1200oC using the SPS HHPD 25 hot furnace.

3.2. Cytocompatibility test

The evaluation of cytotoxicity was done for Ti-15Zr-4Nb-2Ta-0.2Pd, Ti-6Al-4V and CP Ti which were cultured separately with murine fibroblast L929 at a concentration of 70 cells/well and incubated for 7 days at a temperature of 37 °C. The results of the calculated plating efficiency and morphology of cell attachment to each sample surface are presented in figure 2 and Figure 3 respectively. Both commercially pure titanium and Ti-6Al-4V alloy are known for their development of highly stable surface oxide layer especially when exposed to air or to aqueous media which is responsible for bone bonding characteristics of titanium implants, hence this project used both CP Ti and Ti-6Al-4V as benchmarks in order to study and compare the behaviour of Ti-15Zr-4Nb-2Ta-0.2Pd in comparison with both CP Ti and Ti-6Al-4V.

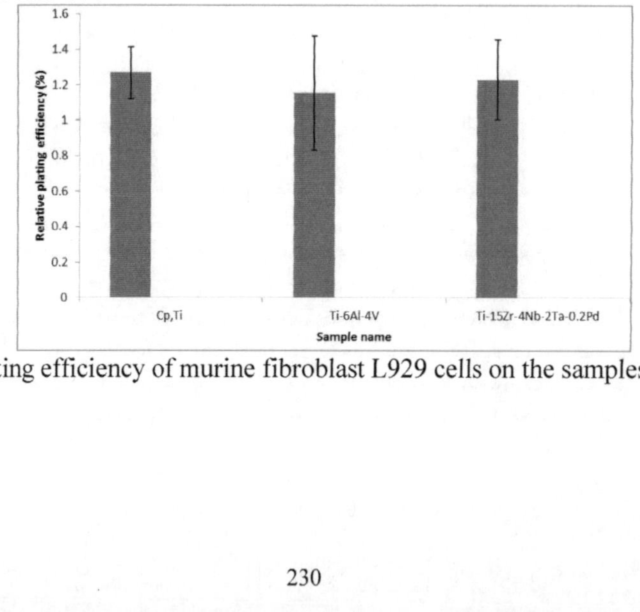

Figure 2. Relative plating efficiency of murine fibroblast L929 cells on the samples.

In the implant and body system, various interactions and reactions may occur including metal ions in the body fluid as well as the transportation to body organs and if the limit of toxicity for the implant is exceeded, this further results in injuries, inflammation and pain for the patient if the implant is found to be toxic rather than compatible. Cell attachment, cytocompatibility or cytotoxicity test serves as an indication as to whether or not an implant used for orthopaedic applications is compatible. Figure 2 shows the relative plating efficiency of all Ti-15Zr-4Nb-2Ta-0.2Pd alloy in comparison with Ti-6Al-4V and CP Ti whereas, Figure 3 shows the morphology of cells attached to each sample surface. No significant difference was observed from Figure 2 due to the fact that cell attachment on all samples as shown in Figure 3 attributed to good cytocompatibility properties with Ti-6Al-4V obtaining the least cell attachment. The relative platting efficiency calculated compliments the results obtained for cell attachment as Ti-6Al-4V still obtained the lowest relative plating efficiency. That said, it is quite obvious that Ti-6Al-4V alloy could be easily substituted with Ti-15Zr-4Nb-2Ta-0.2Pd alloy as it is less toxic and attributed more relative plating efficiency. Furthermore, events after implantation include interactions between biological environments and the artificial material surface or implant, thus, the onset of biological reactions, as well as the particular response by the human body are most critical in orthopaedic device applications.

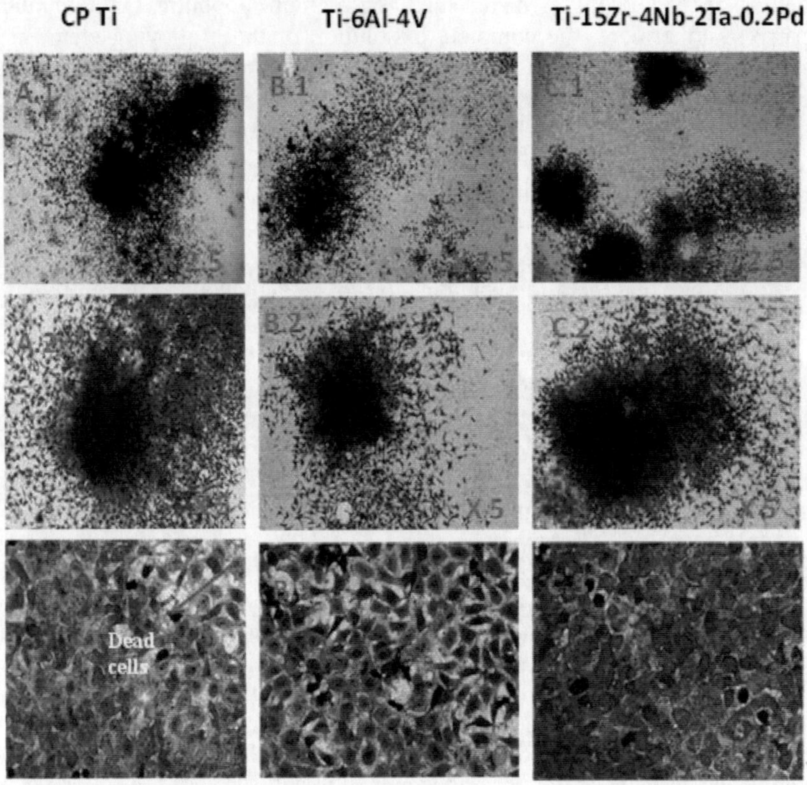

Figure 3. Optical microscope images of colony morphology for the surface of (A) CP Ti, (B) Ti-6Al-4V, (C) Ti-15Zr-4Nb-2Ta-0.2Pd surfaces at X 2.5, 5 and 20 magnification.

The growth of cells on sample surfaces was observed under optical images in Figure 3. Results obtained were in agreement with the calculated plating efficiency presented in Figure 2. Amongst CP Ti, Ti-6Al-4V and Ti-15Zr-4Nb-2Ta-0.2Pd samples surfaces, the murine fibroblast L929 cells well adhered, grew and formed more colonies on CP Ti and Ti-15Zr-4Nb-2Ta-0.2Pd and least attached to the Ti-6Al-4V alloy surface. The cells were well distributed and spread over the samples. Cell attachment relies on the interactions with extracellular substrates in order to maintain the critical cell functions such as cell proliferation. This is further described as the process that results in an increase in the number of cells defined by the balance between cells that survive verses cell loss as a result of cell death. It was easy to distinguish between cells that survived amongst dead cells as cells that survived appeared shinny and were visible to the naked eye after staining whereas dead cells appeared dark and dull on the surface as shown by arrows in A3, B3 and C3 for CP Ti, Ti-6Al-4V and Ti-15Zr-4Nb-2Ta-0.2Pd respectively. More dead cells were observed on the Ti-6Al-4V surface as opposed to both CP Ti and Ti-15Zr-4Nb-2Ta-0.2Pd. Cell attachment also relies on the expression of structural and adhesive proteins as well as cell viability. A decrease in cell viability and cell proliferation greatly reduces the capacity of the cell to integrate with the implant and this would result in bone growth to the implant. Furthermore, cell adhesion and proliferation sturdily depend on the surface microstructure, chemical composition, wettability and surface roughness [6].Ti-15Zr-4Nb-2Ta-0.2Pd alloy and CP Ti exhibited superior cell adhesion and proliferation than Ti-6Al-4V after 7 days of cell culturing.

4. Conclusion

This pioneering work outlines nano-sintered pure titanium (CP Ti) and titanium alloys (Ti-6Al-4V and Ti-15Zr-4Nb-2Ta-0.2Pd) fabricated by spark plasma sintering for evaluation of cytocompatibility characteristics. The samples resulted in high densification and adequate microstructure due to the complete dissolution of the alloying elements such as zirconium, niobium, tantalum and palladium in the titanium matrix. Furthermore, the complete dissolution of those alloying elements resulted in a good combination of the microstructure of fully densified samples. The sintering parameters and sintering technique (SPS) used for this project resulted in a homogeneous microstructure with low surface porosity and contamination. It is well known that higher sintering temperatures and longer holding times constitute to intensive grain growth. SPS is known to attain finer grain sizes which have proven to exhibit better service properties, hence the microstructural parts of the nanosized matrix obtained through milling are the most eligible for meeting today's industrial and biomedical applications. Results obtained suggested Ti-15Zr-4Nb-2Ta-0.2Pd alloy to be used for biomedical applications due to the fact that the alloy contains non-toxic elements and therefore has the ability to replace Ti-6Al-4V which was found to contain toxic elements; Al and V. Overall results obtained where biocompatibility is concerned, Ti-15Zr-4Nb-2Ta-0.2Pd alloy produced by mechanical alloying and spark plasma sintering showed a potential advantage to be used for biomedical applications. In addition, both tantalum (Ta) and niobium (Nb) making up the Ti-15Zr-4Nb-2Ta-0.2Pd alloy are known for their extremely stable oxide layer formed on the alloy surface which exhibits metal ion release and still shows no cytotoxicity effects. For this reason, Ti-15Zr-4Nb-2Ta-0.2Pd alloy can be regarded as a candidate for biomaterial in orthopaedic device application.

Acknowledgments

The authors wish to acknowledge the DST-NRF collaborative Postgraduate Grant, VUT-Academic Exchange Grant, DST-titanium centre of competence (TiCoC) for providing the funding for this project, CSIR, TUT and NIMS Japan for availing their facilities.

References

[1] GEETHA, M., SINGH, A.K., ASOKAMANI, R. & GOGIA, A.K., 2009. Ti based biomaterials the ultimate choice for orthopaedic implants. Progress in material science, 54:397-425.
[2] CORDEIRO, J.M., KUMAR, M & BARAO, A.R. 2017. Is there scientific evidence favoring the substitution of commercially pure titanium with titanium alloys for manufacture of dental implants? Material Science and Engineering C, 71:1201-1215.
[3] LI, Y., WONG, C., XIONG , J., HODGSON, P. & CHO, K.2010. Cytotoxicity of titanium and titanium alloying elements.-Biomaterials & Bioengineering,89(5):493-497.
[4] HUSSEIN, M.A., SURYANARAYANA, C. & AL-AQEELI, N. 2015. Fabrication of nanograined Ti-Nb-Zr biomaterials using SPS. Materials and design, 87: 693-700.
[5] SHRI, D.N.A., TSUCHIYA, K. & YAMAMOTO, A. 2014. Cytocompatibility evaluation and

surface characterization of TiNi deformed by high-pressure torsion. Materials Science and Engineering C, 43: 411-417.
[6] KAZEK-KESIK, ALICJA., KROK-BORKOWICZ, MALGRZATA., PAMULA, ELZBIETA & WOJCIECH, SIMKA. 2014. Electrochemical and biological characterization of coatings formed on Ti-15Mo alloy by plasma electrolytic oxidation. Material Science and Engineering C, 43: 172-181.
[7] HUSSEIN, M.A., SURYANARAYANA, C., ARUMUGAM, M.K. & AL-AQEELI, N. 2015. Effect of sintering parameters on mictrostructure, mechanical properties and electrochemical behavour of Nb-Zr alloy for biomedical applications.-Materials & Design, 83:344-351.

Sintering Performance and Corrosion behavior of biomedical Ti-24Nb-4Zr-8Sn alloy produced using SPS in various simulated body solutions

J Madonsela[1,*], **W Matizamhuka** [1], **R Machaka** [2], **B Shongwe** [3], **A Yamamoto**[4]

[1] Department of Metallurgical Engineering, Vaal University of Technology (VUT), RSA
[2]Titanium Centre of Competence, Materials Science & Manufacturing, Council for Scientific and Industrial Research, Meiring Naudé Road, Brummeria, Pretoria 0185, South Africa
[3] Institute for NanoEngineering Research, Department of Chemical, Metallurgical and Materials Engineering, Tshwane University of Technology (TUT), RSA
[4] Biomaterials Unit, National Institute for Materials Science (NIMS), Japan

* jerman.madonsela@gmail.com,

Abstract. In this study, the Ti-24Nb-4Zr-8Sn (Ti2448) alloy was produced using spark plasma sintering technique and its behaviour in 0.9% NaCl, Hank's solution and cell culture medium were investigated using open circuit potential and potentiodynamic polarization techniques. The results were compared with that of CP Ti and Ti-6Al-4V alloy. Relative densities above 99.0% were achieved for all three systems. CP Ti and Ti64 had both 100% relative density, and Ti2448 showed a slightly lower density of 99.8%. Corrosion results showed that all three materials exhibited good corrosion resistance due to the formation of a protective passive film. In 0.9% NaCl Ti2448 had the highest current density (9.05 nA/cm2), implying that its corrosion resistance is relatively poor in comparison to Ti (6.41 nA/cm2) and Ti64 (5.43 nA/cm2), respectively. The same behaviour was observed in Hank's solution. In cell culture medium, Ti2448 showed better corrosion resistance with the lowest current density of 2.96 nA/cm2 compared to 4.86 nA/cm2 and 5.62 nA/cm2 of Ti and Ti64 respectively. However, the current densities observed are quite low and insignificant that they lie within acceptable ranges for Ti2448 to be qualified as a biomaterial.

1. Introduction

Biomaterials are materials used for making devices that can interact with biological systems to coexist for longer service with minimal failure. They are widely used in the repair, replacement, augmentation of diseased or damaged parts of the musculoskeletal system such as bones, joints and teeth. Titanium alloys are the most suitable candidates used as metallic biomaterials due to their biocompatibility, low Young´s modulus and good corrosion resistance [1]. Ti-6Al-4V is one of the most widely used alloys for orthopaedic implants [2], although it suffers from some drawbacks: adverse health effects arise from the release of Al and V ions into the human body [3] and this alloy exhibits relatively high Young's modulus as compared to that of bone [1]. These limitations are overcome through the development of non-toxic β-stabilizing alloys in order to improve both biochemical and biomechanical compatibility. By favouring the formation of β phase, Young's modulus of the alloy can be decreased to facilitate the load transfer from the implant to the bone, avoiding bone resorption and final loosening of the implant. The Ti-Nb-Zr (TNZ) alloys are particularly interesting since not only do they consist of nontoxic elements but also possess Young's modulus amongst Ti alloys [4]. An improved corrosion resistance was reported in binary Nb-Zr alloy due to Nb content, and an increase in wear resistance in Ti-Nb-Zr-Ta (TNZT) alloy. Moreover, additions of tin (Sn) to these alloys serves to improve their densification and ultimately beta phase formation [6].

2. Experimental Procedure

2.1. Sample preparation and fabrication

The selected alloys for the present study were as follows: commercially pure titanium (grade 2), Ti-6Al-4V and Ti-24Nb-4Zr-8Sn. The alloy composition is given in weight per cent. Appropriate amounts of powder metals of defined purity and particle size (Table 1) were mixed and subsequently tubular mixed (Tubular T2F) for 30 minutes at 320 rpm prior to mechanical alloying. The three alloy systems were then milled using high energy ball mill (Simoloyer CM100, CSIR) under an argon protective atmosphere at room temperature. The grinding media was made of zirconia. The ball to powder ratio was approximately 10:1. Milling was performed at the rotation speed of 500 rpm for 5 hours without the addition of any process control agent.

Prepared powders were consolidated at the Tshwane University of Technology using SPS apparatus (FCT System-GmbH HHP-D5). The powder in a graphite die was heated to 1200 °C at a heating rate of 100 °C/min with a pressure of 50 MPa. The holding time was 10 min. Hussein et al. [5] reported that the heating rate and holding time of 100 °C/min and 10 min respectively minimize grain growth. According to Hussein et al. [5] sintering at 1200 °C allows the complete dissolution of Nb and the attainment of the homogenized nanocrystalline structure with better mechanical properties. Moreover, according to Hussein et al [5], nanostructured alloys have better compatibility compared to their coarse-grained counterparts. These powders were sintered to 40 mm diameter discs and 2 mm thick. The relative density of the compacts was measured by the Archimedes method. The sintered samples for scanning electron microscope (SEM) (FE-SEM JEOL 6000, CSIR) observation was prepared by the standard metallographic method. The polished samples were etched with Kroll's reagent.

Table 1. Characteristics of raw metal powders.

Component	Purity (%)	Particle size (μm)	Density (g/cm^3)	Melting point (°C)
Ti	99.9	44	4.51	1668
Al	99.5	44	2.70	660
V	99.5	44	6.0	1910
Nb	99.8	3	8.57	2648
Zr	99.5	149	6.49	1852
Sn	99.5	0.1	7.30	232

2.2. Corrosion test

The corrosion behaviour of the materials in artificial body fluids [Hanks solution (NaCl 8.0 g/L; KCl 0.4 g/L; KH_2PO_4 0.6g/L; $CaCl_2$ 0.14 g/L; $MgCl_2.6H_2O$ 0.1 g/L; $NaHCO_3$ 0.35 g/L; $NaH_2PO_4.2H_2O$ 0.06 g/L; $MgSO_4.7H_2O$ 0.06 g/L; Glucose 1.0 g/L at pH 7.4), 0.9% NaCl solution and E-MEM + 10% FBS) was carried out using potentiodynamic polarization technique. The specimens with a square surface area (8 x 8 mm) and 2 mm thick were used for tests. All electrochemical tests were conducted using a potentiostat (VersaSTAT3) controlled by a computer (VersaStudio) software for data acquisition. The cell used was a typical three-electrode consisting of the platinum counter electrode (CE), a working electrode (specimen) and silver/silver chloride reference electrode. The tests were performed using scan rate of 0.1667 mV/s and polarized from – 0.25 V to 1.5 V. All the electrochemical experiments were conducted in an incubator and the test temperature was controlled as 37 ± 1°C. Two samples were used for each alloy group in each solution, and 10 ml test solution was used in every electrochemical test. The tests were conducted according to corrosion testing standard (ASTM F2129).

3. Results and Discussion

The average relative densities of the compacts are shown in Fig. 1. Commercially pure titanium and Ti-6Al-4V compacts achieved higher relative densities than Ti-24Nb-4Zr-8Sn. The reason is possibly due to the additions of the higher melting alloying elements such as Nb and Zr that results in incomplete densification.

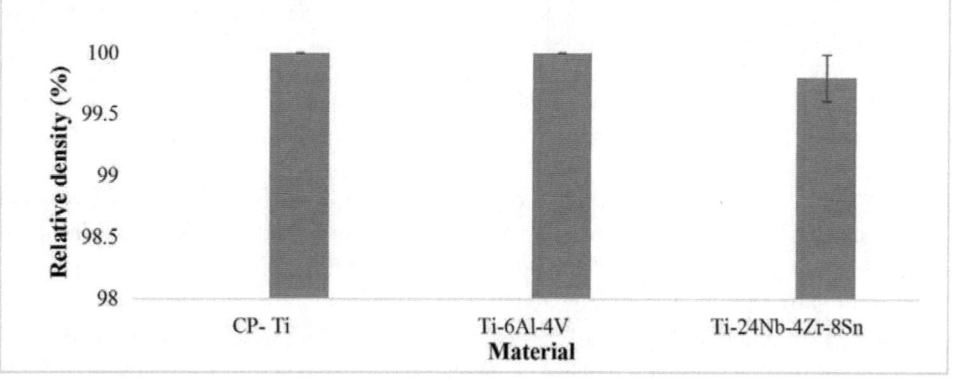

Figure 1: Average relative densities of CP Ti, Ti6Al4V and Ti-24Nb-4Zr-8Sn alloys after SPS.

Figure 2 illustrates the microstructures of the compacts sintered at 1200 °C for 10 min. Refined microstructures can be observed in pure titanium as shown in Figures 2(c) and 2(d). This has been attributed to fast cooling during the SPS sintering cycle as also reported by Falodun et al [12]. Figures 2(a) and (b) show the typical CP Ti and Ti6Al4V microstructures – the α/β phases are more homogeneously distributed as expected. Figure 2(c) however exhibits porosity and an inhomogeneous distribution of the α/β phases in the microstructure of the sintered Ti2448 samples. Xia Li and co-workers [13] also recently observed similar microstructures in their studies of the fabrication of biomedical Ti-24Nb-4Zr-8Sn alloy with high strength and low elastic modulus via powder metallurgy– the insufficient diffusion of elemental powders was highlighted as the possible reasons. Further investigations are currently focused on studying the porosity and the homogenisation of the microstructures.

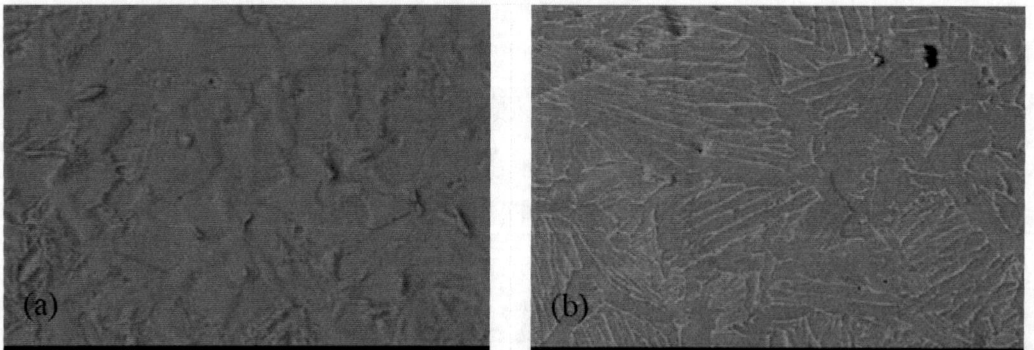

(a) Sintered CP Ti | **(b)** Sintered Ti6Al4V

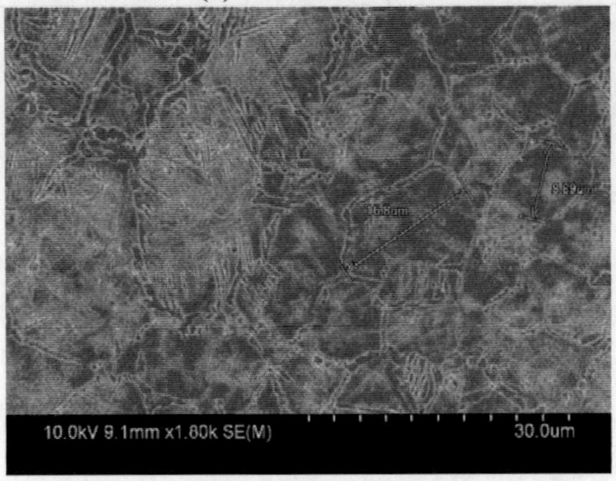

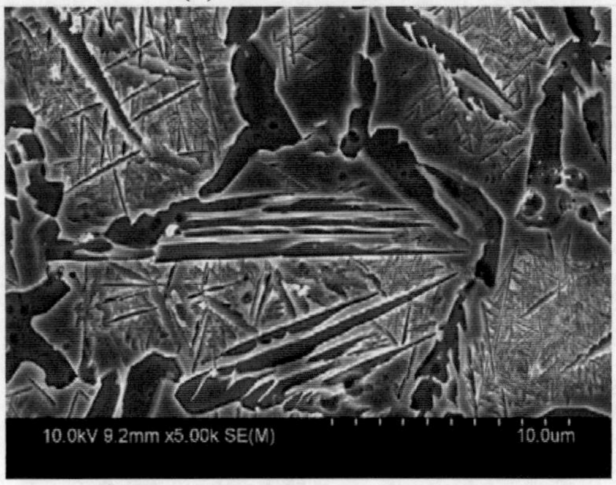

(c) Sintered Ti2448 | **(d)** Sintered Ti2448 at high magnification

Figure 2: SEM micrographs of studied alloys

The potentiodynamic polarization curves for the studied alloys in 0.9% NaCl, Hank's and E-MEM + 10% FBS solutions at 37 ºC are shown in Figures 3(a-c). It is clear all the samples exhibit a typical active-passive characteristic. The corrosion potentials (Ecorr) were estimated from these curves and displayed in Table 2. There is no distinctive difference on the anodic polarization curves of these three alloys, indicating that the same anodic reaction occurred on their surface but with different rates. In 0.9% NaCl Ti2448 had the highest current density (9.05 nA/cm2), implying its corrosion resistance is relatively poor in comparison to Ti (6.41 nA/cm2) and Ti64 (5.43 nA/cm2), respectively. The same behaviour was observed in Hank's solution. In cell culture medium, Ti2448 showed better corrosion resistance with the lowest current density of 2.96 nA/cm2 compared to 4.86 nA/cm2 and 5.62 nA/cm2 of Ti and Ti64 respectively. However, the corrosion rates observed are quite low and insignificant that they lie within acceptable ranges for Ti2448 to be qualified as a biomaterial as provided in the corrosion testing standard (ASTM F2129).

Table 2: Corrosion parameters of the potentiodynamic polarization for the Ti2448, CP-Ti and Ti-6Al-4V alloy in different solutions at 37 °C.

Materials	Solutions	I_{corr} (nA/cm^2)	E_p (V, Ag/AgCl 3M NaCl)
Ti2448	0.9 NaCl	9.0475	-0.253
	Hank's	4.048	-0.327
	E-MEM + 10% FBS	2.4805	-0.351
CP-Ti	0.9 NaCl	6.4105	-0.304
	Hank's	2.290	-0.348
	E-MEM + 10% FBS	4.859	-0.122
Ti6Al-4V	0.9 NaCl	5.427	-0.088
	Hank's	1.977	-0.257
	E-MEM + 10% FBS	5.618	-0.112

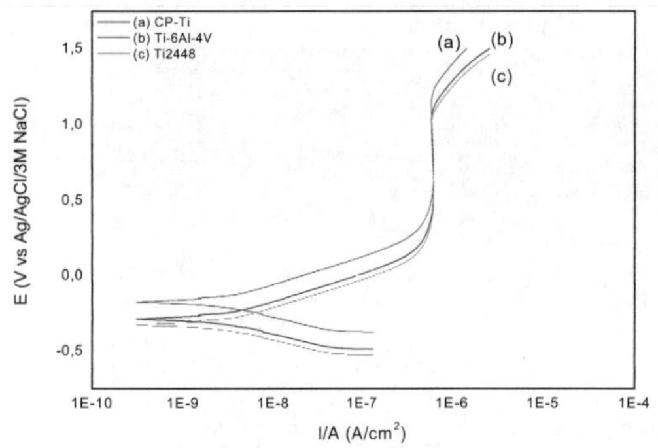

(a) 0.9 wt. % NaCl solution

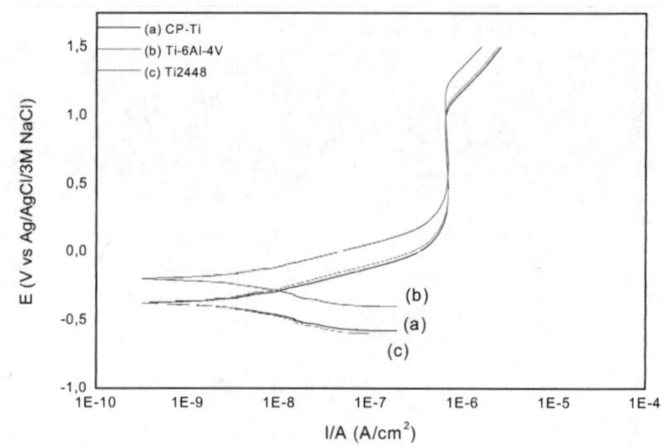

(b) Hanks's solution

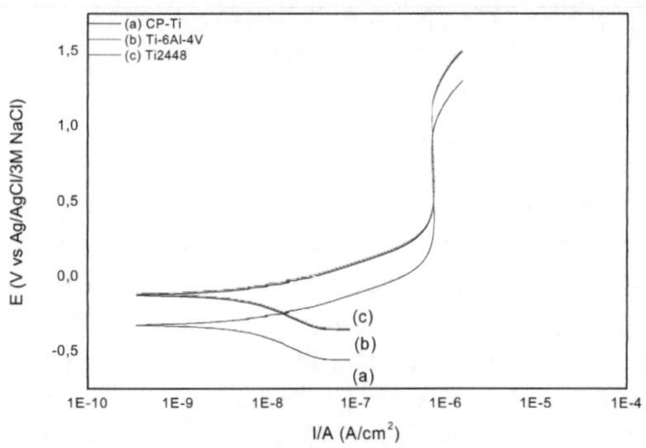

(c) E-MEM + 10% FBS

Figure 3: Potentiodynamic polarization curves in (a) 0.9 wt. % NaCl solution, (b) Hanks's solution and (c) E-MEM + 10% FBS, all conducted at 37 °C.

4. Conclusions

From the material preparation and corrosion behaviour analyses, the following conclusions have been made:

1. Higher relative densities were achieved for the three alloys indicating that the set sintering parameters were favourable.
2. Some instances of undissolved Nb were picked up. These suggest that the milling time was rather insufficient.
3. The Ti228 alloy presented a good passivation performance and its corrosion current density is comparable to that of CP-Ti and Ti-6Al-4V alloy.
4. Ti2448 alloy showed the highest current densities in both 0.9% NaCl and Hank's solution. However, its current density was the lowest in cell culture medium.

References

[1] Geetha M, Singh A K and Asokamani R 2009. *Progress in Materials Science* **54** 397-425.

[2] Long M, Rack HJ 1998 Titanium alloys in total joint replacement – a materials science perspective, *Biomaterials* **19** 1621-39.

[3] Steinemann SG. In: Winter GD, Leray JL, de Goot K, editors. Evaluation of biomaterials. Wiley, Chichester: New York; 1980. 1-33.

[4] Niinomi M 2002 Recent metallic materials for biomedical applications *Met Mater Trans* **33A** 477-86.

[5] Hussein M A., Suryanarayana C, Al-Aqeeli N 2015 Fabrication of nano-grained Ti-Nb-Zr biomaterials using plasma sintering *Materials and Design* **87** 693-700.

[6] Li S J, Cui T C, Li Y L, Hao Y L and Yang R 2008 Ultrafine-grained β-type titanium alloy with nonlinear elasticity and high ductility *App. Phys. Lett,*. **9** 43128

[7] Henriques V A R. 2011 Titanium production for aerospace applications *J. of Aero. Tech. and Man*

[8] Henriques V A R, Campos P P and Cairo C A A 2005 Production of titanium alloys for advanced aerospace systems by powder metallurgy *Materials*

[9] Atli K C , Karaman I , Noebe R D 2014 Influence of tantalum additions on the microstructure and shape memory response of Ti50.5Ni24.5Pd25 high-temperature shape memory alloy *Mat Scie and Eng* A **613** 250-8.

[10] Chen G. Powder metallurgical titanium alloys (TiNi and Ti-6Al-4V): injection moulding, press-and-sinter, and hot pressing, ResearchSpace@ Auckland, 2014.Kon M, Hirakata L.M.,

[11] Asaoka K. Porous 2004 Ti-6Al-4V alloy fabricated by spark plasma sintering for biomimetic surface modification *J Biomed Mater Res B Appl Biomater* **68** 88-93.

[12] Falodun O E, Obadele B A, Oke S R, Maja M E and Olubambi P A 2017 *IOP Conf. Ser.: Mater. Sci. Eng.* 272 012029

[13] Xia Li, Shulong Ye, Xini Yuan, Peng Yu 2018 Fabrication of biomedical Ti-24Nb-4Zr-8Sn alloy with high strength and low elastic modulus by powder metallurgy, *J. of Alloys and Compounds*

IOP Conf. Series: Materials Science and Engineering **430** (2018) 012038 doi:10.1088/1757-899X/430/1/012038

Investigation of abnormal corrosion of 10.5 – 12.5 chromium ferritic stainless steel used to fabricate railway coal wagons

T C Mamphekgo[1,*], V J Matjeke[1] and K Pillay[1]

[1] Transnet Engineering, Research & Development, Private Bag X 528, Kilnerpark, South Africa, 0127

[*]Tefelo.mamphekgo@transnet.net

Abstract: The 10.5 – 12.5 per cent chromium ferritic stainless steel has attained success as a material used to manufacture and refurbish coal wagons. It was designed to be a low-cost material making use of the minimum chromium content required to impart acceptable corrosion resistance. There was a sudden occurrence of uniform corrosion of coal wagons manufactured between the year 2015 and 2016. This study focuses on investigating abnormally corroded wagons that are manufactured using the ferritic stainless steels. Furthermore, the exposure and salt spray tests were conducted on different consignments. The optical microscope was used to conduct a metallographic examination and chemical analysis was conducted to compare different consignments ferritic stainless steel. The root cause analysis on the plate from in-service wagon indicated that the material was indeed ferritic stainless steel and the surface roughness test showed some presence of pits on the surface of the plate. The experimental results of the different consignments indicated that one plate was dark grey which corroded quite significant during corrosion testing, whilst the other was bright grey which showed a lower corrosion rate.

1 Introduction

Ferritic stainless steel with 10.5-12.5 percentage chromium is recognised as the utility stainless steel consisting of good weldability and mechanical properties [1, 2].The most commonly used alloying elements used in stainless steel making include chromium and nickel this contributes greatly to the corrosion resistance by forming a thin invisible protective and adherent oxide layer on the surface of the metal [3, 4]. Amongst many other applications, this utility stainless steel has achieved great success as a material of construction in the manufacturing of coal wagons for the railway industry in the last three decades [1, 2].

The first wagons manufactured out of this stainless steel were placed in-service in the year 1985, in the South African railways [2]. Generally, the lifespan of these coal wagons extends to over 30 years due to its excellent corrosion and abrasion resistance [8], however, the wagons that were manufactured between the year 2015 and 2016 showed a rather interesting red oxide layer soon after exposure to the in-service conditions. Considering that prior to this period there have not been any records of this abnormal corrosion on the coal wagons, this discovery was rather abnormal and intriguing as it is scientifically known that the stainless steel does not show such type of oxide in its nature.

A plate was then sectioned from one of the wagons which were put in-service for further examinations. Furthermore, two consignments of ferritic stainless steels plates were supplied and from this, it was also noted that the one consignment consisted of dark grey coloured ferritic stainless steel and the other one consisted of bright grey stainless steel. This raised an alarm as the ferritic stainless steel supplied in the past was bright grey in colour, therefore an investigation was also conducted to further understand this difference in colour of the ferritic stainless steel. The aim of this investigation is to establish the root cause of the sudden abnormal corrosion of the ferritic stainless steel. The findings

IOP Conf. Series: Materials Science and Engineering **430** (2018) 012038 doi:10.1088/1757-899X/430/1/012038

of this investigation will assist in identifying inferior stainless steel for the manufacturing of the coal wagons.

2 Methodology

The plate sectioned from the in-service wagon was subjected to visual examination to study the surface defects, chemical analysis to determine the type and grade of the ferritic stainless steel, surface roughness test to measure the size of the surface defects and microscopical analysis to examine the microstructure of the material. Two plates from different consignments were taken from the stores and tested (see figure 8 and 9). The bright grey and dark grey plates were marked A and B respectively. Test samples were sectioned and prepared for chemical analysis, microscopical examination, corrosion testing and x-ray diffraction analysis. Test samples were also subject to corrosion testing such as exposure test and salt spray test. Corrosion tests were carried out under a controlled laboratory environment to assess the corrosion properties of the plates test sample. The exposure test was done in accordance with specification ASTM G31-72 and G1-90 to compare the corrosion properties of the two plates. Test samples for the exposure test were prepared and immersed in a solution of 3.5% NaCl for 336 hours at room temperature. The test samples were then removed, washed, rinsed with ethanol, oven dried and weighed. Furthermore, two test samples were tested in the salt spray cabinet according to ASTM B117 where a consistent salt spray of 3.5% NaCl was applied.

3 Root cause analysis

This section focuses on the investigation of the causes of the abnormal corrosion on the coal wagons. Figure 1 below shows ferritic stainless steel plate that was used on the wagon.

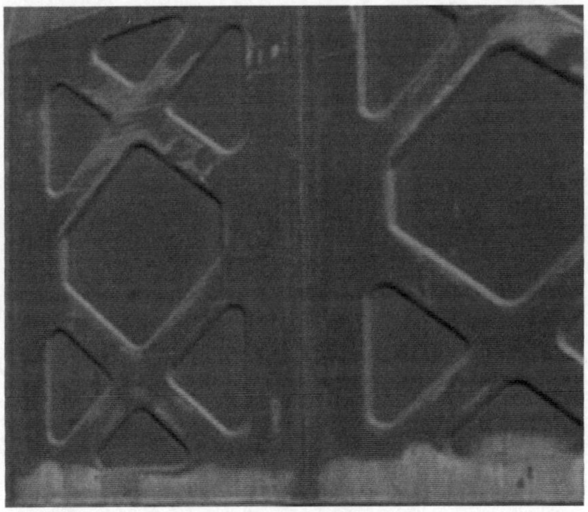

Figure 1. Corroded ferritic stainless steel plate as obtained from the in-service wagon.

3.1 Surface roughness

The surface roughness of the in-service plate was analysed using a confocal laser microscope. Figure 2 and 3 shows the contour and 3D profile of the sample as obtained. This figure clearly indicates the presence of pitting and corrosion attack on the sample obtained from the in-service plate. Figure 4 and 5 shows the parameters obtained from the roughness and waviness measurements taken along the bulk of the sample as shown by the extracted profile. The skewness, R_{sk}, is indicative of the sample consisting of many valleys or pits since it is a negative value. Furthermore, the arithmetic average roughness, R_a, was obtained to be 0.775 μm.

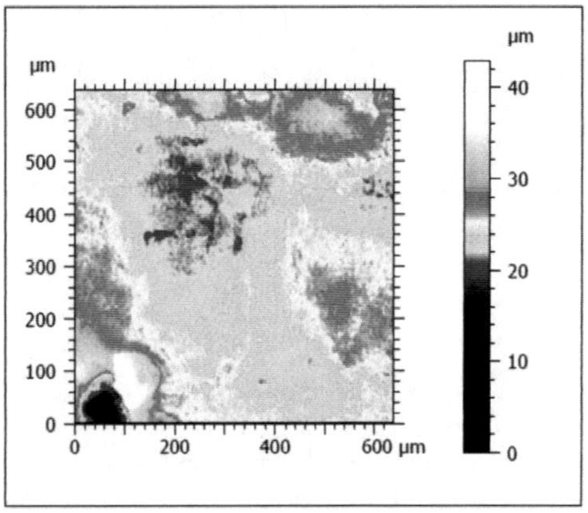

Figure 2. Contour profile of sample showing different heights.

Figure 3. 3-D surface profile as obtained from confocal laser microscope.

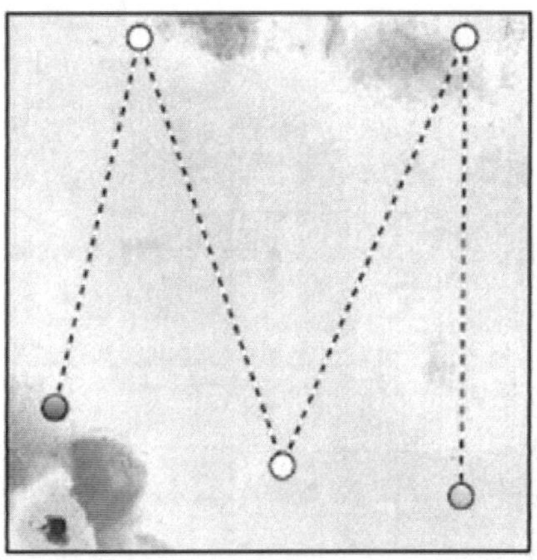

Figure 4. Extracted profile showing lines along which measurements were taken.

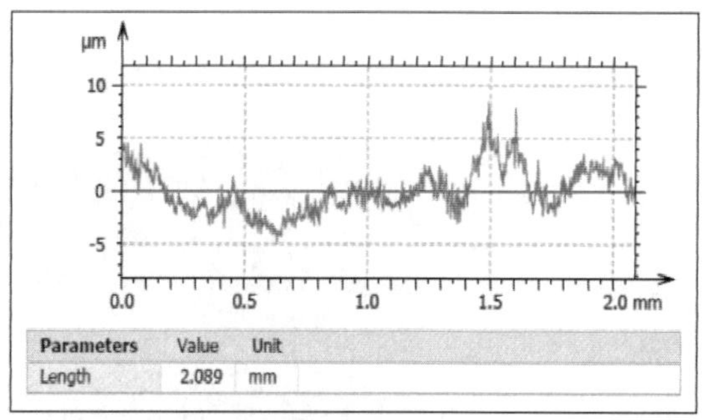

Figure 5. Longitudinal profile showing measurements taken along the extracted profile showing length and depth.

3.2 Microscopical examination

The microstructures of the plate that corroded in service is shown in figure 6 below. The general microstructure of plate was ferritic as expected from the 10.5 - 12.5 percentage chromium stainless steel.

Figure 6. The microstructure of the in-service plate showing fine ferritic grains.

3.3 Chemical properties

The chemical analysis results of the plate are tabulated in Table 1 below. The sample had a chromium content of 11.3 per cent, this indicated that the corroded plate was indeed a 10.5 – 12.5 per cent chromium ferritic stainless steel.

Table 1. Elemental composition for the plate obtained from the in-service coal wagon.

%C	%Si	%Mn	%P	%S	%Cr	%Ni	%Ti	%N
0.012	0.521	0.782	0.032	0.008	11.33	0.685	0.203	0.010

3.4 X-ray diffraction analysis

The test sample from the in-service wagon plate was subjected to x-ray diffraction analysis to determine the crystallographic structure and the phases of the material. The XRD test results indicated the sample consisted of a significant amount of BCC iron phase, titanium carbides and chromium carbide phases, as shown in figure 7 below.

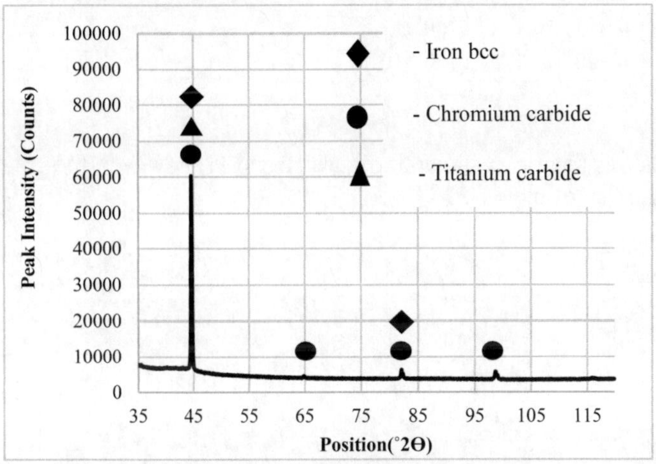

Figure 7. The x-ray diffractogram for the in-service plate.

4. Experimental Results

Figure 8 and 9 shows the plates in the as-received conditions from the stores. It was observed that plate B had red rust on the surface even before being put in service.

Figure 8. Plate A in the as-received condition from the stores.

Figure 9. Plate B in the as-received condition from the stores.

IOP Conf. Series: Materials Science and Engineering **430** (2018) 012038 doi:10.1088/1757-899X/430/1/012038

4.1 Chemical properties

The chemical analysis of the two plates is shown in table 1. The test results had shown that the chemistry of the plates was comparable, however, the nitrogen content of plate A was much higher than that of plate B.

Table 2. The chemical composition of the plate A and B.

%Elements	Plate A	Plate B
Carbon	0.039	0.016
Silicon	0.956	0.800
Manganese	1.034	0.870
Phosphorus	0.031	0.024
Sulphur	0.0023	0.0005
Chromium	11.04	11.25
Nickel	0.714	0.743
Titanium	0.313	0.217
Nitrogen	0.320	0.010

4.2 Microscopical examination

The general microstructure of both plates was ferritic as shown in figure 10 and 11, however, the grains of plate A was observed to be coarser than that of plate B.

Figure 10. The microstructure of plate A showing a coarse ferritic grains

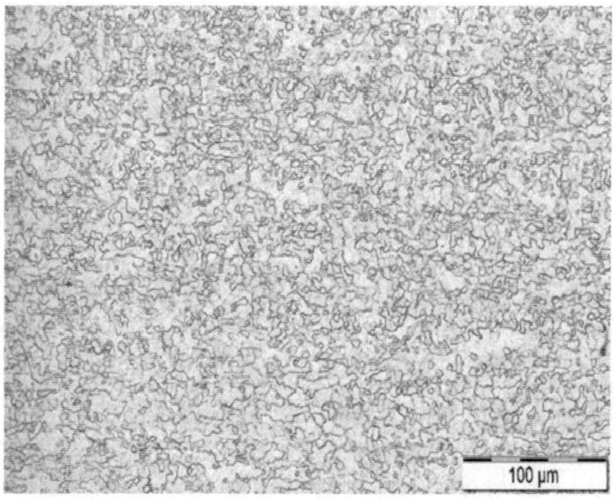

Figure 11. The microstructure of plate B showing fine ferritic grains

4.3 Corrosion test results

4.3.1 Exposure Test. Figure 12 and 13 shows the exposure test samples. From the test results, it was clearly evident that plate A has lower corrosion properties compared to the plate B, as plate B was corroding at a much higher rate than plate A. After the test samples were washed and the weight measured, the corrosion rate was calculated to be of an average rate of 0.03735mm/per year for plate B, while plate A corroded at a lower average rate of 0.00965mm/per year.

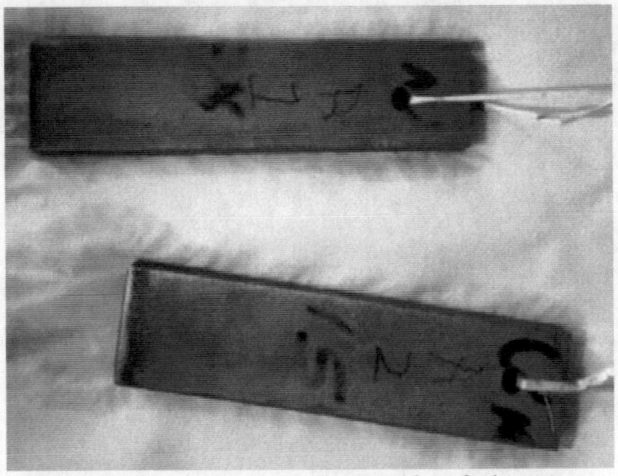

Figure 12. Exposure test sample of plate A.

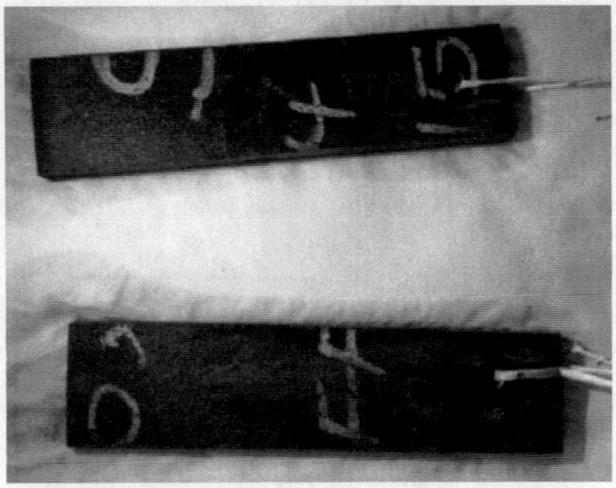

Figure 13. Exposure test sample of plate B.

4.3.2 Salt Spray Test. The sample from plate A did not show any significant corrosion after the salt spray test (see figure 14), however, the sample from plate B showed much more significant corrosion as can be seen by the corrosion products on the surface of the plate B.

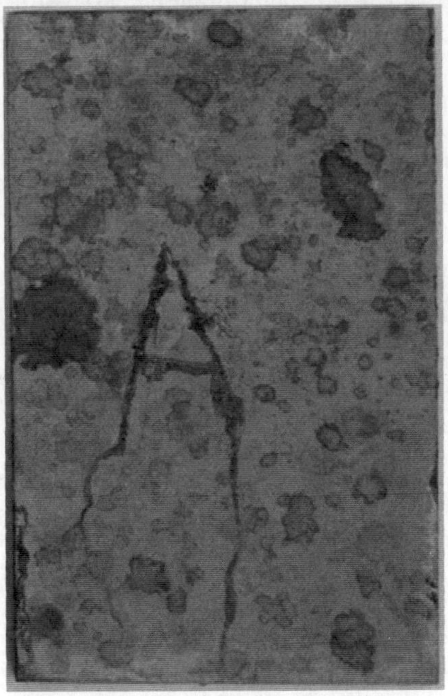

Figure 14. Salt spray test sample of plate A.

Figure 15. Salt spray test sample of plate B.

4.4 X-ray diffraction analysis

The test samples from plate A and B were subjected to x-ray diffraction analysis after salt spray testing, to determine the crystallographic structure and the phases of the material. The XRD test results indicated that all the three test samples consisted of a significant amount of BCC iron phase, titanium carbides and chromium carbide phases (see figure 16 and 17).

IOP Conf. Series: Materials Science and Engineering **430** (2018) 012038 doi:10.1088/1757-899X/430/1/012038

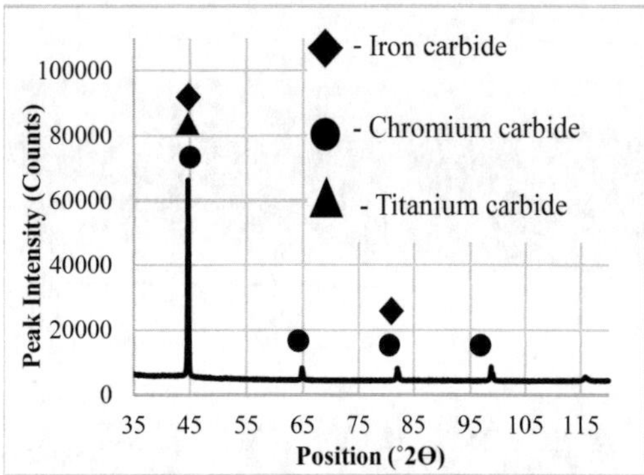

Figure 16. X-ray diffractogram for plate A test sample.

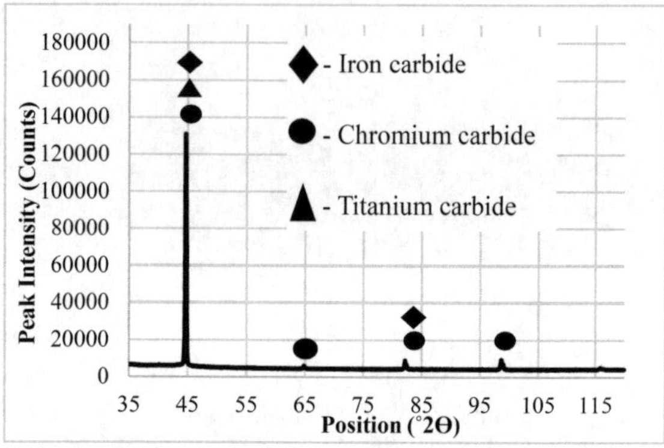

Figure 17. X-ray diffractogram for plate B test sample.

5. Discussions and Conclusions

From the root-cause analysis of the plate sectioned from the in-service coal wagon, it was observed that the plate had a dark grey layer on the surface, this is due to the formation of a thick dark oxide layer which forms during hot rolling of the ferritic stainless-steel sheets [12]. This oxide layer contains depleted chromium layers which render it poor in corrosion properties [13], the layer is usually removed by means of pickling the ferritic stainless steel in inorganic acids such sulphuric and hydrofluoric acid after the rolling and heat treatment processes [12, 14].

It was quite evident that the plate was not pickled after rolling processes as it was showing a red oxide on the after being put in-service for a short period of time. The chemical analysis of the in-service plate had shown that it was a ferritic stainless as the chromium content was 11,33 per cent, whilst the general microstructure was ferritic with fine grains. The surface roughness analysis indicated that there was a presence of pits on the surface of the plate, which was due to pitting corrosion. Pitting corrosion was due to the accumulation of chloride and sulphate ions on the surface of the coal wagon and led to the breaking down of the oxide layer and the initiation of corrosion [8, 9]. The experimental results of the two ferritic stainless steel taken from the stores shown that the two plates had comparable chemistry, however the titanium and nitrogen content of test sample A was slightly higher than that of test sample B.

The microstructures of both test sample A and B were ferritic, however, test sample A showed significantly coarse grains as compared to the more refined microstructure of test sample B. The test sample B consist of low amount of titanium in solid solution as seen from the chemical analysis results, this together with low annealing temperatures of about 880 to 925 degree Celsius, promotes faster recrystallisation of the grains of the material and at the same time promotes the migration of the titanium atoms to the grain boundaries. The titanium atoms then restrict the growth of the grains by the effect of pinning [15]. The XRD analysis results had shown that all the test samples consisted of chromium carbides (Cr7C3), titanium carbides (TiC) and BCC iron phases. Titanium is added to the stainless steel during steel making because of its high affinity for carbon than chromium, therefore it reacts with the carbon in the stainless steel to form TiC thereby preventing the sensitisation effect which is due to the formation of Cr7C3 [11,15]. However, the Cr7C3 carbides cannot be eliminated completely as there is a high amount of chromium in the ferritic stainless steel, hence the presence of small amount of this phase on the XRD results.

References

[1] Beckitt,C, "Sweet tastes of success for 3Cr12 utility stainless steel," *Anti-corrosion methods and materials,* **40**, no. 2, pp. 9-11, 1993

[2] SASSDA,[Online].Available:https://www.sassda.co.za/wp-content/uploads/2013/10/3CR12INACTION.pdf. [Accessed 03 July 2018].

[3] Yuan F,2012. [Online]. Available: http://www.intechopen.com/books/pitting-corrosion/corrosive-effects-of-chlorides-on-metals. [Accessed 25 August 2016]

[4] Outokumpu OJY, "Outokumpu OJY," 2013. [Online]. Available: http://www.adatum.com. [Accessed 22 March 2017]

[5] Ahmad Z, "Types of Corrosion: Materials and environment," in *Principles of Corrosion Engineering and Corrosion Control*, Butterworth, IchemE, 2006, p. 146

[6] Bornmyr A, Winkler F, Winkler J Bohler Welding Ghmb, "Avesta finishing chemicals," 2009. [Online]. [Accessed 20 March 2017]

[7] Chong P.H., Liu Z., Skeldon P. and Crouse P., "Characterisation and corrosion performance of laser-melted 3Cr12 steel," *Applied surface science,* pp. 362-368, 2005.

[8] Thomas C.R, "SX 3CR12 - a material for abrasion-corrosion," *Journal of the South African Institute of Mining and Metallurgy,* **81**, no. 10, pp. 298 - 302, 1981

[9] Shi P, Shi H, Liu C and Jiang M "Effect of pickling process on removal of oxide layer on the surface of ferritic stainless steel," *The Canadian journal of metallurgy and material science,* 2017

[10] Yue Y, "Descaling behaviour of 430 hot-rolled stainless steel in HCl-based solution," *Journal os iron and steel research, international,* **3**, no. 23, pp. 190-196, 2016.

[11] Crookes R, "Pickling and passivating stainless steel," *Materials and applications series,* vol. 4, no. **2**, 2007

[12] Kuzucu V, Aksoy M and Korkut MH "The effect of strong carbide-forming elements such as Mo, Ti, V and Nb on the microstructure of ferritic stainless steel," *Journal of materials processing technology,* no. **82**, pp. 165-171, 1998.

[13] "Hilti corporation," May 2015. [Online]. Available: http://www.hilti.com. [Accessed 22 March 2017]

[14] Frankel GS, "Pitting Corrosion of Metals: A Review of the Critical Factors," *Journal of the Electrochemical Society,* **145**, no. 6, pp. 2186-2198, 1998.

[15] de Almeida Reis Tanure L.P, de Alcantara CM, De Oliveira TR, Santos DB and Gonzalez BM "Microstructure, Texture and Microhardness Evolution during annealing heat treatment and Mechanical behavior of the Niobium-stabilized Ferritic stainless steel ASTM430 and Niobium-Titanium- Stabilized ferritic stainless steel ASTM439:-a comparative study," *Materials Research,* **6**, no. 20, pp. 1650-1657, 2017

[16] Skalarska-Smialowska Z, Pitting and crevice corrosion, NACE international, 2005

Investigation of corrosion behaviour of carbon steel in simulated soil solution from anodic component of polarisation curve

Phumlani Mjwana[1, 2, a] Mandla Mahlobo[1, 2, b] Obadele Babatunde[1, c] Philippe Refait[2, c] Peter Olubambi[1, e]

[1] Centre for Nanoengineering and Tribology, University of Johannesburg, Corner of Siemert and Beit Streets, Johannesburg, Gauteng, 2094 South Africa
[2] Laboratoire des Sciences de l'Ingénieur pour l'Environnement (LaSIE), UMR 7356 CNRS-Univ. La Rochelle, Bât. Marie Curie, Av. Michel Crépeau, F-17042 La Rochelle cedex 01 France

[a] mjwanap@gmail.com

Abstract. Marine structures and buried pipelines are constantly exposed to extreme conditions with elements such as temperature variation, soil ion content, dissolved metals, etc posing a continuous threat on the integrity and lifespan of the structures. Quantification of these elements has been limited to studying only a few of the parameters, with the resulting corrosion behaviour not fully understood. The purpose of this study was to investigate the corrosion behaviour of carbon steel in simulated soil solution using the anodic component attained from modelled polarisation curves. The behaviour of the metal/electrolyte interface was studied using non-invasive in-situ electrochemical techniques at OCP. A combination of voltammetry, electrochemical impedance spectroscopy with X-ray diffraction analysis of the mineral layer were utilised. The expression for the anodic component of the current as a function of potential was attained from $\log|j_a|$ vs. potential plots. Mathematical modelling of the experimental polarisation curves was done using *OriginPro Data Analysis & Graphing Software*. Results showed that at OCP, lepidocrocite, carbonated green rust, calcite and aragonite were found as the corrosion process involved anodic and cathodic zones. Voltammetry around OCP (VAOCP) and linear polarisation resistance (LPR) showed the necessity for corrosion protection as the adopted electrochemical system resulted in a progressively corrosive environment. Kinetics and mechanism of the anodic and cathodic process indicated corrosion processes similar to aerated soil conditions.

1. Introduction

Buried pipelines are continuously exposed to harmful and threatening elements in the immediate surrounding soil environment such as soil moisture, temperature variation, pH, soil ion content, dissolved metals, etc. all of which could prove detrimental to the service life of buried pipelines [1]. To protect these pipelines from external corrosion, a combination of organic coatings and cathodic protection (CP) is used. According to the EN 12954: 2001 standard, a residual corrosion rate lower of equal to 10 μm/y must be maintained. Application of CP in immersed conditions containing calcium and magnesium results in the formation of calcareous deposits. While this layer has been shown to have a beneficial effect for CP [2], the role has never been quantified

Cathodic protection (CP) is one of the most common and effective means for corrosion prevention in offshore metallic facilities and buried carbon steel structures [3,4]. CP leads to scaling of which, in saline fluids transportation industries such as oil recovery, power generation and water desalination may lead to major problems [5]. There remains a beneficial effect of the scaling in CP in terms of corrosion protection [6]. The scaling deposit often contain calcium carbonate in the form of calcite and/or aragonite with brucite precipitating at low applied potentials.

It has been demonstrated that high temperature, more cathodic potentials and increased interfacial pH are the real driving forces of the precipitation of brucite [6]. It was further shown that the brucite layer acts as an inhibitor for the $CaCO_3$ aragonite deposition as the initial scaling process occurs between the brucite crystals. Instantaneous crystallization of $CaCO_3$, independent of applied potential, was shown by optical microscope in a study in artificial water [7]. The same study illustrated the inhibiting effect of SO_4^{2-} of which modified the calcite morphology under cathodic polarization. One of the most crucial parameters that influences the metal/electrolyte residual corrosion behaviour is O_2 at the vicinity of the electrode.

The present study focuses on the interactive behavior of carbon steel and O_2 in a corrosive environment and studies the corrosion behavior at the metal/electrolyte interface

2. Experimental procedure

The S235JR carbon steel was used in the present study is commonly used for pipelines and construction of buildings and bridges [8]. Its average composition (weight %) was: 98.94% Fe, 0.097% C, 0.49% Mn, 0.020% P, 0.09% Si, 0.05% Cr, 0.11% Ni, 0.16% Cu, and 0.042% S. Cylindrical coupons (15 mm diameter, 6 mm thick) were cut from the same sheet, polished with silicon carbide grade P600, rinsed with acetone and rapidly dried in an air flow. Coupons were connected by welding to copper wire, embedded in epoxy resin and studied at the end of each experiment for analysis of deposit layer and corrosion products. Only one surface of the coupon, with total area 1.77 cm^2, was exposed to the simulated soil solution. A single coupon was immersed in 750ml simulated solution in each electrochemical cell with platinum plate and saturated calomel electrode (SCE) used as the counter and reference electrode respectively. The solution was gently stirred and the beaker top left uncovered and thus exposed to the atmosphere.

Table 1. Mineral composition of Electrolyte 1 as prepared on the day of each experiment

Compounds	**Composition (*mmol/l*)**
NaCl	10
$CaCl_2 \cdot 2H_2O$	2
$MgCl_2 \cdot 6H_2O$	1
$Na_2SO_4 \cdot 7H_2O$	1.5
$NaHCO_3$	2

The simulated soil solution was prepared using analytical grade reagents. All experiments were conducted in soil simulated solution containing main species present in soil as determined by ASTM norm D1141 (American Society for Testing Materials) represented in Table 1. Prepared using analytical grade reagents, Electrolyte 1 (Table 1) was used as the calcareous deposit forming solution. Electrolyte 1 was used for all experiments unless otherwise stated.

Voltammetry around OCP (VAOCP) could be used as a non-destructive technique in order to monitor the evolution with time of the residual corrosion behavior [9]. It includes a moderate potential range (±50 mV in the present study) which is assumed sufficient to have a negligible damage on the metal/electrolyte interface. The fitting procedure of the experimental polarization curve used for the present study was adapted from Akkouche et al. [10] in order to determine the controlling corrosion kinetics [11]. A scan rate $dE/dt = 0.1\ mVs^{-1}$ was used.

In order to obtain the polarization curves around OCP, the potential was swept from OCP to OCP +50mV (forward scan) and then from OCP +50mV to OCP –50mV (reverse scan). During VAOCP experiments potential range was ±50mV unless stated otherwise. For CP experiments the above mentioned procedure necessarily modifies the metal/electrolyte interface. For this reason, the forward scan gives the true nature of the metal/electrolyte interface. The corrosion potential (E_{corr}) was directly read on the log|j| vs. E_{IRfree} curve while corrosion current density (j_{corr}), the Tafel coefficients β_a and β_c, and the limiting current density (j_{lim}) were obtained from the computer fitted curve. Polarization curves were attained at day 2 and day 7.

A true representation of the corrosion behavior was obtained from the anodic component of the modelled polarization curves. For this study, the anodic component was compared for forward scans at day 7 by extrapolating the current density value at the applied potential for the respective potential [12]. This would give a true reflection of the residual corrosion rate (V_{CR}). It remains true that cathodic polarization protected the electrode from corrosion. There was a residual corrosion behavior that may be observed and quantified [12,13]. As a basis and point of reference, the anodic component was also evaluated for modelled polarization curves at OCP.

X-ray diffraction (XRD) experiments were carried out at room temperature with an INTEL XRD Equinox 6000 in a step mode scanning using Co-Kα

3. Results and discussion

With the many parameters involved in the electrochemical system, the path travelled by current is one of the most crucial. Ohm's law takes into account the electric current flow between two points and the resistance of the path. With resistance being one of the major components, the Ohmic drop or IR drop has to be taken into account because the measured potential is influenced by the resistance of the simulated soil, metallic path and/or corrosion products. Electrochemical impedance spectroscopy (EIS) was conducted to examine the high frequency behaviour at OCP and thus determine the IR drop. The system OCP ($E_{OCP} = -0.69V_{SCE} \pm 0.02V_{SCE}$) was monitored for 7 days and the system allowed to reach steady state before EIS was conducted. The impedance data (Figure 1) reveals a R_e value of 110 Ω.

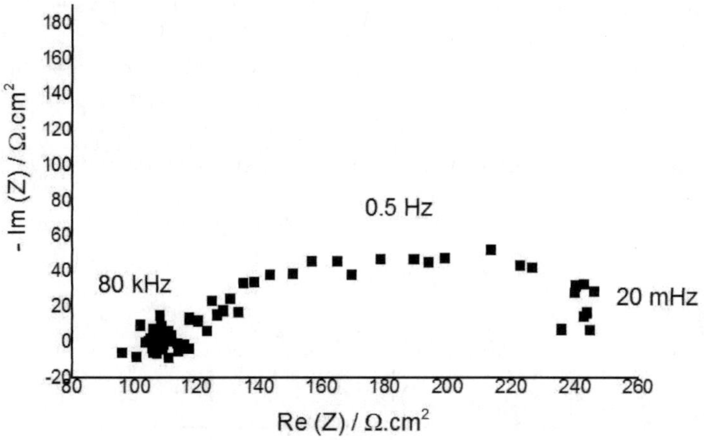

Figure 1. Nyquist plot at OCP

3.1. *Polarisation resistance*

Linear polarisation resistance (LPR) was performed at day 2 and day 7 to assess the corrosion behaviour of the electrode at the metal/electrolyte interface [9]. Polarisation resistance (R_P) could be determined by finding the gradient of the current density vs potential line at zero current (Figure 2). At current density zero there was no current flowing and the respective potential was taken as the free corrosion potential (E_{corr}). It was at this point where R_P was best represented. Therefore the inverse of the gradient was the polarization resistance with units Ohm squared centimetre ($\Omega.cm^2$). R_P values determined were 1301.7 $\Omega.cm^2$ and 1210.8 $\Omega.cm^2$ for day 2 and day 7 respectively. These values were a direct indication of the resistance as to polarisation of the metal, in the respective electrolyte, against corrosion with respect to time. It can be deduced that the decrease of R_P indicated a reduction of corrosion resistance with time. Therefore a higher corrosion rate was expected and thus a greater need for corrosion protection. This also has a direct link to the increase of potential observed in Figure 2 which points towards formation of corrosion products.

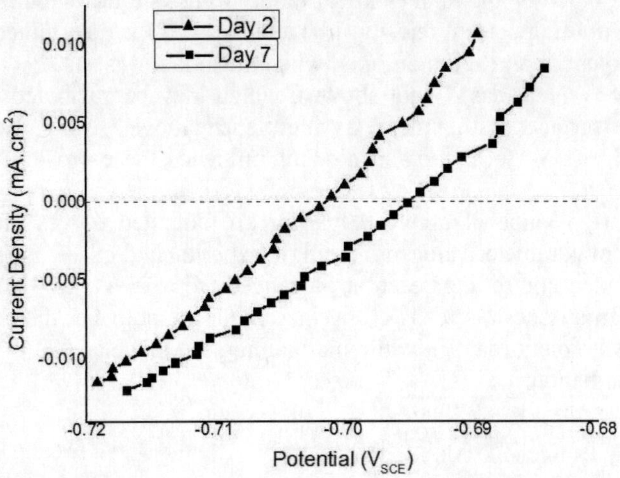

Figure 2. Linear polarisation resistance plot for day 2 and 7 of OCP experiment

3.2. Open circuit potential

The free corrosion potential (E_{OCP}) for the system was determined by allowing the electrochemical cell to reach a steady state (see Figure 3). No current or potential was applied. The potential starts off at a high value of –0.48 V_{SCE}. This potential corresponds to the oxide film formed as a result of iron oxidation in the atmosphere. As the oxide film dissolves, the potential decreases. The potential reaches its lowest point after a single day of –0.70 V_{SCE}. Over the seven day period the potential plateaus at an averages value of –0.69 V_{SCE}.

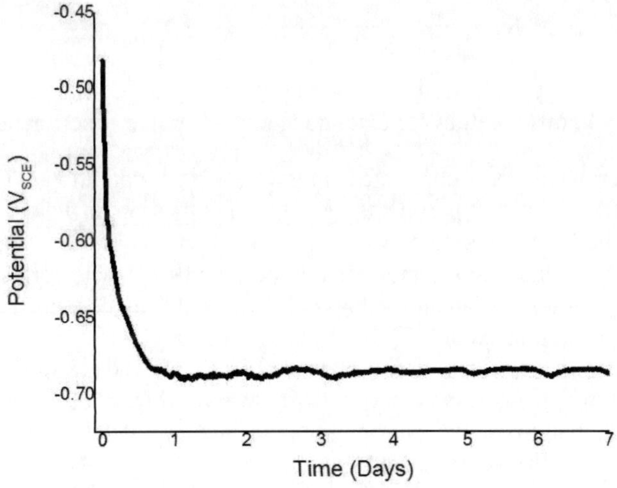

Figure 3. OCP evolution monitored over 7 days.

3.3. Voltammetry and corrosion rates

The kinetics and mechanism of the anodic and cathodic processes at OCP was done using VAOCP with potential range of ±50 mV [10]. VAOCP is able to get information about the electrochemical system while inducing minimal damage to the metal/electrolyte interface. The controlling kinetic model of the electrochemical system had to be determined according to the best computed fit when compared to the experimental data. Kinetic models include charge transfer (tafel), mixed activation/diffusion (mixed), diffusion control (diffusion) and a combination of H_2O and O_2 reduction control (hydroevolution) [10]. The anodic controlling mechanism was assumed to be charge transfer and this is an assumption that is carried throughout the study. This can be observed from Figure 4, and the modelled curves to follow, that the anodic experimental and computed curve are best fitted with a straight line, which indicates Tafel law behaviour.

The parameters determined by the mathematical model were as follows; current density (j_{corr}), tafel coefficients (β_a and β_c), limiting current density (j_{lim}) and E_{corr}. E_{corr} is determined graphically from the log current density vs potential plot and then fixed when the model was run. E_{corr} increases from – 0.70 V_{SCE} at day 2 to – 0.68 V_{SCE} at day 7 (not shown), which may be attributed to the formation of a corrosion layer on the surface of the metal as mentioned above. Although anodic and cathodic potentials were applied, the VAOCP method did not influence the corrosion of steel because E_{corr} values were close to E_{OCP}.

With a Chi Square (R^2) value of above 0.999 for all modelled curves, this method of kinetic modelling gave a range of parameter values for a single experimental curve. For example, day 7 curve parameter values for mixed controlling reaction βa ranged from 15 V^{-1} to 33 V^{-1}, with β_c deviating inconsistently as well. This issue was resolved by increasing the potential range for VAOCP to ±100 mV at the end of the experiment (day 7), while maintaining the range at ±50 mV at day 2 in order to limit the interfacial perturbation.

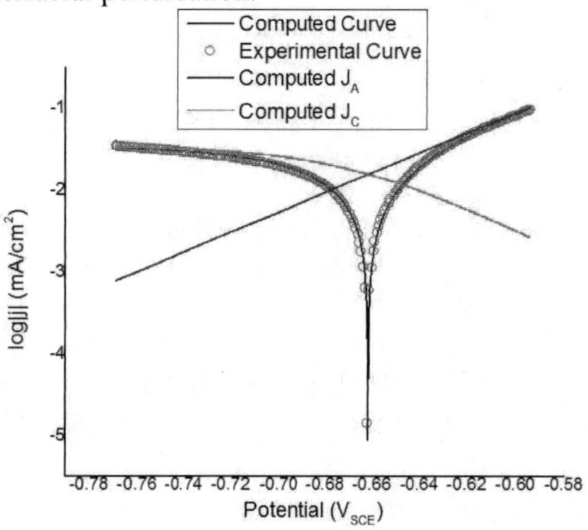

j_{corr} (mA/cm²)	0.021 ± 0.002
β_a (V^{-1})	28 ± 1
β_c (V^{-1})	-25 ± 4
j_{lim} (mA/cm²)	-0.040 ± 0.006
E_{corr} (V_{SCE})	-0.69 ± 0.01

Figure 4. Computer fitted polarisation curve with potential range ±100 mV at day 7 with averaged parameter values for extended potential range experiments.

Three sets of experiments were conducted under these new conditions. The cathodic controlling mechanism deduced for all OCP experiments was mixed activation/diffusion control, while the Tafel behaviour of the anodic branch mentioned is shown in Figure 4. The β_a value was fixed between a specific range while taking into account previous work done [9,10]. The corresponding table in Figure 4 gives the overall summarized parameter values. This translates to a corrosion rate (V_{corr}) of 248 ± 29 µm/yr. This correlates to aerated conditions in soil environment [10].

From observing the cathodic component behaviour (Figure 4), it tends towards a horizontal plateau. Therefore, the controlling mechanism was very close to pure diffusional control. The O_2 reached the electrode at a slow rate, but a pure diffusion regime was not completely attained. With time corrosion products formed, blocking the active electrode surface and thus increasing the E_{corr}. This was also consistent with the slight decrease in j_{corr} between day 2 and day 7 (not shown).

XRD diffractograms for coupons at OCP (Figure 5) show presence of lepidocrocite (L) and carbonated green rust (GR) with chemical makeup γ-FeO(OH) and $Fe^{II}_4Fe^{III}_2(OH)_{12}.CO_3$ respectively. GR is usually found in aerated conditions as illustrated by [12]. This points towards the availability of O_2 at the electrode vicinity. Aragonite (A) and calcite (C) were detected with Aluminum (Al). Al was observed because the sample holder is made up of Aluminum. Aragonite and calcite indicates the presence of active cathodic zones. This would explain why the electrode surface is not completely covered with L and GR, instead there are anodic and cathodic zones. The electrolyte allows easy formation of A and C and therefore these two species should be expected when cathodic polarization is applied.

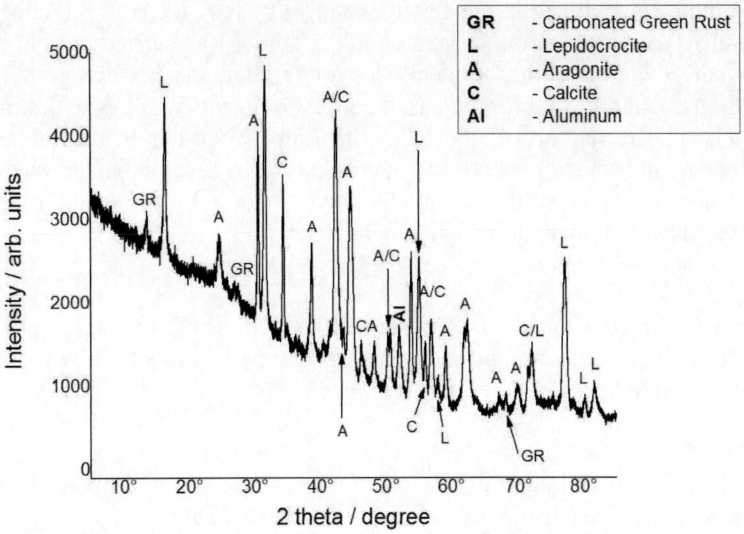

Figure 5. XRD diffractogram of products on coupons at OCP at the end of the experiment.

3.4. *Anodic component*

A true representation of the interfacial corrosion behaviour was obtained from the anodic component of the modelled polarisation curves. For this study the anodic component was compared for forward scans at day 7 by extrapolating the current density value at OCP [12]. As a basis and point of reference, the anodic component was also evaluated for modelled polarisation curves at OCP.

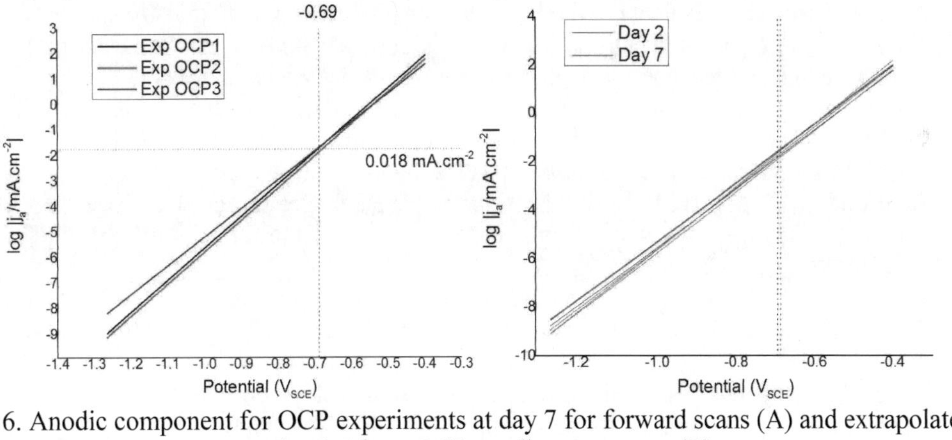

Figure 6. Anodic component for OCP experiments at day 7 for forward scans (A) and extrapolated current density from OCP anodic component (B)

The E_{OCP} was found to be –0.69 V_{SCE} of which the extrapolated anodic current density was 0.018 mA.cm^{-2} (see Figure 6A). This translates to a corrosion rate of 213 μm/yr. As a point of reference for cathodic protection anodic component analysis [12], the residual corrosion rate was predicted for applied cathodic potential –0.7 V_{SCE} to –1.2 V_{SCE} from the current density vs. potential curves (log$|j_a|$ vs. E) in Figure 6. It can be deduced that when cathodic potential is applied at –0.8 V_{SCE}, the corresponding current density would be 0.008 mA/cm^2 and thus a residual corrosion rate of 14 μm/yr.

Interfacial corrosion behaviour can be directly related to the change in β_a values. This may be observed from the gradient of the anodic component as the gradient increases with time (Figure 6B). This behaviour may be due to the formation of corrosion products. The average EOCP for day 2 and 7 (–0.693 and –0.685 V_{SCE} respectively), represented by the vertical dotted lines, further validates this assumption. With time corrosion products form, blocking the active electrode surface and thus increasing the E_{corr}. The O_2 reaches the electrode at a slow rate but a pure diffusion was not completely attained. This is also consistent with the slight decrease in j_{corr} with time (not shown).

4. Conclusion

With the many parameters involved in the electrochemical system, the path travelled by current is one of the most crucial. Voltammetry around open circuit potential was carried out in order to understand interfacial behaviour without altering the metal/electrolyte interface. Formation of corrosion products on the metal surface resulted in slightly increased corrosion potential values with time. This may point towards the effects of perturbation of interface with time. A similar behaviour was observed with increased β_a values as they were observed to increase with time as depicted by anodic component graphs. The availability of O_2 at the metal/electrolyte interface was noted to be limited as the polarization curves showed limited O_2 diffusion behaviour.

Acknowledgement
This study is part of the collaborative Masters study at the University of Johannesburg and University of La Rochelle, carried out by Phumlani Mjwana and financially supported by the National Research Foundation (NRF) - South Africa

References
[1] Bahadori A. 8 - Commissioning of Cathodic Protection Systems. Cathodic Corrosion Protection Systems. Gulf Professional Publishing; 2014. 355-399
[2] Shi W, Lyon SB. Investigation using localised SVET into protection at defects in epoxy coated mild steel under intermittent cathodic protection simulating inter-tidal and splash zones. 2017 *Prog Org Coatings*. **102**:66–70
[3] Neville A, Morizot AP. Calcareous scales formed by cathodic protection - An assessment of characteristics and kinetics. 2002 *J Cryst Growth*. **243**(3–4):490–502.
[4] Barbalat M, Lanarde L, Caron D, Meyer M, Vittonato J, Castillon F, et al. Electrochemical study of the corrosion rate of carbon steel in soil: Evolution with time and determination of residual corrosion rates under cathodic protection. 2012 *Corros Sci*.;55:246–53
[5] Gabrielli C, Keddam M, Khalil A, Rosset R, Zidoune M. Study of calcium carbonate scales by electrochemical impedance spectroscopy. 1997 *Electrochim Acta*.
[6] Karoui H, Riffault B, Jeannin M, Kahoul A, Gil O, Ben Amor M, et al. Electrochemical scaling of stainless steel in artificial seawater: Role of experimental conditions on CaCO3and Mg(OH)2formation. 2013 *Desalination*. **311**:234–40.
[7] Ben Amor Y, Bousselmi L, Bernard MC, Tribollet B. Nucleation-growth process of calcium carbonate electrodeposition in artificial water-Influence of the sulfate ions. 2011 *J Cryst Growth*. **320**(1):69–77
[8] Kossakowski PG. Prediction of Ductile Fracture for S235JR Steel Using the Stress Modified Critical Strain and Gurson-Tvergaard-Needleman Models. 2012 *J Mater Civ Eng* **24**(12):1492–500.
[9] Rahal C, Masmoudi M, Abdelhedi R, Sabot R, Jeannin M, Bouaziz M, et al. Olive leaf extract as natural corrosion inhibitor for pure copper in 0.5 M NaCl solution: A study by voltammetry around OCP. 2016 *J Electroanal Chem* **769**:53–61
[10] Akkouche R, Rémazeilles C, Jeannin M, Barbalat M, Sabot R, Refait P. Electrochimica Acta In fl uence of soil moisture on the corrosion processes of carbon steel in arti fi cial soil : Active area and differential aeration cells 2016 *Electrochim Acta* **213**:698–708.
[11] Mahlobo MGR, Mjwana P, Tladi MNA, Obadele BA, Olubambi PA, Refait P. Evaluation of cathodic protection performance of carbon steel pipeline buried in soil: A review. In: Mechanical and Intelligent Manufacturing Technologies (ICMIMT), 2018 IEEE 9th International Conference on. IEEE; 2018. p. 37–43.
[12] Nguyen Dang D, Lanarde L, Jeannin M, Sabot R, Refait P. Influence of soil moisture on the residual corrosion rates of buried carbon steel structures under cathodic protection. Electrochim Acta. 2015;176:1410–9
[13] Barbalat M, Caron D, Lanarde L, Meyer M, Fontaine S, Castillon F, et al. Estimation of residual corrosion rates of steel under cathodic protection in soils via voltammetry. 2013 *Corros Sci* **73**:222–9
[14] Xiao-gang LI. Effect of Environmental Factors on Electrochemical Behavior of X70 Pipeline Steel in Simulated Soil Solution. 2009 *J Iron Steel Res Int* **16**(6):52–7.

IOP Conf. Series: Materials Science and Engineering **430** (2018) 012040 doi:10.1088/1757-899X/430/1/012040

Flow-induced corrosion behaviour of low alloy steel in the presence of mono-ethylene glycol

Sunday Aribo[1, 2], Mantsha Sesheweni[1], Olufemi Sanumi[1], Adekunle Ogunbadejo[2], Oladeji Ige[1], Peter Apata Olubambi[1]

[1]Centre for Nanoengineering and Tribocorrosion, School of Mining, Metallurgy and Chemical Engineering, University of Johannesburg, South Africa
[2]Department of Metallurgical and Materials Engineering, Federal University of Technology, Akure, Nigeria

Email: aribosunny@yahoo.com

Abstract. The flow-induced corrosion behaviour of welded low ally steel in 1 wt. % NaCl environment containing monoethylene glycol at 0°C and 25 °C respectively, was investigated. A modified submerged impinging jet rig was adopted to simulate the low temperature environment. Varying speeds between 1 and 3 m/s were used to study the degradation mechanism of the welded steel. Also, the ratio of MEG: 1 wt. % NaCl was kept at 1:4 for the MEG/NaCl solution. Static corrosion condition was also studied and compared with the flowing conditions. Results indicated that the alloy degraded with observable weight loss in pure 1wt. % NaCl at both temperatures whereas there was very negligible weight loss in all the samples in MEG/NaCl solution.

1. Introduction

It is well documented that flow can aggravate corrosion susceptibility of steels by accelerating mass transfer rates of the reacting species and by destroying the protective film formed on the surface of the steel [1]. The localized attack on the film is said to be due primarily to high near-wall turbulence and the stress developed in the protective film. Other authors argued that mechanical stress and chemical dissolution acting simultaneously are responsible for iron carbonate removal during flow accelerated corrosion [2]. The scale breakdown can happen for a variety of reasons, many of them related to fluid flow [2]. It is hypothesized that following the scale damage, a galvanic effect is established between the scale covered surfaces (cathode) and the scale-free surface (anode) leading to propagation of localized attack.

Meanwhile, as reported by Aribo et al. [3], hydrate formation is a flow assurance problems in wet-gas transportation in deep water wells at high pressure and low temperature [4]. One way to resolve this flow assurance challenge is the addition of thermodynamic and hydrate inhibitor such as the ethylene glycol [4, 5]. In the course of doing the job of dehydration, mono-ethylene glycol (MEG) has also been found to inhibit corrosion [6-8].

Monoethylene glycol (MEG) has also been reported to cause an increase in density and viscosity of the aqueous solution [9-10]. Addition of monoethylene glycol (MEG) to wet gas influences CO2 corrosion by lowering the dew point of water which cause a decrease in Top-of-the-Line Corrosion [9, 11]. Also, there is a decrease in corrosion rate at the bottom of the line when with addition of monoethylene glycol to water phase [11-12]. MEG has also been reported [13] to adsorb on steel surface affecting the anodic reaction of the corrosion process. This report shows that monoethylene glycol inhibits CO2 corrosion.

This research is however an attempt to study the behavior of monoethylene glycol in interaction with chloride environment at a low temperature under dynamic conditions. This is because most literature reported the inhibitive performance of MEG under static corrosion conditions. One other innovation of this work is that very low temperature flowing condition is adopted.

2. Method

Welded low alloy steel (sectioned into HAZ, parent metal, Weldment) with chemical composition shown in Table 1 was adopted for this research. A submerged impinging jet reported by Aribo et al. [14] was modified by incorporating a compartment around the reservoir to house ice blocks (Figures 1). 1 wt. % NaCl electrolyte and 1 wt. % NaCl and MEG in ratio 4:1 to MEG were used as the corrosive environments. Experiments were conducted at 25°C and 0°C, respectively. Meanwhile temperature of 0°C was achieved by using the adapted rig in Figure 1 with the reservoir placed inside a plastic cooler. Ice block was then packed around the reservoir containing the electrolyte. A digital thermometer was used to monitor the temperature of the electrolyte. Each experiment was conducted for 1 hour and the weight loss determined by a digital weighing balance. Meanwhile flowing speed was varied between 1- 3 m/s. All experiments were conducted in triplicate and the average results reported.

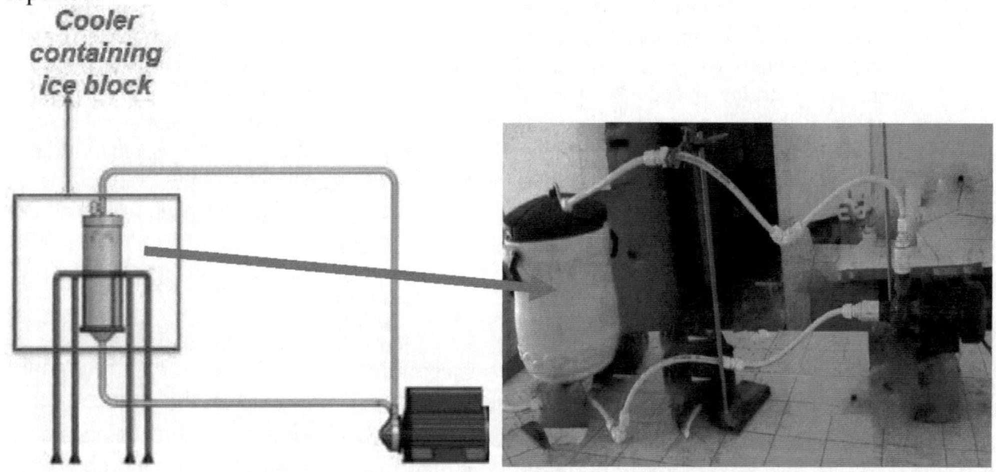

Figure 1: Adapted submerged impingement jet rig

Table 1: Chemical composition of the low alloy steel

	C	Mn	Si	P	S	Cr	NI	Mo
Wt.%	0.18	1.3	0.3	0.25	0.10	0.30	0.30	0.12

3. Results

3.1. Weight loss in NaCl environment at 0°C and 25 °C.

Figures 2a and 2b Show the weight loss analysis of the samples in 1 wt. % NaCl at 0°C and 25°C. It is observed that at static condition, the weld metal exhibited low weight loss while the parent metal and heat affected zone showed substantial weight loss. When the materials were subjected to flow conditions, all the zones of the welded metal showed increase weight loss with increase speed. As expected, flowing conditions enhanced more material loss for all materials. However the weld metal showed higher resistance to material degradation compared with HAZ and weld metal.

3.2. Weight loss in MEG/NaCl environment at 0°C and 25 °C.

Figures 3a and 3b indicate a very negligible weight loss in all samples in the presence of MEG/NaCl as compared to NaCl at 0 °C and 25 °C in Figure 2. It can be observed that all the zones of the welded alloy showed negligible weight loss at static condition. The HAZ has the highest loss in weight while the weldment showed the highest resistance. The decrease in the weight loss of the weld metal may likely be due to the presence of high corrosion resistant alloying elements that are likely to be present in the filler metal used during the welding operation.

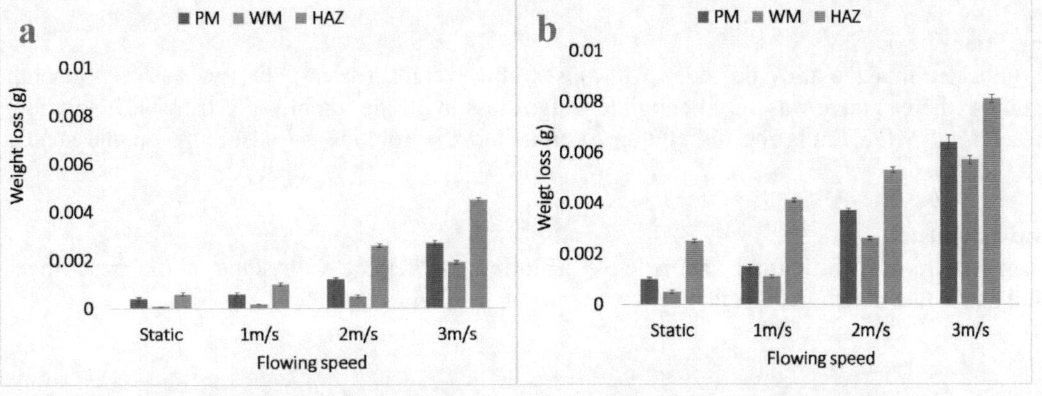

Figure 2: Material loss in 1 wt. % NaCl at 0°C (a) and 25°C (b)

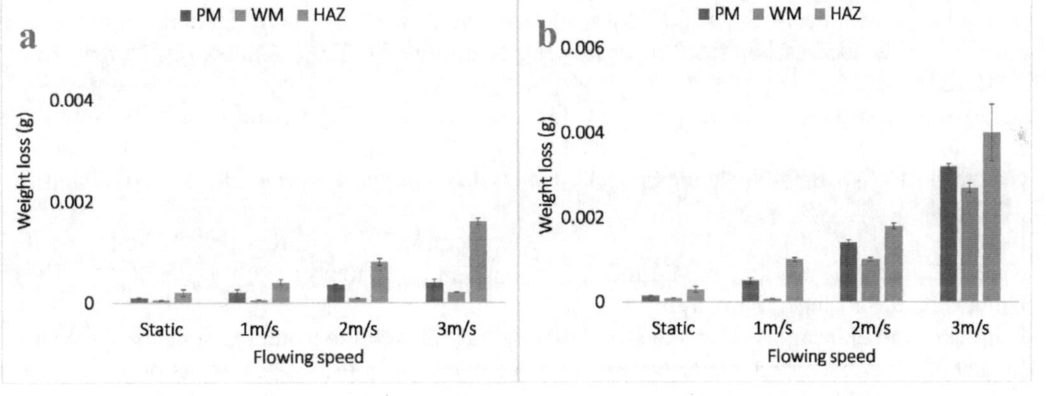

Figure 3: Material loss in 1 wt. % NaCl/MEG at 0°C (a) and 25°C (b)

4. Summary of Discussion

Monoethylene glycol (MEG) has been reported to increase the viscosity of water and hence lower the diffusivity of active species causing corrosion [6, 11]. It has also been argued that solution polarity which increases the solvation energy of iron ions is decreased with addition of MEG [6]. Lower solution polarity and hence lower solvation energy impair the activation and driving force for corrosion activity. Aside this other authors have also argued that MEG acts like anodic inhibitor blocking the anodic sites. This invariably reduces iron dissolution and hence lower corrosion rate [13]. Other authors also reported that solution conductivity is significantly decreased and hence higher Ohmic drop is expected with addition MEG [16]. With all the illustrations above it is not surprising that the corrosion susceptibility of all the sections of the welded alloy decreased with addition of 20 wt. % of MEG to the NaCl solution.

Influence of flow rate on Materials degradation is a subject area that has been widely reported. Higher hydrodynamic influences corrosion rate by moving the corrosion species towards reaction area and moving the corrosion products away from the surface of the anode. Hence higher corrosion rate is expected at higher flowing rates [1].

Effects of temperature on the corrosion behaviour of the alloy in the presence of MEG can be thus explained. It has been argued that thermal degradation of MEG occurs at higher temperatures [1]. It is expected that at 25°C, the thermal oxidation of MEG may result in the formation glycolic, oxalic and formic acids which are corrosive agents capable of instigating corrosion [1]. Aside this it is a known fact that corrosion species become more aggressive as temperature of the electrolyte increases.

5. Conclusions

Results indicated that the alloy degraded with observable weight loss in pure 1wt. % NaCl at both temperatures whereas there was very negligible weight loss in all the samples in MEG/NaCl solution. The presence of MEG limits the dissolution of iron into the solution by adsorbing on the steel's surface.

Acknowledgements:

Thanks to **Ezekhume Emmanuel and Adeoye Abiodun** for helping with some of the experiment setup and data collection in the laboratory

References

[1] Schmitt G and Horst Meier M 2006 Fundamental aspects of CO2 metal loss corrosion – part11: Influence of deferent parameters on CO2 corrosion mechanism *61st Annual Conf. & Expo, Corrosion,* Germany

[2] Alawadhi K, Aloraier A S, Joshi S, Alsarraf J and Swilem S 2013 Investigation on preferential corrosion of welded carbon steel under flowing conditions by EIS *J. Mater.Eng.Perform* **22** (8) 2403-10

[3] Aribo S, Adeyeye O D, Olusegun S, Ige O O, Ogunbadejo A S, Igbafen E O and Olubambi P A 2017 Inhibitive performance of sida acuta and monoethylene glycol on the under deposit corrosion of UNS S30403 stainless steel, *Int. J. Adv. Manuf. Technol,* DOI: 10.1007/s00170-017-0624-3

[4] Ekawati D 2011 Effect of temperature, bicarbonate, and MEG concentration on pre-corroded carbon steels. Master's Thesis University of Stavanger, Norway

[5] Brustad S, Løken K and Waalmann J G 2005 Hydrate prevention using MEG instead of MeOH : Impact of experience from major Norwegian developments on technology selection for injection and recovery of MEG," in *Offshore Tech. Conf. paper no.17355*

[6] Gulbrandsen E and Morard J 1998 Why does glycol inhibit CO_2 corrosion, *Corrosion Conf. and Expo paper no 98221* (Houston, TX)

[7] Gregg M R., Sharp A and Bartrip K 2003 Corrosion inhibitor developments for offshore gas gathering systems on Canada's atlantic coast. *Corrosion Conf. and Expo Paper no 03332* (Houston, TX)

[8] Pojtanabuntoeng T, Salasi M and Gubner R. 2014 The influence of mono ethylene glycol (MEG) on CO2 corrosion of carbon steel at elevated temperatures (80 to 120 °C) *Corrosion Conference and Expo, Paper no 4176* (Houston, TX)

[9] Oyevaar M H, Morssinkhof R W J and Westerterp K R 1989 Density, viscosity, solubility, and siffusivity of CO_2 and N_2O in Solutions of diethanolamine in aqueous ethylene glycol at 298 K *J. Chem. Eng.Data* **34** 77–82

[10] Kan A T, Lu H and Tomson M B 2013 Effects of monoethylene glycol on carbon dioxide partitioning in gas/monoethylene glycol/water/salt mixed systems *Ind.Eng.Chem.Res.* **49** (12) 5884–90

[11] Mendez C, Singer M, Camacho A, Hernandez S, Nesic S 2005 Effect of acetic acid , pH and

MEG on the CO_2 top of the line Corrosion *Corrosion Conference and Expo*, *Paper no* 05278 (Houston, TX)
[12] DeWaard C, Lotz U, and Milliams D E 1991 Predictive model for CO2 corrosion engineering in wet natural gas pipelines *Corr.Eng* **47**(12) 976–85
[13] Gulbrandsen E and Morard J H 1991 Why does glycol inhibit CO2 corrosion *Corrosion Conference and Expo*, *paper no*221 (Houston, TX).
[14] Aribo S, Fakorede A, Ige O, Olubambi P 2017 Erosion-corrosion behaviour of aluminium alloy 6063 hybrid composite *Wear* **376-377PA** 562-8
[15] AlHarooni K, Barifcani A, Pack D, Gubner R and Ghodkay V 2015 Inhibition effects of thermally degraded MEG on hydrate formation for gas systems *J. Pet. Sci.Eng* **135** 608- 17
[16] Zhang G A, Xu L Y, Cheng Y F 2008 Mechanistic aspects of electrochemical corrosion of aluminum alloy in ethylene–glycol–water solutions', Electrochim. Acta **53** 8245–52

IOP Conf. Series: Materials Science and Engineering **430** (2018) 012041 doi:10.1088/1757-899X/430/1/012041

A comparative corrosion study of titanium strips produced by wrought and direct powder rolling processes

N Mothopeng[1,2], N Maledi[2], M Maminza[1] and S Chikosha[1,*]

[1] Light Metals, Materials Science & Manufacturing, Council for Scientific and Industrial Research, Meiring Naudé Road, Brummeria, Pretoria 0185, South Africa
[2] School of Chemical and Metallurgical Engineering, University of the Witwatersrand, Private Bag 3, South Africa, 2050

*SChikosha@csir.co.za

Abstract. Titanium is a light metal that has a high strength to weight ratio compared to steel. It has good corrosion resistance and is biocompatible. However, due to its high costs, its use has been limited mainly to the aerospace and medical industries. Direct powder rolling (DPR), a powder metallurgy (PM) method, is being developed with the aim of reducing the cost of producing titanium flat mill products. Critical to the success of the DPR produced products will be their ability to match or exceed the properties and performance behaviour of the equivalent products produced by wrought metallurgy. Titanium strips that are 89 % and 98 % dense were produced by DPR and their corrosion behaviour was compared to the wrought produced titanium strips. The corrosion tests were carried out in 1.75 wt.%, 2.0 wt.% and 3.5 wt.% NaCl solutions kept at 23 °C using potentiodynamic polarization. The polarization curves showed that the wrought strips were the most corrosion resistant, followed by the DPR-98 % then the DPR-89 %. The influence of pores on the corrosion behaviour was high, therefore, effort needs to be made to reduce the pores of the DPR samples.

1. Introduction

Titanium (Ti) is generally known for its high strength and resistance to corrosion in a wide variety of environments [1] due to its ability to form a passive oxide film on its surface when exposed to oxidising environments. [2]. The corrosion resistance increases with the increasing thickness of the passive film [3]. Initially, titanium was predominantly used in the aerospace industry due to its light weight and high strength. Presently, due to its remarkable properties, titanium and titanium alloys have managed to pave their way into the petrochemical and power generation industries as condensers, heat exchangers [4],[5] and consumer products such as jewellery [6],[7]. Titanium products, though favourable, are considered very expensive due to raw materials and manufacturing costs [4],[7]. South Africa (SA) is the 2nd largest producer of ilmenite ($FeTiO_3$) which is the ore used to produce titanium. Nonetheless, SA does not produce any high-value semi-finished products such as Ti ingots, and/or mill products, powder etc. Mill products consist of plates, sheets, slabs, billets, bars, tubes and make up 75% of the global market share.

The Council for Scientific and Industrial Research (CSIR), through its Titanium Centre of Competence (TiCoC) in partnership with various universities, other science councils and private sector is mandated to add value to titanium through beneficiation of locally produced titanium ore in South Africa. This initiative involves the development of an innovative method of producing titanium metal in powder form. Therefore, it is equally important to develop powder metallurgy (PM) processing techniques that can convert the produced titanium powder into high-value mill products, to test their properties and performance against the wrought commercial grade equivalent for various application environments.

Direct powder rolling (DPR) of titanium powder is being developed mainly because the technique has the potential to offer cost benefits by reducing fabrication steps and has high material utilization [9]. The DPR process can be used to produce flat mill products such as strips, sheets and plates. The DPR process involves a roll compaction stage followed by a sintering stage or roll compaction followed by hot rolling and other subsequent finishing stages [10]. The objective of this paper was to compare the electrochemical corrosion behaviour of DPR produced titanium strips with the equivalent grade of wrought produced commercially pure (CP) Ti strips in a NaCl solutions

2. Experimental procedure

2.1 As received sample characterization

The developmental DPR strips were supplied by CSIR as Ti strips of different densities (Sample A-89 % and Sample B-98 %). The wrought strips (Sample C) were received from SAETRA (PTY) LTD as the supplier of ATI Mill product as CP Ti Grade 2 strips according to the ASTM B265.13A standard. The samples were sectioned and cut using water jet cutting. The samples were then hot mounted and prepared metallographically for microstructural analysis. The as-polished samples were etched with Kroll's reagent (46 ml distilled water + 3 ml nitric acid+ 1 ml hydrofluoric acid) for 5 seconds. The polished and etched surfaces were analysed on the Leica DMI 5000M optical microscope and image analysis was performed on the Olympus IMS software using the intercept method according to ASTM E112-13. Another set of samples were cut and the structural analysis was performed using the PANalytical X'Pert PRO X-ray diffraction (XRD) with a Cu-Kα diffractometer using λ = 1.54062 Å radiation at an accelerating voltage of 45 kV, a current of 40 mA. The samples were scanned at 2θ from 20 ° to 120 ° angles at a scan rate of 0.0065.

2.2 Electrochemical testing

The 3.5 wt.%, 2 wt.% and 1.75 wt.% NaCl solutions were prepared using distilled water. The solutions were incubated at room temperature (23°C ± 1 °C) for 24 hours before the tests. Corrosion testing specimens were prepared by attaching a copper wire to a surface of the samples using an aluminium tape and cold mounted in resin. The exposed surface of the samples during potentiodynamic testing was 0.071cm^2. The experiment was carried out according to ASTM G3-89 2010 by means of the electrochemical test conducted on the Autolab (NOVA version 2.1.0) potentiostat, with the electrolyte kept at room temperature (23 °C ± 1 °C) using a thermostat controlled bath. A silver chloride electrode (Ag/AgCl) was used as the reference electrode (RE), graphite as the counter electrode (CE) and the samples as the working electrode (WE). The tests were conducted at a potential range of -1 V to +2 V at a scan rate of 5 mV/s. All measurements were carried out at least triplicated in deaerated solutions for the repeatability of data.

3. Results and discussion

3.1 XRD results

The structural analysis of samples is shown in figure 1. The XRD analysis confirms that the materials are unalloyed titanium, with a hexagonal structure with a = 2.9500 Å, b = 2.9500 Å and c = 4.6860 Å lattice parameters.

IOP Conf. Series: Materials Science and Engineering **430** (2018) 012041 doi:10.1088/1757-899X/430/1/012041

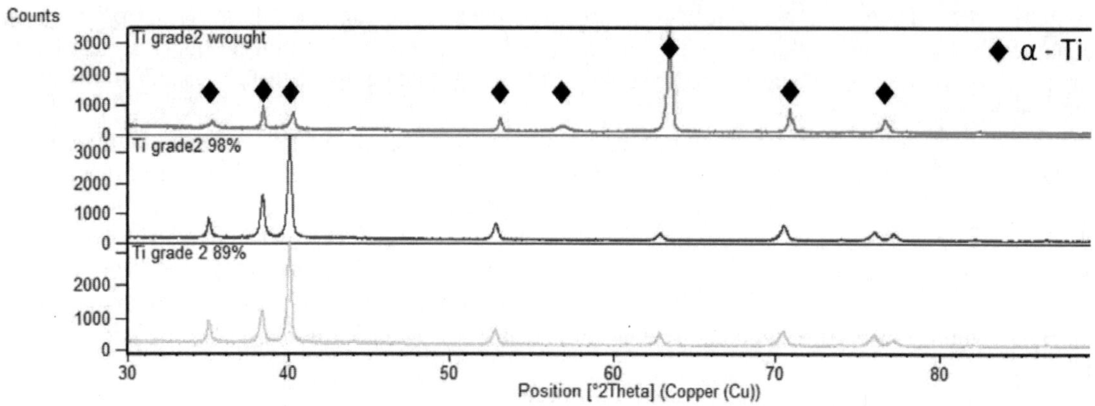

Figure 1. The XRD patterns for the grade 2 titanium strips.

3.2 Composition of samples

Table 1. The composition of the grade 2 Ti samples according to suppliers.

	Ti	Fe	O_2	N_2	C	H_2	Residuals
DPR: Sample A-89% (wt.%)	bal	-	0.22	0.02	-	-	-
DPR: Sample B-98% (wt.%)	bal	-	0.24	0.01	-	-	-
Wrought ASTM B265-13a: Sample C (wt.%)	bal	0.11	0.10	0.004	0.1	-	-
ASTM B265-15: Ti grade 2 (wt.%)	**bal**	**≤0.30**	**≤0.25**	**≤0.03**	**≤0.08**	**≤0.015**	**≤0.3**

In order for the titanium strip to be classified as grade 2, it must meet the ASTM B265 standard for strips, sheets and plates. Table 1 gives the composition requirements for grade 2 according to the latest ASTM B265-15 standard. Based on the grading and compositional analysis provided by suppliers, both DPR and wrought samples were classified as titanium grade 2.

3.3 Optical microscope results

The microstructures of the three titanium samples analysed using an optical microscope are shown in figure 2. From the optical microscope, it can be seen that sample A-89 % has the most pores followed by the sample B-98 % than the wrought sample C. The difference in the number of pores observed confirms the differences in the densities of the samples. After analysing the images, it was found that the pores in sample A-89 % occupy 6.8 % of the total area, pores in sample B-98 % occupy 2.6 % and sample C-wrought occupy a negligible value.

The images in figure 3 show the grains of all the samples. The DPR samples show large grains of approximately 52 μm and 60 μm average length for the DPR-89 % and DPR-98 % samples respectively (figure 3a and b). The wrought sample shows smaller equiaxed grains of approximately 27 μm average length (figure 3c).

3.4 Electrochemical results

Figure 4(a-c) shows the polarization curves of the samples in 1.75 wt.%, 2 wt.% and 3.5wt.% NaCl solutions respectively. All the curves show active (A), transmissive (T) and passive (P) regions. The titanium oxidation reaction is predominant in the active region. The oxide spreads over the entire surface suggesting the presence of a passive film. Figure 4a shows that DPR B-98 % and wrought CP Ti grade 2 are more corrosion resistant relative to the DPR A-89 % sample. The cathodic reactions of the samples are different, the wrought sample always shows a transmissive region. The tendency for the transpassive behaviour increases with a decrease in porosity. All three potentiodynamic polarization curves show that the wrought sample gave better corrosion resistance to corrosion in the three concentrations of NaCl. The acceleration of the formation of the passive region depends on the chlorides concentration. The passive region forms readily in more concentrated solutions than in dilute concentrations [11].

Figure 2. Optical micrographs of a) DPR A-89% b) DPR B-98% and c) wrought C produced samples.

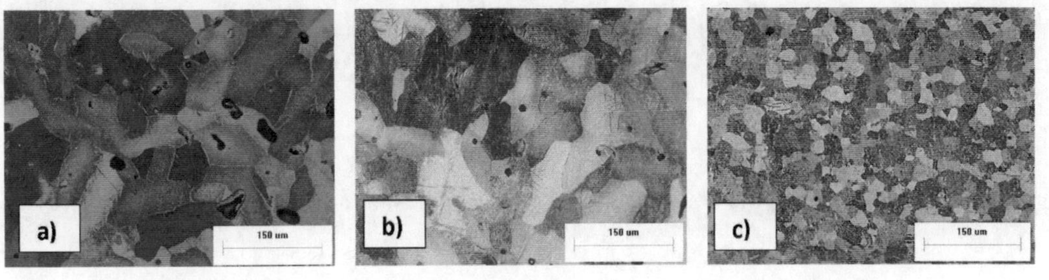

Figure 3. Optical micrographs of a) DPR A-89% b) DPR B-98% and c) wrought C produced samples.

Figure 5a presents the values of the corrosion rates per year for all the samples in different NaCl solutions. The corrosion rates of the samples increase with increasing concentration of NaCl and the i_{corr} of the samples increase with increasing concentration of NaCl, this is consistent with the known tendency of chloride ions to attack the surface [11]. Increasing the concentration of NaCl increases the chloride ion (Cl^-) concentration which causes the partially protective film to break down and leads to the instability of the metal. The highest corrosion rate was found to be 2.2mm/year for the DPR A-89 % sample in 3.5 wt.% NaCl with the lowest being the wrought sample at 0.2 mm/year in the 1.75 wt.% NaCl. The general behaviour observed in Figure 5a is that the wrought C samples have a lower tendency to corrode, followed by the DPR B-98 % then the DPR A-89 %. The reason for this behaviour can be explained using the porosity of the three samples. Blackwood and Chooi [12] reported that the corrosion rate of porous titanium is significantly higher than that of highly compact titanium with little or no porosity. Although the non-ideal passive behaviour in porous samples has not been fully explained yet, Blackwood et al. [13] found that it may be related to the difference in electrode potential with the depth of the pores.

The grain size is considered to have an influence on the corrosion resistance of a material. However, the texture and number of pores have a higher effect of the corrosion behaviour than the grain size [14]. Hoseini et al [14] investigated the effect of grain size and texture on the corrosion behaviour of CP Ti and found that the sample with larger grains was more corrosion resistant relative to a sample with smaller grains. The findings by Hoseini et al [14] are applicable provided all the other factors such as porosity, composition etc. are constant. For this paper, other factors are not constant therefore it was difficult to study the effect of the grain size on the corrosion rate.

IOP Conf. Series: Materials Science and Engineering **430** (2018) 012041 doi:10.1088/1757-899X/430/1/012041

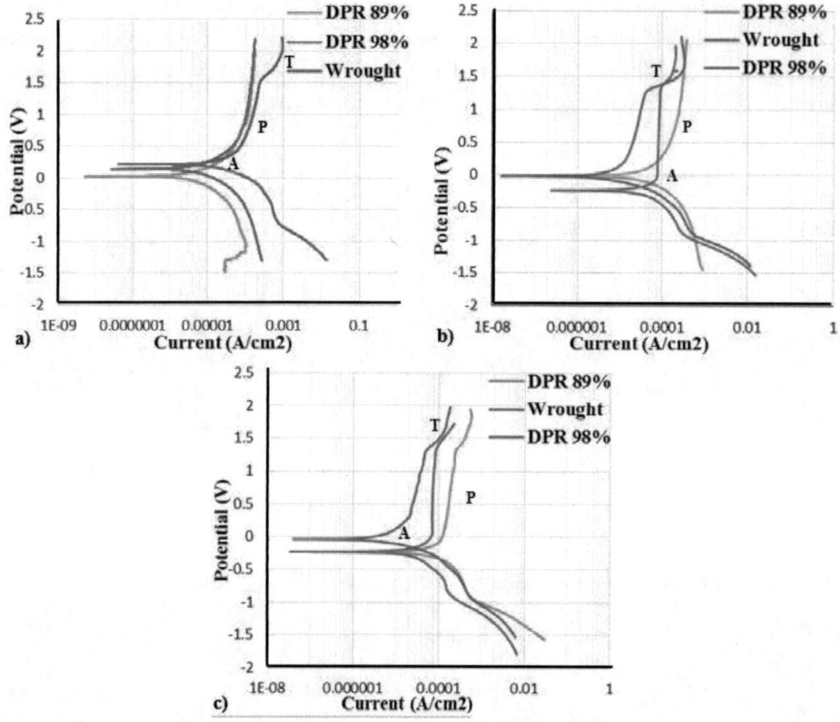

Figure 4. Potentiodynamic polarization curves for the samples in (a) 1.75 wt.%, (b) 2 wt.%, (c) 3.5 wt.% NaCl solutions.

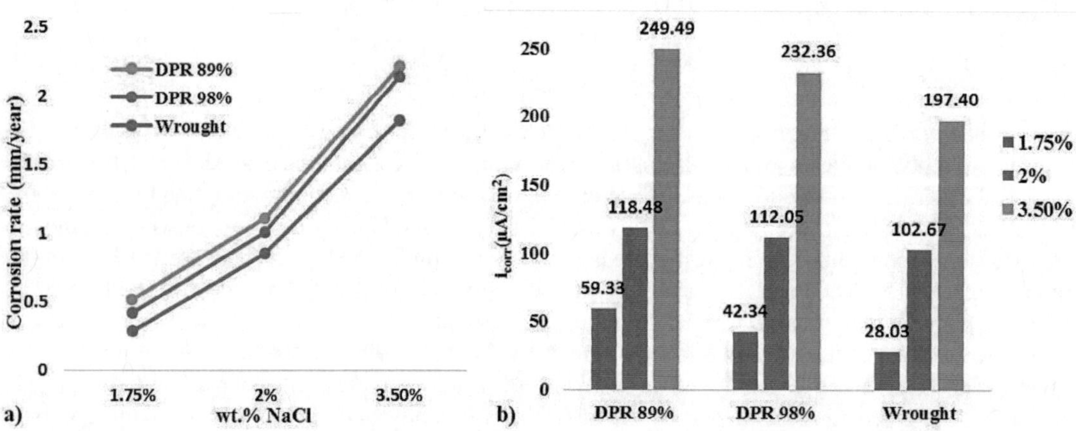

Figure 5. The a) corrosion rates and b) current densities of DPR A-89 %, DPR B-98 % and wrought C samples.

4. Conclusions

According to metallurgical characterization and electrochemical results from the tests conducted, the following conclusions were made:

- The DPR B-98 % produced Ti grade 2 strips had better corrosion resistance compared to the DPR A-89 % dense strip. This shows that it is very critical that the DPR produces full density samples for corrosion applications.

- The wrought sample had a low corrosion rate compared to the two DPR produced Ti strips. Improvement of the porosity of the DPR samples is required in order to match or exceed the corrosion behaviour of the wrought sample.
- It can be concluded, therefore, that further work is required to reduce the porosity of the DPR produced strips to improve their density.

Acknowledgements
The authors would like to acknowledge the Council of Scientific and Industrial Research (CSIR) for sponsoring the work and the University of the Witwatersrand for providing some of the laboratory equipment.

References

[1] Z. B. Wang, H. X. Hu, Y. G. Zheng, W. Ke, and Y. X. Qiao, Comparison of the corrosion behavior of pure titanium and its alloys in fluoride-containing sulfuric acid, *Corros. Sci.*, **103**, pp. 50–65, 2016
[2] I. Cvijović-Alagić, Z. Cvijović, S. Mitrović, V. Panić, and M. Rakin, Wear and corrosion behaviour of Ti-13Nb-13Zr and Ti-6Al-4V alloys in simulated physiological solution, *Corros. Sci.,* **53**, no. 2, pp. 796–808, 2011
[3] R. Singh and N. B. Dahotre, Corrosion degradation and prevention by surface modification of biometallic materials, *J. Mater. Sci. Mater. Med.,* **18**, no. 5, pp. 725–751, 2007
[4] S. J. Oosthuizen, Titanium: the innovators' metal-Historical case studies tracing titanium process and product innovation, J. *South African Inst. Min. Metall.*, **111**, pp. 781–786, 2011
[5] N. Mundhenk, P. Huttenloch, B. Sanjuan, T. Kohl, H. Steger, and R. Zorn, Corrosion and scaling as interrelated phenomena in an operating geothermal power plant, Corros. Sci., **70**, pp. 17–28, 2013
[6] M. Peters, T*itanium and Titanium Alloys*. 2003
[7] R. Thomas, Titanium in the geothermal industry, *Geothermics,* **32**, no. 4, pp. 679–687, 2003.
[8] J. G. Banker and M. R. Scaturro, Titanium for Secondary Marine Structures, *in International Titanium Association,* 1996, pp. 1–10
[9] B. Wu, Z. Pan, S. Li, D. Cuiuri, D. Ding, and H. Li, The anisotropic corrosion behaviour of wire arc additive manufactured Ti-6Al-4V alloy in 3.5% NaCl solution, *Corros. Sci.,* **137**, no. August 2017, pp. 176–183, 2018
[10] D. Cantin et al., *Direct Powder Rolling (DPR) of Titanium. Light Metals Flagship,* Orlando, Florida, 2010
[11] M. C. Zhao, M. Liu, G. L. Song, and A. Atrens, Influence of pH and chloride ion concentration on the corrosion of Mg alloy ZE41, *Corros. Sci.,* **50**, no. 11, pp. 3168–3178, 2008
[12] D. J. Blackwood and S. K. M. Chooi, Stability of protective oxide films formed on a porous titanium, *Corros. Sci.*, **44**, no. 3, pp. 395–405, 2002
[13] D. J. Blackwood, A. W. C. Chua, K. H. W. Seah, R. Thampuran, and S. H. Teoh, Corrosion behaviour of porous titanium-graphite composites designed for surgical implants, *Corros. Sci.*, **42**, no. 3, pp. 481–503, 2000
[14] M. Hoseini, A. Shahryari, S. Omanovic, and J. A. Szpunar, Comparative effect of grain size and texture on the corrosion behaviour of commercially pure titanium processed by equal channel angular pressing, *Corros. Sci.*, **51**, no. 12, pp. 3064–3067, 2009

IOP Conf. Series: Materials Science and Engineering **430** (2018) 012042 doi:10.1088/1757-899X/430/1/012042

The direct observation of surface martensite formation upon cooling to temperatures close to ambient in a heat treated AISI 301 stainless steel

T W Mukarati[1,*], R J Mostert[1] and C W Siyasiya[1]

[1] University of Pretoria, Private bag X20 Hatfield 0028, South Africa

[a] tmcarati@gmail.com

Abstract. Martensitic transformation, either athermal or strain-induced, in metastable AISI 301 austenitic steel is of high technical interest as this facilitates manipulation of mechanical properties. This work is part of a project aimed at identifying the influence of variables that determine the athermal α′-martensitic transformation in this steel in order to ensure consistent martensite formation and optimized mechanical properties. The influences of initial austenitic grain size and surface concentrations of interstitial elements (C, N) on martensite start temperature were investigated. The surface concentrations of both carbon and nitrogen were decreased through decarburisation and the formation of martensite upon cooling from ambient temperatures was directly observed using a cryogenic Scanning Electron Microscope (cryo-SEM). It was found that martensite formed on the heat-treated surfaces at much higher temperatures than that expected for AISI 301 steel. The observations were confirmed using electron backscatter diffraction (EBSD). Other analytical techniques including glow-discharge optical emission spectroscopy (GDOES), and Neutron Diffraction were carried out to explain the observations.

1. Introduction

The AISI 301 austenitic stainless steel is lean in alloying elements and hence leads to a greater susceptibility to formation of strain-induced martensite (SIM) upon cold deformation. The metastable austenitic microstructure can transform to martensite through either cold deformation or thermal cooling [1], [2]. Martensitic transformation, is of great interest and crucial as mechanical properties can be boosted greatly with this phase transformation [3].

Understanding microstructural evolution is the key in developing a comprehensive constitutive model for material behaviour [4] for this steel type. The optimization of the microstructure is needed to control the mechanical properties [5] and to ensure cast-to-cast consistency of such properties.

Carbon and nitrogen are two interstitial elements which are known to thermodynamically stabilize the austenite phase and also suppress strain-induced martensitic transformation upon cold deformation [5]. Figure 1 shows that nitrogen has a stronger austenite stabilization effect compared to carbon and an addition of 0.1% (mass) nitrogen, results in the decrease of M_s temperature by more than 300 Kelvin for an austenitic stainless steel.

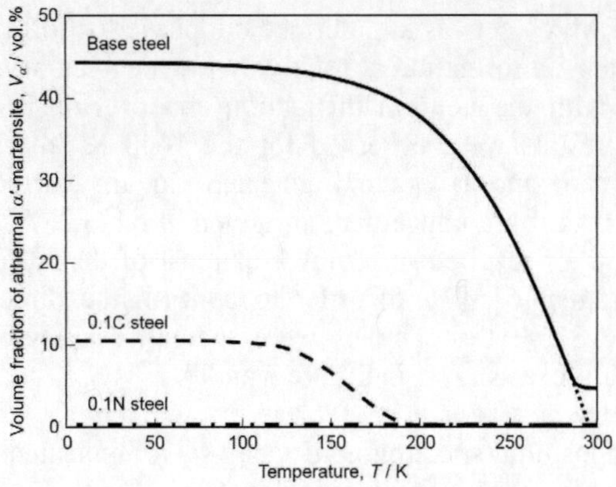

Figure 1. Changes in volume fraction of athermal α'-martensite upon cryogenic cooling [5] for a Fe- 18% Cr- 8% Ni base steel showing the influence of additions of carbon and nitrogen.

2. Materials and methods

The starting materials were hot rolled, annealed and pickled AISI 301 stainless steel plates of 4.44 mm thickness. As-received samples were cut into three samples of 300mm x 30mm. Two samples were heat treated at 1448 K and 1648 K in air for 10.5 minutes and were both air cooled. The three samples (15 mm x 15 mm surface area): as-received, heat treated at 1448 K and 1648 K were electropolished and characterized in the unetched condition using cryogenic SEM (Zeiss Crossbeam 540 FEG SEM), which is equipped with cooling facility using liquid nitrogen. In order to observe microstructural changes, the same spot of the sample surface was analyzed using the in-lens detector of the microscope as the temperature was lowered. The samples were thereafter analyzed using optical microscopy and electron back scattered diffraction.

Analysis such as GDOES was done on the plant and laboratory annealed samples to determine the change in surface concentration of interstitial elements (C, N). Interrupted cryogenic neutron diffraction analysis was done on the as-received sample with 15 analysis temperatures used between ambient and 3 K. EBSD analysis was carried out using EBSD Oxford instrument at an accelerating voltage of 20kV and a working distance of 10 mm. The samples for cryogenic SEM and EBSD were prepared by electropolishing using a Struers Lectropol-5 equipment and A3 electrolyte (600 ml methanol + 360 ml butoxyethanol + 60 ml perchloric acid) and operating voltage of 35 V at room temperature. No etching was done.

3. Results and discussion

The chemical composition (in wt. %) of the as-received hot rolled, annealed and pickled AISI 301 coil is given in Table 1.

The equilibrium thermodynamic stability of the phases of this steel was calculated as a function of temperature ranging from 1220 K to 1740 K, using Thermo-Calc software, as shown in Figure 2. Figure 3 and Table 2 demonstrate the strong effect of the annealing temperature on austenite grain size, with an increase from 1353 K (Fig 3a) to 1448K (Fig. 3. B) resulting in more than doubling the grain size. Figure 4 relays the Neutron diffraction spectra before and after cooling the as-received sample to 4 K in a cryogenic neutron diffraction facility, showing no observed crystallographic changes due to the cooling. The interrupted in-situ neutron diffraction results observed between 300 K 3 K did not identify any transformation of the austenite of the as-received sample. Figure 5 and 6 present SEM images obtained during in-situ observation of the as-polished surface with cryogenic cooling of the 1448 K laboratory annealed sample. Accordingly, a small quantity of martensite could be observed at room temperature, but the martensite fraction increased significantly upon cooling to 243 K (b) and 125 K (c). Figure 6 present SEM images taken at high magnification (5000 times) showing very thin parallel lines (ε-martensite) which acts as an intermediate phase for formation of α′-martensite. These in-situ observations of martensite formation at relatively low degrees of undercooling were obviously surprising and in contrast with the neutron diffraction and previously published, [5], (Fig. 1, for example) results. Similar results were observed for the 1648 K laboratory annealed sample. The GDOES analyses of the plant and laboratory annealed (in air) surface chemistry, as far as the interstitial element concentrations are concerned, are given in table 2. Accordingly, it is clear that the laboratory heat treatments in air resulted in removal of most of the surface carbon and a substantial portion of the surface nitrogen [6], [7]. In order to confirm the direct observation of martensite formation in the cryogenic SEM, the samples were thereafter analysed using EBSD and optical microscope at room temperature, as shown in Figure 7 and 8.

Table 1. Chemical composition (spectrographic analysis) of annealed and pickled AISI 301 as supplied (mass %).

C	Mn	P	S	Si	Cr	Ni	N_2	Co
0.02	1.57	0.023	0.003	0.41	17.41	6.6	0.119	0.12
Cu	**Mo**	**Sn**	**Ti**	**Al**	**Nb**	**O**	**B**	**Fe**
0.09	0.05	0.009	0.009	0.005	0.009	0.0074	0.0017	Balance

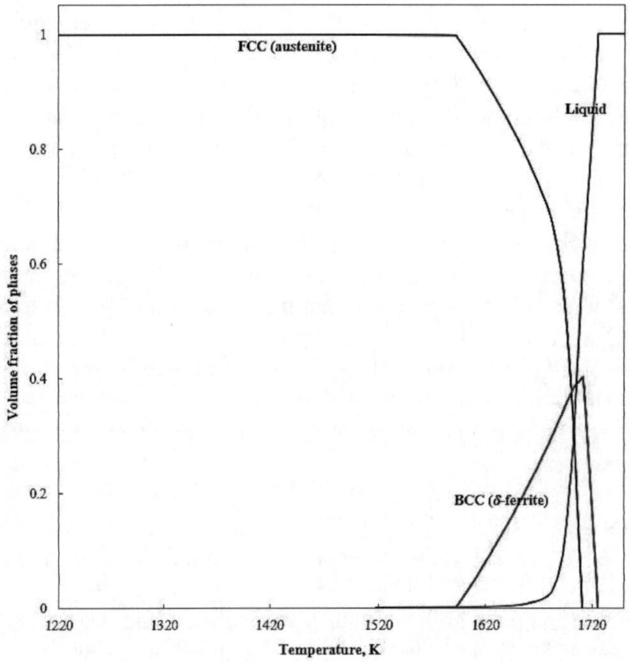

Figure 2. Thermo-Calc for the given chemistry in Table 1.

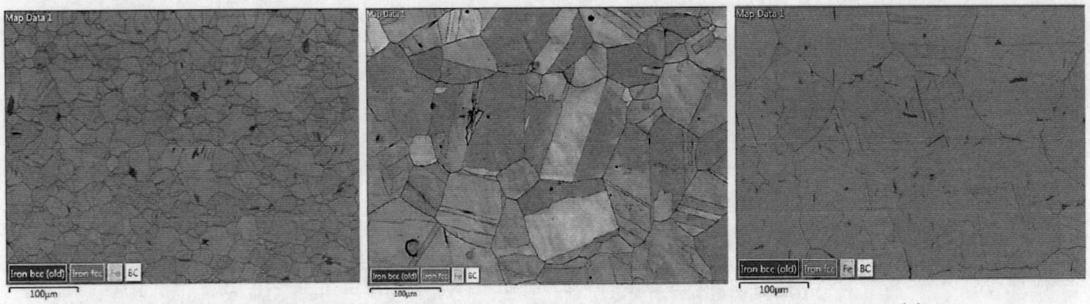

(a) (b) (c)

Figure 3. EBSD images of (a) as-received hot rolled, annealed and pickled AISI 301 stainless steel (b) heat treated at 1448K and (c) heat treated at 1648K, all taken before cryogenic work.

Table 2. The surface chemical composition of carbon and nitrogen of the as-received and laboratory annealed samples, as obtained using GDOES.

	As received (plant annealed at 1353K)	**Annealed at 1448 K**	**Annealed at 1648 K**
Carbon (%wt)	0.0174	0.0030	0.0038
N_2 (%wt)	0.1190	0.0574	0.0603
*Initial austenitic grain size (μm).	25.0	59.5	102.7

*The initial austenitic grain sizes were determined using Tango software of the EBSD.

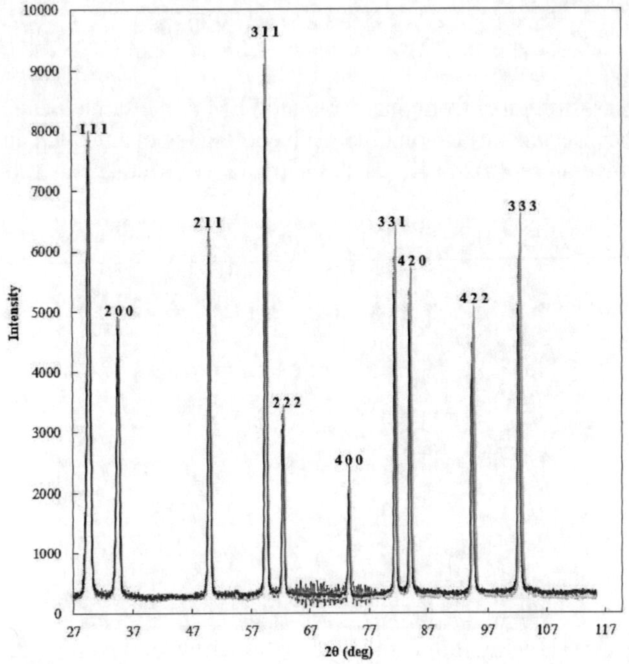

Figure 4. Neutron diffraction analysis of steel in 'as-received' condition at 300 K, 250 K, 245 K, 242 K, 238 K, 235 K, 233 K, 230 K, 227 K, 224 K, 220 K, 200 K, 150 K, 10 0K, and 3 K, with the spectra unchanged throughout the temperature range.

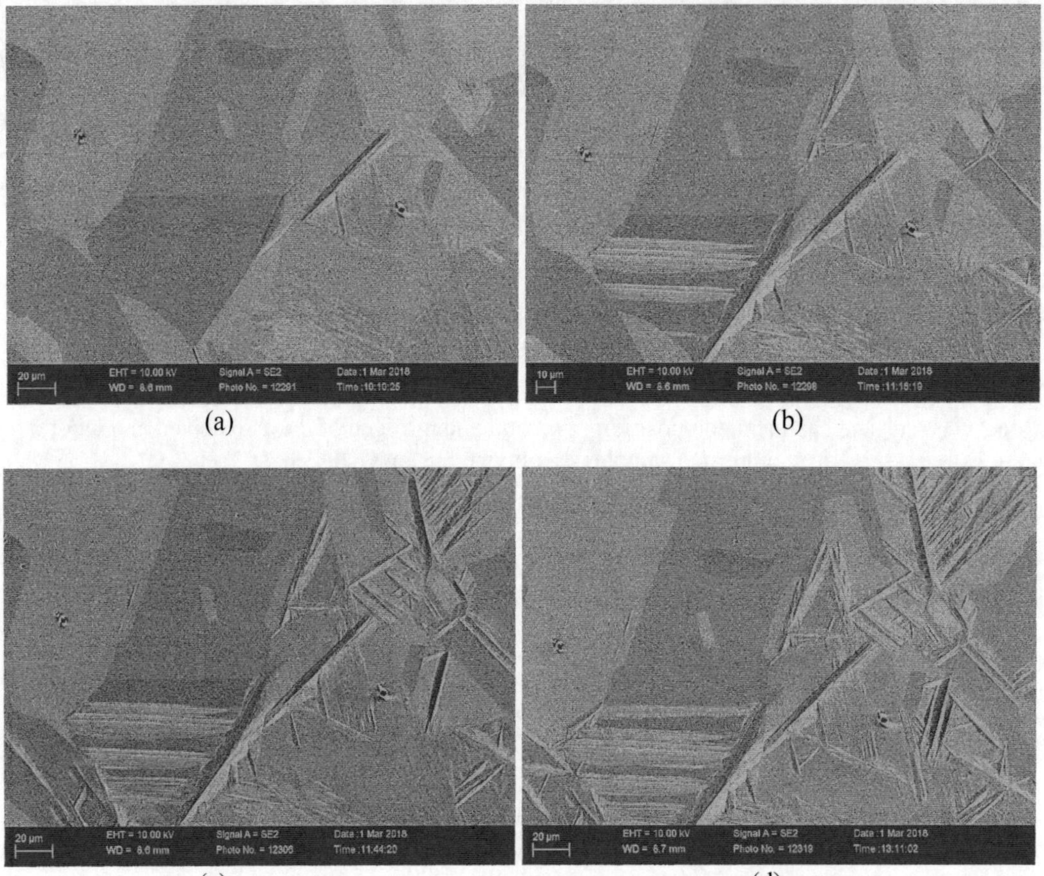

Figure 5. Images of microstructures from the cryogenic SEM at the same location showing formation of α′-martensite of the affected surface on cryogenic cooling (steel annealed at 1448 K after) (a) at room temperature, (b)243 K, (c) 233 K (d) 125 K, all taken at 1000 mag.

Figure 6. Microstructure showing thin parallel lines, (ε-martensite) which acts as an intermediate phase for formation of α′-martensite at 125 K at high magnification.

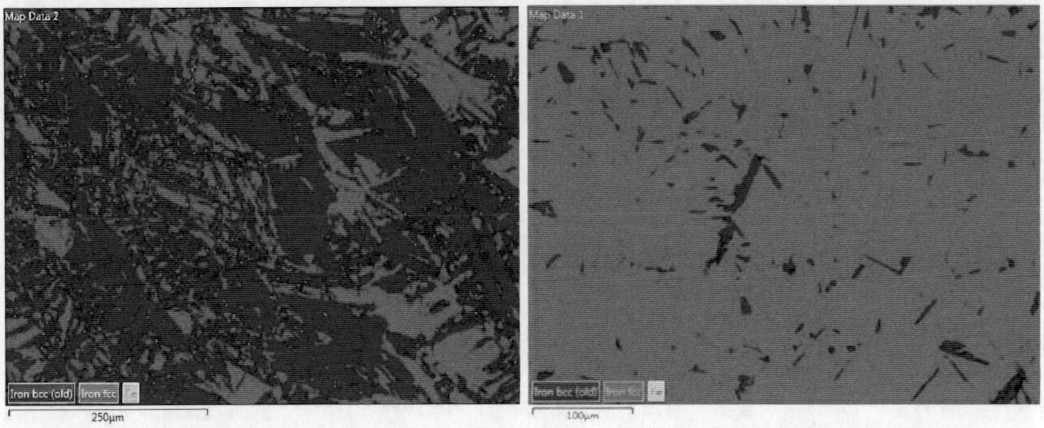

(a) *(b)*

Figure 7. EBSD images of steel annealed at (a) 1448 K, (b) 1648 K after cryogenic cooling to 123 K.

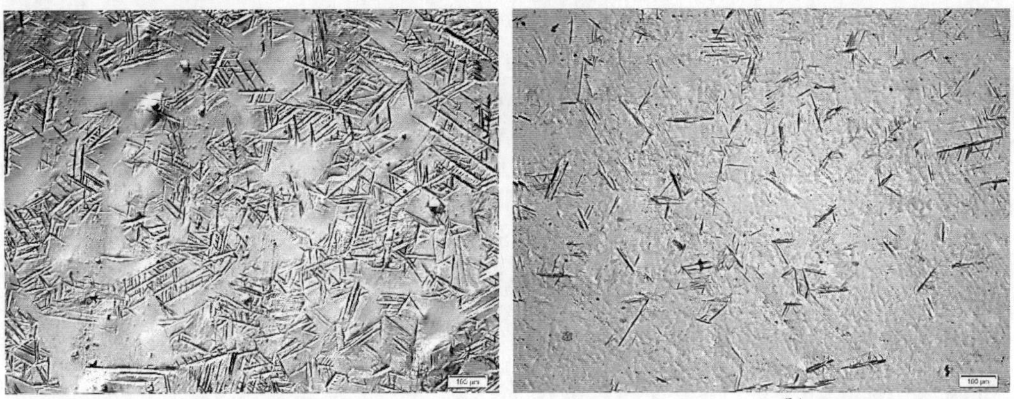

(a) *(b)*

Figure 8. Optical images of steel annealed at (a) 1448 K, (b) 1648K after cryogenic cooling to 123 K.

The 'as-received' steel had been plant annealed at 1353 K in a continuous annealing furnace. For this base steel, no α'-martensite transformation was observed during cryogenic cooling and neutron diffraction analysis cooling the sample down to a temperature as low as 4 Kelvin. Samples which were laboratory annealed at 1448 K and 1648 K have however shown α'-martensite developing at relatively elevated temperatures during cryogenic cooling using a cryogenic SEM. Specific spots of the electropolished samples were selected using the in-lens detector of Zeiss Crossbeam 540 FEG SEM whilst the temperature was lowered by liquid nitrogen. Coarse α'-martensite were observed to form at temperatures of 243 K and lower. At high magnification, very thin parallel lines were observed. This was believed to be ε-martensite which acts as an intermediate phase for the $\gamma \rightarrow \alpha'$ transformation. According to the published results, Ms for this steel [5] in the as-received condition cannot be observed above zero Kelvin. The direct observation of martensite development on the electropolished surfaces at much higher temperatures after heat treatment in air at 1448 K and 1648 K, is believed to be due to localised decarburization and denitriding during laboratory annealing in air, as demonstrated by the GDOES analyses performed. The calculated Ms temperature for the surface chemistries of the laboratory annealed (1448K and 1648K annealed samples) were found to be much higher than that of the bulk composition; 184.5 K and 173.5 K respectively for the two annealing temperatures using the Eichelman and Hull equation [8]. The observed Ms temperatures of the laboratory annealed surfaces were in the order of 243 K and 228 K respectively, and it is believed that the deviation of actual from calculated Ms temperatures are probably due to grain size effects with larger grain sizes leading to higher Ms-temperatures. Martensitic transformations are autocatalytic as martensite laths partition the grain on reaching the grain boundaries thereby inducing the formation of other laths. Martensitic transformation of decarburized and denitrified surfaces in industrial plants can therefore accelerate bulk martensitic transformation during subsequent straining and lead to variable mechanical properties in metastable stainless steels.

The Thermo-Calc model demonstrated that, for the bulk composition, formation of δ-ferrite occurs from around 1573 K. EBSD images of the steel annealed at 1648 K showed a significant percentage of δ-ferrite, approximately 11.4% (volume fraction of 0.11). Formation of δ-ferrite (bcc) from austenite (fcc) implies a significant amount of interstitial elements such as nitrogen and carbon released into the solution which will then stabilize austenite further. This effect was observed by a smaller volume fraction of α'-martensite forming during cryogenic cooling as compared to the steel annealed at 1448 K where there was no formation of δ-ferrite.

4. Conclusions

For the metastable AISI 301 stainless steel investigated, the formation of α'-martensite on the surface is possible at temperatures close to ambient, due to decarburization and denitriding of the surface. In a production environment, this unexpected martensitic transformation could cause variable responses to strain hardening due to the autocatalytic nature of the austenite to martensite transformation. An inert gas atmosphere is therefore recommended during plant annealing to prevent changes in surface chemistry through decarburization and denitriding.

Acknowledgements

The authors are grateful to Columbus Stainless Steel company for financial support as well as the material, AMI-FMDN, administered by Mintek for supporting this work, Laboratory for Microscopy and Microanalysis (University of Pretoria) and NECSA (South Africa) for their service and assistance.

References

[1] J. Man *et al.*, "Microstructural changes during deformation of AISI 300 grade austenitic stainless steels: Impact of chemical heterogeneity," *Procedia Struct. Integr.*, vol. 2, pp. 2299–2306, 2016.

[2] M. Wendler *et al.*, "Anomalous stabilization of austenitic stainless steels at cryogenic temperatures," *Mater. Sci. Eng. A*, vol. 675, pp. 415–420, 2016.

[3] X. Yang, S. Sun, and T. Zhang, "The mechanism of bcc α' nucleation in single hcp ε laths in the fcc $\gamma\rightarrow$ hcp $\varepsilon\rightarrow$ bcc α' martensitic phase transformation," *Acta Mater.*, vol. 95, pp. 264–273, 2015.

[4] L. E. Murr, K. P. Staudhammer, and S. S. Hecker, "Effects of Strain State and Strain Rate on Deformation-Induced Transformation in 304 Stainless Steel : Part I1 . Microstructural Study," *Metall. Trans. A*, vol. 13, pp. 627–635, 1982.

[5] T. Masumura, N. Nakada, T. Tsuchiyama, and S. Takaki, "The difference in thermal and mechanical stabilities of austenite between carbon- and nitrogen-added metastable austenitic stainless steels," *Acta Mater.*, vol. 84, pp. 330–338, 2015.

[6] M. Wendler, M. Hauser, O. Fabrichnaya, L. Krüger, A. Weiß, and J. Mola, "Thermal and deformation-induced phase transformation behavior austenitic – martensitic cast stainless steels," *Mater. Sci. Eng. A*, vol. 645, pp. 28–39, 2015.

[7] J. Mola, R. Roman, and K. Buxiao, "High tensile ductility in a conventionally martensitic Ni- and Mn-free stainless steel made austenitic by full dissolution of interstitial alloying elements," *Eur. Stainl. Steel Conf. Proc. Graz Austria. 8th*, 2015.

[8] M. Moallemi, A. Najafizadeh, A. Kermanpur, and A. Rezaee, "Effect of reversion annealing on the formation of nano / ultrafine grained structure in 201 austenitic stainless steel," vol. 530, pp. 378–381, 2011.

IOP Conf. Series: Materials Science and Engineering **430** (2018) 012043 doi:10.1088/1757-899X/430/1/012043

Effect of putrefied quenching oil on the transformation behaviour of railway spring steel.

V.J. Matjeke[1,2], G. Mukwevho[1], J.W. van der Merwe[2,3],

[1] Transnet Engineering, Research & Development, Private Bag X 528, Kilnerpark, South Africa, 0127

[2] School of Chemical and Metallurgical Engineering, University of the Witwatersrand, Johannesburg, Private Bag X3, Wits, South Africa, 2050

[3] DST-NRF Centre of Excellence in Strong Materials, University of the Witwatersrand, Johannesburg, Private Bag X3, Wits, South Africa, 2050

E-mail: Goodness.Mukwevho@transnet.net

Abstract. Bogie springs are safety critical components that also provide a smooth ride to the rail vehicle by absorbing vibrations and shock loads during the movement of the train. This study investigates the cause of premature failure of springs. These springs were manufactured from a high strength steel grade, SAE J404 grade 5160. The investigation process focused on heat treatment processes and failure mechanisms. Oil properties were also scrutinised by conducting quench tests. The overall findings of the investigation highlighted the detrimental effect of a degraded quenching oil on the fatigue life of the springs. Implementing certain measures from these findings will increase the spring lifespan and contribute towards safe operation of the trains.

1. Introduction

Bogie springs are integral in stabilising and keeping railway fleet on track. They furthermore, give smooth ride to the rail vehicle by absorbing vibrations and shock loads during the movement of the train [1]. In the last century, efforts have been made to produce high performance springs. This is done precisely in an effort to increase the loads and lifespan on the springs. Prominence has been put on improving the strength whilst preserving ductility, toughness and fatigue properties by improving the heat treatment processes and alloy improvement [2].

Premature failures of the springs has been a nightmare for engineers and scientist worldwide since rail inception [3, 4, 5]. The purpose of this study is to investigate primary causes of spring premature failure and evaluation of manufacturing process. The causes of failure of springs are usually due to material, design, manufacturing process and load factors [6]. Generally, springs are designed to withstand cyclic loading over their lifetime. Although the lifetime of the spring is largely determined by loading conditions and outside factors such as corrosion and mechanical damages plays a role in the reduction of lifespan [7]. There are many factors determining the spring life, the service record of the springs in South Africa indicates that generally the springs last for a period of 25 years operating in similar conditions as the failed spring.

2. Background

A spate of helical bogie spring failures has caused grave concern for railway operators in South Africa. The springs are manufactured using hard drawn steel rods that are hot coiled into spring shape followed by heat treatment and shot peening. The springs are normalised and coiled at 850 - 880 °C followed by austenitizing for an hour at 860 - 880 °C and quenching in oil at 40 °C. The quenching oil properties are summarised in Table 1. The quenching process is then followed by tempering at 430 – 450 °C for 3 hours.

The spring that is investigated is one of the many springs that failed in-service and during the fatigue testing. The spring recorded 203800 cycles before complete failure at a frequency of 3Hz. The cyclic and mean load were 13.044 and -43.48 kN respectively. The springs are fitted on to the heavy haul locomotives. The required microstructure is tempered martensite with minimum hardness of 50 Rockwell C. The springs are manufactured out of SAE J404 grade 5160 material with the rod diameter ranging from 25 to 35 mm.

Locomotive bogies consist of 12 dual spring sets, i.e. four dual springs per axle. The locomotive carries a maximum load of 22 tons per axle. The load is transmitted directly from the bolster to the subframe to effect compression to the springs. This load is a far less than the spring rated load. Bogie springs are often inspected and monitored for fatigue cracks only during a maintenance intervention cycle due to accessibility difficulties. The intervention period is every four years, during heavy lifting. The first section of this study covers one of the investigation of spring failure that is representative of numerous similar failures, whilst the second part focuses on quenching experimental exercise.

Table 1. Physical and chemical properties of the quenchant

Properties	
Appearance	Liquid
Flash point	>204 °C
Density	0.854 g/cm^3
Pour point	-15 °C
Viscosity	20 m^2/s @ 40 °C

2.1 Visual examination

The fracture surfaces of the failed spring used in this investigation is shown in Figures 1 and 2. The failure of these springs can be described as fatigue failure that originated on the surface. The failed spring fracture surface in Figure 1 revealed dull area followed by approximately 13 percent fatigue beach marks from the dark area before brittle failure. Figure 2 revealed that the dark dull area which appeared as a pre-existing crack that oxidised during heat treatment. The dark area is identified as the initiation point of the fatigue crack.

Figure 1..Complete spring fracture surface

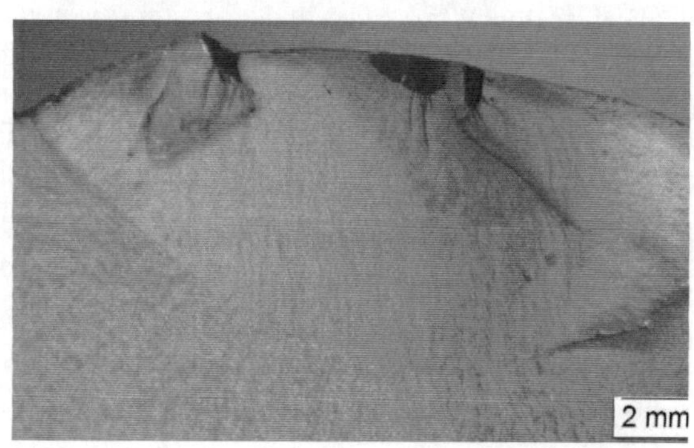

Figure 2. Cropped image from the point of origin

2.2 Scanning electron microscope (SEM)

The origin of the fatigue crack was examined by means of scanning electron microscope. The initiation point featured a mixture of transgranular and intergranular cracking typical of quench cracking. The SEM image of the origin of the crack is presented in Figure 3. The surface of the origin was covered with what appears to be heat related iron oxide. Oxidation occurred during further heat treatment process (tempering)

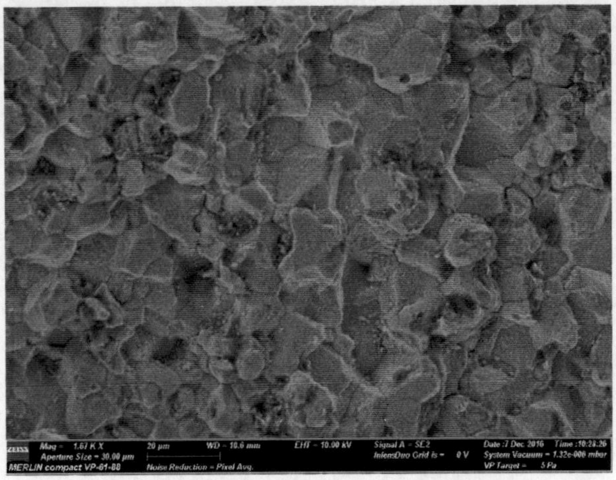

Figure 3. Failure origin SEM image

2.3 Chemical composition

The chemical composition was conducted by means of optical emission spectrometry in element weight percentage. The chemical composition results are presented in Table 2. The composition of the failed spring complied with the requirements of SAE 5160. The steel is classified as hardenable high carbon steel.

Table 2. Chemical composition of the failed spring

Element	**%C**	**%Si**	**%Mn**	**%P**	**%S**	**%Cr**
Failed spring	0.60	0.28	0.84	0.013	0.004	0.78
SAE J404 5160	0.56-0.64	0.15-0.35	0.75-1.00	0.04 max	0.03 max	0.70-090

2.4 Oil quench test

The oil sample that was used to produce the failed springs was collected from the quench tank for quench test analysis. An additional sample of the unused oil was also taken for analysis. The oil samples were taken in separate clean containers for testing. The oil samples were tested using portable quenchant test system (IVF smart quench). The quench tests results are presented in Table 3. Figurative illustrations are shown in Figures 4 and 5. The used oil showed deterioration and as a result had a fast cooling rate at the martensitic start range. The harmonisation of SAE 5160 continuous cooling temperature (CCT) curve shown in Figure 6 and the convection curve shows full martensitic transformation. The CCT curve in Figure 6 was generated using the JMatPro software. The cooling rates of the used oil was severe at critical temperatures of 280 and 300°C.

Table 3. Quench test results

Property	Unit	New oil	Used oil
Maximum cooling rate	°C/s	61.6	78.9
Temperature at Maximum cooling rate	°C	526.0	515.9
Temperature at start of boiling	°C	619.9	615.3
Temperature at start of convection	°C	331.9	240.1
Colling rate at 300 °C	°C/s	7.3	19.0
Time to 600°C	s	10.9	9.8
Time to 400°C	s	15.8	13.1
Time to 300°C	s	41.3	15.1
Time to 200°C	s	49.8	27.2

Theta 1	°C	616.3	612.3
Theta 2	°C	398.6	326.3
HP- IVF (Oil)		271.2	756.3

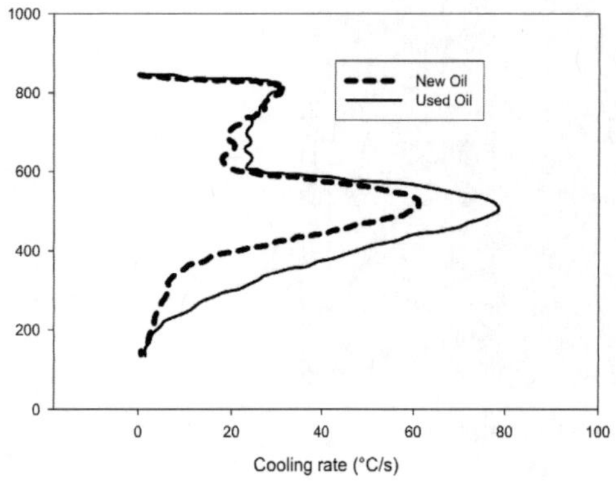

Figure 4. Oil properties cooling rates.

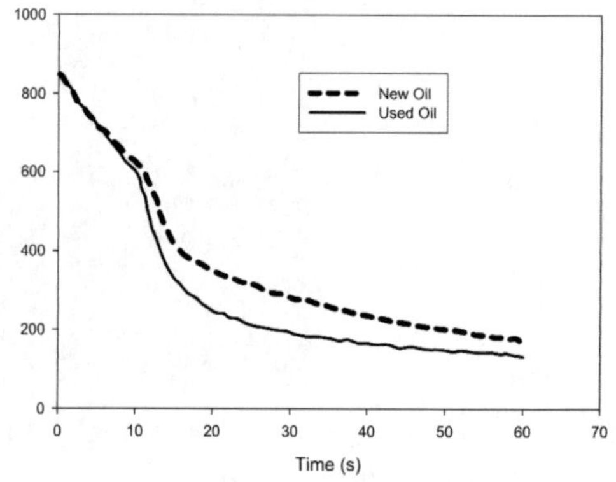

Figure 5. Oil properties convection curves

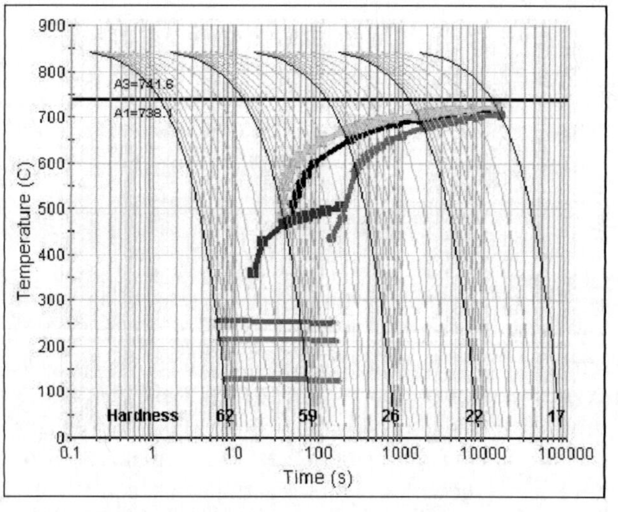

Figure 6. SAE 5160 CCT diagram

3. Heat treatment experiment

For purposes of validating the earlier discussed findings, hardening experiments were undertaken in order to simulate the heat treatment process. Sample material of SAE 5160 was heated to 860°C and quenched into the new and used oil respectively.

Figure 7. Microstructural microstructure Quenched with used oil

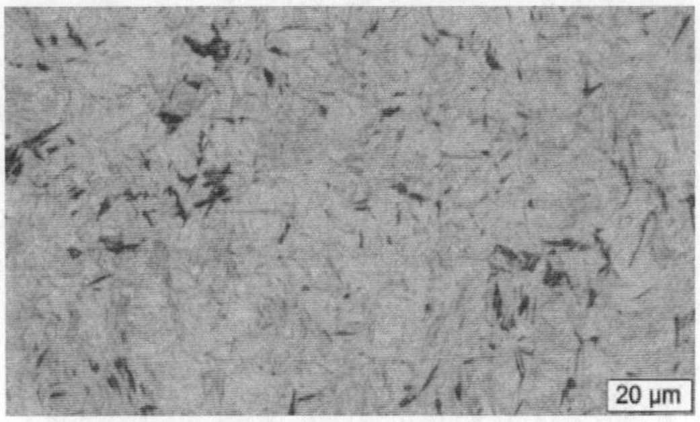

Figure 8. Microstructural microstructure quenched with new oil

3.1 Microscopic examination

The microstructural images are shown in Figures 7 and 8. The used oil microstructure revealed a fully transformed martensite with a crack running through. On the other hand, the microstructure of the new oil reveal matrix of martensite and bainite formations.

3.2 Hardness results

The hardness profile measurements were carried out using micro-Vickers hardness tester. The hardness average and profile results are presented in Table 4 and Figure 9 to 10. The used oil revealed higher hardness when compared to the new oil. The hardness profile was done from one edge to the other, at 1mm interval.

Table 4. Average hardness measurements

Quenchant	Average	Standard deviation	Standard Error
New Oil	780.9	23.3	4.6
Used Oil	866.8	13.9	2.8

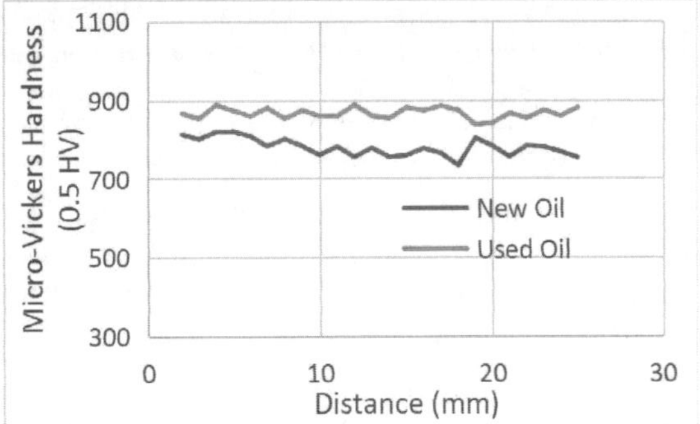

Figure 9. Hardness profile measurement

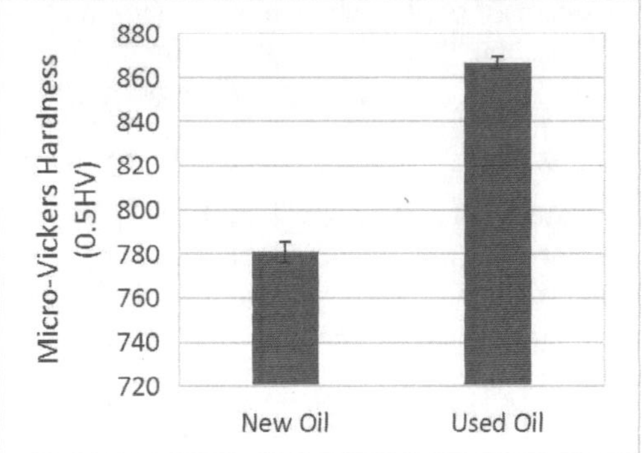

Figure 10. Hardness measurements average

3.3 X-ray diffraction(XRD) results

XRD analysis was conducted on the samples that were quenched with the used oil and new oil. The XRD spectrums are shown in Figure 11, Figure 12. The quantitative analysis showed that the sample quenched with used oil had more martensite transformation than the one quenched with new oil see Table 5. This confirms that the deterioration of oil has an impact on the cooling rates. Filters were used on the diffraction. Phase quantification was conducted using Rietveld.

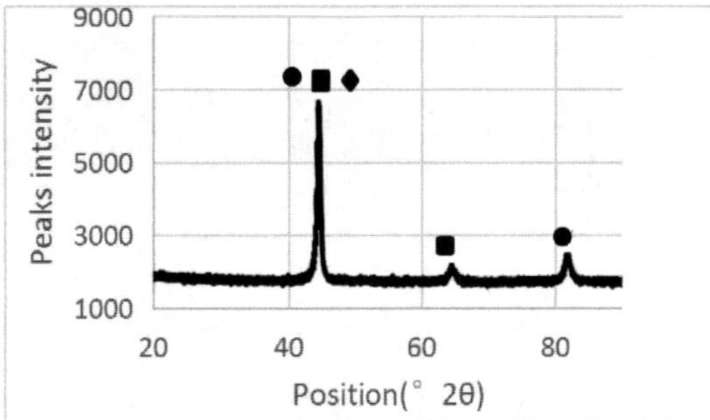

Figure 11. XRD spectrum, ● -Martensite, ■ -Iron Carbide,◆-Ferrite on used oil sample

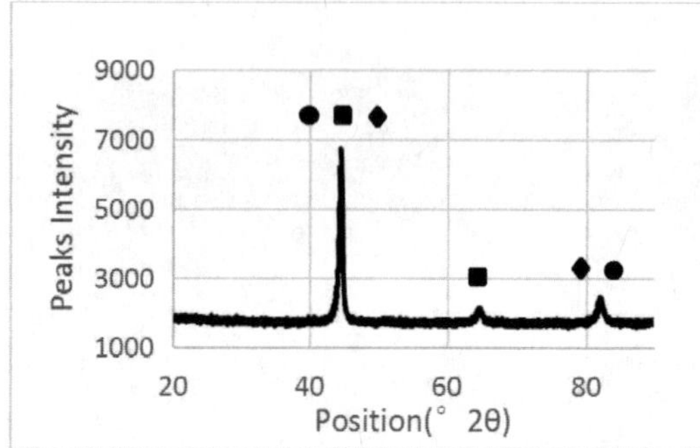

Figure 12. XRD spectrum, ● -Martensite,■ Iron Carbide,◆-Ferrite on new oil sample

Table 5. Quantified phase analysis

Phase	Used Oil (%)	New Oil (%)
Martensite	77.7	67.4
Fe_3C	21.1	31.7
Ferrite	1.1	0.9

4. Discussion

The failure of the spring was attributed to the pre-existing cracks that occurred during the production of the springs. The pre-existing crack was positively recognized as quench crack, that developed during the phase transformation. The crack occurred due to uneven rapid cooling of the spring introduced by the degraded oil. The main cause of quench cracks is severe cooling rate and subsequent development of tensile stress on the surface of the spring [8]. The tensile stresses are increased by volumetric martensite expansion during phase transformation. The cracks are often fine on the surface, detectable by dye penetrant testing. The internal stresses that cause cracking can be reduced by selecting a quenchant with slower cooling rates and still promote low martensite transformation [9]. Many researchers have studied the characteristics and behaviour quenchants, however, the effect of the quenching media on specific steels is still lacking [10]. There are numerous grades of quenching oils with different properties and quenching capabilities.

5. Conclusion

The failure of the spring was caused by the existence of a surface quench crack that acted as a precursor for the fatigue crack and subsequent failure. It has been proven that the internal stresses formed can be reduced by selecting quenching media that promote a relatively slow martensite transformation. It is important that the quench oil properties are regularly monitored in order to avoid using degraded oil for quenching operation. Inadequate heat treatment process resulted in deficient microstructures and subsequently poor in-service performance of the springs.

6. Reference

[1] Baek SG, Shin B, Lee SW, Choi YS, Kim J, Koo JC 2013 *Optimization of high speed EMU suspension parameters for vibration reduction*. J Mech Sci Technol;**27**:305–11.

[2] Podgornik B, Torkar M, Burja J, Godec M, Senčič B 2015 *Improving properties of spring steel through nano-particles alloying*. Mater Sci Eng A;**638**:183–9.

[3] Mukhopadhyay NK, Das SK, Ravikumar B, Ranganath VR, Ghosh Chowdhury S 1997 *Premature failure of a leaf spring due to improper materials processing*. Eng Fail Anal;**4**:161–70.

[4] Zhu Y, Wang Y, Huang Y 2014 *Failure analysis of a helical compression spring for a heavy vehicle's suspension system*. Case Stud Eng Fail Anal;**2**:169–73.

[5] Das SK, Mukhopadhyay NK, Kumar BR, Bhattacharya DK 2007 *Failure analysis of a*

passenger car coil spring. Eng Fail Anal;**14**:158–63.

[6] Pyttel B, Brunner I, Kaiser B, Berger C, Mahendran M 2014 *Fatigue behaviour of helical compression springs at a very high number of cycles - Investigation of various influences*. Int J Fatigue;**60**:101–9.

[7] Artaraz F, Sánchez Beitia S 1991 *An unsuitable residual stress state in train springs originated by shot peening*. Int J Fatigue;**13**:165–8.

[8] Lingamanaik SN, Chen BK 2014 *Microstructural and thermo-mechanical analysis of quench cracking during the production of bainitic-martensitic railway wheels*. Eng Fail Anal;**40**:25–32.

[9] Ravi Kumar B, Bhattacharya DK, Das SK, Chowdhury SG 2000 *Premature fatigue failure of a spring due to quench cracks*. Eng Fail Anal;**7**:377–84.

[10] Chen N, Han L, Zhang W *et al. Enhancing mechanical properties and avoiding cracks by simulation of quenching connecting rods*. *Mater Lett* 2007;**61**:3021–4.

IOP Conf. Series: Materials Science and Engineering **430** (2018) 012044 doi:10.1088/1757-899X/430/1/012044

Effect of heat treatment on strength and ductility of 52CrMoV4 spring steel.

VJ Matjeke[1,2,3], G Mukwevho[1], AM Maleka[1] , JW van der Merwe[2,3]

1. Transnet Engineering, Research & Development, Private Bag X 528, Kilnerpark, South Africa, 0127
2. School of Chemical and Metallurgical Engineering, University of the Witwatersrand, Johannesburg, Private Bag X3, Wits, South Africa, 2050
3. DST-NRF Centre of Excellence in Strong Materials, University of the Witwatersrand, Johannesburg, Private Bag X3, Wits, South Africa, 2050

[a]Goodness.Mukwevho@transnet.net

Abstract. In this study we examined the behaviour of the 52CrMoV4 spring steel at various tempering temperatures and soaking times. The hardening and quenching process was set up to achieve a minimum hardness of 800 HV by selecting an appropriate quenching medium. The microstructure was evaluated optically and with scanning electron microscopy (SEM) and the mechanical properties were characterised by hardness and tensile tests. An excellent combination of strength and ductility was achieved on conventional alloyed steel by adjusting heat treatment process parameters. The strength and ductility of the material was found to be comparable to that of advanced high strength steel (AHSS).

1 Introduction

The manufacturing of heavy duty springs from steel requires an exact understanding of the heat treatment process and material behaviour. Although the presence of alloying elements can improve the mechanical properties, they also escalate the cost of the steel [1]. Therefore, the selection of the steel and manufacturing process must be robust in such a way that the spring will exhibit excellent strength and ductility with minimal alloying elements. Furthermore, it is important to select a quenching media that will harden the steel without forming quench cracks or distortion. This is a function of the selected heat treatment process, the quenching media and the alloying elements in order to provide acceptable mechanical properties. The heat treatment process alters the microstructure and strength of material, which will directly influence fatigue life [2].

Spring manufacturing requires materials with good formidability as well as strength in order to improve performance [3]. The majority of automotive springs are made from medium carbon high strength steels [4]. To increase the strength of the steel is simple, however, it is difficult to get a combination of high strength and ductility. Generally, an increase in strength results in a loss of ductility [5]. This study will focus on the heat treatment process, microstructure and mechanical properties of spring steel. The manufacturing process of the springs entails high temperature coiling, hardening and tempering [6].

This study will investigate whether a heat treatment process can make a conventional 52CrMoV4steel compete with advanced high strength steel (AHSS). The strength and hardenability of the 52CrMoV4 steel is improved by additions of elements such as chromium, molybdenum and vanadium. Molybdenum provides solution strengthening in steel during heat treatment, while chromium and vanadium forms carbides that improve the hardness of the material during tempering [7]. Molybdenum also forms fine carbides precipitates that increases the strain hardening effect [8]. On the other hand vanadium has been used as a micro -alloying element for high strength low alloy steels since the 1950s [9]. It improves the strength and hardness by formation of vanadium carbides and grain refinement [10]. With the effort to increase the strength, precaution is required to avoid temper embrittlement [11]. The strength is a critical factor in determining the fatigue life of a material.

2 Experimentation and material

A 52CrMoV4 steel rods were cut into 25 X 320 mm sections. The samples were placed in the furnace when it stabilised at 860 °C, thereafter the samples were soaked and austenitised at 860 °C temperature for an hour. The samples were quenched in oil with the quenching properties as shown in Table 1. Three tempering temperatures were used: 410, 450 and 470 °C; and two tempering times for each temperature, namely: 2 and 3 hours. The continuous cooling curve (CCT) of the steel were simulated using JmatPro software by inserting the chemical composition of the spring steel. The chemical composition was determined with a Bruker optical emission spectrometer. The oil quenching ability was analysed with a portable quenchant test system (IVF smart quench). The transverse section of the rod was cut, ground and polished to a 1 µm surface finish for metallographic examination. The polished surface was etched with a 3% Nital reagent. The microstructure was characterised with the optical microscope and SEM, while the penetrant test (PT) was used to determine the presence of quench related flaws. Micro hardness profile measurements were conducted across the transverse section of the steel rod from one edge to the other edge at a 1 mm interval using a 5 gf load for the micro Vickers hardness test. Rietveld analysis was used to quantify X-ray diffraction (XRD) results to characterise phases.

Table 1. Quenching oil properties

Properties	Oil
Appearance	Liquid
Fire point (°C)	220
Flash point (°C)	204
Density (g/cm^3)	0.854
Pour point (°C)	-15
Viscosity (mm^2/s @ 40 °C)	20

3 Results and discussion

3.1 Chemical Analysis

The chemical composition of the spring steel is presented in Table 1. The chemical analysis confirmed that the steel is a micro-alloyed 52CrMoV4 steel containing 0.19 wt% Mo and 0.11 wt % V. The addition of Mo and V alloying element improves the hardenability of the steel.

Table 2. Chemical composition of the steel in weight percentage

Element	% C	% Si	% Mn	% P	% S	% Cr	% Ni	% Cu	% Mo	% V
Mass %	0.55	0.28	0.92	0.012	0.007	1.05	0.01	0.01	0.19	0.11

3.2 Quench Test Analysis

The oil quench test results are shown in Figure 1. The oil showed complex cooling behaviour at various temperature stages. The harmonisation of the continuous cooling temperature (CCT) curve and the oil cooling rates observed in Figures 1 and 2, predicts full martensite transformation. The CCT curve revealed martensite start to be 265 °C, whilst the cooling rate of the quenching medium was approximately 6.7 °C/s.

3.3 Microstructural Examination

The microstructure of the as-quenched steel rods is presented in Figures 3 and 4. The general microstructure of the as quenched spring steel was martensitic. There was no evidence of cracks or quenching related flaws. The XRD analysis of the as quenched steel revealed the structure to be a mixture of martensite (79%), iron carbide (19.8) and ferrite (2%). The XRD spectrum is shown in Figure 5. The as tempered microstructures are presented in Figures 6 to 11. The general microstructure was tempered microstructure. Figures 6 to 11 revealed the decomposed unstable martensite to tempered martensite. The SEM analysis images are shown in Figures 12 to 17. Figure 12 showed rod shaped iron carbide (Fe_3C) layered around ferrite that retained the high-density dislocation of martensite. Figures 13 and 14 showed ferrites with low density of dislocation and the mixture of rod-shaped and spherical iron carbides (Fe_3C). Figures 15 to 17 presented a structure of tempered martensite which is predominately made up of spherical and rod-shaped carbides. There is a presence of needle like carbides as well. The observation on the samples tempered 2 hours was a microstructure with rod shaped carbides whilst the samples tested for 3 hours showed pronounced spherical carbides.

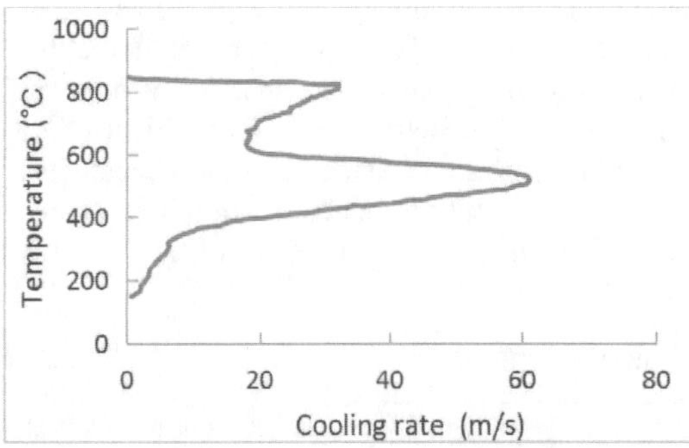

Figure 1. Quenching media cooling rate plot

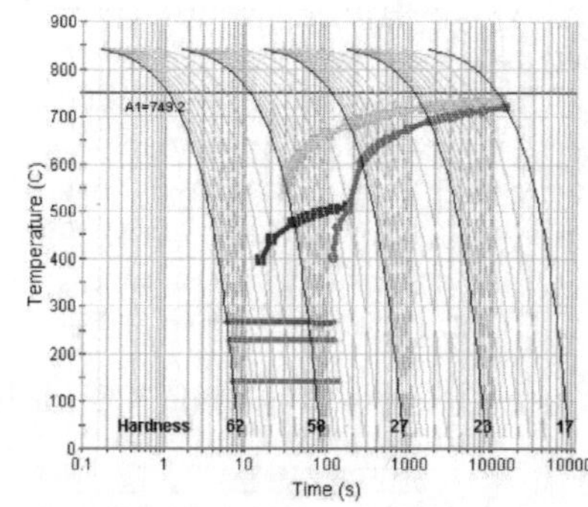

Figure 2. Extrapolated CCT curve for 52CrMoV4

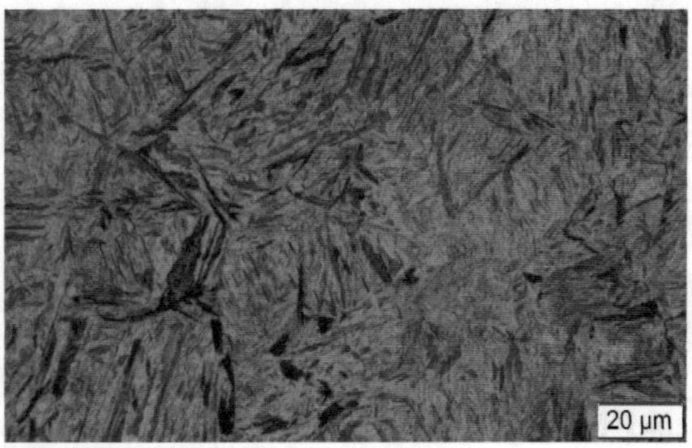

Figure 3. The as-quenched martensite

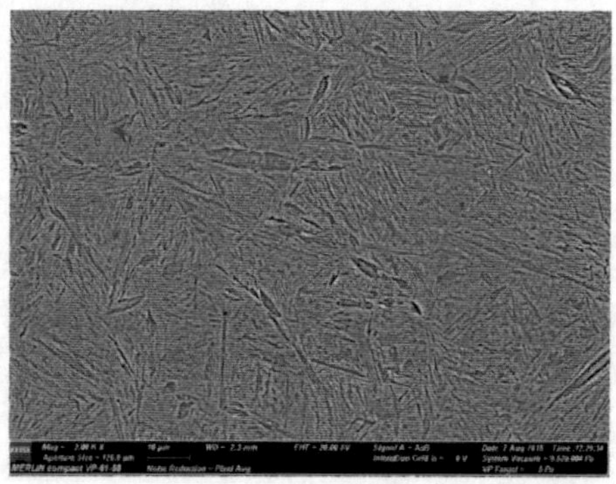

Figure 4. SEM image of the as-quenched martensite

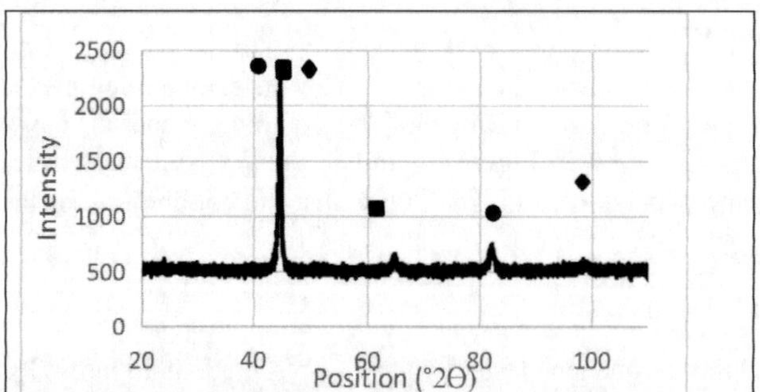

Figure 5. XRD spectrum, showing ● -Martensite ■ -Iron Carbide,◆-Ferrite

Figure 6. Microstructure tempered at 410°C for 2 hours

Figure 7. Microstructure tempered at 410°C for 3 hours

Figure 8. Microstructure tempered at 450°C for 2 hours

Figure 9. Microstructure tempered at 450°C for 3 hours

Figure 10. Microstructure tempered at 470°C for 2 hours

Figure 11. Microstructure tempered at 470°C for 3 hours

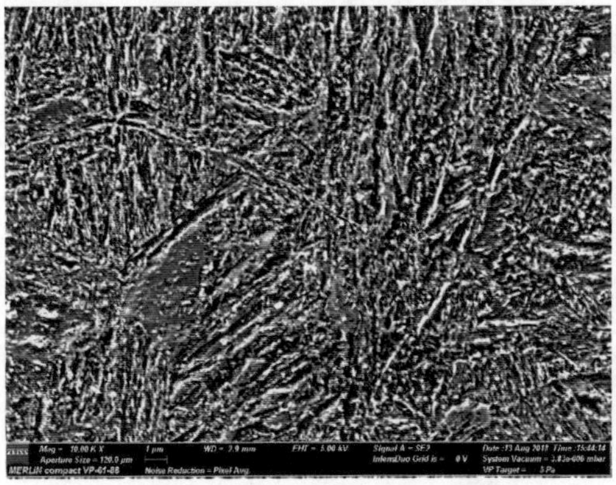

Figure 12. Microstructure tempered at 410°C for 2 hours

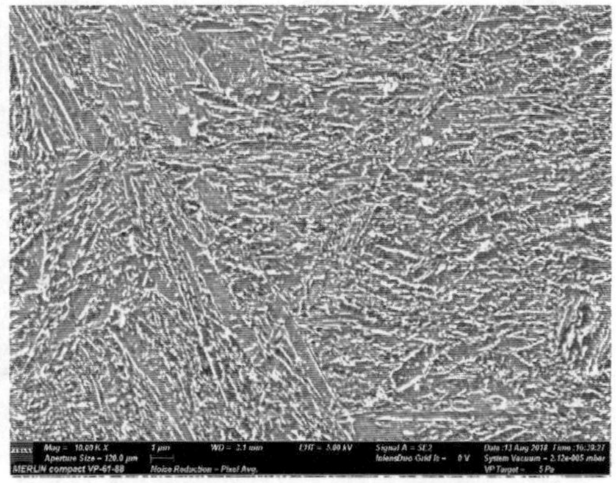

Figure 13. Microstructure tempered at 410°C for 3 hours

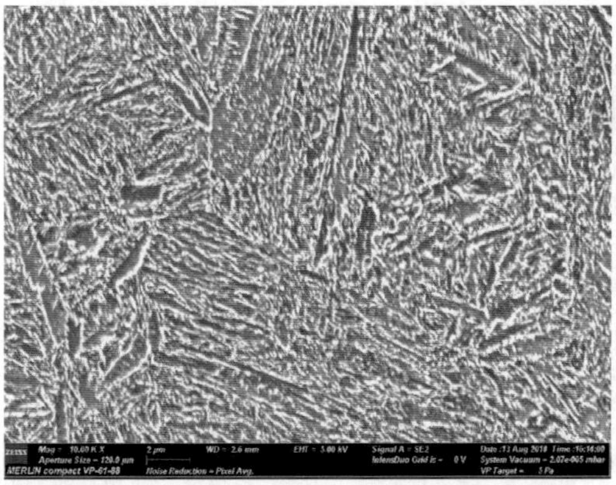

Figure 14. Microstructure tempered at 450°C for 2 hours

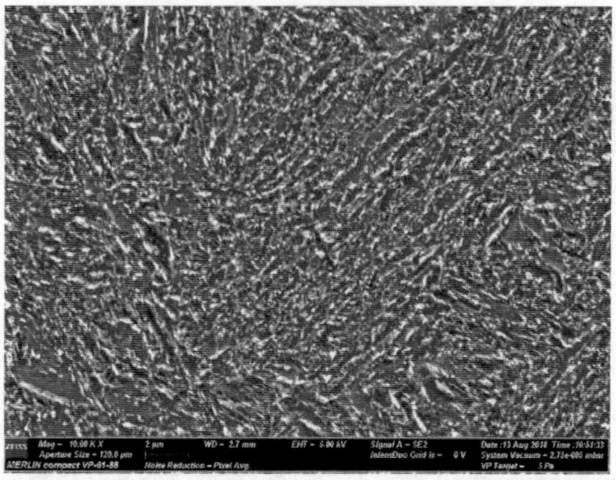

Figure 15. Microstructure tempered at 450°C for 3 hours

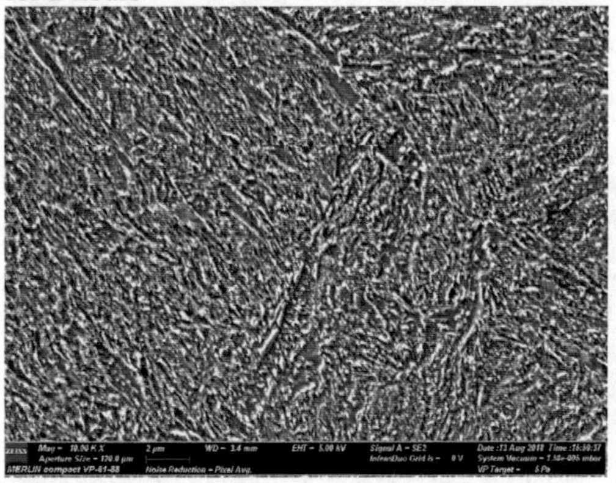

Figure 16. Microstructure tempered at 470°C for 2 hours

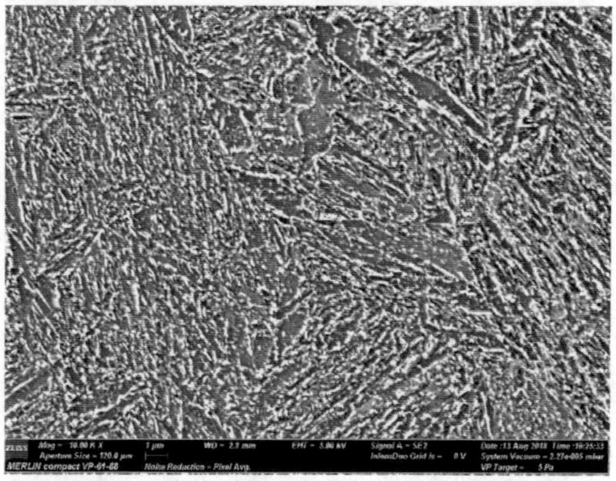

Figure 17. Microstructure tempered at 470°C for 3 hours

3.4 Mechanical properties

The as-quenched and tempered microhardness results are illustrated in Tables 3 and 4 The hardness generally decreased with the increase in tempering temperature, however the sample that were tempered for 3 hour showed evidence of secondary precipitation hardening when compared to the material tempered for 2 hours (see Figure 13). The secondary hardening was as a result of coalesce of Mo_3C and V_4C_3. On the other hand, the strength decreased, and ductility increased with the increase of tempering temperature. The tensile results are shown in Figure 14 and Table 6. The elastic region of the tensile test sample showed elastic behaviour or spring effect until yielding point.

Table 3. The as quenched hardness measurements

Average	Standard Deviation	Standard Error
804	15	3.03

Table 4. The tempered hardness profile average

Measurements	Average	Standard Deviation	Standard Error
410°C 2hrs	521	12.4	2.6
410°C 3hrs	524	17.3	3.6
450°C 2hrs	475	12.8	2.7
450°C 3hrs	491	15.5	3.2
470°C 2hrs	482	14.6	2.9
470°C 3hrs	486	12.7	2.6

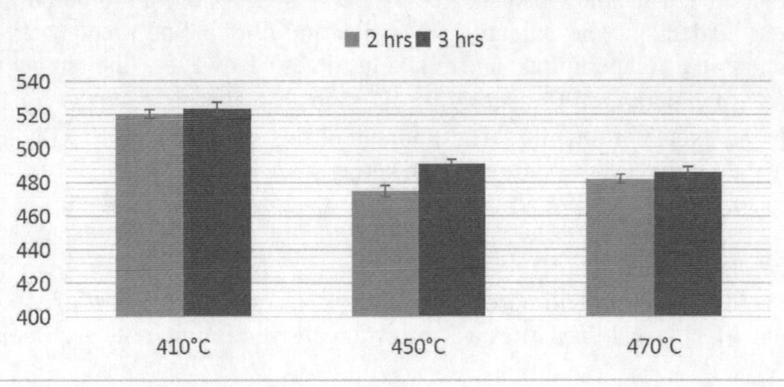

Figure 18. Hardness measurements for various tempering temperatures and times

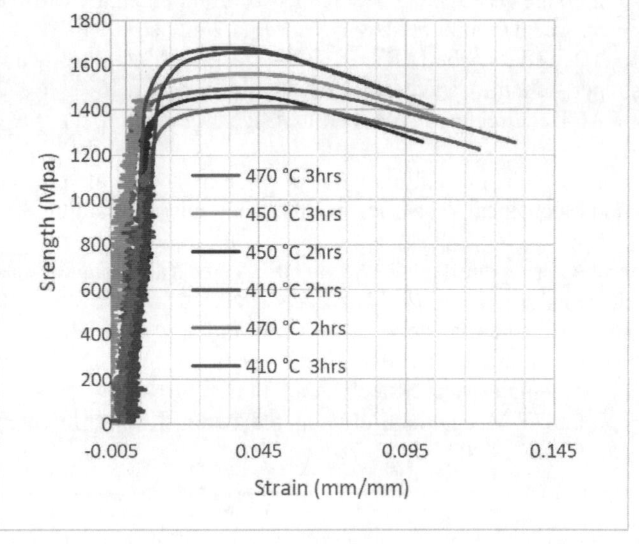

Figure 19. Tensile curves of quenched and tempered steel.

Table 5. Tensile strength and total strain of quenched and tempered steel.

Tempering Parameters	Ultimate Tensile Strength (MPa)	Yield Strength (MPa)	Strain (%)
410°C 2h	1674	1633	9
410°C 3h	1655	1608	10
450°C 2h	1462	1424	11
450°C 3h	1558	1516	12
470°C 2h	1412	1343	12
470°C 3h	1495	1420	13

4 Conclusions

The presence of molybdenum and vanadium influenced the mechanical properties by forming fine carbides precipitates that introduced secondary precipitation hardening. The optimum strength and ductility of the spring steel was improved by subjecting the material to various tempering temperatures and soaking time. The study also showed that tempering time is an important variable in influencing precipitation hardening. The amount of Mo-rich and C-rich fine precipitates (carbides) varied at different tempering temperatures and soaking times. However, the amount carbides increased with increase in temperature. Tempering of 52CrMoV4 material at 470°C for 3 hour has provided high strength and excellent ductility. The findings of this scientific study will assist spring manufacturers optimise the mechanical properties of the spring steel.

5 Reference

[1] Kostryzhev A, Singh N, Chen L, Killmore C, Pereloma E. Comparative Effect of Mo and Cr on Microstructure and Mechanical Properties in NbV-Microalloyed Bainitic Steels. *Metals (Basel)* 2018;**8**:134.
[2] Fragoudakis R, Savaidis G, Michailidis N. Optimizing the development and manufacturing of 56SiCr7 leaf springs. *Int J Fatigue* 2017;**103**:168–75.
[3] Manuel F, Cerda C. Third Generation Advanced High Strength Steels via Ultrafast Heating., 2017.
[4] Ardehali Barani A, Ponge D, Raabe D. Refinement of grain boundary carbides in a Si-Cr spring steel by thermomechanical treatment. *Mater Sci Eng A* 2006;**426**:194–201.
[5] Li Z, Pradeep KG, Deng Y, Raabe D, Tasan CC. Metastable high-entropy dual-phase alloys overcome the strength–ductility trade-off. 2016:4–11.
[6] Fedin VM, Borts AI. Heat treatment and thermoplastic treatment. *Met Sci Heat Treat* 2009;**51**:544–52.
[7] Kostryzhev A, Singh N, Chen L, Killmore C, Pereloma E. Comparative Effect of Mo and Cr on Microstructure and Mechanical Properties in NbV-Microalloyed Bainitic Steels. *Metals (Basel)* 2018;**8**:134.
[8] Vinet L, Zhedanov A. Pricipal effects of Mo in HSLA steels and cross effects with microalloying elements. *Int Semin Appl Mo Steels* 2010:1–18.
[9] Sage AM. Vanadium in structural steels. *J S Afr Inst Min Met* 1984;**84**:138–46.
[10] Kannam M. The microstructure, hardness, impact toughness, tensile deformation and final frature of four specialty high strength steels. 2011:111.
[11] Zia-Ebrahimi F, Krauss G. Mechanisms of tempered martensite embrittlement in medium-carbon steels. *Acta Metall* 1984;**32**:1767–78.

IOP Conf. Series: Materials Science and Engineering **430** (2018) 012045 doi:10.1088/1757-899X/430/1/012045

The influence of full and partial austenitization temperatures on the quench and partition heat treatment process for an advanced high strength steel

V Kurup[1], **C W Siyasiya**[1], **R J Mostert**[1] and **J Wicks**[1,2]

[1] Department of Metallurgy and Material Science, University of Pretoria, South Africa
[2]Arcelor Mittal, Vanderbijlpark, South Africa

[*]vinod.kurup@up.ac.za

Abstract. The effect of full and partial austenitization on the microstructural evolution of a medium carbon steel with 2.2 mass %Si was studied during the quench and partitioning heat treatment process. The quench and partition process is designed to ensure that carbon partitions to retained austenite to obtain martensite and carbon- stabilized retained austenite. The full and partial austenitization temperatures were established after determining the Ac_3 and Ac_1 temperature from dilatometer runs. Heat treatment was carried out in a B*ä*hr Dilatometer and replicated in salt bath furnaces. Phase transformations were studied by dilatometry, scanning electron microscopy and EBSD analysis. Hardness tests were carried out to evaluate the mechanical properties. The results indicate that full austenitization treatment at 940 ^{0}C resulted in thin films of retained austenite along with a lath martensite structure. The partial austenitization at 800 ^{0}C resulted in plate-like martensite with blocky retained austenite. The full austenitization gave rise to a higher volume fraction of retained austenite of about 22% which led to better mechanical properties.

1. Introduction

The average use of advanced high strength steel (AHSS) in car body applications has significantly grown over the last 20 years [1]. Numerous methods for obtaining third-generation AHSS have been proposed. The quench and partitioning process (Q&P) is considered as one of the alternatives for manufacturing AHSS [2] [3][4] [5]. During the Q&P heat treatment, austenite or austenite-ferrite mixture is quenched to an optimal quench temperature (QT) between M_S and M_f to get the desired fraction of martensite and retained austenite. This is followed by a partitioning heat treatment (PT) where the carbon from supersaturated martensite is transported to untransformed retained austenite to stabilize it. The methodology to determine optimum quench temperature has been reported elsewhere [5]. The original Q&P process is based on two important concepts: (a) constrained carbon equilibrium (CCE) and (b) athermal martensite transformation. The CCE condition implies that interstitial carbon partition from martensite to retained austenite and there is no phase boundary motion [6][5].

The Q&P steels are medium carbon steels with a composition similar to TRIP type steel [7]. The carbon content is kept low in order to avoid welding related problems. Si is added to prevent the formation of carbide precipitates in order to ensure that sufficient carbon is available to stabilize the retained austenite at room temperature [8]. Mn is an austenite former and also helps in increasing hardenability of the steel. [9]. In addition to Si, Al is also added to suppress the carbide formation [10].

Santofimia et al found that by increasing the Mn content above 3.5 mass% the initial microstructure after hot rolling consisted of martensite and polygonal ferrite. The application of Q&P heat treatment to such microstructures resulted in good mechanical properties. [11]

When the steel is subjected to a partial austenization, the desired amount of ferrite and austenite is obtained by controlling the inter-critical temperature. The subsequent Q&P process changes the volume fraction and components of the microstructure. Partial annealing followed by the Q&P process has been reported for high Al steel. [11]

Therefore, in this work, a medium carbon steel with a high silicon content of 2.2 mass% was subjected to Q&P process after both full and partial austenitization. The effect of partial and full austenization was evaluated by the evolution of microstructure and hardness test. The optimal Q&P process after both partial and full austenization was studied.

2. Experimental Procedure

The experimental steel used in the present work had the nominal composition of 0.27C- 2.8Mn- 2.2Si-0.5Al-0.75Cr and balance Fe. All composition values are in mass percentage. The steel was induction melted and cast into an ingot. It was reheated to 1200 °C to homogenize and then hot rolled and forged into 13 mm diameter rods. Si and Al were added to ensure prevention of carbide formation so that enough carbon remains in martensite to stabilize the austenite during the Q&P heat treatment. Two different sets of heat treatments were carried out, see Figure 1.

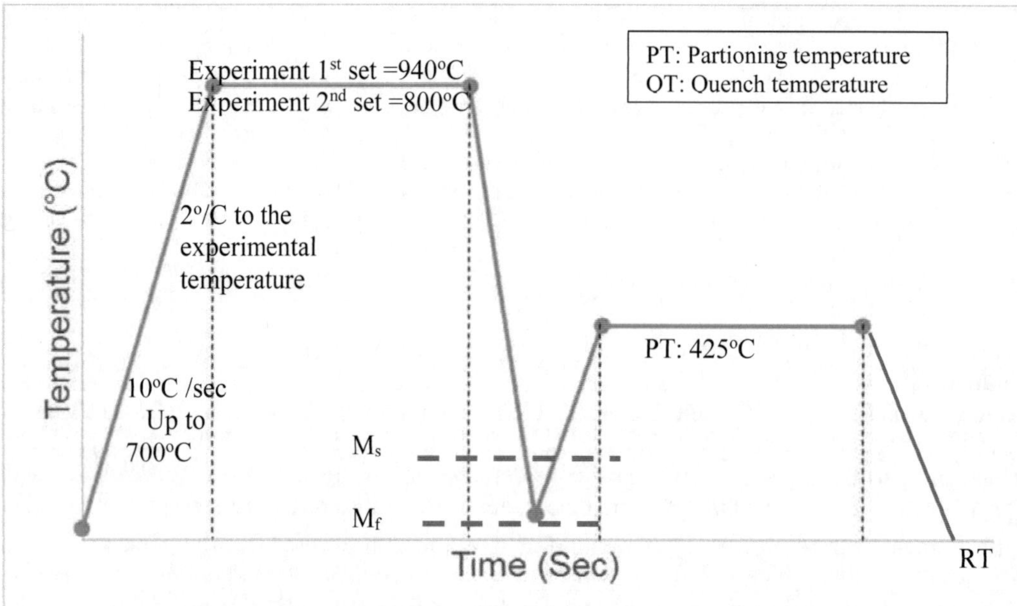

Figure 1. Q&P heat treatment cycle for the 1st and 2nd set of experiments

The Bähr Dilatometer 850D™ was used to measure the Ms and critical temperatures Ac_1 and Ac_3. The tests were carried following ASTM A-1033-04 [12]. Thermodynamic analysis was carried out using Thermo-Calc 2017b computer software for phase analysis at different austenitization temperatures [13]. Figure (2) shows the temperature–phase fraction phase diagram generated from Thermo-Calc [13].

Heat treatments were carried out in the dilatometer machine. The first set of tests comprised of the heating sample above the A_{c3} temperature of 940 °C and thereafter quenching using helium to the calculated optimum quench temperature as determined by the Speer model [5]. The samples were then reheated to the partitioning temperature of 425 °C and isothermally held for 200 s before quenching to room temperature. The quenching temperature was varied 40 ^{0}C above and below the calculated optimum quench temperature.

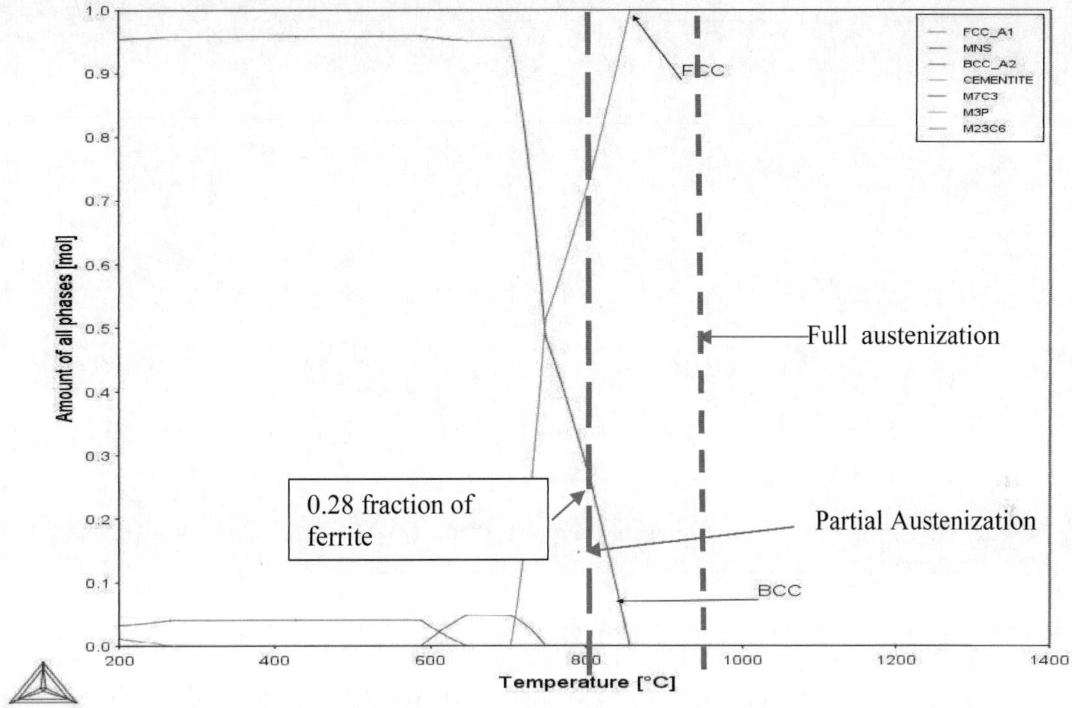

Figure 2. Temperature – volume phase fraction of the alloy as calculated by Thermo-Calc[13]

In the second set, the samples were heated in the two-phase region at 800 ^{0}C, between A_{c3} and A_{c1}, quenched to the optimum quench temperature, and partitioned as 1st set. The partial austenitization induces ferrite formation and also increases the carbon content of the austenite.. The optimal quench temperature was theoretically calculated for both sets of experiments using the respective Ms temperatures and thermodynamic data.

The heat-treated samples were analyzed using x-ray diffraction (XRD).. The samples were analyzed using a PANalytical X'Pert Pro powder diffractometer in θ–θ configuration with an X'Celerator detector and variable divergence and fixed receiving slits with Fe filtered Co-Kα radiation (λ=1.789Å). The phases were identified using X'Pert Highscore Plus computer software. The phase quantification was carried out using the Rietveld method [14]. The sample with the highest amount of retained austenite was analyzed by scanning electron microscope (SEM) and electron backscatter diffraction (EBSD). The step size used for SEM-EBSD was 60 nm. Vickers hardness test was carried out to evaluate the hardness of the alloy under different heat treatment conditions. The hardness test was done on all samples with a load of 500 grams. Ten hardness tests were taken lengthwise in the middle of each sample in order to obtain more accurate average hardness values.

3. Results and Discussion

3.1. Critical temperatures

The A_{C1}, A_{C3} and $M_{S.}$ temperatures of the steels are shown in Table 1. Figure 3 shows change in length versus temperature graphs used for the estimation of Ms temperature.

Table 1. Critical temperatures obtained from dilatometer experiments

Critical temperatures	Full Austenitization at 940 °C	Partial Austenitization 800 °C
A_{C1}	765	765
A_{C3}	855	855
M_s	370	275

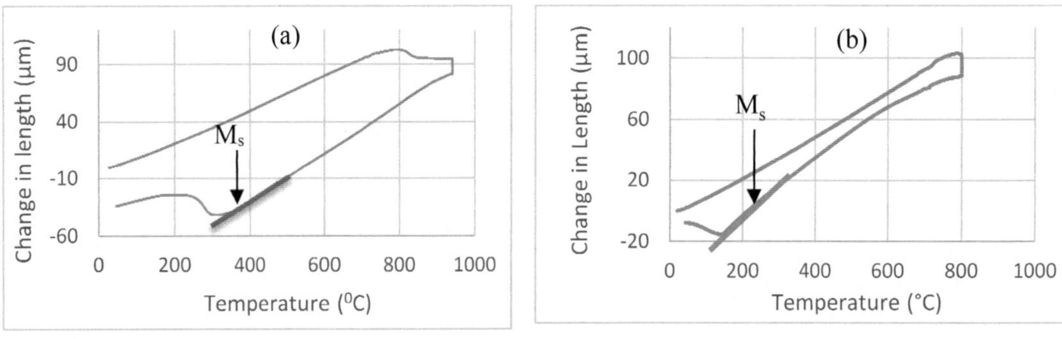

Figure 3: Change in length versus temperature graph (a) M_s temperature of steel after full austenitization at 940 °C. (b) M_s temperature of steel partially austenitized at 800 °C

The optimal quench temperature calculated using the Speer method [5] which use the KM equation that gives the fraction of martensite at the quench temperature [15]:

$$f_m = 1 - e^{-\alpha(Ms-QT)} \quad (1)$$

where f_m the fraction of martensite transformed and Ms is the martensite start temperature and QT is the quench temperature. α is the rate parameter in this case taken to be 0.011 [15] The Ms Equations, which were used are as follows:

$$M_s = 545 - 330 \times C + 2 \times Al - 14 \times Cr - 23 \times Mn - 7 \times Si \quad (2)$$
$$M_s = 565 - 600(1 - e^{-0.96 \times C}) - 31 \times Mn - 13 \times Si - 10 \times Cr \quad (3)$$

The exponential equation (3) and equation (2) were used to calculate the theoretical Ms temperature for the steel austenitized at 940 and 800 °C respectively. The theoretical optimal quench temperature for the fully austenitized sample was found to be 225 °C and for the partially austenized sample, it was 183 °C. The optimal quench temperature is important in the design of Q&P heat treatment cycles as this temperature determines the volume fraction of martensite formed and the potential level of carbon enrichment of austenite during the partitioning process. This carbon enrichment increases the thermal stability of austenite and it's M_S temperature decreases which leads to less fresh martensite on second quenching.

3.2. XRD measurements

The XRD analysis of the as-cast structure revealed the presence of martensite and 8% retained austenite. Figure 4 shows the XRD pattern of the alloy with partial austenitization at 800 °C and quenched to 183 °C.

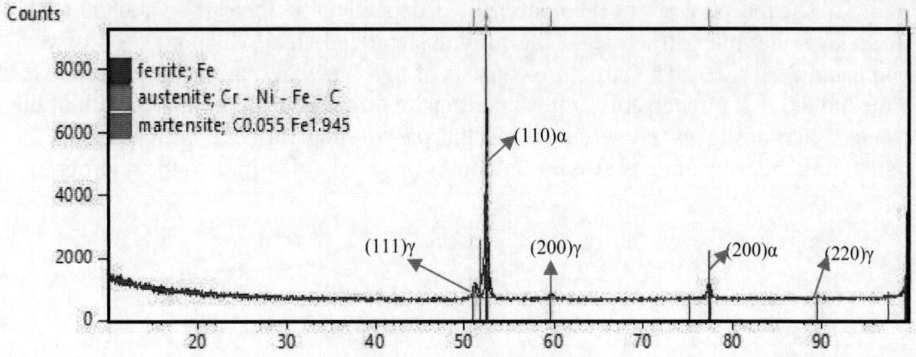

Figure 4. XRD pattern of alloy heat treat at 800⁰C and quenched at 183⁰C

Table 1. Volume fraction of retained austenite for different Q&P heat treatment, measured by XRD

Partial austenitization 800 °C		**Full austenitization 940 °C**	
Quench temperature (°C)	**Retained austenite (%)**	**Quench temperature (°C)**	**Retained austenite (%)**
143	20.3	185	17.3
183	11.5	225	22.4
223	9.9	265	19.6

Table 2 shows the amount of retained austenite at each quench temperature under partial and full austenization condition. For steel with full austenization, the optimal quench temperature of 225 0C resulted in maximum retained austenite of 22.4%. However, in the case of the partially austenized sample, maximum retained austenite was obtained at 143 °C rather than the calculated 183 °C and was found to be 20.3%. This may be because not enough martensite was formed to supply carbon atoms to stabilize the retained austenite. Therefore, the unstable retained austenite transformed into fresh martensite during the second quenching stage.

From the optimal temperature calculations and XRD results in Table 2, the effect of quench temperature on the Q&P heat treatment is shown in Figures 5 and 6. It can be seen that as the quench temperature increased, the volume fraction of martensite decreased while that of austenite increased. Along with that, carbon concentration in austenite also decreased immensely. The theoretical carbon content in the retained austenite was higher in the steel which was partially annealed. From the iron-carbon phase diagram[13], it can be seen that carbon content in the austenite in the two-phase region is more, which points to the high carbon content in the retained austenite, which was partially austenized. The dots in figures 6 and 7 indicate the experimental retained austenite results from XRD.

3.3 Microstructural characterization

Scanning electron microscope (SEM) was used to study the Q&P heat-treated samples.

Figures 7 (a) and (b) shows coarse and rounded martensite structure along with the ferrite phase from partially austenitized steel. The ferrite phase can be distinguished by the absence of carbides in it. The difference in morphology of martensite between partial and full austenitization is attributed to the higher carbon content in the former which leads to more of plate-like martensite than the lath-like structure of the latter. For the steel which was heat treated at 940 °C with Ms temperature of 370 °C, when quenched to 225 °C, exhibits a higher driving force (ΔT). This leads to a high nucleation rate resulting in the finer structure of martensite. On the contrary, the higher carbon content martensite of the partially austenitized steel leads to lower Ms temperature of 275 °C, see Figure 3. Therefore, when quenched to 183 °C, the ΔT is lower which leads to lower nucleation rate resulting in coarse plate-like martensite. As the quench, the temperature decreases to 143 °C the ΔT increases so it results in a finer structure in comparison to the steel quench to 183 °C. The morphology of the martensite is dependent on the carbon content of the austenite. Figure 7 (c) shows the microstructure of the fully austenitized steel which consists of very fine lath martensite with some carbides. The structure consists of two types of martensite 1) secondary martensite which formed in the first quench and 2) fresh martensite formed during a final quench of the heat treatment. Secondary martensite is tempered during the partitioning process, which shows the presence of carbides in it. Retained austenite is also present but is not easily identified in the micrograph.

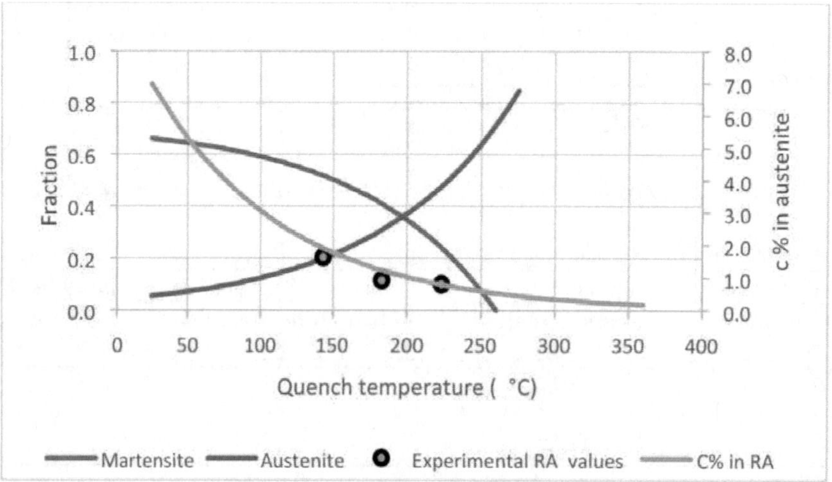

Figure 5. Fraction of martensite and austenite at various quench temperatures for steel with partial austenitization at 800^{0}C

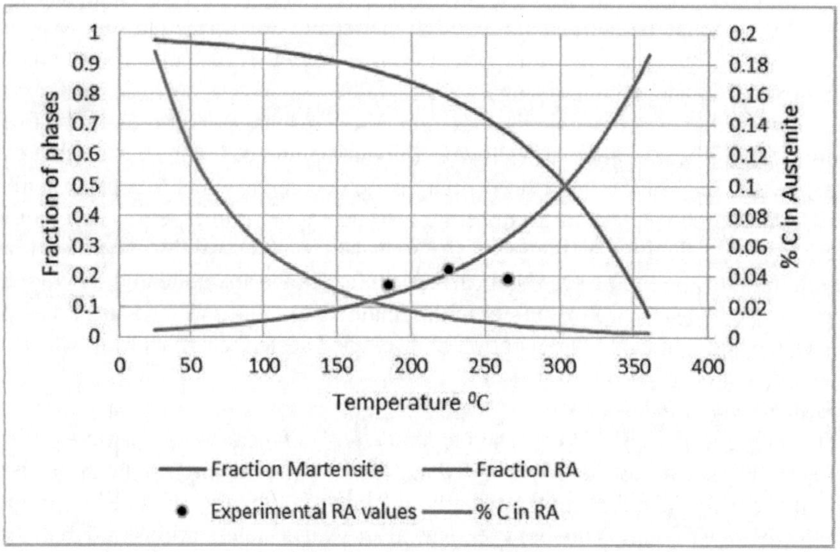

Figure 6. Fraction of martensite and austenite at various quench temperatures for with full austenitization at 940^{0}C

The Retained austenite though present cannot be easily distinguished from the SEM images. Therefore, SEM -EBSD was carried out to see the size, morphology and distribution of retained austenite. Figures 8 (a) and (b) show the EBSD of steel that underwent full austenitization and partial austenitization respectively. It can be seen that some fraction of retained austenite is in the form of thin films and is distributed between the martensite lath phase while the coarse blocky retained austenite is situated along the prior austenite grain boundaries. On the contrary, for the partially austenitized steel, the retained austenite is mainly coarse and blocky. Longer partitioning time was necessary to stabilize the blocky retained austenite. As a result, less retained austenite is obtained after the Q&P process in partially austenitized steel.

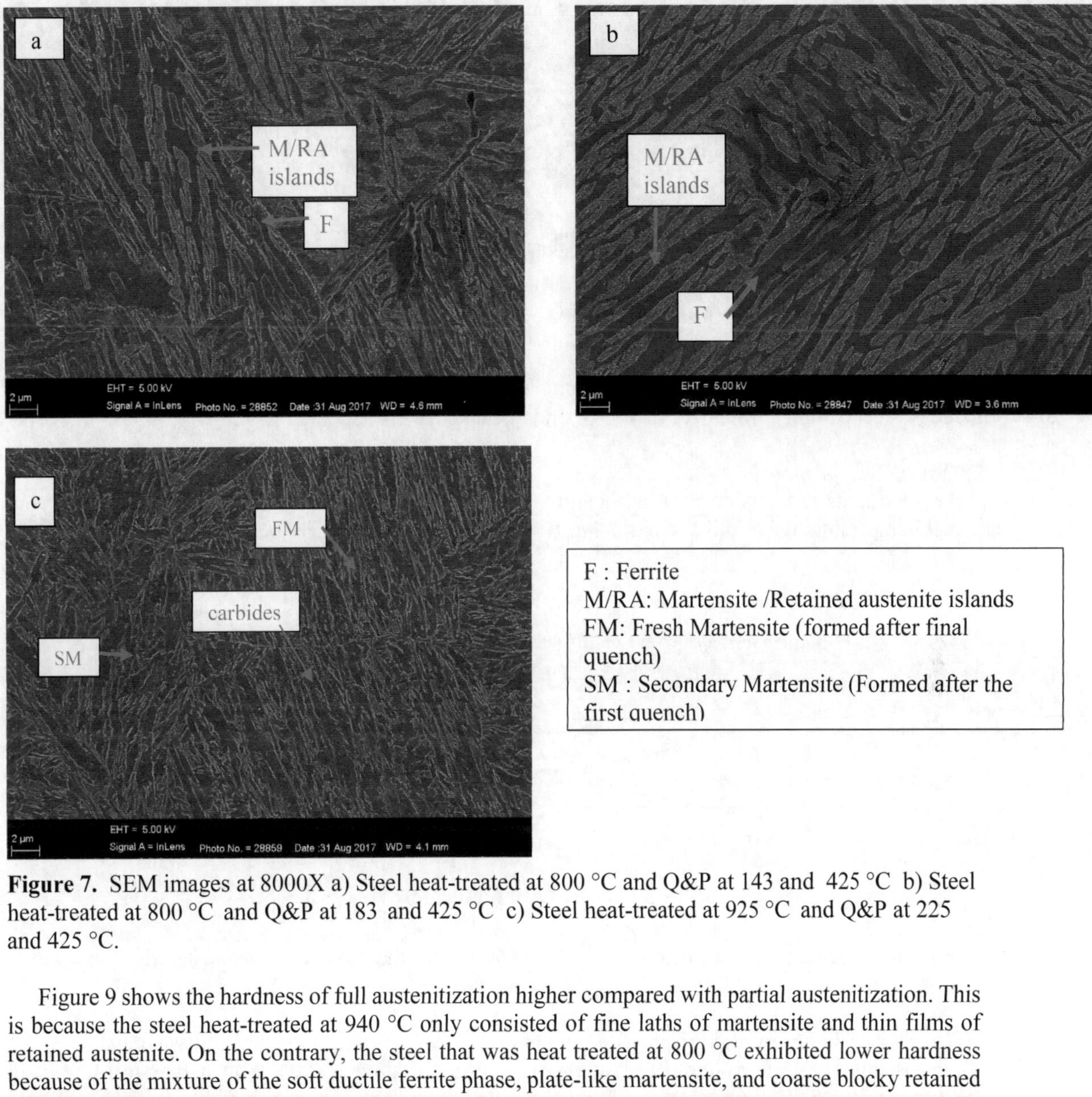

Figure 7. SEM images at 8000X a) Steel heat-treated at 800 °C and Q&P at 143 and 425 °C b) Steel heat-treated at 800 °C and Q&P at 183 and 425 °C c) Steel heat-treated at 925 °C and Q&P at 225 and 425 °C.

Figure 9 shows the hardness of full austenitization higher compared with partial austenitization. This is because the steel heat-treated at 940 °C only consisted of fine laths of martensite and thin films of retained austenite. On the contrary, the steel that was heat treated at 800 °C exhibited lower hardness because of the mixture of the soft ductile ferrite phase, plate-like martensite, and coarse blocky retained austenite. For fully austenitized steel, the hardness is insensitive to quench temperature. For partially austenitized steel, the hardness increased with an increase in quench temperature. This is attributed to the amount of the softer retained austenite which is at its maximum volume fraction at the quench temperature of 143 °C, i.e. 20.3%. In order to elucidate the discrepancy of maximum %RA between the predicted optimal quench temperature of 183 °C and the one observed at a quench temperature of

143 °C in the partially austenitized steel, the thermal profiles of the partitioning heat treatment were analysed, see Figure 10.

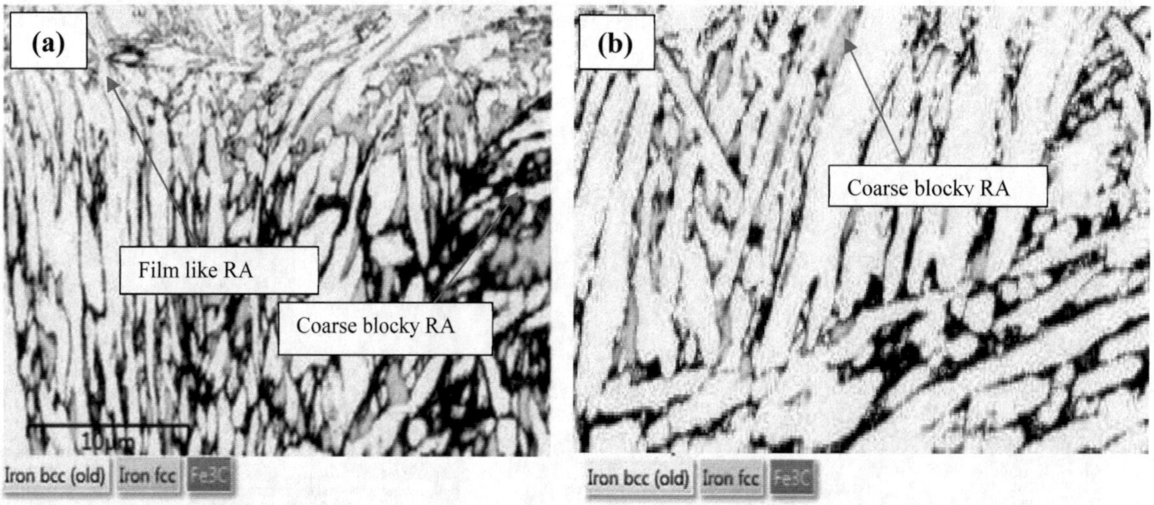

Figure 8. (a) The EBSD image of steel austenitized at 940 °C and quenched to 225 °C, (b) The EBSD image of steel partially austenized at 800 ^{0}C and quenched at 143 ^{0}C

3.4 Mechanical properties

Vickers hardness test was done on all heat-treated samples. The average hardness values at different quench temperature under full and partial austenitization are shown in Figure 9.

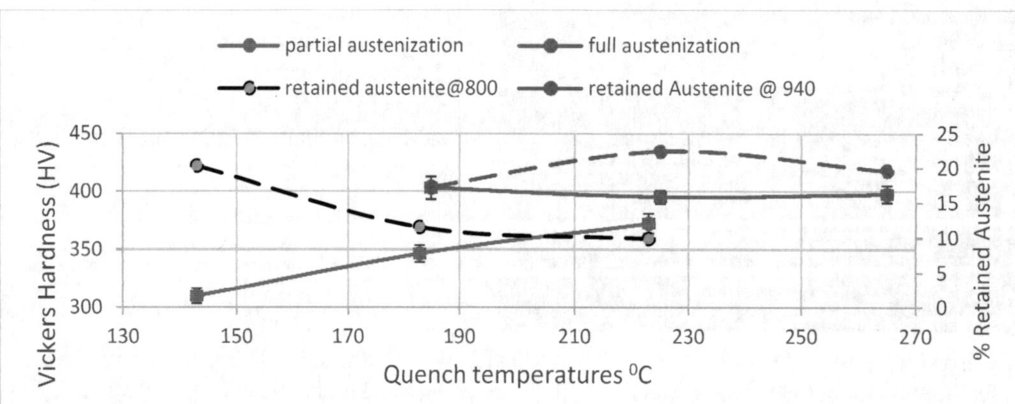

Figure 9. Hardness test at various quench temperature for the full and partial austenitisation heat treatment.

The expansion observed during partitioning in Figure 10 (a) is attributed to the phase transformation of the austenite (FCC) to bainite (BCC) after partial austenitization at 800 °C, quenching to 1430C and partitioning at 425 °C. During bainite formation, there is carbon rejection, which is taken up by the austenite phase, and this leads to the increased stability of the retained austenite. Evidence of bainite formation can be seen from the microstructure in Figure 11. After the final quench, no expansion is observed in Figure 10(a), which implies that no fresh martensite was formed.

No significant expansion was observed during partitioning in Figure 10 (b) after partial austenitization at 800 0C, quenching to 1430C and partitioning at 425 0C but the end of final quench shows expansion which is attributed to fresh martensite formation. This implies that the retained austenite obtained during the first quench was not fully stabilized during partitioning. This unstable austenite finally decomposed into fresh martensite. Hence, the lower amount of retained austenite at 1830C than at 1430C. The same argument applies to the full austenitization, see Figure 10 (c).

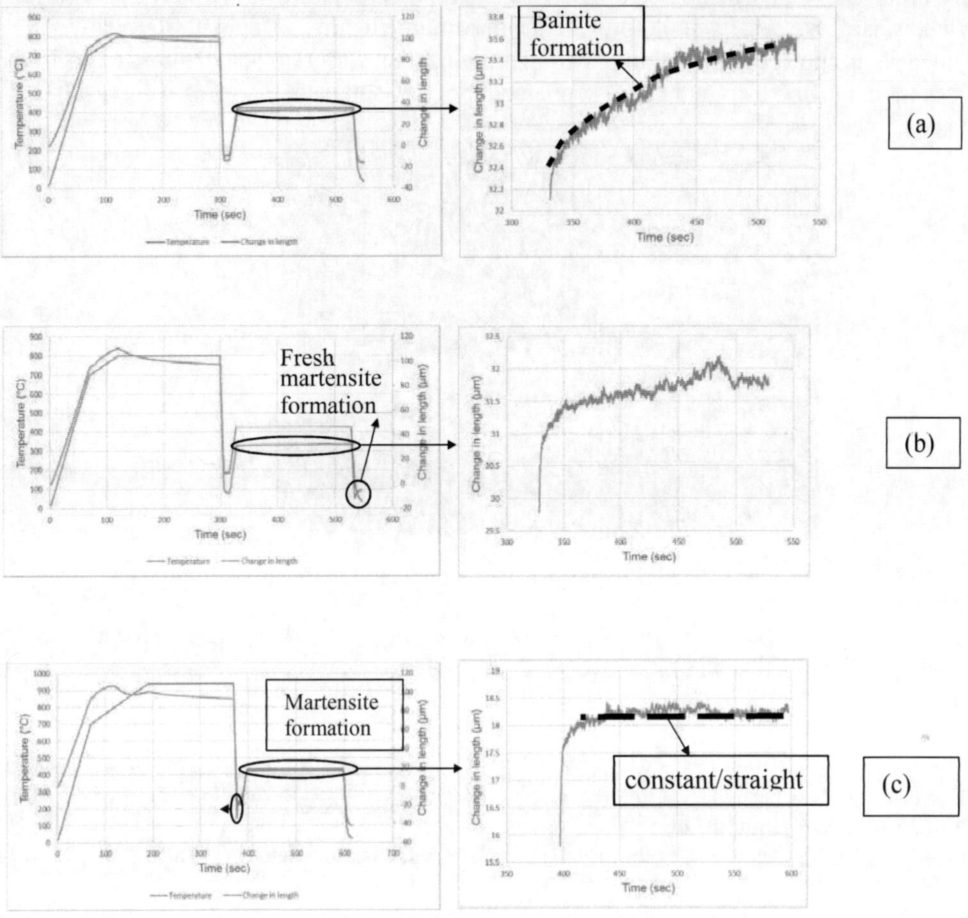

Figure 10. Q&P heat treatment (a) partially austenitized at 800 °C, quenched to 143^{0}C and partitioned at 425 °C, (b) partially austenitized at 800 °C, quenched to 183 °C and partitioned at 425 °C and (c) fully austenitized at 940 °C, quenched to 225 °C and partitioned at 425 °C.

4. Conclusions

- Q & P process increases the amount RA from 8 % in the as-cast alloy to 22 % in heat treated condition which shows that partitioning step help in stabilizing austenite. Full and partial austenitization before the Q&P heat treatment alters the morphology of martensite from thin lath-shaped to the coarser plate-like morphology because of the different initial carbon content in austenite.
- XRD results show maximum RA (22.4%) at 225 ^{0}C for full austenitization which confirms with optimum temperature calculation based on the CCE condition. The XRD results for the partial austenitization shows maximum RA (20.3%) at 143 ^{0}C instead of 183 ^{0}C as calculated from optimum temperature calculation. This deviation is because of bainite formation as seen from the thermal profiles of the different Q&P heat treatments.

- EBSD images show mainly thin film-like RA in case of full austenization while you have coarse blocky type RA after partial austenization. Since the size affects the stability of RA less RA is obtained in case of partial austenizatiion
- In partial and full austenization, steel with maximum retained austenite has the lowest Vickers hardness, which would provide the maximum ductility along with strength.
- Full austenization is preferred over partial austenization for Q&P heat treatment due to the fineness in the microstructure and more amount of RA obtained

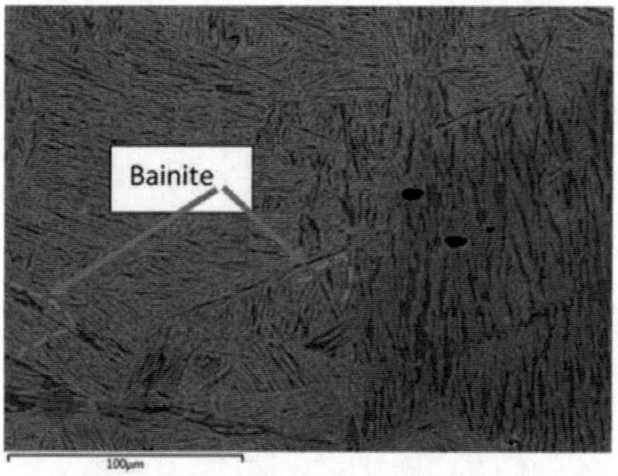

Figure 11. SEM image of sample quenched at 143 ^{0}C and partitioned at 425 ^{0}C

Acknowledgements
The authors would like to thank Ferrous Metal Development Network (FMDN), Mintek under Advanced Metal Initiative for financial support.

References

[1] Lesch C, Kwiaton N, Klose F B 2017 "Advanced High Strength Steel (AHSS) for automotive applications tailored by smart application." Steel Research International (Wiley-VCH Verlag GmBh &Co.KGaA,Weinheim) 88 (10).
[2] Edmonds D V, He K, Rizzo F C, Cooman B C, Matlock D K and Speer J G 2006. *Mat. Sci. and Eng. A* 25-34.
[3] Kim D H, Speer J G, Kim H S and de.Cooman B C 2009 *Met. and Mat. Trans A* **40A** 2048-2060.
[4] Wang L, Speer JG 2013 *Metallogr. Microstruct. Anal.* 268-281.
[5] Speer J, Matlock D K, De Cooman B C, Schroth J G. 2003 *Acta Materialia* **51** (9): 2611-2622.
[6] Clarke A J, Speer J G, Miller M K, Hackenberg R E, Edmonds D V, Matlock D K, Rizzo F C, Clarke K D, De.Moor E 2008 *Acta Materialia* **56** (1): 16-22.
[7] Matsumura O, Sakuma Y,Takechi H 1987. *Scr.metall.* **21** (1301).
[8] Santofimia M J, Zhao L, Petrov R, Kwakernaak C, Sloof W G, Sietsma J 2011 *Acta Materialia* **59** (15): 6059-6068.
[9] Sakuma Y, Matsumura O,Takechi H 1991 *Met. Trans. A* **22** (2): 489-498.
[10] Santofimia M J, Zhao L, Petrov R and Sietma J 2008 *Mat.Char.***59** (12): 1758.
[11] Santofimia M J, Zhao L, Petrov R and Sietma J. 2010. *Mat. Char.* **10** (61): 937.
[12] ASTM A1033-04, Standard Practice for Quantitative Measurement and Reporting of Hypoeutectoid Carbon and Low-Alloy Steel Phase Transformations, ASTM International, West Conshohocken, PA, 2004, www.astm.org
[13] www.thermocalc.com.
[14] Rietveld H M 1988 *Australian Journal of Physics* **41**: p.113.
[15] Marburger R E, D.P.Koistinen 1959 *Acta Metallurgica* **7** 59-60

Evaluation of machining performance of pineapple filler based reinforced polymer composites using abrasive water jet machining process

Jagadish[1], Kapil Gupta[2*], Maran Rajakumaran[3]

[1] Department of Mechanical Engineering, National Institute of Technology Raipur-India

[2] Department of Mechanical and Industrial Engineering Technology, University of Johannesburg, Doornfontein-2028, Johannesburg, Republic of South Africa

[3] Department of Mechanical Engineering, National Institute of Technology Silchar, Assam, India

*Corresponding Author- kgupta@uj.ac.za

Abstract. In the present scenario, polymer composites are being used in various applications such as defence, building construction, aerospace, and packaging etc. due to their unique properties. Machining of polymer composites using conventional machining methods is extremely difficult and expensive. Therefore, non-conventional machining technologies have been explored. This paper presents the results of experimental investigation conducted on abrasive water jet machining of pineapple filler based reinforced polymer composites. A total of nine experiments have been conducted based on Taguchi's robust design of experiment technique considering four input parameters such as stand of distance, working pressure, nozzle speed, and abrasive grain size. Material removal rate and surface roughness parameters are considered as machinability indicators. The optimum machinability indicators obtained are material removal rate- 71 g/s, and average surface roughness- 0.13 μm. The current research identifies abrasive water jet machining process as a sustainable substitute of conventional processes for machining pineapple filler based reinforced polymer composites.

1. Introduction

Composites have become inevitable in present scenario due to their excellent specific characteristics such as high strength, lightweight and better corrosion resistance. Composites especially polymer composites are made of matrix (epoxy resin) and reinforcement (fibers) materials. Among the various polymer composites, natural fiber/filler reinforced polymer (NFRP) composites are gaining significant importance in different sectors such as automotive, aircraft industries, furniture's, packaging and construction due to their advantages over synthetic fiber [1]. Furthermore, NFRP composites are eco-friendly, recyclable, biodegradable and abundantly available. Examples of natural fiber composites that are used in various applications include jute, ramie, cotton, banana, sisal, wood, flax, bamboo, hemp, areca etc. [1].

Among the different natural fillers, the most commonly used natural fillers are pineapple leaf based fillers (PLF). PLF is widely grown in the entire region of India such as Assam, Tripura, Meghalaya, Manipur, West Bengal, Gujarat, Karnataka, and Kerala etc. [2]. Besides, many fruit bearing plants, the pineapple plant bears fruit only once in its lifetime, beyond which the plants are left to die and decay as an agro waste. Thus, instead of letting the plant to decay as an agro waste they can be used for some potential applications by extraction of fibers/fillers from the leaves. The extracted leaves can be further utilized in many applications such as fabric, automobile, furniture and sports industries as an alternative to synthetic fibers/fillers [3]. Furthermore, a synthetic fiber such as glass is available in high density, non-degradable, costly and cause environmental issues. To overcome these difficulties, synthetic fibers can be replaced with PLF; this offers lower weight, less costly, biodegradable and environmentally friendly. Several studies have been carried out on determination of physical and mechanical properties of PLF based polymer composites and compared the results with glass fibers. Furthermore, with upcoming demand of pineapple fillers based polymer (PFRP) composites, machining, especially cutting of these materials with different shape with high superior finish, and higher productivity at minimum machining costs and maximum machining effectiveness, have become an important issue to investigate the machinability behavior of these materials. In addition, these composites are subjected to extensive machining operations before actual use.

Beside, its advantages and applications; due to the anisotropic and inhomogeneous properties, the machining of PFRP composites by conventional process (such as, drilling, milling, sawing, and grinding etc.) is very difficult and a costly endeavor [4] and also produces environmental burdens during machining. Hence, in order to overcome these limitations non-traditional machining processes are preferred [4]. Since, the PFRP composites are of non-conductive nature; the electrical discharge machining (EDM), wire-electrical discharge machining (WEDM), and electrochemical machining (ECM) processes cannot be used. On the other hand, the laser cutting can be used but this process suffers from the problem of a large heat-affected zone, which results in melting of PFRP composites due to lesser thermal conductivity. Furthermore, ultrasonic machining (USM) can be used for machining of PFRP composites, but limited to metal matrix composites and ceramic matrix composites. This process also possesses work piece and tool size limitations [5] and it is a slow process as well [6]. In spite of these problems, non-traditional machining processes are considered as hazardous process because in which large amount of harmful elements in the form of solid, liquid, and gaseous wastes are discharged; resulting in serious occupational health and environmental issues[7]. To overcome these limitations, abrasive water jet machining (AWJM) is used for machining of PFRP composites. AWJM process is capable of machining ductile and electrically non-conductive materials with better material removal rate [8, 9]. Another added advantage of AWJM process is that it doesn't require any extra coolant and lubricants as water by itself acts as a tool as well as coolant and lubricant which makes it an eco-friendly and green manufacturing process. However, there is a scarcity of work on machining of PFRP composites using AWJM process. This study aims to explore capability of AWJM process to machine PFRP composites with high productivity and surface quality. A detailed investigation on the effect of AWJM parameters such as stand-off distance, work pressure, nozzle speed, abrasive grain size on machinability indicators namely material removal rate, and average surface roughness has been done. Optimum values of process parameters have also been obtained for the best values of the aforementioned machinability indicators. The subsequent sections details the various stages of this experimental study.

2. Material and Method

2.1. Workpiece preparation

The present study utilizes pineapple leaf filler (PLF) as reinforcement material with density of 1520 kg/m^3. Initially the pineapple leaves are collected and soaked in water for a span of approximately 2 weeks. After the stipulated time period, due to bacterial fermentation the gummy portion of leaf surrounding the fiber becomes soft and swollen [10, 11]. Now with the aid of blunted object the gummy portion is manually scraped out leaving behind the fiber. The fiber is then thoroughly rinsed in distilled water and dried under the sun or oven. The fiber is then converted into powder form for mixing with matrix. The chief constituents of PLF are cellulose, lignin and ash [12, 13]. The matrix used is epoxy (Araldite LY 556) with density of 1160 kg/m^3 and a corresponding hardener (HY951) with density of 950 kg/m^3, which are mixed in the ratio of 10:8 by weight [13]. To this mixture, 3% of PLF is added and mechanically stirred until a homogenous mixture is obtained [13]. Thereafter the mixture is poured in to the mold kept inside the vacuum glass chamber and allowed to cure for 24 to 48 hrs at room temperature. The obtained specimen of size 180 mm x 150 mm x 5 mm is then taken for further machining. The pineapple leaf fiber along with the prepared composite and the machined specimen are shown in Figure 1.

2.2. *Experimental procedure*

In the present work four independent process variables, which significantly affect the performance of the process such as SoD, WP, NS, and AGS are selected with three levels each. Even though the full factorial array covers all the possible combination of factors (LP= 34 =81), it is a costly and time consuming process, especially when the number of process variables are more [14]. Hence, in the present study, Taguchi orthogonal array (OA) is used to reduce the number of experiments without considerably affecting the accuracy of the results. The minimum number of experimental runs calculated using Taguchi orthogonal array is found to be nine (9) [1+4 (3-1) = 9] for four parameters with three levels. The values of the process parameters and their corresponding levels are shown in Table 1. Machining of PLFP composite has been carried out using CNC abrasive water jet cutting machine (Figure 2) manufactured by DARDI International Corporation, China. Throughout the experiment the discharge rate and orifice diameter are fixed at 2.31 l/min and 0.25 mm respectively with maximum possible water pressure being limited to 3800 bar. Square holes of size 20 mm X 20 mm have been cut during the experiment from the PFRP composite of dimension 180 mm × 140 mm × 6 mm.

Figure 1. (a) Extracted pineapple leaf fiber (b) pineapple leaf filler (c) prepared composite (d) machined specimen.

Table 1. Details of input process parameters

Input parameters	Unit	Symbol	Level 1	Level 2	Level 3
Stand of Distance	mm	SoD	1	2	3
Work Pressure	MPa	WP	100	125	150
Nozzle Speed	mm/min	NS	100	200	300
Abrasive Grain Size	mesh	AGS	70	80	90

Figure 2. Experimental setup.

The experiment repeated thrice for each set of parameter conditions as per the orthogonal array and the average values were taken for calculation of material removal rate (MRR), and surface roughness (SR) (see Table 2) in order to minimize experimental errors and obtain accurate results [14]. The surface roughness is measured using surface profilometer (Make: Tokyo Seimitsu Co. Ltd. Model: Hanhysurf E-35B) and the MRR is calculated by measuring the weight of the composite before and after machining and the time taken for machining [15] using the Eq. (1).

$$MRR = \frac{W_B - W_A}{T} \quad (1)$$

where, W_A is the weight of the composite after machining and W_B is the weight of the composite before machining, while T is the time taken for machining a 20 X 20 mm hole.

Table 2. Experimental results corresponding to nine combinations of AWJM parameters

	Input parameters				Output parameters	
Ex. No.	SoD (mm)	WP (MPa)	NS (mm/min)	AGS (mesh)	MRR (g/s)	SR (μm)
1	1	100	100	70	03.251	0.150
2	1	125	200	80	12.652	0.143
3	1	150	300	90	28.927	0.172
4	2	100	200	90	24.956	0.130
5	2	125	300	70	47.853	0.135
6	2	150	100	80	18.6251	0.185
7	3	100	300	80	66.771	0.160
8	3	125	100	90	24.784	0.102
9	3	150	200	70	55.551	0.178

3. Results and discussion

The effect of input parameters such as SoD, WP, NS and AGS on MRR is shown in Figure 3. From the graph, it is observed that there is an increasing trend for MRR with increase in SoD and NS. The reason for this can be attributed to the fact that as SoD increases the area over which the water impacts increases and eventually more material gets removed, whereas the nozzle speed increases the kinetic energy of the abrasive particles present inside the nozzle increases and hence more energy is imparted to the workpiece for material removal mechanism [16]. Besides, SoD and NS, other factors did not influence MRR much (see Figure 3). In the case of WP, MRR decreases as WP increases from 100 MPa to 125 MPa and at 150 MPa, the MRR increases considerably. As the WP at higher level increases, it increases the kinetic energy of the abrasive particles inside the nozzle, which results in higher MRR as observed in Figure. 3. Moreover, a marginal decrement of MRR is observed (see Figure. 3) as AGS increases from 70 mesh to 90 mesh. This is mainly because, as the mesh size increases particle size of abrasives decreases and small abrasive grain cover less area when bombarded on the composite surface resulting in reduced MRR. Thus, it is observed that, parameter SoD and NS are found to be most significant while parameter AGS is least significant on the AWJM responses during PFRP composites machining. Moreover, parameter WP is found to be moderate influence on the responses. From the analysis, the optimal combinations of input parameters for higher MRR are SoD (3 mm, level 3), WP (150 MPa, level 3), NS (300 mm/min, level 3) and AGS (70 mesh, level 3).

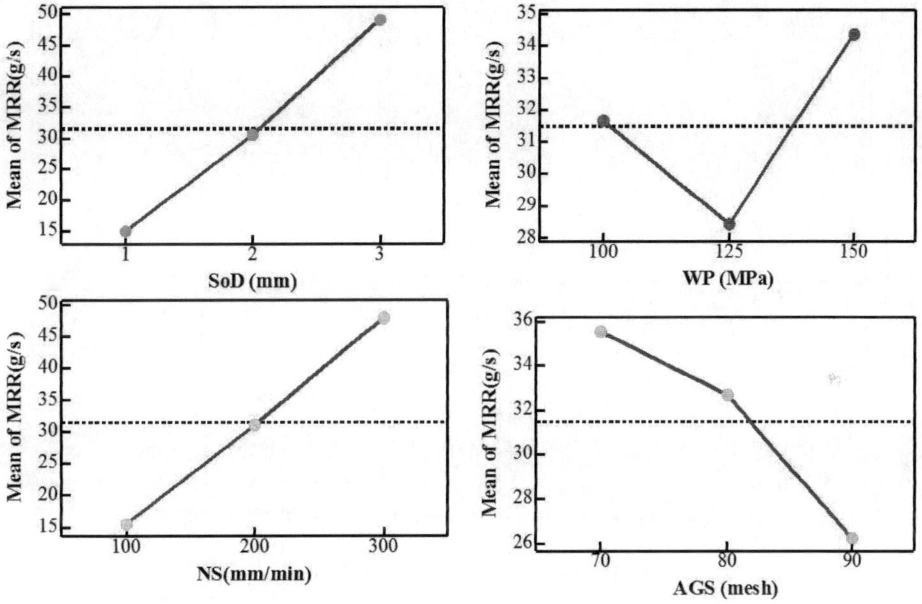

Figure 3. Effect of process parameters on MRR.

On studying the effect of input parameters on surface roughness (Figure 4), it is observed that surface roughness decreases with increase in SoD. It is because as SoD increases, the jet diameter increases due to air drag, which ultimately reduces the kinetic energy density of the jet at the point of impact with the workpiece [16]. Hence, low SoD gives high kinetic energy and better surface finish. With increase in NS, surface roughness increases because of the fact that as NS increases, the kinetic energy of abrasive particles increases and more material is removed from the workpiece, which causes an increase in surface roughness [12]. In case of WP, surface roughness first decreases (0.146 μm to 0.126 μm) with increase of WP and then increases (0.126 μm to 0.178 μm). This is because at low WP, the penetration is less and hence less SR due to less material removal and as WP increases the SR increases as observed at 150 MPa due to higher penetration and hence higher SR. On the other hand, the effect of AGS on the SR found to be moderate i.e. increase of SR from (0.155 μm) at 70 mesh to (0.162 μm) at 80 mesh then it drastically decreased to 0.134 μm at 90 mesh. This is because, as the AGS (mesh size) is increased (70 to 90 mesh) mean grain size of the abrasives gets reduced to smaller size, resulting in lesser material removal and smoother surface as observed in Figure.4. Therefore, it is desirable to have higher values of mesh size for abrasives to get the smoother SR in the case of PFRP composite. Thus, parameter WP and NS found to be most significant while parameter SoD is least significant on the AWJM responses during PFRP composites machining. From the analysis, the optimal combination of input parameters for lower SR are SoD (3 mm, level 3), WP (125 MPa, level 2), NS (100 mm/min, level 1) and AGS (90 mesh, level 3).

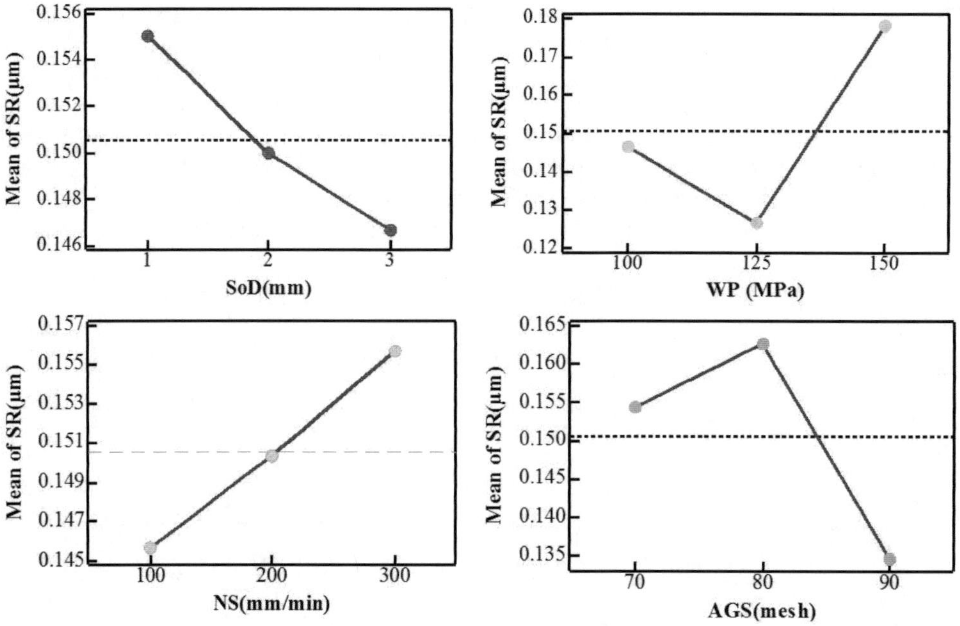

Figure 4. Effect of process parameters on SR

At last, validation of the optimum results obtained from parametric analysis is carried out via confirmatory tests and results are depicted in Table 3. From the results, it is observed that, results for MRR and SR are found to be comparable and acceptable with experimental results.

Table 3. Results of confirmation tests.

Response Parameters	Exp. Results	Optimum Parameters				Confirmatory Test Result
		SoD (mm)	WP (MPa)	NS (mm/min)	AGS (mesh)	
MRR	66.77	3	150	300	70	71.07
SR	0.143	3	125	100	90	0.131

4. Conclusion

The present paper reports an experimental investigation carried out to machine PFRP composites by AWJM process. Effects of process parameters on MRR and SR have been studied. Nozzle speed is the most significant in case of surface roughness, while stand-off-distance and nozzle speed are more significant in the case of material removal rate. The optimum machinability indicators obtained are MRR- 71 g/s, and average roughness- 0.13 μm. The optimal setting for optimum MRR is found to be- SoD (3 mm, level 3), NS (300 mm/min, level 3), WP (150 MPa, level 3) and AGS (70 mesh, level 1); while optimum settings for SR are SoD (3 mm, level 3), WP (125 MPa, level 2), NS (100 mm/min, level 1) and AGS (90 mesh, level 3). Based upon the outcomes of the present study, it can be concluded that AWJM process is a vital substitute of conventional processes for machining PFRP composites with high productivity, surface quality and sustainability.

References

[1] Sanjay MR, Arpitha GR, Naik LL, Gopalakrishna K, Yogesha B 2016 *Nat. Resour.* 7 108-114 DOI: 10.4236/nr.2016.73011
[2] Kannojiya R, Gaurav K, Ranjan R, Tiyer NK, Pandey KM 2013 *J Environ Res Dev.* **7** 1385–1390
[3] Begum K, Islam MA 2013 *Res J Eng Sci.* **2(3)** 46–53
[4] Konig W, Wulf C, Gral P, Willerscheid H 1985 *CIRP Annals* **34** 537–548 DOI: 10.1016/S0007-8506(07)60186-3
[5] Teti R (2002) Machining of composite materials. CIRP-Ann Manuf Technol, 51(2):611-634
[6] Arifur R, 2012 *Springer*, Switzerland.
[7] Sivapirakasam SP, Mathew J, Surianarayanan M 2011 *Expert Sys Appl.* **38(7)** 8370-8374
[8] Ahmad J 2009 *Handbook of Machining of Polymer Matrix Composites* (Boston: Springer) DOI: 10.1007/978-0-387-68619-6
[9] Phokane TC, Gupta K, Gupta MK 2018 *Part C J Mech Eng Sci* DOI: 10.1177/0954406217747913
[10] Adam A, Yahya A 2016 *ARPN J. Eng. Appl. Sci* **11** 161–165
[11] Siva N, Paul JJ 2015 *Int. Conf. on Recent Advancement in Mechanical Engineering & Technology* p 243
[12] Kannojiya R, Gaurav K, Ranjan R, Tiyer NK, Pandey KM 2013 *J. Environ. Res. Dev.* **207** 1385–1390
[13] Asim M, Khalina A, Jawaid M, Nasir M, Zahra D, Ishak MR, Enamul Hoque M 2015 *Int. J. Polym. Sci.* 1–16 DOI: 10.1155/2015/950567
[14] Montgomery DG 2009 *Design and analysis of experiments* (New Delhi: John Willey & Sons)
[15] Lemma E, Chen L, Siores E, Wang J 2002 *Compos. Struct.* **57** 297–303 DOI: 10.1016/S0263-8223(02)00097-1
[16] Mhamunkar MM 2016 *Int J. Adv. Res. Sci. Engg. Techno.* **3** 1640–1646

IOP Conf. Series: Materials Science and Engineering **430** (2018) 012047 doi:10.1088/1757-899X/430/1/012047

Synthesis of a Novel Amphiphilic Nano-Chitosan Material

W M R Matshe[1], L Tshweu[1], M O Balogun[1]*

[1]Biopolymer Modification & Therapeutics Lab, Materials Science & Manufacturing, Council for Scientific and Industrial Research, Meiring Naude Road, Brummeria, Pretoria 0001, South Africa

*mbalogun@csir.co.za

Abstract. Chitosan (CS) is a semi-synthetic polysaccharide derived from the most abundant naturally occurring polymer, chitin. It has found applications in fields as diverse as health, cosmetics, agriculture, environmental remediation etc. While biodegradability is important for non-toxic polymer-based material, stable and mechanically resilient chitosan is needed for harsh environmental applications. We synthesized a nano-chitosan derivative (LA-CS-Suc) with comparable thermal stability as the parent polymer. Synthesis involved chemically linking an organic acid to CS via carbodiimide chemistry. LA-CS-Suc was characterized by FT-IR, TGA, DSC and DLS. Cellular toxicity was also carried out. Successful conjugation was confirmed by FT-IR spectroscopy. Thermal stability was comparable to parent polymer. LA-CS-Suc had particle sizes ranging from 215 to 320 nm and they showed no cellular toxicity.

1. Introduction

Chitin is the major macromolecule in the exoskeleton of crustaceans, insects, and fungi. CS is obtained by deacetylation of chitin. Greater than 50% deacetylated chitin is classified as CS. CS has wide applications because it is biodegradable, biocompatible and non-toxic.[1] These properties along with its amenability to chemical modifications make it an ideal starting polymer for the development of environmentally friendly materials.[2]

Many chemical modification strategies of CS have involved substitution on the C-2 primary amine—a readily reactive nucleophile. Less reactive, but also available for chemical derivatization, are the C-3 and C-6 hydroxyl groups.

Zhong et al explored the electrostatic potential of maleic-CS nano-fibres in measuring proton current flow.[3] They demonstrated a solid-state chitosan-based ionic flow modulator that shows potential in biomedical applications. CS-based material have also shown potential in fuel cell applications as a stable, mechanically resilient and environmentally safe component for the modulation of cationic and anionic conductivities.[4, 5]

Here, we report the chemical synthesis and characterization of polycationic and polyanionic nano-CS which retain the mechanical properties of the parent CS polymer.

2. Materials and Methods

Chitosan of molecular weight 10 kDa and >90%deacetylation was obtained from D. B. Fine Chemicals, South Africa. All other chemicals were purchased from Sigma-Aldrich. Ethanol (EtOH, 99.5%), and acetone (99.5%) were purchased from Merck. Distilled and deionized water (dH_2O) was used in experiments. Ultraviolet spectrum was measured by using the Mettler Toledo UltraViolet5Bio (UV5Bio). Malvern Zetasizer Nano ZS was used to determine the hydrodynamic particle size of the material. Infrared absorption spectra were measured using the Perkin Elmer Spectrum 100 Fourier Transport Infrared (FTIR) Spectrometer. The Mettler Toledo thermogravimetric analysis (TGA) SDTA851^{e} was used to obtain the degradation thermogram of the material. Differential scanning calorimetry (DSC) Q2000 TA Instrument was used to determine the melting point of the material.

2.1 Synthesis of Lauryl-CS:
CS (2.0 g, 11.2 mmol, 1.0 eqv) was completely dissolved in aqueous acetic acid solution (1%, v/v, 100 ml) and lyophilized. The dried chitosan was then dissolved in dH_2O (200 ml) and DMF (40 ml) was added to the solution slowly without precipitating the polymer. To a stirring solution of LA (120 mg, 0.56 mmol, 0.05 eqv) in DMF (10 ml) was added EDC·HCl (233 mg, 1.4 mmol, 0.15 eqv) and NHS (207 mg, 1.8 mmol, 0.15 eqv) and the reaction was left to run for 2 hours. The EDC-activated LA solution was added dropwise to the stirring solution of lyophilized chitosan and the reaction was left to run for 24 hours. The LA-conjugated CS was precipitated with cold EtOH and centrifuged at 6000 rpm for 20 min, twice. After the final wash residual solvent was removed *in vacuo* to obtain the pure lauryl-CS (LA-CS).

2.2 Synthesis of Succinylated LA-CS:
LA-CS was suspended in a solution of dH_2O (100 ml), DMF (40 ml), made alkaline using TEA. Succinic anhydride (223 mg, 2.24 mmol, 0.20 eqv) was added to the stirring solution and the reaction was left for 24 hours. LA-CS-hemisuccinic acid (LA-CS-Suc) was precipitated using cold EtOH and centrifuged at 6000 rpm for 20 min, twice. The residual solvent was removed *in vacuo* and the product was then placed in an oven at 40 °C to dry.

2.3 Thermal studies:
Differential scanning calorimetry (DSC) and thermogravimetric analysis (TGA) were conducted for all derivatized materials to assess decomposition rate, moisture content and melting point temperature. DSC comprised of a heating-cooling-heating cycle from -60 °C to 300 °C at a rate of 10 °C/min. Nitrogen gas flow was set to 25 ml/min. TGA of derivatized and non-derivatized CS were subjected to temperatures of ambient to 400 °C.

2.4 Particle size and surface charge:
Particle size and surface charge analyses were conducted by dissolving 1 mg each of CS, CS-LA, LA-CS-Suc in 1 ml of 1% acetic acid (for CS), 1 ml of 4% acetic acid (for CS-LA), 1 ml of PBS (pH = 7.4) and 1 ml of NaCl (1.0 M) (for LA-CS-Suc). Sodium bicarbonate was added to the CS-LA-Suc to enhance its aqueous solubility. All the solutions were stirred for 20 min at room temperature and analysed using refractive index of 1.330 and absorption value of 0.001.

2.5 Cell toxicity:
A haemolytic assay was used to determine the toxicity of LA-CS-Suc to red blood cells (RBCs). Non-derivatized CS was used as reference control. A 2% (v/v) RBC solution was prepared. LA-CS-Suc and CS were suspended in PBS at 10 mg/ml stock suspension (pH 7.4). The LA-CS-Suc and non-derivatized were serially diluted from the 10 mg/ml stock solution. A 1% (v/v) solution of Triton X-100 was used to obtain 100% haemolysis. 1200 μl of each sample was added to 300 μL of RBC solution. The suspensions were then incubated with shaking at 37 °C for 2 hours. Thereafter, all samples were centrifuged at 5000 rpm for 3 min at 20 °C. The supernatants (1 ml) were withdrawn and analysed on a UV spectrophotometre at 570 nm for haemoglobin content. The results obtained were expressed as percent haemolysis. Zero percent haemolysis was taken as the absorbance of the supernatant of the RBC incubated in PBS (pH = 7.4) solution, while 100% haemolysis was expressed as the absorbance of the supernatant of RBC treated with 1% solution of Triton X-100.

3. Results and Discussion

CS was successfully modified with LA using carbodiimide chemistry to yield the product LA-CS (Figure 1). LA-CS was further acetylated with the dicarboxylic acid succinic acid with the reagent succinic anhydride. The EDAC and anhydride were used because they are less harsh on the polymer than alternative coupling reagents like acid chlorides. Yields of up to 70% were obtained in these reactions.

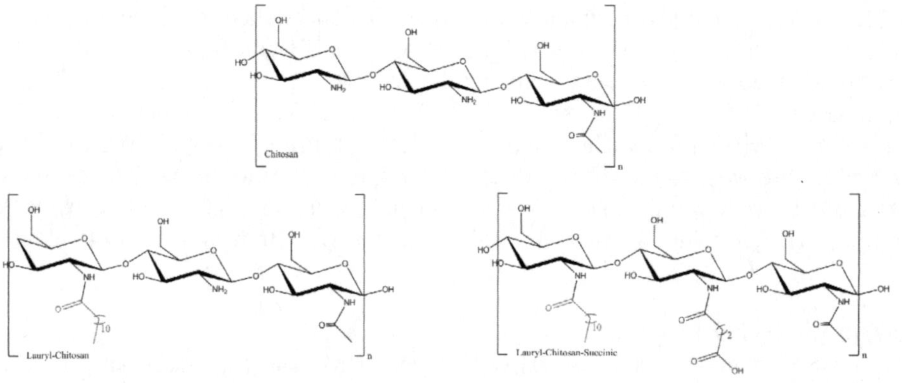

Figure 1. Chemical structures of chitosan and its chemical derivatives.

3.1 Characterization

Confirmation of successful product formation was obtained by FT-IR spectroscopy (Figure 2). Transmission peaks at 1640-1655 cm^{-1} for both modified chitosan correspond to the carbonyl (C=O) stretching as a result of amide bond. The peak around 3350 cm^{-1} corresponds to the stretching of the sugar monomers' hydroxyl groups (O-H). The occurrence of peaks 2929 cm^{-1} for LA-CS-Suc and 2978 cm^{-1} for CS-LA is a result of C-H stretching for LA and succinic acid. The absence of peak 1770 cm^{-1}, a characteristic peak of an ester bond, supports N-acylation of CS with LA and succinic acid.

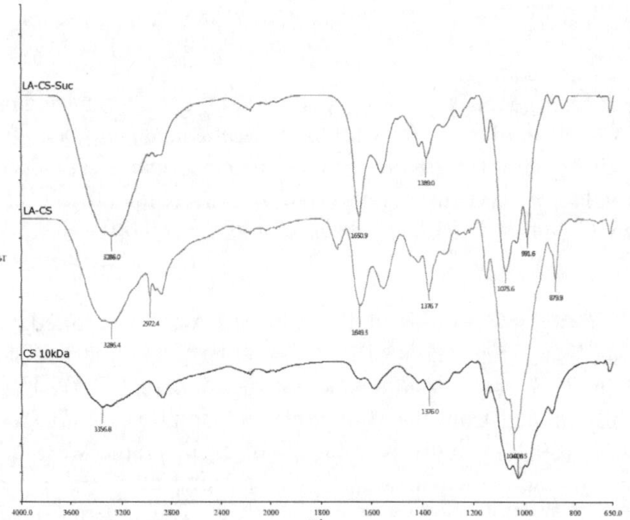

Figure 2. FT-IR spectrum of chemically modified chitosan with LA and succinic acid. CS-LA: CS derivative of LA. LA-CS- Suc: CS derivative of lauric and succinic acid

Thermal stability of non-derivatized and derivatized CS was confirmed by TGA and DSC (Figure 3 and 4). The initial reduction in weight of 2-9% for succinic acid, CS, LA-CS and LA-CS-Suc was proposedly due to moisture content. LA showed near 100% weight reduction within 50 °C. Succinic

IOP Conf. Series: Materials Science and Engineering **430** (2018) 012047 doi:10.1088/1757-899X/430/1/012047

acid showed a greater weight loss between 200-230 °C. CS's greatest thermal degradation appeared between 270-320 °C with over 30%. Similar percent reduction in weight loss up to 70 °C was observed for LA-CS-Suc. CS (> 80 °C) and LA-CS-Suc (>325 °C) experienced a gradual reduction in weight loss. This weight loss patterns are primary due to degradation, depolymerisation and dehydration of acylated units of the polysaccharides. Although thermal stability for LA-CS-Suc seems to be significantly less compared to CS, the multiple degradation patterns of LA-CS-Suc proposes a change in transition state, as a function of temperature. This is supported by the DSC thermogram of LA-CS-Suc (Figure 4).

3.2 Thermal studies

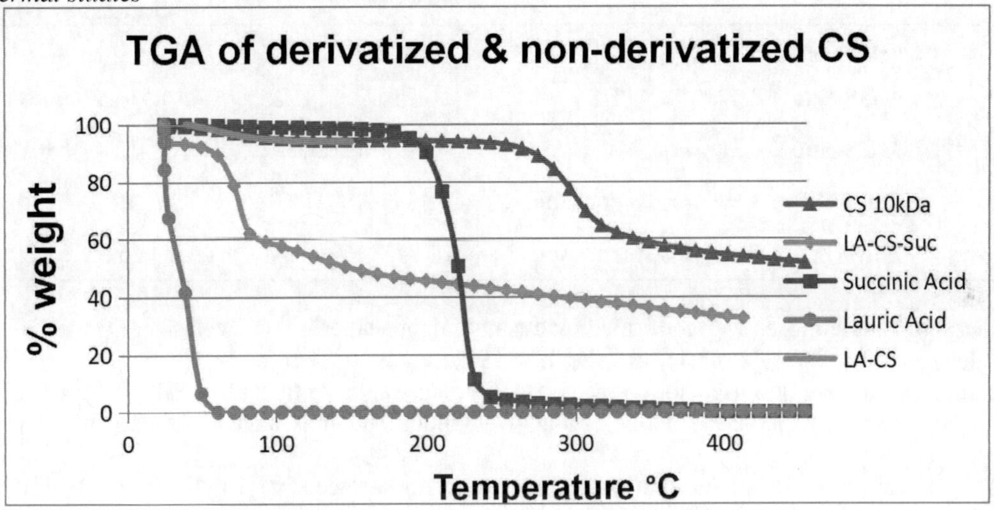

Figure 3. TGA of CS, LA-CS- Suc, lauric and succinic acid.

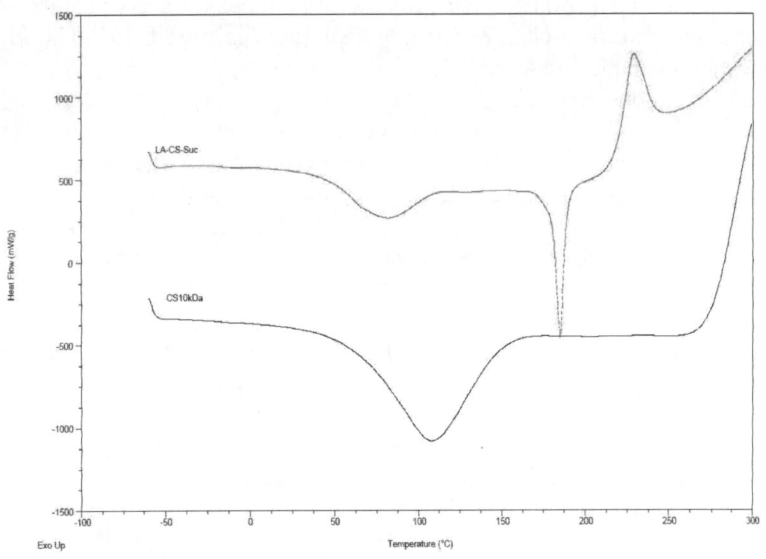

Figure 4. DSC thermogram of CS10kDa and LA-CS-Suc

DSC thermal decomposition of LA-CS-Suc shows two distinct peaks at proximately 175 °C and 230 °C. This thermogram suggests two ideas of thermal degradation of LA-CS-Suc. The peak around 175 °C proposes crystallinity of the modified CS. The exothermal peak around 230 °C suggests decomposition of this crystalline conformation. These observations as a function of temperature, suggests LA-CS-Suc undergoes change in transition states from an ordered to an unordered conformation (crystalline to amorphous). This proposes energy absorption and release characteristics of LA-CS-Suc with the increase in temperature. Similar peaks are not observed for CS10kDa.

3.3 Particle size

Table 1. Particle size and surface charge of CS derivatives in various aqueous diluents.

Sample Type	Diluent	Average particle size (d.nm ± SD)	Average Particle surface charge (± SD)
LA-CS-Suc	dH_2O	320.4 ± 28.43	-20.77 ± 0.40
LA-CS-Suc	PBS	216.9 ± 17.49	-14.93 ± 1.84
LA-CS-Suc	NaCl	257.6 ± 5.71	-12.47 ± 0.53
LA-CS	4% Acetic Acid	371.6 ± 21.49	19.3 ± 3.40
CS 10kDa	1% Acetic Acid	1078.67 ± 5.44	26.93 ± 1.20

Synthesis of nano-chitosan was confirmed by dynamic light scattering on a zetasizer (Table 1). LA-CS was determined to be 371 nm while the LA-CS-Suc was smaller at between 216 and 320 nm depending on the solution used for analysis. LA-CS-Suc was investigated in PBS, 1.0 M NaCl, and deionized water with particle size increasing over about 100 nm range. Non-derivatized CS was observed at over 1 μm.

A significant property of the particles was stability in surface charges. Non-derivatized CS is a polyelectrolyte with net cationic charge in acidic solutions. Its pKa is 6.5. This explains the high positive charge observed. However, this charge is reduced when LA is conjugated to CS. This is in part due to some of the $-NH_2$ groups been converted to amides after conjugation to LA. There is a switch to strong anionic surface charge when succinic acid is introduced. The high charge values confirm the formation of stable nanoparticles.

CS in aqueous solutions has a uniform distribution of functional groups capable of forming hydrogen bonding with water. This thermodynamically stable linear conformation is disrupted by the introduction of the hydrophobic LA. LA constrains the polymeric backbone to adopt a particulate conformation that sequesters the aliphatic chain into the core, minimizing exposure to the aqueous environment (Figure 5). The charged amines of LA-CS and carboxylates of LA-CS-Suc result in repulsion between particles. This effectively prevents particle aggregation and enhances discrete nanoparticle formation.

The anionic charge and size of LA-CS-Suc was found to be very stable in solutions of different ionic salt strengths. In deionized water the charge and particle size were greatest. This could be attributable to a near total lack of counter ions which would stabilize individual particles. However, the presence of Na^+Cl^- ions in the 1 M NaCl solution provides small counter ions that separate and stabilize particles. Further, in the phosphate-buffered solution a much bigger PO_4^{3-} complex interferes with particle aggregation and reduces the particle size even more.

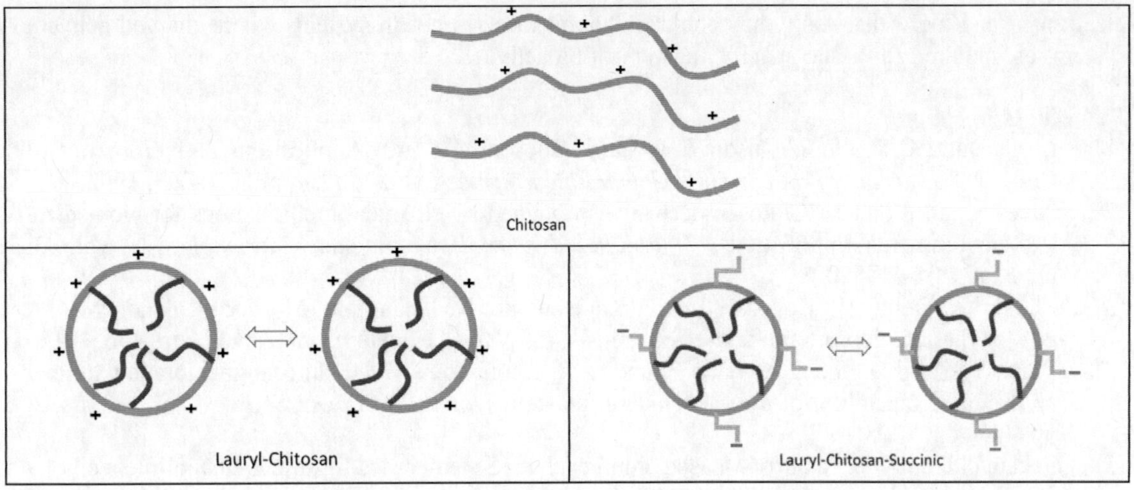

Figure 5. Schematic representation of the proposed conformations of chitosan (CS), Lauryl-chitosan (LA-CS), and Lauryl-Chitosan-Succinic acid (LA-CS-Suc).

3.4 Cell toxicity

The toxicity of the LA-CS-Suc derivatives was investigated with RBCs. Haemolysis is known to be related to high cationic charge.[6] Because LA-CS has a similar cationic charge to CS we only investigated LA-CS-Suc. There was no significant difference observed between the haemolysis of CS and LA-CS-Suc at concentrations up to 10 mg/ml (Figure 6). The extent of haemolysis obtained was less than 10% and similar to the degree of haemolysis seen with the PBS negative control. This clearly indicates that the synthetic derivatization of CS did not reduce its safety profile.

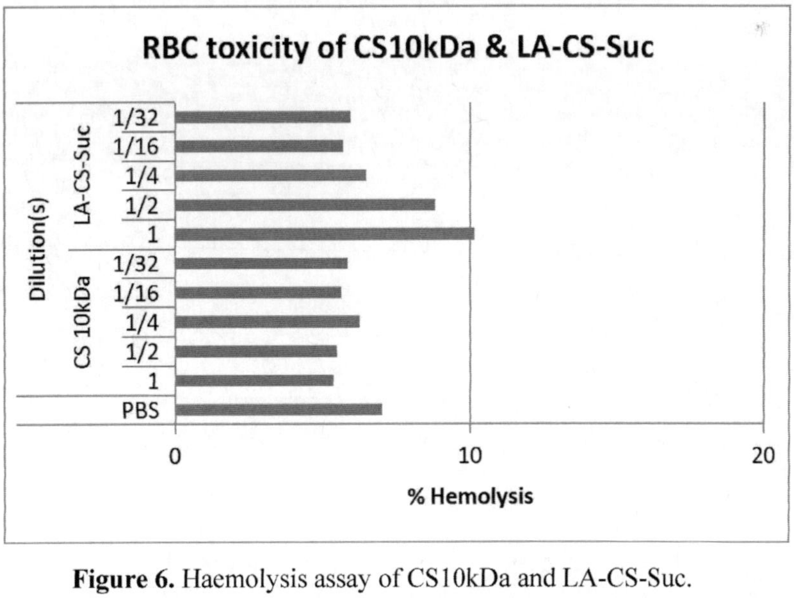

Figure 6. Haemolysis assay of CS10kDa and LA-CS-Suc.

4. Conclusion

In this work we successfully synthesized and characterized cationic and anionic nano-CS derivatives. These particles did not lose the attractive properties of the parent chitosan like thermal stability and low toxicity. It is envisaged that the increased charge to particle ratios of these particles will make them attractive as green, biodegradable electrolytes. This could potentially make them applicable in

IOP Conf. Series: Materials Science and Engineering **430** (2018) 012047 doi:10.1088/1757-899X/430/1/012047

electronics, and fuel cells. Also, they could find application in health systems where charged delivery systems are required for tissue-specific delivery of bioactives.

References

[1] Li Q, Dunn E T, Grandmaison E W, and. Goosen M F A, "Applications and Properties of Chitosan," *Journal of Bioactive and Compatible Polymers,* vol. 7, no. 4, pp. 370-397, 1992.

[2] Alves N and Mano J, "Chitosan derivatives obtained by chemical modifications for biomedical and environmental applications," *International journal of biological macromolecules,* vol. 43, no. 5, pp. 401-414, 2008.

[3] Zhong C, Deng Y, Roudsari A F, Kapetanovic A, Anantram M, and Rolandi M, "A polysaccharide bioprotonic field-effect transistor," *Nature communications,* vol. 2, p. 476, 2011.

[4] Wang Y, Yang D, Zheng X, Jiang Z, and Li J, "Zeolite beta-filled chitosan membrane with low methanol permeability for direct methanol fuel cell," *Journal of Power Sources,* vol. 183, no. 2, pp. 454-463, 2008.

[5] Feketeföldi *et al.* B, "Chitosan-based anion exchange membranes for direct ethanol fuel cells," *J Membr Sci Technol Google Scholar,* 2016.

[6] Duncan R and Gaspar R, "Nanomedicine (s) under the microscope," *Molecular pharmaceutics,* vol. 8, no. 6, pp. 2101-2141, 2011.

IOP Conf. Series: Materials Science and Engineering **430** (2018) 012048 doi:10.1088/1757-899X/430/1/012048

Synthesis, characterization and antimicrobial activities of quaternary chitosan-based materials

Zamani Cele[1,*], Lindokuhle Ndlandla[1], Anou Somboro[2], Daniel Gyamfi[2] and Mohammed Balogun[1]

[1] Biopolymer Modification & Therapeutics Lab, Polymers and Composites, Materials Science & Manufacturing, Council for Scientific and Industrial Research, Meiring Naude Road, Brummeria, Pretoria 0001, South Africa

[2] Pharmaceutical Sciences, School of Health Sciences, University of KwaZulu-Natal, Westville Campus, Durban 4000, South Africa

*ZCele1@csir.co.za

Abstract. The emergence of new pathogenic strains together with continuous rise of antimicrobial resistance and the death of new antibiotics in the clinical pipeline raise an urgent call for the development of potent antimicrobial agents. There has been growing interest in the use of new agents, such as antimicrobial polymers, as alternatives for therapy and disinfection. Cationic chitosan derivatives, N-(2-hydroxypropyl)-3-trimethylammonium chitosan chlorides (HTCC), have been widely studied as potent antibacterial agents. However, their systemic structure-activity relationship (SAR), activity toward drug resistant bacteria and fungi, and mode of action are very rare. Herein, we investigated antibacterial efficacies of the HTCC polymer derivatives against multidrug resistant bacteria, including clinical isolates. The polymers were found to be active against a variety of bacterial pathogens (MIC = 32−128 μg/mL). Interestingly, these polymeric materials were active against Streptococcus Gram-positive bacteria, which are a predominant cause of hospital-acquired infections (HAIs).

Introduction

The emergence of antimicrobial resistance (AMR) and the steady decline of new antibiotics in the clinical pipeline is an increasing global health concern for humans and agriculture [1]. AMR causes a high minimal inhibitory concentration (MIC) of the current antibiotics, resulting in a decreased susceptible range for treatment and adverse outcomes which then calls for an urgent call for the development of potent antimicrobial agents [2]. Surfaces in clinical settings may act as reservoirs of microorganisms which could in turn lead to the spread of infection upon human contact, by either patients or healthcare workers. In cases where transmission by direct contact predominates, then surface-disinfection technologies should have a major impact in reducing infection rates. An obvious measure is therefore the development of medical devices with antimicrobial properties; either by application of antimicrobial coatings or by adding antibiotics to the material. The overuse of antibiotics has contributed to a rise in bacterial resistance, which then limits their applications in medical devices, leaving us with antimicrobial coatings as our last resort.

Chitosan, a biopolymer that is extracted from crustaceans, fungi and insects [3] is known to exert antimicrobial activity against both Gram-positive and Gram-negative bacteria together with fungi [4]. Unfortunately, the application of chitosan as antimicrobial polymers has been restricted by its lack of solubility in both organic solvents and neutral pH aqueous media. However, chitosan tested at lower pH values has revealed an increased activity against different microbes [5]. Quaternary chitosan derivatives, N-(2-hydroxypropyl)-3- trimethylammonium chitosan chlorides (HTCC), have been widely investigated as potent antimicrobial polymeric materials. A recent report by Jiaul Hoque et al showed that HTTC polymers are active against multidrug resistant bacteria including clinical isolates and pathogenic fungi (MIC = 125-250) and also displayed rapid microbicidal kinetics, killing pathogens within 60−120 min [6]. Micro- and nanomaterials have been recognised as an alternative to antibiotics for the treatment of diseases caused by antimicrobial resistant (AMR) microorganisms [7]. With a variety of metal based nanoparticles (NP), such as ZnO, TiO2, Au, Al, Ag, and Cu NPs, killing the bacteria using one or more mechanisms that damage cellular components, cell membrane, and wall or inhibit DNA synthesis and enzyme activity [8]. However, the application of metallic NPs in treating infectious diseases has been limited due to their potential environment toxicity. Herein, we report the facile synthesis and antimicrobial activity of chitosan derivatives against drug-resistant bacteria and pathogenic fungi. In addition, we also performed structure−activity studies, which is an essential part of the progress toward realizing the full potential of antimicrobial chitosan derivatives.

Experimental Section

Materials, Instruments and Bacterial Isolates

Chitosan (degree of deacetylation ≥93%), Allyl bromide, phthalic anhydride, glycidyl methacrylate, hydrazine hydrate, N-methyl-2-pyrrolidone, glycidyltrimethylammonium chloride (GTMAC), potassium dichromate, silver nitrate and deuterated water were obtained from Sigma Aldrich. All solvents and other chemicals were purchased from local commercial suppliers and were of analytical reagent grade, unless otherwise stated. The water used in all experiments was double distilled. Nuclear magnetic resonance spectra were recorded on a CSIR Inova-400 instrument. Infrared spectra of the chitosan derivatives were recorded on Perkin Elmer Spectrum 100 Fourier-transform infrared spectrometer. Bacterial isolates used in this study were sub-cultured twice from freezer stocks onto Mueller-Hinton agar (MHA) plates and incubated at 37 °C prior to the experiments. All subsequent liquid subcultures were derived from colonies isolated from the agar plates and were grown in cation-adjusted Mueller-Hinton broth (CAMHB) medium.

Synthesis of chitosan derivatives

N-(2-Hydroxy)propyl-3-trimethylammonium chitosan chloride (HTCC). Chitosan (0.3 g) was suspended in double distilled water (20 mL) and the suspension was stirred for 2 h at room temperature. GTMAC (3 equiv) was then added to the polymer solution in three portions at approximately 3 h intervals. The reaction temperature was raised to 85 °C for 24 h after the final addition of GTMAC. The chitosan derivative was precipitated in cold acetone and filtered through a sintered glass funnel and washed with acetone and ethanol copiously. The product was characterized by ^{1}H NMR and FT-IR spectroscopy. The degrees of substitutions (DS) or degrees of quaternization (DQ) in the quaternary chitosan derivatives were determined by Mohr's method of titration. For the derivative **2b,** chitosan (100 mg, 0.62 mmol) was dispersed in distilled water (15 mL) at 85 °C, followed by the addition of GTMAC (1.2 mL) in three portions (0.4 mL each) at 2 h intervals. After 3 h, 1.2 mL of glycidyl methacrylate was added in situ, and the reaction was stirred for additional 18 h. The yellowish, clear reaction solution was precipitated in cold acetone (50 mL). The white product was then collected by filtration and washed freely with acetone and ethanol, yielding **2b**.

N-Phthaloylation of chitosan. The synthesis procedure for the derivative is shown in Scheme 2. Chitosan (300 mg, 1.85 mmol) and phthalic anhydride (827 mg, 5.58 mmol) were suspended in 15 mL of DMF containing 5% v/v water and the mixture was heated to 120 °C while stirring under nitrogen atmosphere for 8 h. The reaction mixture was then cooled to room temperature before adding it to ice

IOP Conf. Series: Materials Science and Engineering **430** (2018) 012048 doi:10.1088/1757-899X/430/1/012048

water. A light brown precipitate was collected by vacuum filtration, washed with cold methanol and air dried for 24 h.

Allylation of N-phthaloyl-chitosan. In the second step, N-phthaloyl-chitosan O-allyl was prepared by reacting N-phthaloyl-chitosan (128 mg, 0.44 mmol) with allyl bromide (212 mg, 1.75 mmol) in DMSO (15 ml) at 80 °C under nitrogen atmosphere for 18 h. The product was then precipitated using a mixture of acetone and ethanol (1:1), collected by vacuum filtration and thoroughly washed copiously with acetone and ethanol.

Deprotection and Quaternization. The N-phthaloyl-chitosan O-allyl was deprotected by using 10% (v/v) ethanolic solution of hydrazine monohydrate at 40 °C and stirred for 24 h. The O-allyl chitosan was then quaternized using the method described above.

Determination of Minimum Inhibitory concentration and Minimum Bactericidal Concentration (MIC/MBC) bacterial isolates. The broth micro-dilution method was employed to determine the minimum inhibitory concentration (MIC) and minimum bactericidal concentration (MBC) values, according to the Clinical and Laboratory Standards Institute (CLSI) guidelines.

Results and Discussion

Synthesis and Characterization

The ability of the C-2 primary amino group of chitosan to be protonated confers a vital distinguishing property onto it, i.e. solubility in aqueous acidic solutions. Cellulose, a closely related polysaccharide which differs structurally only in having a hydroxyl group at this position, is insoluble in aqueous solutions. Chemical modifications on chitosan's primary amine are facile and have extended the applications of the polymer. N-alkyl substitutions are common and result in non-pH dependent quaternary cationic derivatives. The quaternary chitosan derivatives were synthesised by reacting high molecular weight chitosan with GTMAC in aqueous acidic conditions (scheme **1**). Quaternary chitosan 2b was synthesized by using GTMAC as quaternizing agent, which was followed by an *in situ* epoxide ring opening reaction using glycidyl methacrylate.

The impact of protecting the hydroxyl groups on the antimicrobial activity of chitosan was investigated by introducing controlled modifications in the hydroxyl group (Scheme 2). The amino groups of chitosan were protected by N-phthaloylation and the product was confirmed by the FTIR band indicating the presence of phthalimide group. The N-phthaloylation facilitated the solubility of the protected chitosan in organic solvents which in turn was vital for the subsequent modifications. This was then followed by allylation at the C-6 position using allyl bromide in DMSO. The amino groups were then unblocked using hydrazine monohydrate, which was then followed by quaternazition using GTMAC.

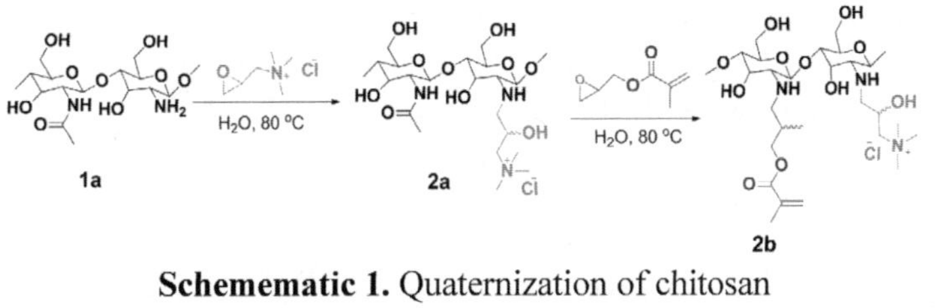

Schemematic 1. Quaternization of chitosan

The chemical modification of chitosan by a particular functional group is usually expressed as the degree of substitution (DS) of chitosan. During the chemical modification and antimicrobial evaluation of chitosan, structure–activity relationship (SAR) studies require precise characterization of the synthesized materials so that the contribution of the structural modification or the covalently attached functional groups on the activity or toxicity can be correctly evaluated. In this study, the NMR and titration method were the only two techniques that were used to calculate the DS.

The DS of the HTCC was determined by Mohr's method of titration. The titration of the chloride ions in HTCC was carried out with aqueous silver nitrate using potassium dichromate as an indicator. DS of HTCC was then calculated by the following equation 1:

$$DS = \frac{V \times c \times 10^{-3}}{V \times c \times 10^{-3} + (W - V \times c \times 10^{-3} \times 314)/162} \times 100\% \qquad (1)$$

Where c (mol/l) is the concentration of $AgNO_3$ solution, V (ml) is the volume of $AgNO_3$ solution, W(g) is the weight of HTTC.

The amount of AgNO3 solution used at the end point during the titration equals the amount of Cl ions present on the HTCC. The DS for **2a** was found to be 0.62 when 3:1 molar ration of GTMAC: to chitosan, while the value increased to 0.98 when using the molar ratio of 6:1, which means that the amino groups of chitosan were fully substituted by quaternary ammonium salt groups.

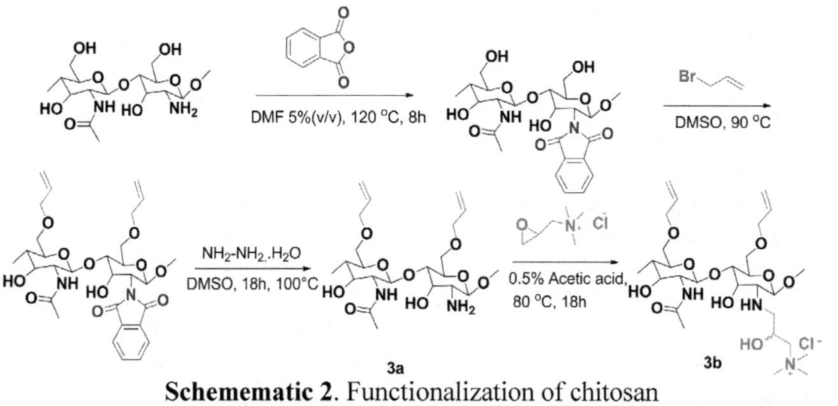

Schemematic 2. Functionalization of chitosan

FTIR studies of chitosan and its derivatives were performed in order to characterize their chemical structure. The presence of the phthalimide group was confirmed by the C=O stretching peaks at 1780 cm^{-1} and 1713 cm^{-1} of N-phthaloyl chitosan, which corresponds to the specific phthalimide group. The additional peaks at 1567 cm^{-1} (N-H deformation and C-N stretching), and 1398 cm^{-1} (coupled C-N and C-O stretching vibration from carbamate group) further confirms the protection of chitosan primary amines. The O-allylation of chitosan was confirmed by the presence of C=C stretch at 1689 cm^{-1}. In addition, the =C-H bending was observed at 940 cm^{-1}. The deprotection of amino groups with ethanolic hydrazine monohydrate was confirmed by the disappearance of =C-H out of plane deformation, C=O stretching and aromatic ring deformation at 1713, 720, and 530 cm^{-1} (figure 3). Finally, the quaternized O-allyl chitosan was characterized by the appearance of C-H bending of trimethylammonium salt group at 1481 cm^{-1} and the disappearance of the characteristic NH_2 bending peak at 1754 cm^{-1}.

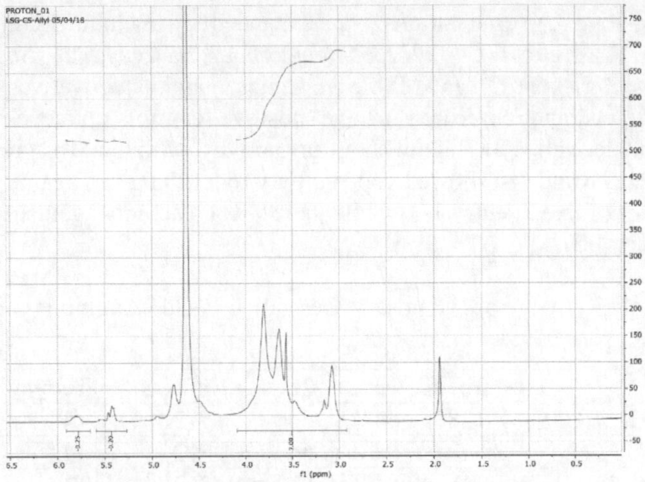

Figure 1. ^{1}H NMR spectrum of O-Allyl chitosan

The ^{1}HNMR spectrum of O-allyl chitosan (Figure 1) was then used to calculate the degree of allylation. The total integral of seven hydrogens of the glucose monomers of the chitosan skeleton is designated as 7.00. The degree of allylation of O-allyl chitosan: **A** equals (A/3H) X 100%, therefore 15%. A refers to total integral of the chiral proton at 5.5 and 5.8 ppm, in which there is only three hydrogens. Degree of allylation = (0.45/3) x 100 = 15%

Antibacterial Activity

We performed the antibacterial assays on the synthesized chitosan derivatives using the Alamar blue method and the MIC (µg/ml) results are summarized in table **1**. Unmodified chitosan (CS) showed no activity at 512 µg/ml according the CLSI guidelines. The quaternized chitosan **2a** showed activity against all Gram-positive bacterial strains (Entry 1-3), including a clinical isolate, *Streptotocus aureus*. Derivatives **2b** and **3b** also showed good antibacterial activity against Gram-positive bacteria, while **3a** was not active at all. Therefore it is evident that the O-allylation of chitosan doesn't improve the bacterial activity of chitosan. The best MIC values were obtained from materials **2a** and **2b** against *Bacillus subtilus* ATCC (32 µg/ml). The Gram-negetive bacterial strains such as *E. coli* ATCC, *Pseudomonas aeroginasae*, *Klebsiella pneumoniae*, *Enterobacter. cloaceae* NDM-1, *E. cloaceae* KPC-2, *E. cloaceae* GES-5, and *S. marcescens* IMP-1 were not susceptible towards the tested chitosan derivative (Entry 6-12). The degree of substitution for **2a** had no effect on the MIC.

Table1. Antimicrobial Activity of chitosan derivatives against Pathogenic Microorganisms

Entry	Strains	2a	2b	3a	3b	CS
1	*S. aureus* ATCC 43300	48	64	NI	64	NI
2	*B. subtilus* ATCC 6051	32	32	NI	64	NI
3	*Salmonella* ATCC	256	256	NI	512	NI
6	*E. coli* ATCC 25922	NI	NI	NI	NI	NI
7	*P. aeroginasae*	NI	NI	NI	NI	NI
8	*K. pneumoniae*	NI	NI	NI	NI	NI
9	*E. cloaceae* NDM-1	NI	NI	NI	NI	NI
10	*E. cloaceae* KPC-2	NI	NI	NI	NI	NI
11	*E. cloaceae* GES-5	NI	NI	NI	NI	NI
12	*S. marcescens* IMP-1	NI	NI	NI	NI	NI

NI= No inhibitory activity at concentration of 512 µg/ml
NB: The highest concentration tested was 512 ug/ml (according the CSLI guideline).

The scope of Gram-positive bacterial strains was expanded by including clinical strains like *S. sanguinis ATCC 10556, S. aureus ATCC 29213, S. epidermidis ATCC 12228, Streptococcus mutants ATCC25175* and *L. rhamnosus ATCC 7469* (Table 2). Chitosan and its derivatives **2a**, **2b** and **3b** were tested against these strains and compared to the positive controls tetracycline and ampicillin. Unmodified chitosan and derivative **3a** did not show any activity, with the MIC greater than 512 μg/ml. The other derivatives (**2a**, **2b** and **3b**) showed good activity (64-128 μg/ml) against the *S. sanguinis, S. aureus* and *S. epidermidis* (Table 2, entry 1-3) while there was no activity against *Strep mutants* and *L. rhamnosus* (Table 2, entry 4-5).

Table 2. Antimicrobial Activity of Chitosan Derivatives and Antibiotics against Pathogenic Microorganisms

Entry	Bacteria	3b	2a	2b	Chitosan	Tetracycline	Ampicillin
1	*S. sanguinis ATCC 10556*	128	64	64	>512	<4	32
2	*S. aureus ATCC 29213*	64	64	64	>512	<4	<4
3	*S. epidermidis ATCC 12228*	128	64	64	>512	32	16
4	*Strep mutants ATCC25175*	>512	>512	>512	>512	32	<4
5	*L. rhamnosus ATCC 7469*	>512	>512	>512	>512	<4	<4

Previously it has been shown that chitosan microparticles interact with outer membrane protein A and lipopolysaccharide that results in bactericidal antimicrobial activity[9]. In the case of these functionalized chitosan derivatives, the minimum bactericidal concentrations (MBCs) were all greater than 3 folds, an indication that the materials are bacteriostatic and not bactericidal.

Conclusion

The water-soluble quaternary chitosan derivatives synthesized and tested against various drug-sensitive and drug-resistant bacterial strains showed improved inhibition of bacterial growth as compared to native chitosan. These derivatives were obtained by chemical introduction of quaternary ammonium salts in the main chain, enhancing the chitosan applicability in a large pH range, opening up a broad range of possibilities. The results of our study show great potential of quaternized chitosan derivatives to be used as safe antimicrobial agents in clinical settings for controlling the spread of infectious diseases caused by both drug sensitive and multidrug resistant microorganisms. The spectrum of activities is currently limited to Gram-positive bacterial strains.

References

[1] Wright G D 2011 *Chem. Commun.* **47** 4055–61
[2] Falagas M E, Tansarli G S, Rafailidis P I, Kapaskelis A and Vardakas K Z 2012 *Antimicrob. Agents Chemother.* **56** 4214–22
[3] Verlee A, Mincke S and Stevens C V 2017 *Carbohydr. Polym.* **164** 268–83
[4] Gortari M C and Hours R A 2013 *Electron. J. Biotechnol.* **16**
[5] Vidar Rúnarsson Ö, Holappa J, Malainer C, Steinsson H, Hjálmarsdóttir M, Nevalainen T and Másson M 2010 *Eur. Polym. J.* **46** 1251–67
[6] Hoque J, Adhikary U, Yadav V, Samaddar S, Konai M M, Prakash R G, Paramanandham K, Shome B R, Sanyal K and Haldar J 2016 *Mol. Pharm.* **13** 3578–89
[7] Moghimi S M 2005 *FASEB J.* **19** 311–30
[8] Samiei M, Farjami A, Dizaj S M and Lotfipour F 2016 *Mater. Sci. Eng. C* **58** 1269–78
[9] Jeon S J, Oh M, Yeo W-S, Galvão K N and Jeong K C 2014 *PLoS One* **9** e92723

IOP Conf. Series: Materials Science and Engineering **430** (2018) 012049 doi:10.1088/1757-899X/430/1/012049

Synthesis and Characterization of Tantalum Oxide Deposited By Metal-Organic Chemical Vapour Deposition (MOCVD)

S O Jeje[1], O O Akinwunmi[2] and M B Shongwe[1]

[1] Institute of NanoEngineering Research, Department of Chemical, Metallurgical and Materials Engineering, Faculty of Engineering and Built Environment, Tshwane University of Technology, Pretoria, South Africa.
[2] Department of Physics and Engineering Physics, Obafemi Awolowo University (OAU), Ile-Ife, Nigeria

*jejesamsonn@gmail.com

Abstract: Tantalum oxide thin-films have rapidly evolved as a potentially important film material in a wide range of industrial applications such as optical coatings, dielectric films, corrosion and heat resistant coatings, dental implants and coronary stents, or prosthesis. The preparation of an improved liquid source metal-organic precursor suitable for the deposition of high-quality tantalum oxide thin films on soda-lime glass using Metal-organic Chemical Vapour Deposition (MOCVD) was reported. Peroxo–hydroxyl–amino tantalum complex precursor was prepared by a method of ligand-exchange reactions between hydroxo–peroxo tantalum complexes and traditional chelating reagents. Preparation of the thin-films was achieved by the pyrolysis of the precursor at four different temperatures (380 °C, 400 °C, 420 °C, 450 °C) with a flow rate of 1.5 dm^3/min for 2 hours deposition period. The deposited films were characterized using Ultraviolet-Visible spectroscopy, Scanning Electron Microscopy (SEM) and Energy Dispersive X-ray spectroscopy (EDX). A direct optical band gap of 4.16 eV – 4.70 eV was obtained from the analysis of the absorption spectrum of the thin-films. SEM micrographs revealed that the deposited film had no regular structure and the layers could be described as amorphous. The EDX results confirmed the presence of Tantalum and Oxygen in the thin-films.

1. Introduction

Tantalum oxides are high dielectric constant material that has reasonably low leakage current and higher breakdown strength [1-3]. Some of its unique properties include high corrosion resistance, piezoelectricity and high refractive index [4]. Thin films of tantalum oxides have triggered considerable interest from the research community, due to their valuable physical properties as well as the promise of compatibility with microelectronics processing [3]. Tantalum oxide thin films have rapidly evolved as a potentially important film material in a wide range of industrial applications such as optical coatings, dielectric films, corrosion and heat resistant coatings, as high-reflective film mirrors for solar cells, charge coupled devices (CCDs), high power laser equipment, dental implants and coronary stents, or prosthesis [2-4].

Tantalum is a very passive metal such that it is almost completely immune to attack by hydrofluoric acid, fluoride ions in acidic solution, free sulphur trioxide and hot/strong alkalis (such as boiling aqueous solutions of metal hydroxides) at a temperature below 150°C [5-7]. This made tantalum suitable for applications in process equipment meant for harsh chemical environments. Good examples of such are surgical tools and implants because of tantalum inertness to bodily fluids [8]. Tantalum has a high affinity for the interstitial gas like oxygen, it forms a dense, non-porous layer of tantalum oxide on the surface that passivates tantalum and gives it its superior corrosion resistance in aqueous media [7].

Tantalum oxide layers employed so far have been deposited by Physical Vapour Deposition (PVD). PVD deposition technique has the limitation of low deposition rate and relative non-conformal coverage in devices and coated tools, which results in non-uniformity of the surface with submicron features and high aspect ratio contacts and via holes. The volatile source of tantalum was the pentahalides. The most common of these was $TaCl_5$ and deposition was through thermal decomposition or hydrogen reduction [9]. Tantalum oxide thin film deposition was accomplished using different techniques such as electron-beam evaporation, ion-assisted deposition, reactive magnetron sputtering and ion beam sputtering [10]. Metal-organic chemical vapour deposition (MOCVD) entails thin solid film formation on a substrate material by a chemical reaction of vapour-phase organometallic precursor that contains a direct metal-carbon bond. MOCVD is preferred to other deposition processes because its step coverage of films had been found to be superior even at a lower temperature. This study is aimed at producing tantalum oxide thin films from peroxo–α-hydroxyl–amino tantalum complex precursor using MOCVD technique and also to describe the optical properties of the tantalum oxide thin films prepared.

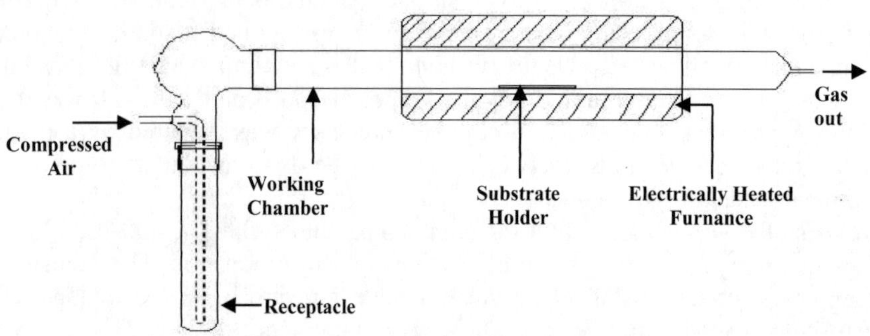

Figure 1. MOCVD thinfilms deposition setup

2. Experimental

The starting materials were tantalum pentachloride ($TaCl_5$), sodium hydroxide (NaOH), hydrogen peroxide (H_2O_2), ammonia (NH_3) solution (all supplied by Sigma Aldrich) and distilled water (H_2O). In a horizontal electrically heated furnace, tantalum pentoxide thin film was deposited on soda-lime glass by metal–organic chemical vapour deposition (MOCVD) technique using peroxo–α-hydroxyl–amino tantalum complex precursor which was prepared using the method of 'Ligand-Exchange Reactions between Hydroxo–Peroxo Tantalum Complexes and Traditional Chelating Reagents' related to the reaction as recorded for Titanium by Masato Kakihana et al. [11]. Equations (1) to (3) show the reaction for the precursor preparation.

$$TaCl_5 + 5HaOH \rightarrow Ta(OH)_5 + 5NaCl \quad (1)$$

$$Ta(OH)_5 + H_2O_2 \rightarrow [Ta(O_2)(OH)_4]^- + H_2O \quad (2)$$

$$[Ta(O_2)(OH)_4]^- + C_{10}H_{16}N_2O_8 \rightarrow peroxo - \alpha - hydroxyl - amino\ acid\ tantalum\ complex \quad (3)$$

The set up for the deposition and the flowchart for the precursor preparation are shown in Figure 1 and Figure 2 respectively. The clear liquid precursor was poured into an unheated receptacle and carrier gas (compressed air) was blown through the liquid precursor at 1.5 dm3/min flow. The compressed airborne precursor was transported into the working chamber maintained at a temperature of 380 °C by an electrically heated furnace for 2 hours; the process was repeated for three other temperatures (400 °C, 420 °C and 450 °C). This was carried out in a fume hood in order to curb problems associated with organic by-products. The thickness of the coating was determined using colour chart approach for thermally grown SiO2 based on reflection theory of visible light. The thin film was characterized using UV visible spectroscopy, Scanning Electron Microscopy (SEM) and Energy Dispersive X-Ray (EDX) to determine the optical, morphological, and compositional analysis of the deposited films respectively.

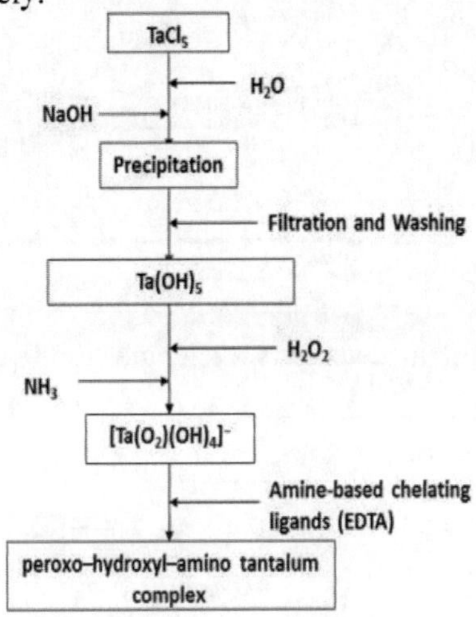

Figure 2. Flowchart for preparation of a water-soluble peroxo–hydroxyl-amino tantalum complex starting from $TaCl_5$

3. Results and Discussion

3.1 Film thickness

Colour chart approach for thermally grown SiO_2 for thickness measurement based on reflection theory of visible light was adopted in this work as it can be used for dielectric films other than SiO_2 or Si_3N_4 [12]. Using calibrated colour charts, thicknesses can be estimated between 10 to 20 nm accuracy and they are useful for oxide films thicker than about 80 nm [12].The estimated value was calculated by using the related expression in equation 4.

$$n_a t_a = n_x t_x \qquad (4)$$

where n_a = refractive index of SiO_2 = 1.46
t_a = SiO_2 film thickness from colour chart
n_x = estimated refractive index of the deposited film =2.275
t_x = unknown film thickness of deposited film

540nm, 560nm, 570nm and 680nm represent SiO_2 film thickness from colour chart at deposition temperature at 380 °C, 400 °C, 420 °C and 450 °C respectively. Equation 4 was applied to calculate films thickness of the deposited Tantalum oxide film at temperature of 380 °C , 400 °C, 420 °C and 450 °C as 346.55nm, 359.38nm, 365.80nm, and 436.40nm respectively. The deposition rate as a function of the deposition temperature indicates that the growth rate for Tantalum oxide films increased with increasing deposition temperature which is in agreement with the work reported by Maeng *et al.* [13]; this can be attributed to the increase of kinetic reaction process in the reaction chamber as the temperature increases.

3.2 Optical characterization of thin films

The UV-visible spectra of the thin films at different temperatures on the soda-lime glass substrate are shown in Figure 3.

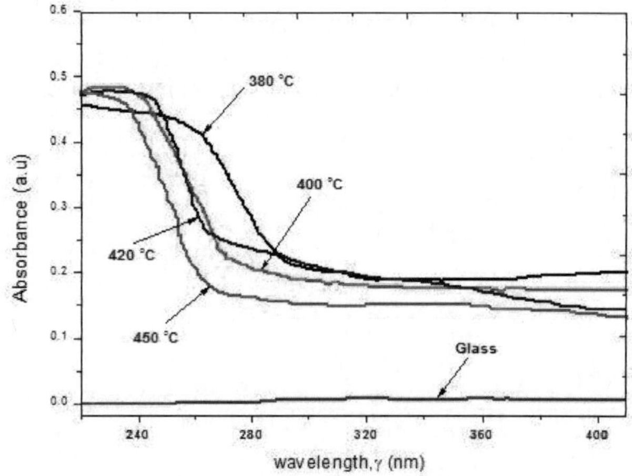

Figure 3. Absorbance versus wavelength for the thinfilms

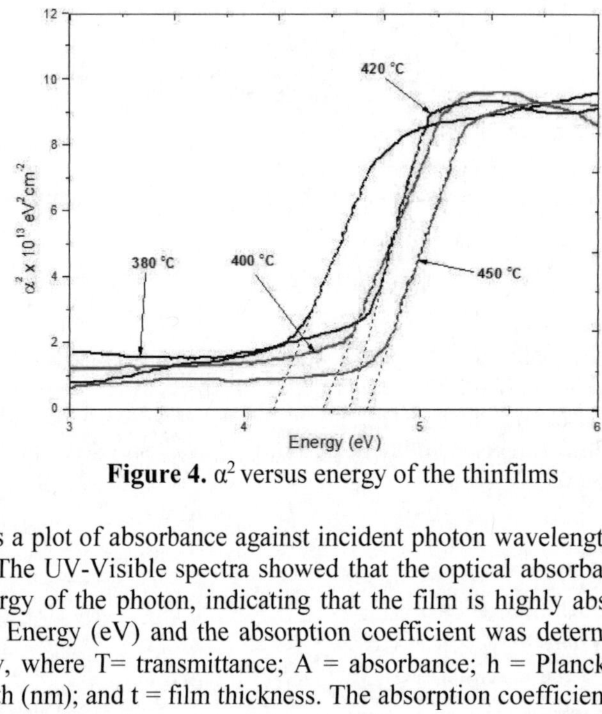

Figure 4. α^2 versus energy of the thinfilms

Each spectrum is a plot of absorbance against incident photon wavelength at normal incidence and room temperature. The UV-Visible spectra showed that the optical absorbance of the films increases with increasing energy of the photon, indicating that the film is highly absorbing in the UV region. The Transmittance, Energy (eV) and the absorption coefficient was determined by equation (4), (5) and (6) respectively, where T= transmittance; A = absorbance; h = Planck's constant; c = speed of light; λ = wavelength (nm); and t = film thickness. The absorption coefficient squared (α2) was plotted against photon energy (eV) in figure 4 to determine the energy band gap of the deposited film. The resulting Tauc plot (figure 4) has a distinct linear regime that indicates the onset of absorption. Thus, extrapolating the linear region to the abscissa yields the energy of the optical band gap of the material.The extrapolations of the linear part of the Tauc plot to the x-axis were 4.16 eV, 4.45 eV, 4.59 eV and 4.7 eV for temperature 380 °C, 400 °C, 420 °C and 450 °C respectively.

$$T = 10\text{-}A \quad (4)$$

$$E = hc/\lambda \quad (5)$$

IOP Conf. Series: Materials Science and Engineering **430** (2018) 012049 doi:10.1088/1757-899X/430/1/012049

$$A = [\ln(1/T)]/t \qquad (6)$$

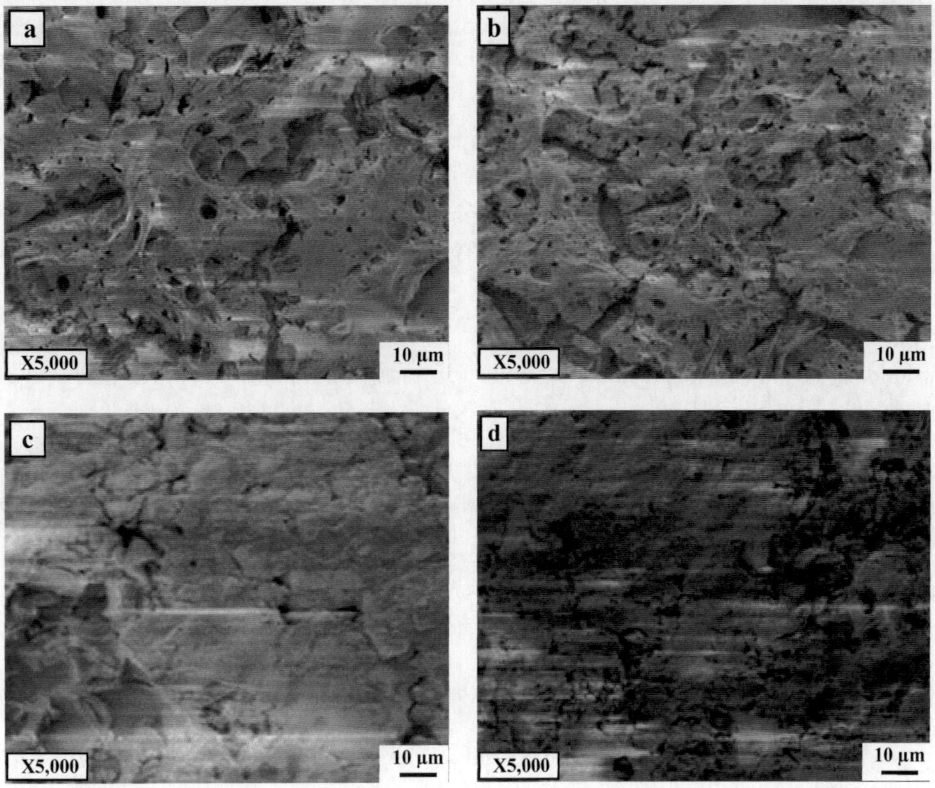

Figure 5. SEM Micrograph of Thin Film at (a) 380°C (b) 400 °C (c) 420 °C and (d) 450 °C

3.3 Morphological and compositional characterization

Figure 5a – d shows the SEM micrograph of the thin films at 380 °C, 400 °C, 420 °C and 450 °C respectively. It is observed that the deposited layers are less of cracks and pores and the grains have no regular structure. The layers can be described as amorphous and the porosity decreases with increment in temperature up to 450 °C. Films that are amorphous do not require lattice matching and epitaxy, therefore, uniform and defect-free multi-layers of amorphous films can be grown even on sharp interfaces down to small individual layer thickness [14]. The EDX spectrum (not shown here) confirmed the presence of Tantalum and Oxygen as the elements present. Other elements (Na, Mg, Si, Ca) in the spectrum are present in the sodalime glass slide used as the thin film substrate. The stoichiometry of the tantalum oxide film could not be ascertained using EDX.

4. Conclusion

A novel single liquid source precursor peroxo–hydroxyl–amino tantalum complex has been synthesized from tantalum chloride and solvents by modifying the method of 'Ligand-Exchange Reactions between Hydroxo–Peroxo Titanium Complexes and Traditional Chelating Reagents'. The tantalum oxide thin films obtained has little or no impurity compound at different deposition temperatures of 380 °C, 400 °C, 420 °C and 450 °C respectively. The thickness of deposited films was found to vary slightly in the range of 346.55 nm to 436.40 nm, therefore the thickness increased with increase in deposition temperature. The direct energy band gap of the deposited films increased from 4.16 eV to 4.70 eV with respect to increasing the deposition temperature from 380 °C to 450 °C.

Acknowledgments
This research was supported by MOCVD laboratory, Physics department, Obafemi Awolowo University, Ile-Ife, Nigeria and Institute of Nano Engineering Research (INER), Tshwane University of Technology, South Africa.

References

[1] Piret F, Singh M, Takoudis C G and Su B L 2008 *Che. Phy. Lett.* **453** 87-91.
[2] Jong-Ho Y and Shi-Woo R 1997 *Thin Solid Films* **292** 324-29.
[3] Vladoiu R, Ciupina V, Mandes A, Dinca V, Prodan M and Musa G 2010 *J. of App. Phys* **108** 093301.
[4] Audier M, Chenevier B, Roussel H, Vincent L, Peña A and Lintanf Salaün A 2011 *J. of Solid State Chemistry* **184** 2033-40.
[5] Cardonne S M, Kumar P, Michaluk C A, and Schwartz H D 1995 *Int. J. of Ref Met and Hard Materials* **13** 187-94.
[6] Patnaik P 2003 *McGraw-Hill Companies* Handbook-of-Inorganic-Chemicals.
[7] Namur R S, Reyes K M and Marino C E B, 2015 *Mat. Research* **18** 91-97.
[8] Rahmati B, Sarhan A D, Basirun W J, and Abas W A B W 2016 *J. of Alloys and Compounds* **676** 369-76.
[9] Levesque A, and Bouteville A 2004 *Chemical Vapor Deposition* **10** 23-28.
[10] Shakoury R, and Willey R R 2016 *Appl Opt* **55** 5353-7.
[11] Kakihana M, Kobayashi M, Tomita K, and Petrykin V 2010 *Bull. of the Chem.l Society of Japan* **83** 1285-308.
[12] Schroder D K 2006 *Semiconductor Material and Device Characterization, Third Edition* (Wiley) 595.
[13] Maeng W J, Park Sang-Joon, and Kim H 2006 *J. of Vacuum Sci. & Tech. B: Microelectronics and Nanometer Structures* **24** 2276.
[14] Gu Z G and Zhang J, 2017 *Coordination Chemistry Review:* Epitaxial growth and applications of oriented metal–organic framework thin films 2017/10/26/.